Advanced Sciences and Technologies for Security Applications

The series Advanced Sciences and Technologies for Security Applications focuses on research monographs in the areas of

- Recognition and identification (including optical imaging, biometrics, authentication, verification, and smart surveillance systems)
- Biological and chemical threat detection (including biosensors, aerosols, materials detection and forensics),

and

- Secure information systems (including encryption, and optical and photonic systems).

The series is intended to give an overview at the highest research level at the frontier of research in the physical sciences.

The editors encourage prospective authors to correspond with them in advance of submitting a manuscript. Submission of manuscripts should be made to the Editor-in-Chief or one of the Editors.

More information about this series at http://www.springer.com/series/5540

Dimitrios P. Nikolelis · Georgia-Paraskevi Nikoleli
Editors

Biosensors for Security and Bioterrorism Applications

Springer

Editors
Dimitrios P. Nikolelis
University of Athens
Athens
Greece

Georgia-Paraskevi Nikoleli
Laboratory of Inorganic and Analytical Chemistry
National Technical University of Athens
Athens
Greece

ISSN 1613-5113 ISSN 2363-9466 (electronic)
Advanced Sciences and Technologies for Security Applications
ISBN 978-3-319-28924-3 ISBN 978-3-319-28926-7 (eBook)
DOI 10.1007/978-3-319-28926-7

Library of Congress Control Number: 2016930265

Printed on acid-free paper

This Springer imprint is published by SpringerNature
The registered company is Springer International Publishing AG Switzerland

Preface

A chemical sensor is a device that transforms the chemical information about a specific sample component to total composition analysis into an analytically useful signal. Chemical sensors usually contain two basic components connected in series: a chemical recognition element ("receptor") and a physicochemical transducer. The recognition system translates the chemical information (i.e., concentration of the analyte) into a chemical or physical output signal. The transducer (i.e., a physical detection system) serves to transfer the signal from the output domain of the recognition element to the electrical, optical, or piezoelectric, etc., domain. A biosensor device is capable of providing specific quantitative analytical information using a biological recognition element (e.g., enzymes, antibodies, natural receptors, cells, etc.), which is retained in direct spatial contact with a transduction element.

Although sophisticated techniques such as chromatography and spectrometry provide accurate and conclusive results, screening tests allow a much higher throughput of samples at a lower cost and with less operator training, so larger numbers of samples can be analyzed. Biosensors combine a biological recognition element (e.g., enzyme, antibody, receptor) with a transducer to produce a measurable signal proportional to the extent of interaction between the recognition element and the analyte compound. The different uses of these devices available today are extremely varied, with food and environmental analysis as an emerging and growing application. The advantages offered by biosensors over other screening methods such as radioimmunoassay, enzyme-linked immunosorbent assay, fluorescence and luminescence immunoassays, with respect to food and environmental analysis, include automation, improved reproducibility, speed of analysis, real-time analysis, but most importantly, the main advantage is that the device can be regenerated and used again, whereas most of the screening methods are only single-use techniques. The main areas of development common to these groups of contaminants include multiplexing, the ability to simultaneously analyze a sample for more than one contaminant, and portability. Biosensors currently have an important role in food safety; further advances in the technology, instrumentation, reagents, and sample handling will surely reinforce this position.

The new biosensor technology has significant technological advantages compared to that of traditional non-sensor-based detection methods. Portable and handheld biosensors, such as dynamic DNA and protein arrays for rapid and accurate detection of pathogens, are typical examples of how biosensor technology can contribute to the defense against bioterrorism. For example, vesicles for use in biosensors have both high specificity and sensitivity, where the vesicles include a receptor specific for the intended analyte and a signal generating component.

A large class of chemical and biological sensors were based on the physical characterization of interfaces. More specifically, electronic (bio)chemical sensing is often related to the characterization of interfaces between ion- and electron-based conductive materials by means of electrical variables such as voltage, current, and charge. Also, recent trends in integrated electronics and the development of nano technology have started a revolution in the field of biosensors allowing the shrinking of very complex electronic systems into millimeter square sizes and this has prompted the development of nanosensors. This would allow implementing complex and sophisticated instrumentation in cheap and portable devices for fast detection of harmful and toxic agents.

The aim of this book is to bring into focus this important research area and advances of biosensors and more specifically those related to the rapid detection of weapons of bioterrorism. The object is related to present advances in the development of portable chemical sensors for the rapid detection of chemical weapons of terrorism; the scope is related to provide a comprehensive review of the most recent research topics most pertinent to the advances of devices that can be used for real-time detections of toxicants such as microbes, pathogens, toxins, nervous gases such as botinilium toxin, Escherichia coli, K. Pneumoniae, sarin, VX, listeria monocytogenes, salmonella, marine biotoxins, staphylococcal enterotoxin B, saxi toxin, gonyautoxin (GTX5), francisella spore virus, bactillus subtilis, ochratoxin.

Biosensors have found a large number of applications in the area of environmental, food, and biomedical analysis. Recent advances include portable devices for rapid detection of insecticides, pesticides, food hormones, toxins, carcinogenic compounds in the environment, such as polycyclic biphenols, etc. Despite public anticipation that biosensors with real-time detection will be able to monitor biological and chemical weapons, the technology has not caught up with the expectations. Presently, biosensors in environmental monitoring stations nationwide can detect compounds like anthrax—but detection can take 12–24 h. The best ones on the market take 20 min. The detection of explosives and especially of biomolecules is an important part of security and defense activities. Technology has enhanced the possibility of combining the functions of drugs and explosives through different biocells in the same analysis module. A testing platform has been developed with a built-in configuration flexibility that allows it to be used for different applications such as chemical, explosives, drug, and biological agents detection. The characteristics of detection are determined by the combination of biocells. It is a new biosensor technology and helps users to combine a number of different detection tasks within the same test, without having to reprogram the instrument.

Biosensors have not yet made a large impact in the area of rapid detection of chemical and biological threats. Biosensors come in thousands of forms and types based on a wide range of physical and chemical principles with varying types of usable outputs. However, the field applications of sensors have been adequate. The diversity of research has been limited mainly to glucose as a mass market. The drawbacks of chemical sensors can be summarized as follows:

a. They are not robust
b. Insufficient for complex analytes
c. Extra laboratory testing is not possible by non-skilled personnel.

Most reports in the literature have suggested that biosensors were at a pre-competitive stage, but highlighted the laboratory proof-of-concept. Presently, we are looking into portable and handheld biosensors, for example, dynamic DNA and protein arrays for rapid and accurate detection of pathogens. A few challenges for biothreat detection had high sensitivity—*detect* very small amounts of pathogens, toxins, and chemical agents; high selectivity—*discriminate* targets from other materials, massively parallel to detect multiple pathogens, minimize false positive, have rapid response, without sample preparation, and inexpensive. To have high spatial resolution, time resolution, selectivity, and sensitivity to chemicals and biosensors, nanowires and potentiometric measurements were used. Techniques which in principle give some nanometer resolution of the area where we want to measure were presented in the literature.

Electrochemical systems based on inhibition of acetylcholinesterase suggest that the detection of nerve agents can be accomplished with fast speed and sensitivity down to fetomolar levels. Detection routes using antibodies and DNA provide many advantages such as high sensitivity of detection, selectivity, and can be used by non-skilled personnel. A wealth of ideas for portability of the sensors was recently presented in the literature. Electroanalytical and optical strategies involving exploiting methods based on the use of immunosensors and genosensors were presented. The combination of screen-printed electrodes with functionalized magnetic beads constitutes a powerful and efficient strategy for the development of disposable magneto-biosensors for the rapid and ultrasensitive detection of many analytes of bioterrorism significance. Magnetic micro- and nanoparticles have a large active surface area which makes possible the immobilization of a high concentration of biomolecules onto the solid phase of the transducer as well as a decrease of matrix effects.

Protein and even cell detection methodologies with interest for various applications were based on nanotechnology (i.e., nanoparticles, nanochannels). Nanoparticle-based immunosensing systems were offered as excellent screening alternatives to sophisticated and high cost equipment that require well-prepared professionals for their use, including data treatment, prior obtaining of final results with interest for further decisions taken in analysis/screening scenarios.

Development of a sensitive and specific biosensor for rapid detection of microorganisms and protein toxins often requires information about the identity of the analyte of interest. On the other hand, recent advances in microbiology and biotechnology have led to the possibility of creating new microorganisms as well as production of new protein toxins with completely or partially unknown DNA or protein sequences. Thus, rapid identification of microorganisms and protein toxins not only enables the detection of the bio-agents but also facilitates the development of highly portable biosensors for use in the field.

The book is targeted at the development of new routes for construction of devices for the rapid detection of chemical warfare toxic agents and therefore for bioterrorism prevention. New trends in methodology are provided, such as advances in microfluidics, which now offer a realistic means for simplified, practical handling with the facility for compressing existing analytical platforms in biosensing. New perspectives for the construction of biosensor devices are described and include novel routes in the high-throughput screening of toxic proteins using immunochemical tools for the construction of electrochemical, optical, and piezoelectric DNA biosensors for rapid detection of weapons of bioterrorism. Recent works with electronic tongues (that is, analytical systems formed by an array of chemical sensors) featuring high selectivity plus a chemometric tool to process a complex multivariate data are also presented. As the generic application covered is related to security, the described systems are those devised to identify and detect explosive compounds. Magneto-actuated biosensors for bacteria and infectious diseases affecting global healthcare are also described in the book.

Since the introduction of modern chemical warfare agents (CWAs) at the beginning of the twentieth century, there has been a continuous interest in the development of robust and reliable analytical tools for the detection of these agents, to provide early alarm in case of terroristic attacks, as well as to monitor their presence in the environment and prevent contamination. Nevertheless, powerful analytical techniques, including chromatographic methods and mass spectrometry are not suitable for field applications and fast early warning, due to the lack of portability, power requirements, long response time, and expensive procedures. In this context, electrochemical (bio)sensors offer advantages in terms of portability, high sensitivity, miniaturization, integration, low cost, and power requirements. The aim of the book is to highlight the important issues of electrochemical (bio)sensors for fast and cost-effective detection of CWAs in the field, considering the main advantages and limitations of this technology, and the latest trends in nanotechnology, lab-on-chip, and functional materials.

There is a growing demand for rapid and reliable methods of determination of microorganism contamination of waters and food products to ensure quality assurance and to improve the healthcare system in general. The majority of the available analytical methods for determination of microorganisms are time-consuming and expensive. In recent years, different approaches have been attempted to develop alternative procedures for determination of microorganisms. A chapter that summarizes the recent achievements in the development of synthetic recognition systems-based devices for monitoring the presence of bacteria, bacteriophages, and viruses, in water and food products is included herein. Molecular imprinting has been most successful in devising relevant synthetic receptors. Application of these recognition systems for determination of microorganisms is described in the present book in detail.

A chapter that is devoted to construction of the immune optical, calorimetric, and piezoelectric biosensors for detection and control of toxic agents such as pesticides, nonylethoxylates, mycotoxins: T2, aflatoxins, patulin, etc., is included. Special attention is given to methods for control of genotoxicity. Cell biosensors are considered as possible types of chemical sensors and their integration into sensor elements is investigated. In general, sensitivity in the field of application of the existing approaches for the control of total toxicity and genotoxicity are thoroughly described.

Due to the structural complexity of marine toxins and the difficulty to produce the corresponding biorecognition molecules, the development of assays and biosensors for their detection has become a challenge. Compared to traditional detection techniques, biosensors can provide advantages in terms of sensitivity, specificity, design versatility, portability, and multiplexed configurations. A chapter that provides a critical overview of the immunosensors, receptor-based biosensors, cell-based biosensors, and aptasensors developed for the detection of palytoxins (PlTXs), brevetoxins (PbTXs), and tetrodotoxins (TTXs) is included. Although only few biosensors for these emerging marine toxins have been described to date, the chapter reflects the promising advances made in this field.

Aptamers are defined as a new generation of nucleic acids which have recently presented promising specifications over antibodies. They can be produced in vitro by **S**ystematic **E**volution of **L**igands by **EX**ponential Enrichment (**SELEX**), and have the ability to recognize selectively and sensitively their targets (protein, toxin, drug, or cell). Thus, they have a wide range of applications in different areas, such as drug delivery, imaging, and biosensing. Accordingly, an increasing number of studies related to aptamer-based sensors "aptasensors" have been introduced in the literature. The recent studies on development of aptasensor technologies, which were applied for toxin detection, have been overviewed.

Mycotoxins such as ochratoxin A and aflatoxins are dangerous food contaminants that usually occur in trace amounts from nanograms to micrograms per gram of food. Therefore, highly sensitive methods are necessary for their detection. Conventional analytical methods such as high-performance liquid chromatography (HPLC) and mass spectroscopy are expensive and time-consuming, therefore biosensor technology is promising for rapid detection of toxicants in the field conditions. Among biosensors, those based on monoclonal antibodies and DNA/RNA aptamers are of special interest, because they provide sensitivity of detection that is better than allowable quantities of toxicants in food. While antibodies are traditional receptors in biosensors, aptamers are novel biopolymers with affinity comparable to that of antibodies. However in contrast to antibodies, aptamers are more stable and the biosensors based on aptamers can be regenerated, allowing their multiple use. A contribution reviews the recent achievements in development of aptamer-based biosensors for detection of selected mycotoxins by electrochemical and acoustic methods.

Recent developments of analytical strategies for determination of potentially hazardous adulterants and allergens achieved to date are presented in the book, highlighting the general considerations and potential prospects for the future. The variety of electrochemical biosensors that have appeared in recent years shows that it is a booming research area with still many challenges, but also great opportunities to develop sensitive, reliable, robust, and cost-effective allergens and adulterants biosensing methodologies.

There is an urgent need for the development of fast and reliable sensors for detection of most toxic compounds that are formed in natural and industrial processes and can have severe consequences on human health. Among many other species, inhibitors of acetylcholinesterase are the focus of many investigations due to a variety of chemical structures and large scale of industrial production. Some anticholinesterase agents, e.g., sarin, soman, and VX gas, were specially developed as chemical warfare with extremely low toxic exposures and lethal consequences for the soldiers and the civil population. Although the accumulated stockpiles of chemical weapons are mostly destructed following the Chemical Weapons Convention, some incidents related to the use of anticholinesterase agents have been reported during the civil wars in Iraq and Syria. The use of homemade sarin in Tokyo subway by Aum Shinrikyo in 1995 is the most known incident related to nerve gases after the Second World War. Meanwhile, threats related to the production and application of anticholinesterase agents by terrorists exist up till today. In addition to chemical warfare, organophosphorus and carbamate pesticides irreversibly inhibit cholinesterase activity and can cause poisoning of agriculture workers and contamination of some foodstuffs. These hazards call for further efforts in the development of appropriate biosensors devoted to detection of anticholinesterase agents in the levels allowing the use of personal protection equipment and hence decrease in the number of potential victims. In this review, progress in the detection of anticholinesterase species based on biosensing technologies is considered with particular emphasis on the results obtained within the past 10 years.

Recently, the diagnosis and treatment of a poisoned person can be done only in specialized centers. Furthermore, currently used clinical methods of intoxication diagnosis are not sufficient for early detection. Conventional laboratory tests based on urine and blood require professional, high-skilled staff, and high cost-equipment as they are arduous and lasting analytical procedures. There is a need to elaborate relatively cheap and easy to use tests, which can simplify and shorten the process of diagnosis of intoxicated patients as well as simplify monitoring of patients from high-risk groups (firemen, miners, security, policemen, soldiers, etc.) having contact with toxic gases. A chapter of the book is focused on novel, early detection sensors for rapid diagnostics of environmental toxicity in the blood of people intoxicated with carbon monoxide.

This book brings together expert scientists with large experience in biosensor technology, some well-recognized environmental analytical chemists, institutes that have large experience in validation, testing, and measurements. A large number of scientists with previous exciting achievements in constructing novel biosensors are authors in the present book. Stimulated paragraphs are focused toward achievements to the construction and application of biosensors to directly monitor toxic agents (i.e., in the field) and set targets for the future development of such devices having pronounced advantages, e.g., portability, cost-effective, real-time, fast response times, etc. Such developments will lead to rapid (near real-time), low cost devices for in situ monitoring of toxic weapons of bioterrorism. It would also serve to solve many problems of detection of analytes that cannot either be detected with the existing instrumentation or due to the high cost of this instrumentation.

The volume brings together contributions from the most eminent international researchers in the field, covering various aspects of work that have not been published in many scientific journals beyond the "state of art" in this field or even commercial units that are available in the market. Very low detection limits, e.g., 10^{-17} M are reported and novel detection schemes for existing or new pathogens, toxins, and other bioterrorism weapons are presented. Inexpensive, robust, portable miniaturized highly selective biosensing systems that were able to detect multi pathogens without sample preparation are the targets of the present review articles; this will give the opportunity to learn new technological schemes that will lead to the construction of devices against bioterrorism that will minimize the risk of weapons of terrorism.

Work on rapid identification of microorganisms and protein toxins by proteomics is an issue that is well presented and the participants of the volume have the opportunity to present their techniques on how to bind the transduction element into the physical sensor unit. Novel techniques of transduction such as nanowiring, nanoparticles, or molecular imprinted polymers are presented and some of these transducers are able to be implanted in the human body. Target analytes detected include a wide range of microbes, pathogens, toxins, nervous gases such as botinilium toxin, Escherichia coli, K. Pneumoniae, sarin, VX, listeria monocytogenes, salmonella, marine biotoxins (such as palitoxins, spirolides, etc), staphylococcal enterotoxin B, saxitoxin, gonyautoxin (GTX5), francisella spore virus, bactillus subtilis, ochratoxin A, cholera toxin, etc. Emphasis is given to the simultaneous analysis of multiple species screen-printed low density microelectrode arrays that were used to develop genosensors to detect Salmonella SSP and staphylococcus aureus. Scanning electrochemical microscopy is used to monitor listeria monocytogenes. Chemical transduction biological elements acting as "receptor" such as antibodies, enzymes, DNA, RNA, lipids, natural, and artificial receptors were immobilized on the physical sensor to recognize the target analyte.

In preparing the book, we have relied on the timely contribution of authors and without their motivation and commitment the publication of this volume would not have been possible. We, thus, extend appreciation to all the authors. We also convey our thanks to Springer for affording us the opportunity to publish this volume.

Athens
2015

Dimitrios P. Nikolelis
Georgia-Paraskevi Nikoleli

Contents

Biosensors for Security and Bioterrorism: Definitions, History, Types of Agents, New Trends and Applications

Georgia-Paraskevi Nikoleli, Stephanos Karapetis, Spyridoula Bratakou, Dimitrios P. Nikolelis, Nikolaos Tzamtzis, Vasilios N. Psychoyios and Nikolas Psaroudakis

Abstract Biosensors are making a large impact in environmental, food, biomedical, and in many other applications. They provide many advantages. in comparison to standard analytical detection methods (i.e., chromatographic techniques) such as minimal sample preparation and handling, faster time analysis, simpler steps of analysis, rapid detection of the analytes of concern, use of non-skilled personnel, and portability for uses in the field applications. The aim of this chapter is to focus on novel research related to the rapid detection of agents and weapons of bioterrorism and provide a comprehensive review of the research topics most pertinent to advancing devices applicable to the rapid real-time detection of toxicants and bioterrorism weapons such as microbes, pathogens, toxins, virus, or nerve gases. The ongoing war on terrorism and the rising security concerns are driving the need for newer faster biosensing devices against bio-warfare agents for both military and civil defense applications. Readers of these review article will learn new schemes of biological weapons that can lead to the construction of devices that will minimize the risk of bio-terrorism.

Keywords Biosensors · Bioterrorism · Virus · Bacteria · Toxins · Nerve gases

---

G.-P. Nikoleli · S. Karapetis · S. Bratakou · N. Tzamtzis · V.N. Psychoyios
Laboratory of Inorganic and Analytical Chemistry, School of Chemical Engineering, Dept 1, Chemical Sciences, National Technical University of Athens, 9 Iroon Polytechniou St., 157 80 Athens, Greece

D.P. Nikolelis (✉)
Laboratory of Environmental Chemistry, Department of Chemistry, University of Athens, Panepistimiopolis-Kouponia, 15771 Athens, Greece
e-mail: dnikolel@chem.uoa.gr

N. Psaroudakis
Laboratory of Inorganic Chemistry, Department of Chemistry, University of Athens, Panepistimiopolis-Kouponia, 15771 Athens, Greece

D.P. Nikolelis and G.-P. Nikoleli (eds.), *Biosensors for Security and Bioterrorism Applications*, Advanced Sciences and Technologies for Security Applications,
DOI 10.1007/978-3-319-28926-7_1

1 Introduction

Recent events have made public health officials acutely aware of the importance to rapidly and accurately detect bioterrorism. Because bioterrorism is difficult to predict or prevent, reliable platforms to rapidly detect and identify bioterrorism agents are important in order to minimize the spread and wide use of these agents and to protect the public health. These platforms must be sensitive, specific, and must also be able to accurately detect a variety of pathogens, including modified or previously uncharacterized agents, directly from complex sample matrices. Recent developments in laboratory prototype devices have been evaluated and were commercialized. Various commercial tests that utilize biochemical, immunological, nucleic acid, and bioluminescence procedures are currently available to identify biological threat agents. Recent developed tests identify bioterrorism agents using DNA aptamers, biochips, evanescent wave biosensors, cantilevers, living cells, and other innovative technologies. This review describes current and developing technologies against bioterrorism and considers challenges to rapid, accurate detect biothreat agents. Although there is no ideal platform, many of these technologies have proved valuable for the detection and identification of bioterrorism agents.

Various tests have been developed to detect and identify biothreat agents. Some of these tests were available before 11 September 2001; other tests have been developed since that time. Although many of these technologies claim to be rapid, accurate, and reliable, few have been evaluated under field conditions for portable uses by non-skilled personnel. This review describes documented current and developing technologies for detection and identification of bioterrorism weapons and addresses the challenges associated with detection in complex sample matrices.

2 Definitions

A bioterrorism attack is the deliberate release of viruses, bacteria, toxins or other harmful agents used to cause illness or death in people, animals, or plants. These agents are typically found in nature, but it is possible that they could be mutated or altered to increase their ability to cause disease, make them resistant to current medicines, or to increase their ability to be spread around the environment. Biological agents can be spread through the air, water, or in food. Terrorists tend to use biological agents because they are extremely difficult to detect and do not cause illness for several hours to several days. Some bioterrorism agents, like the smallpox virus, can be spread from person to person and some, like anthrax, cannot.

Bioterrorism is an attractive weapon because the toxicants are relatively easy and inexpensive to obtain, can be easily disseminated, and can cause widespread fear and panic beyond the actual physical damage. Military leaders, however, have learned that, as a military asset, bioterrorism has some important limitations; it is difficult to employ a biological weapon in a way that only the enemy is affected and not friendly forces. A biological weapon is useful to terrorists mainly as a method of creating mass panic and disruption to a state or a country. However, technologists have warned of the potential power which genetic engineering might place in the hands of future bio-terrorists.

3 History

There have been many reviews written regarding the history, theory, and use of bioterrorism, biothreats, biological weapons, and biological warfare from the 14th century to today. The reader is referred to the following for additional information: Atlas [1], Christopher et al. [2], Hawley and Eitzen [3], Heden [4], Klietmann and Ruoff [5], van Courtland Moon [6], and Tucker [7].

The era of biological weapons was significantly advanced in the 20th century by modern microbiology and multiple international wars. The biological and chemical horrors inflicted during World War I resulted in the drafting of the 1925 Geneva Protocol for the Prohibition of the Use in War of Asphyxiating, Poisonous or Other Gases and of Bacteriological Methods of Warfare. However, many countries that signed the document did so with contingencies in the event of attack by a non-ratifying entity and with the stipulation that the protocol did not prevent investigative research.

Subsequently, Germany, Japan, the Union of Soviet Socialist Republics, and the United States initiated research programs to eliminate the threats of biological weapons [2]. In 1969, under President Richard M. Nixon, the United States began dismantling its offensive biological weapon programs. Henceforth, all biothreat agent research programs in the United States were of a defensive nature, and the 1972 Geneva Convention on the Prohibition of the Development, Production, and Stockpiling of Bacteriological and Toxin Weapons and on Their Destruction was developed and ratified. However, several countries that signed the convention (notably the Union of Soviet Socialist Republics and Iraq) continued offensive research and production of biological agents as recently as the mid-1990s. Additionally, there have been increasingly more subnational terrorist and radical groups that have independently worked on offensive use of biological weapons since the mid-1980s to today [7].

3.1 20th Century

By the time World War I began, attempts to use anthrax were directed at animal populations. Shortly after the start of World War I, Germany launched a biological sabotage campaign in the United States, Russia, Romania, and France. Anton

Dilger was sent to the United States in 1915 carrying cultures of glanders, a virulent disease of horses and mules. Dilger set up a laboratory in his house. He used stevedores working in the docks in Baltimore to infect horses with glanders while they were waiting to be shipped to Britain. Dilger was under suspicion as being a German agent, but was never arrested. Dilger eventually fled to Madrid, Spain, where he died during the Influenza Pandemic of 1918. In 1916, the Russians arrested a German agent with similar intentions. Germany and its allies infected French cavalry horses and many of Russia's mules and horses on the Eastern Front. These actions hindered artillery and troop movements, as well as supply convoys.

In Oregon in 1984, followers of the Bhagwan Shree Rajneesh attempted to control a local election by incapacitating the local population. This was done by infecting salad bars in 11 restaurants, produced in grocery stores, doorknobs, and other public domains with *Salmonella typhimurium* bacteria in the city of The Dalles, Oregon. The attack infected 751 people with severe food poisoning. There were no fatalities. This incident was the first known bioterrorist attack in the United States in the 20th century.

In June 1993 the religious group Aum Shinrikyo released anthrax in Tokyo. Eyewitnesses reported a foul odor. The attack was a total failure, infecting not a single person. The reason for this, ironically, is that the group used the vaccine strain of the bacterium. The spores recovered from the attack showed that they were identical to an anthrax vaccine strain given to animals at the time. These vaccine strains are missing the genes that cause a symptomatic response.

3.2 21st Century

2001—USA and Chile—Anthrax Attacks: In September and October 2001, several cases of anthrax broke out in the United States in the 2001 anthrax attacks, apparently caused deliberately. Letters laced with infectious anthrax were concurrently delivered to news media offices and the U.S Congress, alongside an ambiguously related case in Chile. The letters killed 5 people.

4 Types of Agents

4.1 Category A

These high-priority agents pose a risk to national security, can be easily transmitted and disseminated, result in high mortality, have potential major public health impact, may cause public panic, or require special action for public health preparedness.

4.1.1 Anthrax

Anthrax is a non-contagious disease caused by the spore-forming bacterium *Bacillus anthracis*. An anthrax vaccine does exist but requires many injections for stable use. When discovered early anthrax can be cured by administering antibiotics (such as ciprofloxacin).

4.1.2 Smallpox

Smallpox is a highly contagious virus. It is transmitted easily through the atmosphere and has a high mortality rate (20-40 %). Smallpox was eradicated in the world in the 1970s, thanks to a worldwide vaccination program. However, some virus samples are still available in Russian and American laboratories. Some believe that after the collapse of the Soviet Union, cultures of smallpox have become available in other countries. Although people born pre-1970 would have been vaccinated for smallpox under the WHO program, the effectiveness of vaccination is limited since the vaccine provides high level of immunity for only 3 to 5 years. Revaccination's protection lasts longer. As a biological weapon smallpox is dangerous because of the highly contagious nature of both the infected and their pox. Also, the infrequency with which vaccines are administered among the general population since the eradication of the disease would leave most people unprotected in the event of an outbreak. Smallpox occurs only in humans, and has no external hosts or vectors.

4.1.3 Botulinum Toxin

Botulinum toxin is one of the deadliest toxins known, and is produced by the bacterium *Clostridium botulinum*. Botulism causes death by respiratory failure and paralysis. Furthermore, the toxin is readily available worldwide due to its cosmetic applications in injections.

4.1.4 Bubonic Plague

Plague is a disease caused by the *Yersinia pestis* bacterium. Rodents are the normal host of plague, and the disease is transmitted to humans by flea bites and occasionally by aerosol in the form of pneumonic plague. The disease has a history of use in biological warfare dating back many centuries, and is considered a threat due to its ease of culture and ability to remain in circulation among local rodents for a long period of time. The weaponized threat comes mainly in the form of pneumonic plague (infection by inhalation).

4.1.5 Viral Hemorrhagic Fevers

This includes hemorrhagic fevers caused by the Filoviridae (Marburg and Ebola), and by the Arenaviridae (for example the Lassa fever and the Bolivian hemorrhagic fever). Ebola has fatality rates ranging from 50-90 %. No cure currently exists, although vaccines are in development. Death from Ebola is commonly due to multiple organ failure and hypovolemic shock. Marburg was first discovered in Marburg, Germany. No treatments currently exist aside from supportive care. The arenaviruses have a greatly reduced fatality rate, but a larger presence, chiefly in central Africa and South America.

4.1.6 Tularemia

Tularemia, or rabbit fever, has a very low fatality rate if treated, but can severely incapacitate. The disease is caused by the Francisella tularensis bacterium, and can be contracted through contact with the fur, inhalation, or ingestion of contaminated water or insect bites.

4.2 Category B

Category B agents are moderately easy to disseminate and have low mortality rates.

- Brucellosis (*Brucella* species)
- Epsilon toxin of *Clostridium perfringens*
- Food safety threats (e.g., *Salmonella species*, *E. coli* O157:H7, Staphylococcus aureus)
- Psittacosis
- Q fever
- Staphylococcal enterotoxin B
- Typhus (*Rickettsia prowazekii*)
- Viral encephalitis
- Water supply threats (e.g., *Cryptosporidium parvum*)

4.3 Category C

Category C agents are emerging pathogens that might be engineered for mass dissemination because of availability, easy to produce and disseminate, or may possess high mortality or a major health impact.

Typical category C agents are;

- Nipah virus
- Hantavirus
- SARS
- H1N1 a strain of influenza (flu)
- HIV/AIDS

5 Recent Advances of Biosensors for the Rapid Detection of Weapons of Terrorism

Recent biological terrorism threats and outbreaks of microbial pathogens clearly emphasize the need for biosensors that can quickly and accurately identify infectious agents. The majority of rapid biosensors generate detectable signals when a molecular probe in the detector interacts with an analyte of interest. Analytes may be whole bacterial or fungal cells, virus particles, or specific molecules, such as chemicals or protein toxins, produced by the infectious agent. Peptides and nucleic acids are most commonly used as probes in biosensors because of their versatility in forming various tertiary structures. The interaction between the probe and the analyte can be detected by various sensor platforms, including quartz crystal microbalances, surface acoustical waves, surface plasmon resonance, amperometrics, and magnetoelastics. The field of biosensors is constantly evolving to develop devices that have higher sensitivity and specificity, and are smaller, portable, and cost-effective. The present article describes recent advances in biosensors for applications in the rapid detection of bioterrorism weapons.

Although biosensors are exhibiting double-digit growth rates, they still have to overcome a number of challenges, including the following:

- New research focus less into fundamental research due to impeding introduction of newer applications
- Development of a single biosensor platform with multi-purpose diagnostics capability has restricted biosensor applications
- Numerous problems encountered in successful commercialization of biosensors have encouraged conservative development strategies
- Competition from non-biosensor technologies has hindered revenue growths
- Low rate of technology transfer and lower level of development has deterred the development newer biosensors

Given the current zeitgeist, the market thrust has shifted to biosensors' security capabilities amid the hot topic of biowarfare. A recent report from market research firm In-Stat revealed that the media spotlight on this application may be premature: Despite the public's anticipation that biosensors with real-time detection will be able to monitor biological and chemical weapons, the technology hasn't caught up with expectations. Presently, biosensors in environmental monitoring stations nationwide can detect compounds like anthrax—but detection can take 12–24 h.

6 Planning and Response

Planning may involve the development of biological identification systems. Until recently, most biological defense strategies have been geared to protecting soldiers on the battlefield rather than ordinary people in cities. Financial cutbacks have limited the tracking of disease outbreaks. Some outbreaks, such as food poisoning due to *E. coli* or *Salmonella*, could be of either natural or deliberate origin.

6.1 Preparedness

Biological agents are relatively easy to obtain by terrorists and are becoming more threatening, and laboratories are working on advanced detection systems to provide early warning, identify contaminated areas and populations at risk, and to facilitate prompt treatment. Methods for predicting the use of biological agents in urban areas as well as assessing the area for the hazards associated with a biological attack are being established in major cities. In addition, forensic technologies are working on identifying biological agents, their geographical origins and/or their initial source. Efforts include decontamination technologies to restore facilities without causing additional environmental concerns.

Early detection and rapid response to bioterrorism depend on close cooperation between public health authorities and law enforcement; however, such cooperation is currently lacking. National detection assets and vaccine stockpiles are not useful if local and state officials do not have access to them.

6.2 Aspects of Protection Against Bioterrorism Mainly in the US Include

- **Detection and resilience strategies in combating bioterrorism**. This occurs primarily through the efforts of the Office of Health Affairs (OHA), a part of the Department of Homeland Security (DHS), whose role is to prepare for an emergency situation that impacts the health of the American populace. Detection has two primary technological factors. First there is OHA's BioWatch program in which collection devices are disseminated to thirty high risk areas throughout the country to detect the presence of aerosolized biological agents before symptoms present in patients [8]. This is significant primarily because it allows a more proactive response to a disease outbreak rather than the more passive treatment of the past.

- **Implementation of the Generation-3 automated detection system**. This advancement is significant simply because it enables action to be taken in four to six hours due to its automatic response system, whereas the previous system required aerosol detectors to be manually transported to laboratories [8]. Resilience is a multifaceted issue as well, as addressed by OHA. One way in which this is ensured is through exercises that establish preparedness; programs like the Anthrax Response Exercise Series exist to ensure that, regardless of the incident, all emergency personnel will be aware of the role they must fill. Moreover, by providing information and education to public leaders, emergency medical services and all employees of the DHS, OHS suggests that it can significantly decrease the impact of bioterrorism [8].
- **Enhancing the technological capabilities of first responders**. This is accomplished through numerous strategies. The first of these strategies was developed by the Science and Technology Directorate (S&T) of DHS to ensure that the danger of suspicious powders could be effectively assessed, (as many dangerous biological agents such as anthrax exist as a white powder). By testing the accuracy and specificity of commercially available systems used by first responders, the hope is that all biologically harmful powders can be rendered ineffective.
- **Enhanced equipment for first responders**. One recent advancement is the commercialization of a new form of Tyvex™ armor which protects the first responders and patients from chemical and biological contaminants. There has also been a new generation of Self-Contained Breathing Apparatuses (SCBA) which has been recently made more robust against bioterrorism agents. All of these technologies combine to form what seems like a relatively strong deterrent to bioterrorism. However, New York City as an entity has numerous organizations and strategies that effectively serve to deter and respond to bioterrorism as it comes. From here the logical progression is into the realm of New York City's specific strategies to prevent bioterrorism.
- **BioShield** The accrual of vaccines and treatments for potential biological threats, also known as medical countermeasures has been an important aspect in preparing for a potential bioterrorist attack; this took the form of a program beginning in 2004, referred to as Project BioShield. The significance of this program should not be overlooked as there is currently enough smallpox vaccine to inoculate every United States citizen and a variety of therapeutic drugs to treat the infected. The Department of Defense also has a variety of laboratories currently working to increase the quantity and efficacy of countermeasures that comprise the national stockpile [9]. Efforts have also been taken to ensure that these medical countermeasures are able to be disseminated effectively in the event of a bioterrorist attack. The National Association of Chain Drug Stores championed this cause by encouraging the participation of the private sector in improving distribution of such countermeasures if required [9].

On a CNN news broadcast in 2011, the CNN chief medical correspondent, Dr. Sanjay Gupta, weighed in on the American government's recent approach to bioterrorist threats. He explains how, even though the United States would be better fending off bioterrorist attacks now than they would be a decade ago, the amount of money available to fight bioterrorism over the last three years has begun to decrease. Looking at a detailed report that examined the funding decrease for bioterrorism in fifty-one American cities, Dr. Gupta stated that the cities "wouldn't be able to distribute vaccines as well" and "wouldn't be able to track viruses". He went on to say that movie portrayals of global pandemics, such as Contagion, were actually quite possible and may occur in the United States under the right conditions.

A news broadcast by MSNBC in 2010 also stressed the low levels of bioterrorism preparedness in the United States. The broadcast stated that a bipartisan report gave the Obama administration a failing grade for its efforts to respond to a bioterrorist attack. The news broadcast invited the former New York City police commissioner, Howard Safir, to explain how the government would fare in combating such an attack. He said how "biological and chemical weapons are probable and relatively easy to disperse". Furthermore, Safir thought that efficiency in bioterrorism preparedness is not necessarily a question of money, but is instead dependent on putting resources in the right places. The broadcast suggested that the nation was not ready for something more serious.

6.3 *Biosurveillance*

In 1999, the University of Pittsburgh's Center for Biomedical Informatics deployed the first automated bioterrorism detection system, called RODS (Real-Time Outbreak Disease Surveillance). RODS is designed to collect data from many data sources and use them to perform signal detection, that is, to detect a possible bioterrorism event at the earliest possible moment. RODS, and other systems like it, collect data from sources including clinic data, laboratory data, and data from over-the-counter drug sales [10]. In 2000, Michael Wagner, the Codirector of the RODS laboratory, conceived the idea of obtaining live data feeds from "non-traditional" (non-health-care) data sources. The RODS laboratory's first efforts eventually led to the establishment of the National Retail Data Monitor, a system which collects data from 20,000 retail locations nationwide [10].

The principles and practices of biosurveillance, a new interdisciplinary science, were defined and described in the *Handbook of Biosurveillance*, edited by Michael Wagner, Andrew Moore and Ron Aryel, and published in 2006. Biosurveillance is the science of real-time disease outbreak detection. Its principles apply to both natural and man-made epidemics (bioterrorism).

Data which potentially could assist in early detection of a bioterrorism event include many categories of information. Health-related data such as that from hospital computer systems, clinical laboratories, electronic health record systems, medical examiner record-keeping systems, 911 call center computers, and veterinary medical record systems could be of help; researchers are also considering the utility of data generated by ranching and feedlot operations, food processors, drinking water systems, school attendance recording, and physiologic monitors, among others. Intuitively, one would expect systems which collect more than one type of data to be more useful than systems which collect only one type of information (such as single-purpose laboratory or 911 call-center based systems), and be less prone to false alarms, and this appears to be the case.

In Europe, disease surveillance is beginning to be organized on the continent-wide scale needed to track a biological emergency. The system not only monitors infected persons, but attempts to discern the origin of the outbreak.

Researchers are experimenting with devices to detect the existence of a threat:

- Tiny electronic chips that would contain living nerve cells to warn of the presence of bacterial toxins (identification of broad range toxins)
- Fiber-optic tubes lined with antibodies coupled to light-emitting molecules (identification of specific pathogens, such as anthrax, botulinum, ricin)

New research shows that ultraviolet avalanche photodiodes offer a high gain, reliability and robustness needed to detect anthrax and other bioterrorism agents in the air. The fabrication methods and device characteristics were described at the 50th Electronic Materials Conference in Santa Barbara on June 25, 2008. Details of the photodiodes were also published in the February 14, 2008 issue of the journal Electronics Letters and the November 2007 issue of the journal IEEE Photonics Technology Letters [11].

6.4 Response to Bioterrorism Incident or Threat

Government agencies which would be called on to respond to a bioterrorism incident would include law enforcement, hazardous materials/decontamination units and emergency medical units, if they exist.

The US military has specialized units, which can respond to a bioterrorism event; among them are the United States Marine Corps' Chemical Biological Incident Response Force and the U.S. Army's 20th Support Command (CBRNE), which can detect, identify, and neutralize threats, and decontaminate victims exposed to bioterror agents. US response would include the Center for Disease Control.

Historically, governments and authorities have relied on quarantines to protect their populations. International bodies such as the World Health Organization already devoted some of their resources to monitoring epidemics and have served clearing-house roles in historical epidemics.

Media attention toward the seriousness of biological attacks increased in 2013-2014. In July 2013, *Forbes* published an article with the title "Bioterrorism: A Dirty Little Threat With Huge Potential Consequences" [12]. In November 2013,

Fox News reported on a new strain of botulism, saying that the Centers for Disease and Control lists botulism as one of two agents that have "the highest risks of mortality and morbidity", noting that there is no antidote for botulism [13]. USA Today reported that the U.S. military in November was trying to develop a vaccine to protect troops from the bacteria that cause the disease Q fever, an agent the military once used as a biological weapon.

7 Conclusions

This review has attempted to provide a survey of commercially available and developing technologies for biothreat agent detection. Only technologies that have been evaluated and published have been included. Many other technologies have not been included because insufficient published data were available to ascertain their accuracy and reliability. While an ideal platform has yet to be developed, many of the systems described in this review have proved invaluable in rapidly and accurately identifying biothreat agents. Although the risk of bioterrorism remains, detection technologies will continue to be improved to meet the challenges of this threat.

Security and biodefense is emerging as a strong market for new applications. The new biosensor technology has significant technological advantages when compared to that of the traditional detection methods; for example, vesicles for use in biosensors have both high specificity and high sensitivity (where the vesicles include a receptor specific for the intended analyte and a signal generating component).

We are looking into portable and handheld biosensors, for example, such as dynamic DNA and Protein Arrays for rapid and accurate detection of pathogens. Bioweapons are extremely damaging and efficient. Progress of biotechnology opens also new routes for weaponizing pathogens.

Challenges for biothreat detection are:

1. High sensitive—detect very small amounts of pathogens, toxins, and chemical agents.
2. Highly selective—discriminate targets from other materials
3. Massively parallel to detect multiple pathogens, minimize false positive, have rapid response, without sample preparation
4. Transportable or handheld, robust, simple to operate
5. Inexpensive
6. Adaptable to new biothreats, integrated chemical-biosensor
7. and finally allow the detection of single molecules.

Targets to achieve are:

1. Single RNA molecule detection.
2. Real-time monitoring of RNA hybridization at single molecule level.

3. Real-time monitoring of protein binding to aptamers at single molecule level.
4. Hybridization of synthetic target DNA with anti-anthrax.
5. Selectivity of protein detection, i.e., selectivity of human thrombin detection by anti-thrombin aptamers.

References

1. Atlas RM (2002) Bioterrorism: from threat to reality. Annu Rev Microbiol 56:167–185
2. Christopher G, Cieslak T, Pavlin J, Eitzen E (1997) Biological warfare: a historical perspective. JAMA 278:412–417
3. Hawley RJ, Eitzen EM Jr (2001) Biological weapons: a primer for microbiologists. Annu Rev Microbiol 55:235–253
4. Heden CG (1967) Defences against biological warfare. Annu Rev Microbiol 21:639–676
5. Klietmann WF, Ruoff KL (2001) Bioterrorism: Implications for the clinical microbiologist. Clin Microbiol Rev 14:364–381
6. van Courtland Moon JE (ed) (1991) Biological and toxin weapons: research, development, and use from the middle ages to 1945. Stockholm International Peace Research Institute, Stockholm, Sweden
7. Tucker JB (1999) Historical trends related to bioterrorism: an empirical analysis. Emerg Infect Dis 5:498–504
8. United States. Cong. House. Committee on Homeland Security. Ensuring Effective Preparedness Responses and Recovery for Events Impacting Health Security Hearing before the Subcommittee on Emergency Preparedness, Response and Communications of the Committee on Homeland Security, House of Representatives, One Hundred Twelfth Congress, First Session, March 17, 2011. 112th Cong., 1st sess. HR 397. Washington: U.S. G.P.O., 2012. Print
9. United States. Cong. House. Committee on Homeland Security. Taking Measure of Countermeasures. Hearing before the Subcommittee on Emergency Preparedness, Response and Communications of the Committee on Homeland Security, House of Representatives, One Hundred Twelfth Congress, First Session, April 13, 2011 and May 12, 2011. 112 Cong., 1st sess. HR 397. Washington: U.S. G.P.O., 2012. Print
10. Wagner MM, Espino J et al (2004) The role of clinical information systems in public health surveillance. Healthcare information management systems, 3rd edn. Springer, New York, pp 513–539
11. Avalanche Photodiodes Target Bioterrorism Agents Newswise, Retrieved on June 25, 2008
12. Bell L (2014) Bioterrorism: a dirty little threat with huge potential consequences. Forbes 2013-07-21. Accessed 17 Feb 2014
13. Heitz D (2014) Deadly bioterror threats: 6 real risks. Fox News. 2013-11-02. Accessed 17 Feb 2014

Microfluidics a Potent Route to Sample Delivery for Non-intrusive Sensors

George Kyriacou, Hong Chang, Joseph Gargiuli, Ajay Agarwal and Pankaj Vadgama

Abstract Biosensors offer wide opportunities for threat agent analysis, but practical analytical systems require these sensors to be integrated with pre- and post analytical steps to enable simplified, seamless operation. Perhaps the most important of these steps relates to sample handling and presentation. Advances in microfluidics now offer a realistic means for simplified, practical handling with the facility for compressing existing analytical platforms in biosensing. Small volume handling can not only allow for miniaturisation, but flow at this scale enables a different type of flow profile and the a facility for direct liquid-liquid exchange. Basic flow principles in microflow are presented followed by a description of aqueous/organic flows and how they cab be used both for solute partitioning and in situ membrane formation. The potential value of miniaturised separation membranes is described, including for sample cleanup, handling and biosensor protection. Finally, examples of sensor integration into microfluidic structures are given as pointer towards future developments. Overall, the chapter seeks to rebalance the traditional emphasis on biosensor design by highlighting the importance of controlled sample presentation as a potential route to low maintenance biosensors with improved response characteristics.

Keywords Microfluidics · Laminar flow · Liquid-liquid interface · Reynolds number · Electrochemical biosensor · In situ polymer membrane · Sample presentation · Miniaturisation

---

G. Kyriacou · H. Chang · J. Gargiuli · P. Vadgama (✉)
School of Engineering and Materials Science, Queen Mary University of London, London E1 4NS, UK
e-mail: p.vadgama@qmul.ac.uk

A. Agarwal
CSIR-Central Electronics Engineering Research Institute, Pilani, Rajasthan, India

D.P. Nikolelis and G.-P. Nikoleli (eds.), *Biosensors for Security and Bioterrorism Applications*, Advanced Sciences and Technologies for Security Applications,
DOI 10.1007/978-3-319-28926-7_2

1 Introduction

The miniaturisation of sensors has been driven by evermore sophisticated, high precision fabrication platforms available through centralised microelectromechanical foundries. Through this MEMs capability, an ever expanding repertoire of controlled geometry sensor surfaces, high density precision arrays and a kaleidoscope architectures have come about, never thought possible in the early, pioneering days of biosensor development [1, 2]. Much of this is centred on the machining of silicon, but with the advent of polymeric materials technologies, many using silicon as the forming template, it has also been possible to leverage, soft polymer structures in sensor design [3]. Perceived advantages have been the creation of devices that are small enough to be non-intrusive, requiring the smallest of samples and able to be intergrated into larger assay components through mechanical adaptability. Such considerations highlight the fact that a functional sensor is not itself the full answer to an analytical need, but must be a part of a cascade of functions from sampling, sample presentation/separation through to signal processing and presentation. Seamless sensor integration into a system is already well understood in the case of medical monitoring, particularly if measurement is to be undertaken in vivo [4]. However, systems having a small analytical 'footprint' for non-intrusive measurement are also needed in other sectors needing extra laboratory testing. This certainly applies to system deployment for bioterrorism/security where practical application demands a versatile capability as regards both location and stand alone operation.

Along with sensor miniaturisation, miniaturised sample presentation is also important; too little attention has been give to the fact that the way in which a sample is presented (dilution, flow, reagent content etc.) can make all the difference between a viable and non-viable system. Clearly, any fluid sample, but especially a liquid, can be controlled across a range of physicochemical variables; beyond sample diluent control, the volume requirement can be controlled, flow rate manipulated, calibrants interposed (one advantage of the early FIA approaches) and critically surface shear and turbulance at the sensing surface manipulated. This last aspect has been shown to be a critical influence not just on analyse flux to a sensor, but on the deposition of colloidal materials likely to foul a sensor surface. This chapter focuses on microfluidics developments as a research field in its own right and then on the way it serve as a sample presentation methodology for miniaturised sensors. As such only a part of the uTAS paradigm is covered; sample prove design and the use of miniaturised sample cleanup and chromatographic components are excluded. However, a sensor and microfluidic combination goes a long way towards underpinning the uTAS ideal.

2 Microfluidics: Basic Features

Analytical structures and devices operating at the micrometer scale are fabricated on either top down approach or bottom up principles, now well established [5]. For top down fabrication, a larger structure is machined without atomic or molecular control to the smaller dimension, while in the bottom up approach atomic scale manipulation is used to build up the larger structure. Whilst this especially applies to nano devices, the principles are the same for micron scale design. The inherent benefit of miniaturisation is the combination with other complex sub-components to create a monolithic structure that is simple in design, ergonomically efficient and appropriate for use by a non-skilled operator [6].

The governing principle around microfluidic sample management is embodied in the Reynolds number (Re). This defines the ratio of inertial versus viscous force during flow:

$$\mathrm{Re} = \frac{\mathrm{DV}\rho}{\mu} \tag{1}$$

Re is dimensionless and here, D is channel diameter (m), V is fluid velocity (m/s), ρ is fluid density (kg/m^3) and μ fluid viscosity ($N \cdot s/m^2$). In low dimension channels, the viscous force dominates and the tendency is then for turbulent flow to switch over to laminar flow. At approximately <3000 Re turbulent flow is avoided and the key analytical implication is that intermolecular exchange and solute transport are determined by Fick's Laws alone. Indeed, much of microfluidics research is aimed at overcoming the barrier to mixing by attempting to engineer turbulence in miniature flows, e.g. to effect more efficient sample and reagent mixing. However, in the present chapter, the special benefits of non-mixing will be highlighted. A particular flow model of value for sample delivery is that of dual, linear fluid flow. This is relatively easy to generate, though the situation becomes marginally more complicated at the inception of flow convergence, where turbulence and local recirculation, along with eddy currents can combine to create unwanted physical mixing [7].

3 Dual Flow for Mobile Liquid-Liquid Interfaces

The dual flow model has been reported by us, based on bifurcated flow channels with convergence of independent flow at the starting point of a simple linear flow system. (Fig. 1). Various designs have been tested empirically for both the angle of convergence and the method flow drainage [8].

Our flow channels avoided dependence on a MST foundry, and were simply formed by adhesive bonding of cut glass plates, which had the advantage of permitting optical microscopy of flow, not possible with Si and avoided the solvent incompatibility of transparent polymers (vide infra). Ultimately, a single channel drainage design, rather than the dual drainage shown in Fig. 1 proved the most effective and convenient. Figure 2 shows a practical manifestation with a design for an integrated electrochemical wire sensor. What the design allows is for a mobile fluid layer to be sustained between sample and sensor. In this way, a colloid containing or contaminated sample can be separated from the sensor surface avoiding the dangers of surface biofouling and associated sensor drift, permitting calibrant/reagent to be introduced without interruption of the sample flow and permitting modulation of the concentration of solute achieved at the sensor surface [9].

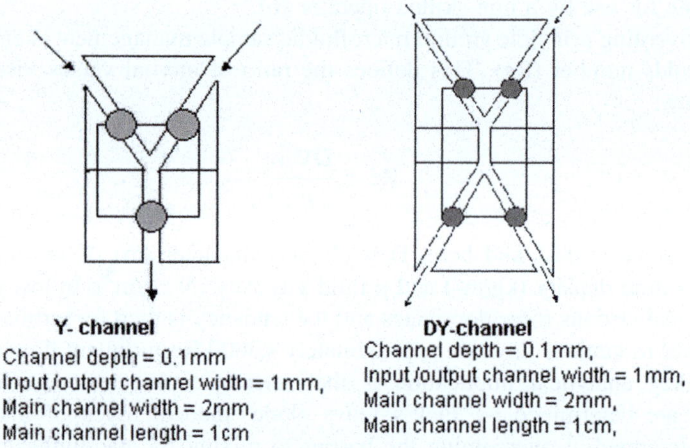

Fig. 1 Glass fabricated flow channels for laminar flow and incorporated distal fluid flow drainage design. The structures are configured to accept parallel fluid flow from micro flow pumps. The *shaded circles* indicate the location of pipette tips positioned vertically for fluid entry from external tubing. The glass construct allows for transillumination and for carriage of organic solvent. (From Ref. [8] with permission)

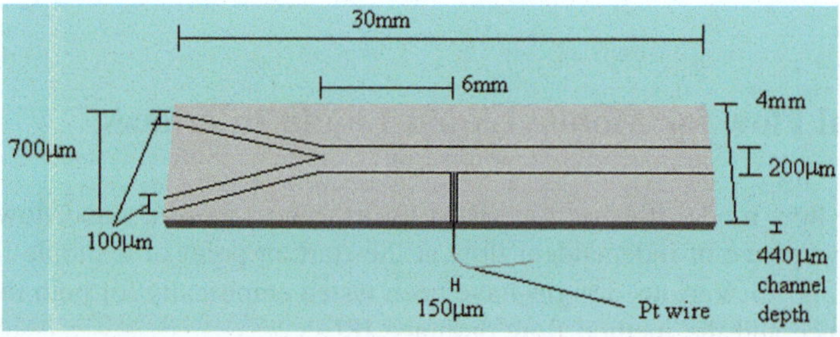

Fig. 2 A flow cell with integrated wire electrode for protecting the sensor tip using a contacting analyte, receiving, flow containing only buffer or reagent, while the non-contacting sample flow stream is kept physically away from the tip, enabling sensor protection from cellular and colloidal components liable to foul the electrode surface

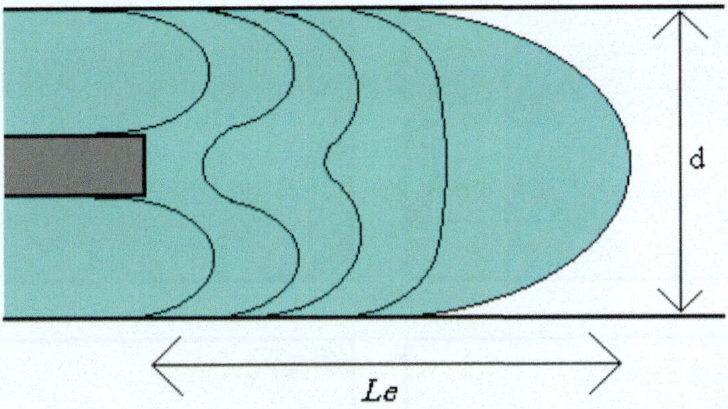

Fig. 3 Flow velocity profile as it develops from the point of dual flow contact on either side of a rectangular splitter palate. The initial curves represent dual parabolic flows, which over distance develop into merged and then a single parabolic flow with maximum velocity at the centre of the channel. The position of the single flow profile is defined as the Langhaar distance (Le)

Detailed consideration of flow and overall design are key to retention of this stable liquid-liquid interface and fluid 'coating' over the sensor. Importantly to adhere to laminar flow principles (Eq. 1) it is not necessary to have a micro dimension flow channel. Indeed, the latter has distinct disadvantages in that surface fouling of the channel wall is more likely to lead to channel blocking. At initial convergence of any such flows, there are inevitably two independent parabolic flow profiles, with their highest velocity, as in any pipe flow, at the centre and the lowest flow rate close to the channel wall converging to a no-slip boundary at the channel wall. Subsequently, a single, fully developed flow profile results attained at a channel location defined as the Langhaar distance (*Le*):

$$Le = 0.058.\ \mathrm{Re}.\ \mathrm{d} \tag{2}$$

where d is the characteristic length of the channel, which for a particular channel flow geometry can be taken as the hydraulic diameter—non-tubular flow. The flow profile convergence is illustrated in Fig. 3. The dimensions of this convergence is important because it determines the type of transport to the sensor surface and the expected response magnitude.

The velocity of flow at the liquid-liquid interface in the middle of the channel is inevitably the same for both liquids and depending upon the time available for their interaction along the channel, Fickian, concentration gradient driven diffusion directs solute transport orthogonally to the flow direction. This can be visualised in Fig. 4, where amylose in one flow solution meets an iodine indicator in the second flow stream. An iodide—amylose complex forms, is readily seen as a line along the length of the fluid interface, but importantly, its formation is determined by the

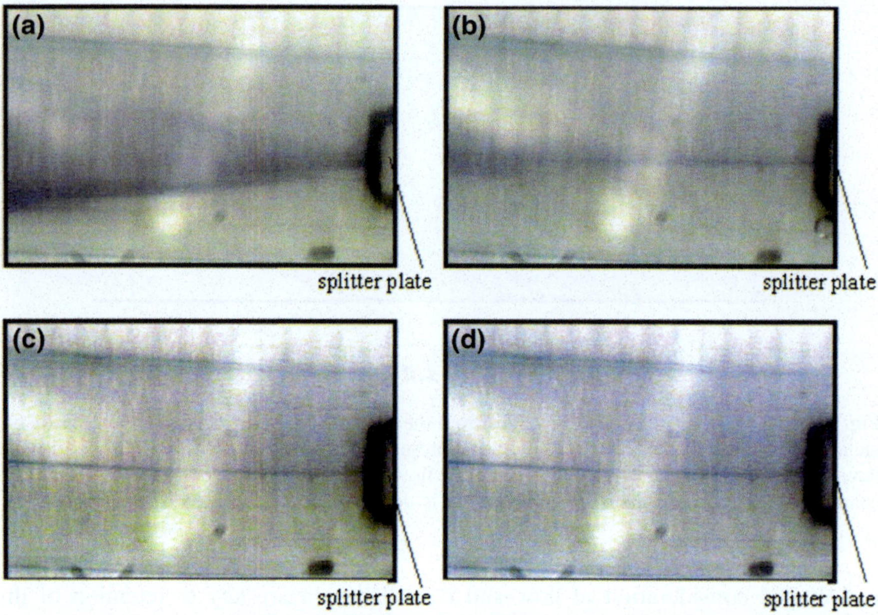

Fig. 4 Digital images of left to right flow near a splitter plate and the creation of flow dependent interaction zones, respectively between amylose loaded and iodide loaded flow streams. Flow is from *right* to *left*, and shows the interfacially generated amylose-iodide complex, known to be visible at sub-micomolar levels. **a** 10 μl/min per inlet, **b** 50 μl/min per inlet, **c** 100 μl/min per inlet, **d** 999 μl/min per inlet

degree of contact at the interface, dictated by the flow rate; higher flow permits only a reaction at the interface and slow flow permitts a widening interaction to give a pyramidal zone of extended diffusive cross channel interactions.

4 Transport Phenomena Under Flow Conditions

The protective principle of dual laminar protective flow, described above, rests on the fact that surface active fouling agents are largely colloidal and cellular, whilst molecules of interest are typically micro solutes. The diffusion coefficients of a micro solute, being at least an order of magnitude greater than the effective diffusion value for a colloid or cell ensures that just a finite liquid layer forms a screen for protecting the sensor surface. Of course, part of the protective feature relates to the fact that the liquid layer is mobile and so a laterally diffusion colloid/cell is swept along the channel before reaching the sensor, so access to the sensing surface (Fig. 2) is much greater than that a fouling agent. The location of an intramural sensor along the flow channel is thus of importance because there is the balance of the increased time available for lateral transport of target molecule versus that of a

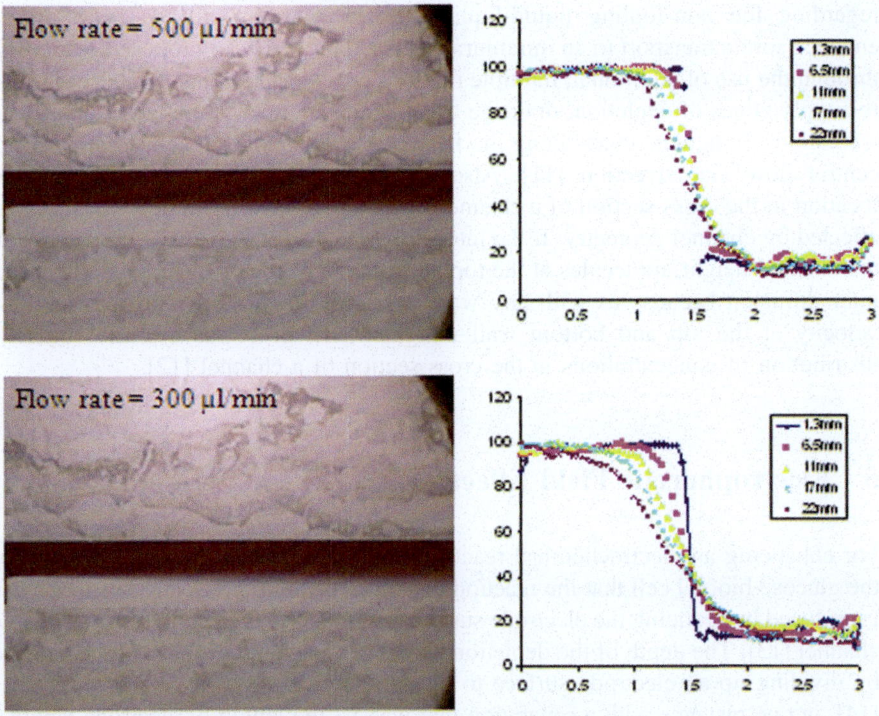

Fig. 5 A cross-channel densitometry analysis showing the location of the intense colour of permanganate solution as permanganate from such a solution diffuses laterally across the direction of flow into an aqueous receiving stream. The image represents a stop-flow experiment with loss of the sharp liquid-liquid interface as the permanganate diffuses across. Quantitative profiling was achieved by *grey* scale imaging. The horizontal axes is in mm and the key refers to distance along the flow channel

colloid, and so the sensor should not be too far down the channel, whilst positioning too near the flow entry means it is within the Languor distance. A quantitative evaluation of the diffusion process for a micro solute was undertaken using permanganate solution with stopped flow (Fig. 5); the concentration drop to the non-permanganate side becomes less distinct over time as lateral diffusion progresses; this apples to both micro- and macro-solutes, and of course the concentration profile itself offers a route to determining the diffusion coefficient value of a solute within a given sample matrix.

Laminar flow has been exploited in other sectors. Thus, in the case of electrochemical fuel cell design [10], non-mixing fuel and oxidant streams have been used in order to avoid the need for a membrane separator. Proposed advantages parallel those for sensing; a single channel simplifies packaging and sealing requirement, problems of membrane damage are avoided, fouling and permeability issues cease to be a worry and there is greater compatibility with Lab on a Chip design with elimination of one structural component. Arguably, whatever the advantages regarding the non-fouling liquid-liquid interface, it may still be necessary to enhance solute transport to an intramural sensor. This has been proposed, variously, through the use of additional, multiple inlets to provide additional solute sources, to re-supply zones of depletion, drainage of depleted solution via multiple outlets and a creation of angular ridges along the flow channel to enhance solute supply from central flow via advection [11]. Also, solute residence time at any particular location in the cross-section of a channel is not just a function of flow velocity, but affected by channel geometry. If the inter-diffusive distance is small compared with the channel height, molecules at the top and bottom of the channel have longer for inter diffusion because as with any wall proximity, there is a much slower flow velocity at the top and bottom wall margins. This gives the typical ‘butterfly’ distribution of concentrations in the cross-section of a channel [12].

5 Electrophoretic Field Effects

For enhancing an electrochemical reaction itself, it has been shown in the case of the glucose biofuel cell that the reaction depletion layer at the electrode surface can be reduced by reducing the electrode surface length along the length of a micro flow channel [13]. The depth of the depletion zone over an electrode can also be reduced by dividing up an electrode surface to create spaced a array of smaller electrodes [14], in one instance with a separation that was ×3 the length of electrode length a distinct enhanced power density was achieved at a microbiofuel cell [15]. This type of realignment will have exactly the same advantage for sensing, and may contribute to accessing low target molecule levels.

The differential diffusion of proteins and other solutes has been used as a means for their purification [16]. If the receiving stream in a dual flow channel has an affinity for these proteins, as has been observed in the case of a system for cell membrane proteins, then both an enhancement of concentration and purification by the receiving stream becomes possible [17]. This principle could be adapted to low level micro solute monitoring if the appropriate partitioning interface can be provided for the receiving stream.

Lateral application of an electric field induces electrophoretic transport of large charged molecules [18, 19], either as a basis for removing them from the assay sample or for increasing their concentration; this has analogy with field flow fractionation (FFF). A typical technical arrangement involves electrodes recessed behind a gel phase facing the channel, thereby avoiding flow disturbance at the electrode whist achieving electrical connectivity to the sample. Quite a dense, narrowed, highly localised distribution of a mobile protein phase can thereby be achieved. This topic has been considered recently [20] and may have value for sample manipulation during sensor based monitoring. Such driven solute transfer between two distinct solvent phases may be asymmetric, thus transport from a PEG solution to a dextran receiving stream occurs readily, but that in the reverse direction is constrained and non-linear, showing threshold effects [19]. In this context, the lateral transfer of protein was not seen to be pH, and therefore charge, dependent, but partition coefficients seemed to be important and a developed interfacial double layer at the liquid-liquid boundary will affect cross-phase mobility. A phosphate buffer was described as generating a high double layer potential leading to accumulation of protein at the interface whilst a Tris combination with BES buffer lowered the interfacial potential and thereby protein distribution [21, 22]. A rather different situation occurs with DNA fragments. Here, an apparent energy minimum is attained at the liquid interface, promoting electrostatically driven adsorption via an electrophoretic field. The complexity of the phenomenon is indicated by the promoted detachment of DNA fragments with increase of the electrical field gradient; this appeared to be thermally promoted, with detachment of fragments occurring in a size dependent manner. It appears that the local electrostatic field strength is substantially greater than any applied field, but the latter affects the local field by both a narrowing the thermodynamic potential energy well and a reduction of the height of the barrier to thermal displacement [23]. The much higher diffusivity of a protein versus DNA also, potentially, will help to separate these components for measurement and such dual liquid components could additionally allowing some size based discrimination of DNA prior to sensing.

6 Cell Resolution

When red and white cells from blood are brought into contact with a flowing liquid-liquid interface, the red cells tend to diffuse out into the bulk phase whereas the white cells are retained because of their larger size and lower mobility; the energy minimum experienced by the cells is also a factor and is determined by cell surface area, greater for white cells [24, 25]. Discrimination is also described in general between different nucleated cell types and between live/dead cells [26, 27]. So cell loaded samples can be manipulated in microfluidics for say accessing cell sub-types for toxin measurement or for avoiding cell and sensor contact.

7 Dual Flow Heterogeneity

A dual flow that involves high and low viscosity liquids shows significant flow asymmetry. At low Re in particular, the more viscous fluid occupies the larger cross-section of the flow channel, compressing the less viscous fluid (Fig. 6), with net flow rate in each fluid being the same; in this way, also, the balance of pressures across the flow interface is maintained [28, 29]. A solvent juxtaposition with inevitably different viscosities has been studied using a combination of phenol/chloroform/isoamyl alcohol and water [30]. The liquids are immiscible, but yet flow stability, and avoiding flow breakup, was a challenge, with surfactant

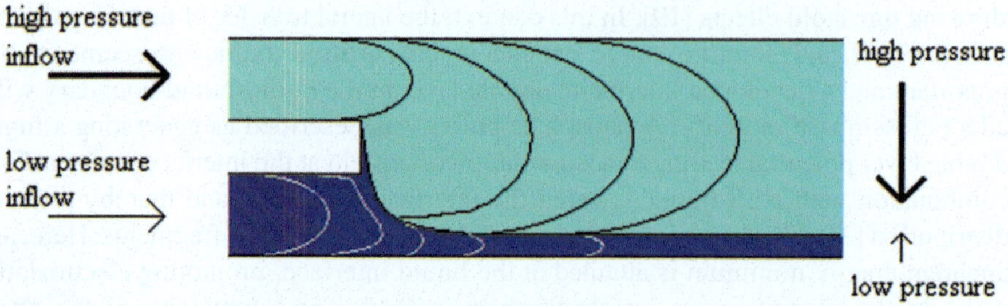

Fig. 6 A typical asymmetric width distribution for dual liquid-liquid flows unseparated by a physical barrier where a high and a low viscosity flow is juxtaposed. The asymmetry can be resolved by controlled, asymmetric pumping of flow

required to avoid initial interfacial curvature and later droplet formation. In this system, protein could be precipitated at the interface, but here it was not sufficiently to be able to completely remove protein that had been on the aqueous side. Where, however a more diffuse, miscible interphase is developed, solvent is present on both sides, determined by mutual solvent molecule diffusion; the result is a wider zone for protein accumulation and a more pure aqueous phase can be generated; the same effect is seen with DNA. Thus a DNA product of cell lysis could be extracted on a continuous basis.

A range of dual organic—aqueous flows have been investigated [31, 32], e.g. for sample extraction and increased solute concentration [33, 34]. Here, however, special flow units are usually required with etched glass typically used to give solvent resistance and allow optical monitoring. Examples of extraction include that of cobalt into a toluene phase [35], amphetamine like agents from urine into a chlorobutane phase [36] and ion pair extraction into chloroform [37]. One reported alternative to glass that simplified fabrication has been UV curable thiolene and a perfluoropolyether [34]; this was respectively used to establish the lipophilicity of caffeine through extraction and as a means of removing radioactive copper from aqueous solution as a possible system for radionuclide separation. In a further example, strychnine was extracted and then back extracted into a chloroform phase [38], and for a carbamate pesticide assay a sequential micro flow system achieved initial generation of a naphthol hydrolysis product, then a coloured diazonium adduct which was continuously extracted into a butanol phase for assay [39]. A further advancement of dual phases into triple parallel phases has been reported by Tokeshi et al. [35]. Here, sample mixing, reaction and extraction took place in a single chip and was defined as continuous flow processing (CFCP), illustrated, for example, for Co^{2+} with extraction of a 2-nitroso-1-naphthol complex from an aqueous phase into a non-polar xylene phase where the complex dissociated, for release of the free Co^{2+} into an acidic trapping phase.

8 In Situ Polymer Membrane Formation

Microfluidics has provided a versatile platform for sample extraction and assay as described above, recently reviewed by Atencia and Beebe [40]. Provided the balance of flows is established, stable, centralise flow patterns can be achieved. However, most descriptions above utilise the above type of direct liquid-liquid interfaces. There is a concern, however if flow disruption occurs, arguably during long term use a disruption to the flow dynamics could destabilise the co-linear flow and lead to measurement instability. Rather less work has been done to incorporate formal barrier structures that could both control solute transport and also stabilise flow. We have specifically addressed this challenge, along with other Groups, using flow balanced organic-aqueous flows to create reactive interfaces for generating polymeric barriers directly in situ. Rapid flow proves an advantage here as it promotes flow stabilisation, whereas slow flow is known to lead to segmentation [41]. Creation of distinct contralateral hydrophobic and hydrophilic flow surfaces also can provide stabilising pinning contacts for an organic and aqueous liquid, respectively. One hydrophobic surface modifier is organosilane on glass [42]; this has proved useful in protecting flow stratified, non-segmented flow. However, the approach does require a selective pre-deposition of the silane surface modifier, and a rather complex, ridged, channel design was necessitated to avoid channel crossover during surface treatment. Different fabrication materials for either side avoid this, e.g. PDMS and glass [43], but here the initial fabrication becomes more demanding. However, such balanced flows have not been used for interfacial membrane formation or synthesis, and focus essentially on extraction [44].

In our studies we used glass only fabricated Y-channels for hosting both solvent and aqueous flows, which considerably simplified construction. Stabilisation of flow and centralisation of the flow interface appeared to be achievable just through differential control flow control, avoiding viscosity induced flow asymmetry flow (Fig. 6). Initially, solvent flow was studied in an essential impermeable flow model system where the polymer generating reaction was guaranteed to be ultra-fast; the flows here delivered reactants for the formation of nylon 6,6 [8] based on a duality of water/xylene. The interface between the solvents moves naturally towards the xylene phase because of the lower solvent viscosity, but was countered by differential flow. The nylon system does not really serve as a convenient solute permeable membrane, except for gas molecules, and so an alternative fast reaction needed to be developed to create a membrane that had permeability and whose permeability and structure could be modified as a continuum, with the potential to host immobilised bioreagents, e.g. as a component in a bioreactor or a sensing system.

Accordingly, protein chemical cross linking was used to create the new class of self-standing membranes. This, of course, is the classic mode of protein immobilisation for supported protein layers for biosensors,. However, the standard cross linker reaction rates are too slow and in micro channels lead to channel blockage. By instead using a diacylchloride crosslinker (Fig. 7) an ultrafast interfacial

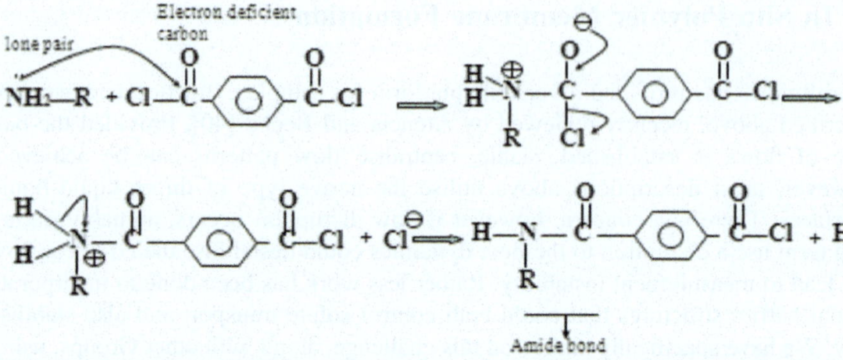

Fig. 7 Amide bond formation at a potential basic amino acid side chain. The rapid tetrephtaloyl chloride crosslinking reaction used for protein cross linking in micro channel. The reaction mixture comprises flow balanced organic and aqueous flow to avoid flow asymmetry.; the strength of this ensures that it is self standing and adherent to the top and base of the channel

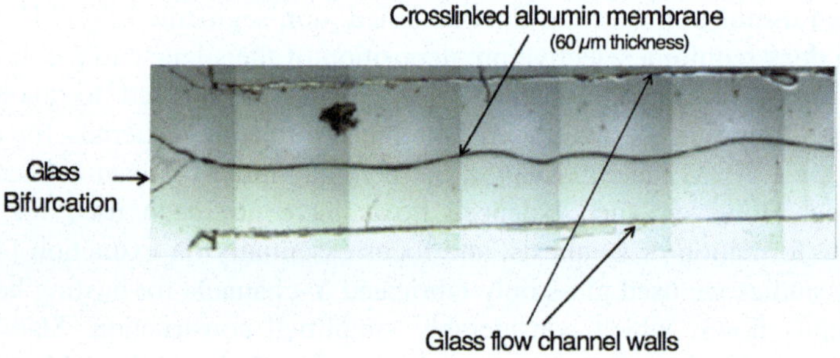

Fig. 8 The reaction interphase generated by the cross linking reaction in Fig. 7 leads to a self-limiting phase whereby a protein mesh limits further diffusive mixing. The edge of the self-limiting membrane is shown, and has the strength to be self supporting while also strongly adherent to the top and base of the glass flow channel

crosslinking occurs and a central, self standing, self limiting membrane results that has a cross linked mesh, overall thickness and permeability that can be tuned by controlling reactant concentrations, flow rates and choice of starting protein [45].

Additionally, a charge based separation is feasible, based on choice of protein isoelectric point and the operating pH for the solute transfer. A typical asymmetric flow was 400 uL/min water for the protein and 700 uL/min xylene for the cross linker in a 2 × 0.1 mm flow channel. Figure 8 shows a centralised, suspended membrane from cross linked albumin. A further benefit of this is the opportunity for using counterflow to accelerate solute transfer, and the operation of organic and other high partitioning solutes for analyte pre-concentration.

With protein based membranes, there are, of course, the readily available amino acid side groups for further facile immobilisation of bioreagents, including enzymes and antibodies. In the case of the former, not only can the enzyme be part of the assay chain, but it could be used to break down interferent species. Antibodies, alternatively, should permit immunosensing formats.

Nominally, a protein phase is biocompatible, though this would need to be determined on a case by case basis. By contrast, with the protein membrane hosting an RGD motif through co-retained fibronectin, laminin, vitronectin and other tissue matrix proteins, cell adhesion can be deliberately facilitated. The potential is there in that case for immobilised cells to serve as a toxin reporter, and a new embodiment of cell based biosensors could be achieved. Monitoring of such cells could be through localised sensor arrays or through optical tracking of the morphology of cell monolayers. A surrogate means of doing this is with electrode arrays in each channel for AC impedance spectroscopy. Cell layer retention on the side of the membrane opposite to the sample would allow delivery of effective growth medium and enable cell survival in a sample stream that no longer needed to be itself mixed with nutrient.

Considerable innovation is represented by work on interposed barrier structures, though the literature remains limited. Hydrogel barriers using, for example, polyacrylamide [46] and alginate [47] have been reported, though such gel phases with their high water content would have limited mechanical integrity, so would be used as a broad interphase and may lack the molecular discrimination of dense membranes. The environmental responsiveness and swelling of such gels could, however, have reactive cell and reagent reservoir release, and perhaps even serve as addressable valves. A rather more membrane like structure has been reported, made using alkali-acid gradients across a flow for the generation of chitosan membranes; these were shown to be able to resolve particulates and antibodies from micro solutes [48]. Nylon membranes of only ~ 10 um thickness were formed in situ in an earlier study using an X-shaped micro channel, alternative to our Y-shaped flow [8]; interestingly this ultra thin barrier had permeability to H_2O_2 [32]. A further polyamide membrane has been reported, formed using a xylene carrier at a water interface [49] and a palladium complex membrane formed from a reaction between a phosphine polymeric ligand and a palladium compound [50]. Overall, the interest in in situ membrane formation, including our own, was stimulated by the work on microfluidics by Whiteside's Group [51, 52].

Use of preformed membranes is certainly an alternative, perhaps easier, approach, though a less elegant solution to sample separation for microfluidics, and the smaller the channel diameter, the more difficult to achieve. However, in one report a simple commercial microporous membrane was sandwiched between two micro flow chambers and used to discriminate low molecular weight peptides from a protein containing biological mixture for later SERS analysis [53]. For blood plasma separation, a microporous polysulphone membrane was used where vertical deployment of the membrane avoided cell sedimentation onto the membrane so avoiding membrane blockage [54].

9 Sensor Incorporation in Microfluidics

Dual flow microfluidics offers the opportunity not only for sensor protection but for reagent loading into the receiving stream; this could include bio- as well as non-biobioreagents. In a simple model with amperometric detection of glucose using glucose oxidase we generated stable sensor responses (Fig. 9). So despite a continuous flow of solution enzyme orthogonal to the diffusion of glucose, a ~ 130 um Pt working electrode retained in the channel wall with no barrier membranes operated reliably; there would have been little concentration variation across the low diameter measurement surface. Importantly, the response is extended to well above the typical functional glucose km of an immobilised enzyme (~6 mM), and reflected solution enzyme km (>30 mM). This is likely to be due to two factors, the diffusion limitation to glucose transfer across the liquid stream and the continuous re-supply of oxygenated solution through ongoing flow. We have also found linear responses with combined enzyme and substrate in the supply stream. In principle it would be possible to operate with reagent requiring enzymes such as the dehydrogenases to broaden the analytical repertoire. Also, if an immunoreaction takes place rapidly as is the case for some homogeneous competitive immunoassays, then

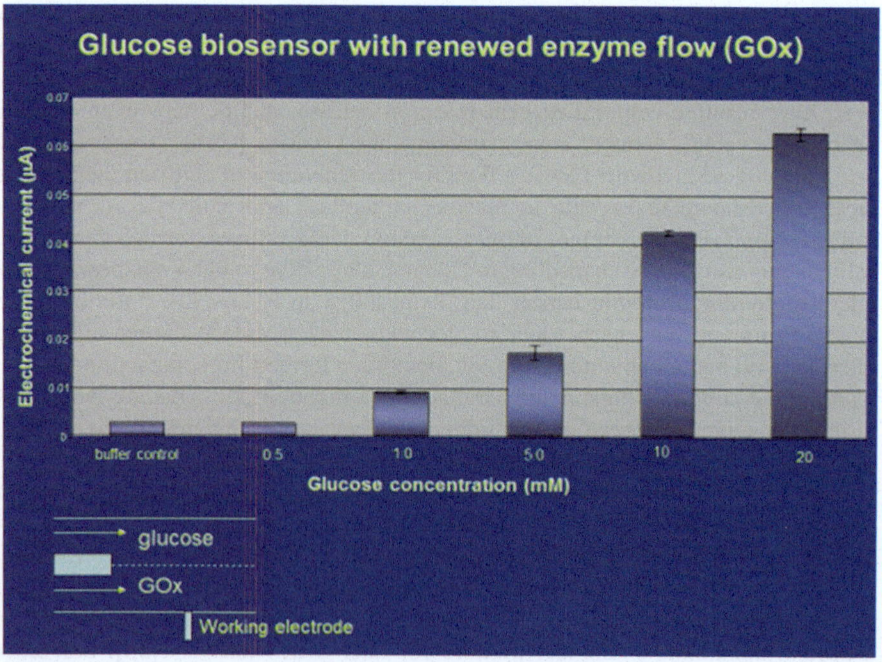

Fig. 9 Extended linearity state responses of a Pt wire working electrode to the H_2O_2 product of a glucose oxidase catalysed reaction using glucose oxidase dissolved in the flowing stream in the flow juxtaposed to the electrode surface, while glucose is present in the non-contacting sample stream

with the continuous re-supply of immunoreagent in the receiving stream, a quasi-continuous readout for immunosensing might be achieved for continuous tracking with immunosensors.

Microfluidics integrated biosensors based on enzymes have been reviewed recently [55], where the various alternatives of particle loaded, wall- and sensor immobilised configurations were highlighted. At the opposite end of the measurement spectrum, addressing multistep sample preparation procedures, a labelled DNA modified electrochemical sensor was integrated into a microfluidic device to registering hybridisation initiated change in the labelled DNA; however, this reagents configuration also included an on chip PCR amplification [56]. In a similar strategy, arrays of electrochemical sensors and heaters have produced for quantitative PCR utilising metal printed circuit board engineering [57].

For toxin detection other than metabolic disruption to cells, inhibition of enzymic reactions has provided a versatile routed for measurement. Thus, a microfluidic chamber incorporating an SPE and AChE [58] has been used for sarin detection. Amperometric detection of a thiocholine product; served as the reporter end point here. By an optical approach an aptamer was used to bind to polychlorinated biphenyls [59]. Here, an Ag based SERS detector surface functionalised with aptamer was integrated into a PDMS microfluidic channel, enabling assay of small sample volumes, whilst also providing for an aqueous interface that was more representative of a biological sample. SPR provides the ultimate surface biosensing technique, and localisation of such detection surfaces (LSPR) [60] has allowed multiple, simultaneous measurements on small sample volumes enabled through parallel microfluidics delivery. The general area of optical biosensors combined with microfluidics is thus receiving increased attention, and that in the near future advances in micro lens technology, organic LEDs and new materials for thin film photodiodes and high density optical pathway devices will lead to miniaturisation of optics with fluidics [61].

More adventurous bio-equivalent electrochemical configurations in microfluidic systems are being reported. Drug and chemical toxicity has been determined, for example, by using a polarised electrode to substitute for natural electron donation to immobilised Cyt P450; this powered oxidative catalysis by this liver enzyme system and enabled evaluation of resulting genotoxic metabolites from chemical agents. DNA damage here was registered by square wave voltammetry at a RuPVP coated electrode [62]. Nanowires offer a further route to low intrusion sensing, complementing microfluidics. In one study, the micro channel itself was use to direct flow for the formation of nanowire patterns [63]. As a departure from solid state flow channels, a soft gel micro channel has been reported with incorporated carbon nanowire arrays of 20 um diameter comprising core-shell titanium carbide with carbon; this was used for monitoring NO release from cultured endothelial cells [64]. Whilst not part of a microfluidic structure, an individually addressable, compressed array of 110 gold micro electrodes for NO/peroxinitrite detection shows the likely detection scale up that could be achieved for future microfluidics [65]. Miniaturisation and array compression with nano sensors has been reviewed [66], indicating the diversity of platforms possible including impedance measurement, potentiometric ion monitoring and nano-scale surface modification for improved bioreagent loading [66]. Extreme miniaturisation of ion selective electrodes, whilst retaining robustness, remains a challenge, and whilst the ISFET model is usable with appropriate MEMs engineering, there is now interest in wire type structures based on metal such as a Pt device where an ionophore coating was used for K^+ and Ca^{2+}, and iridium oxide for pH. Thin film sensors as have been designed for PDMS micro channels [67]. So diverse range of integrated structures is emerging, and is likely to enable multi parameter ion monitoring perhaps as future hyphenated systems with voltammetric and optical detection [68].

Immunosensors have also been under investigation at mininaturised levels, and a substantial range of surface immobilisation chemistries and flow designs are reported [69]. As with other measurement modalities the drive for such research is miniaturisation, multiple analysis and improved response to background ratios. What is really needed is a re-engineering of the affinity binding chemistry itself so either the limit of sensitivity can be enhanced, or the dynamics of binding reversal can be accelerated to enable continuous immunosensor based tracking in microfluidics. However, the scene is set for microfluidics to at least simplify the multiple reagent flows, binding and separation steps in immunoassay to allow remote use by non-technical operators.

10 Conclusions

A broad range of technical achievements with microfluidics and sensors has been covered. Whilst these descriptions do not themselves amount to a comprehensive overview of uTAS, they do provide information about the key elements of detection and sample/reagent transfer at smaller scales. This is qualitatively distinct from macroscale fluid movement, highly exploitable, and the likely focus of future investigations. This is likely to move on to nanoscale fluid handling where wall effects dominate even more, so allowing greater materials based solute and fluid manipulation. The flow emphasis in this chapter attempts to presage an era when continuous sampling and operation will be the norm for decentralised measurement. The eventual goal is continuous monitoring, at least as a way of setting up alerts for threat agents and for identifying trends. We have seen considerable advances here with respect to small molecules, but there remains a need to move on to (bio-)macromolecular measurement with, if not continuous, then quasi-continuous measurement with rapid, repeat assays enabled through rapid microfluidic handling. Here also it can be envisaged that if arrays can be compressed into such miniature systems, then microfluidic delivery can become a part of high throughput arrays. This will become easier for a future when miniaturisation technologies shift from the micro to the nano scale as a norm through advances materials and engineering science. It is possible that health care monitoring will be the initial focus of attention, but the follow on to threat agent monitoring can only thereby be accelerated.

Acknowledgments Generous support from the EPSRC, BBSRC and UKIER is gratefully acknowledged.

References

1. Ino K (2015) Microchemistry- and MEMS-based integrated electrochemical devices for bioassay applications. Electrochemistry 83:688–694
2. Wu CC, Lee GB, Chen MH, Luo CH (2005) Micromachined oxygen gas sensors for microscopic energy consumption measurement systems. J Med Eng Technol 29:278–287
3. Sawada E, Kazawa H, Yoshida Y, Iwasaki K, Mitsubayashi K (2006) A flexible and wearable glucose sensor based on functional polymers with Soft-MEMS techniques. Biosens Bioelectron 22:558–562
4. Zhou RJ (2005) Greenberg, Microsensors and microbiosensors for retinal implants. Front Biosci 10:166–179
5. Rios A, Zougagh M (2015) Modern qualitative analysis by miniaturized and microfluidic systems. Trac-Trend Anal Chem 69:105–113
6. Kaprou G, Papadakis G, Kokkoris G, Papadopoulos V, Kefala I, Papageorgiou D, Gizeli E, Tserepi A (2015) Miniaturized devices towards an integrated Lab-on-a-chip platform for DNA diagnostics. In: Van Den Driesche S (ed) Bio-Mems and medical microdevices
7. Braff WA, Bazant MZ, Buie CR (2015) Inertial effects on the generation of co-laminar flows. J Fluid Mech 767:85–94
8. Gargiuli J, Shapiro E, Gulhane H, Nair G, Drikakis D, Vadgama P (2006) Microfluidic systems for in situ formation of nylon 6,6 membranes. J Membr Sci 282:257–265
9. Kyriacou G, Vadgama P, Wang W (2006) Characterization of a laminar flow cell for the prevention of biosensor fouling. Med Eng Phys 28:989–998
10. Shaegh SAM, Nguyen NT, Chan SH (2011) A review on membraneless laminar flow-based fuel cells. Int J Hydrogen Energy 36:5675–5694
11. Yoon SK, Fichtl GW, Kenis PJA (2006) Active control of the depletion boundary layers in microfluidic electrochemical reactors. Lab Chip 6:1516–1524
12. Kamholz AE, Weigl BH, Finlayson BA, Yager P (1999) Quantitative analysis of molecular interaction in a microfluidic channel: The T-sensor. Anal Chem 71:5340–5347
13. Zebda A, Renaud L, Cretin M, Innocent C, Ferrigno R, Tingry S (2010) Membrane less microchannel glucose biofuel cell with improved electrical performances. Sens Actuat B Chem 149:44–50
14. Lim KG, Palmore GTR (2007) Microfluidic biofuel cells: the influence of electrode diffusion layer on performance. Biosens Bioelectron 22:941–947
15. Münchow G, Schönfeld F, Hardt S, Graf K (2008) Protein diffusion across the interface in aqueous two-phase systems. Langmuir 24:8547–8553
16. Meagher RJ, Light YK, Singh AK (2008) Rapid, continuous purification of proteins in a microfluidic device using genetically-engineered partition tags. Lab Chip 8:527–532
17. Hu R, Feng XJ, Chen P, Fu M, Chen H, Guo L, Liu BF, Rapid J (2011) highly efficient extraction and purification of membrane proteins using a microfluidic continuous-flow based aqueous two-phase system. Chromatogr A 1218:171–177
18. Stichlmair J, Schmidt J, Proplesch R (1992) Electroextraction—A novel separation technique. Chem Eng Sci 47:3015–3022
19. Münchow G, Hardt S, Kutter JP, Drese KS (2007) Electrophoretic partitioning of proteins in two-phase microflows. Lab Chip 7:98–102
20. Hardt S, Hahn T (2012) Microfluidics with aqueous two-phase systems. Lab Chip 12:434–442
21. Haynes CA, Carson J, Blanch HW, Prausnitz JM (1997) Electroststic potentials and protein partitioning in aqueous 2-phase systems. AIChE J 37:1401–1409

22. Theos CW, Clark WM (1995) Electroextraction-2-phase electrophoresis. Appl Biochem Biotechnol 54:143–157
23. Hahn T, Hardt S (2011) Size-dependent detachment of DNA molecules from liquid-liquid interfaces. Soft Matter 7:6320–6327
24. SooHoo JR, Walker GM (2009) Microfluidic aqueous two phase system for leukocyte concentration from whole blood. Biomed Microdevices 11:323–329
25. Bresme F, Oettel M (2007) Nanoparticles at fluid interfaces. J Phys: Condens Matter 19:1–33
26. Nam KH, Chang WJ, Hong H, Lim SM, Kim DI, Koo YM (2005) Continuous-flow fractionation of animal cells in microfluidic device using aqueous two-phase extraction. Biomed Microdevices 7:189–195
27. Tsukamoto M, Taira S, Yamamura S, Morita Y, Nagatani N, Takamura Y, Tamiya E (2009) Cell separation by an aqueous two-phase system in a microfluidic device. Analyst 134:1994–1998
28. Kim BJ, Liu YZ, Sung HJ (2004) Micro PIV measurement of two-fluid flow with different refractive indices. Meas Sci Technol 15:1097–1103
29. Xia HM, Wang ZP, Koh YX, May KT (2010) A microfluidic mixer with self-excited 'turbulent' fluid motion for wide viscosity ratio applications. Lab Chip 10:1712–1716
30. Reddy V, Zahn JD (2005) Interfacial stabilization of organic-aqueous two-phase microflows for a miniaturized DNA extraction module. J Colloid Interf Sci 286:158–165
31. Hisamoto H, Horiuchi T, Tokeshi M, Hibara A, Kitamori T (2001) On-chip integration of neutral ionophore-based ion pair extraction reaction. Anal Chem 73:1382–1386
32. Hisamoto H, Shimizu Y, Uchiyama K, Tokeshi M, Kikutani Y, Hibara A, Kitamori T (2003) Chemicofunctional membrane for integrated chemical processes on a microchip. Anal Chem 75:350–354
33. Ciceri D, Perera JM, Stevens GW (2014) The use of microfluidic devices in solvent extraction. J Chem Technol Biotechnol 89:771–786
34. Goyal S, Desai AV, Lewis RW, Ranganathan DR, Li H, Zeng D, Reichert DE, Kenis PJA (2014) Thiolene and SIFEL-based microfluidic platforms for liquid-liquid extraction. Sens Actuat B-Chem 190:634–644
35. Tokeshi M, Minagawa T, Uchiyama K, Hibara A, Sato K, Hisamoto H, Kitamori T (2002) Continuous-flow chemical processing on a microchip by combining microunit operations and a multiphase flow network. Anal Chem 74:1565–1571
36. Miyaguchi H, Tokeshi M, Kikutani Y, Hibara A, Inoue H, Kitamori T (2006) Microchip-based liquid-liquid extraction for gas-chromatography analysis of amphetamine-type stimulants in urine. J Chromatogr A 1129:105–110
37. Tokeshi M, Minagawa T, Kitamori T (2000) Integration of a microextraction system on a glass chip: ion-pair solvent extraction of Fe(II) with 4,7-diphenyl-1,10-phenanthrolinedisulfonic acid and tri-n-octylmethylammonium chloride. Anal Chem 72:1711–1714
38. Tetala KKR, Swarts JW, Chen B, Janssen AEM, van Beek TA (2009) A three-phase microfluidic chip for rapid sample clean-up of alkaloids from plant extracts. Lab Chip 9:2085–2092
39. Smirnova A, Shimura K, Hibara A, Proskurnin MA, Kitamori T (2007) Application of a micro multiphase laminar flow on a microchip for extraction and determination of derivatized carbamate pesticides. Anal Sci 23:103–107
40. Atencia J, Beebe DJ (2005) Controlled microfluidic interfaces. Nature 437:648–655
41. Kuban P, Berg J, Dasgupta PK (2003) Vertically stratified flows in microchannels. Computational simulations and applications to solvent extraction and ion exchange. Anal Chem 75:3549–3556
42. Hibara A, Iwayama S, Matsuoka S, Ueno M, Kikutani Y, Tokeshi M, Kitamori T (2005) Surface modification method of microchannels for gas-liquid two-phase flow in microchips. Anal Chem 77:943–947
43. Xiao H, Liang D, Liu GC, Guo M, Xing WL, Cheng J (2006) Initial study of two-phase laminar flow extraction chip for sample preparation for gas chromatography. Lab Chip 6:1067–1072

44. Assmann N, Ładosz A, Rudolf von Rohr P (2013) Continuous Micro Liquid-Liquid Extraction. Chem Eng Technol 36:921–936
45. Chang H, Khan R, Rong Z, Sapelkin A, Vadgama P (2010) Study of albumin and fibrinogen membranes formed by interfacial crosslinking using microfluidic flow. Biofabrication 2(Art. No: 035002)
46. Orhan JB, Knaack R, Parashar VK, Gijs MAM (2008) In situ fabrication of a poly-acrylamide membrane in a microfluidic channel. Microelectron Eng 85:1083–1085
47. Braschler T, Johann R, Heule M, Metref L, Renaud P (2005) Gentle cell trapping and release on a microfluidic chip by in situ alginate hydrogel formation. Lab Chip 5:553–559
48. Luo X, Berlin DL, Betz J, Payne GF, Bentley WE, Rubloff GW (2010) In situ generation of pH gradients in microfluidic devices for biofabrication of freestanding, semi-permeable chitosan membranes. Lab Chip 10:59–65
49. Zhao B, Viernes NOL, Moore JS, Beebe DJ (2002) Control and applications of immiscible liquids in microchannels. J Am Chem Soc 124:5284–5285
50. Uozumi Y, Yamada YMA, Beppu T, Fukuyama N, Ueno M, Kitamori T (2006) Instantaneous carbon-carbon bond formation using a microchannel reactor with a catalytic membrane. J Am Chem Soc 128:15994–15995
51. Kenis PJA, Ismagilov RF, Takayama S, Whitesides GM, Li SL, White HS (2000) Fabrication inside microchannels using fluid flow. Acc Chem Res 33:841–847
52. Kenis PJA, Ismagilov RF, Whitesides GM (1999) Microfabrication inside capillaries using multiphase laminar flow patterning. Science 285:83–85
53. Perozziello G, Candeloro P, Gentile F, Coluccio ML, Tallerico M, De Grazia A, Nicastri A, Perri AM, Parrotta E, Pardeo F, Catalano R, Cuda G, Di Fabrizio E (2015) A microfluidic dialysis device for complex biological mixture SERS analysis. Microelectron Eng 144:37–41
54. Liu C, Mauk M, Gross R, Bushman FD, Edelstein PH, Collman RG, Bau HH (2013) Membrane-based sedimentation-assisted plasma separator for point-of-care applications. Anal Chem 85:10463–10470
55. Mross S, Pierrat S, Zimmermann T, Kraft M (2015) Microfluidic enzymatic biosensing systems: a review. Biosens Bioelectron 70:376–391
56. Hsieh K, Ferguson BS, Eisenstein M, Plaxco KW, Soh HT (2015) Integrated electrochemical microsystems for genetic detection of pathogens at the point of care. Acc Chem Res 48:911–920
57. Tseng H-Y, Adamik V, Parsons J, Lan S-S, Malfesi S, Lum J, Shannon L, Gray B (2014) Development of an electrochemical biosensor array for quantitative polymerase chain reaction utilizing three-metal printed circuit board technology. Sens Actuat B-Chem 204:459–466
58. Tan HY, Loke WK, Nam-Trung N, Tan SN, Tay NB, Wang W, Ng SH (2014) Lab-on-a-chip for rapid electrochemical detection of nerve agent Sarin. Biomed Microdevices 16:269–275
59. Fu C, Wang Y, Chen G, Yang L, Xu S, Xu W (2015) Aptamer-based surface-enhanced raman scattering-microfluidic sensor for sensitive and selective polychlorinated biphenyls detection. Anal Chem 87:9555–9558
60. Ruemmele JA, Hall WP, Ruvuna LK, Van Duyne RP (2013) A localized surface plasmon resonance imaging instrument for multiplexed biosensing. Anal Chem 85:4560–4566
61. Ligler FS (2009) Perspective on optical biosensors and integrated sensor systems. Anal Chem 81:519–526
62. Wasalathanthri DP, Mani V, Tang CK, Rusling JF (2011) Microfluidic electrochemical array for detection of reactive metabolites formed by cytochrome P450 enzymes. Anal Chem 83:9499–9506
63. Chen J, Zu Y, Rajagopalan KK, Wang S (2015) Manufacturing a nanowire-based sensing system via flow-guided assembly in a microchannel array template. Nanotechnology 26
64. Li L-M, Wang X-Y, Hu L-S, Chen R-S, Huang Y, Chen S-J, Huang W-H, Huo K-F, Chu PK (2012) Vascular lumen simulation and highly-sensitive nitric oxide detection using three-dimensional gelatin chip coupled to TiC/C nanowire arrays microelectrode. Lab Chip 12:4249–4256

65. Quinton D, Girard A, Kim LTT, Raimbault V, Griscom L, Razan F, Griveau S, Bedioui F (2011) On-chip multi-electrochemical sensor array platform for simultaneous screening of nitric oxide and peroxynitrite. Lab Chip 11:1342–1350
66. Wei D, Bailey MJA, Andrew P, Ryhaenen T (2009) Electrochemical biosensors at the nanoscale. Lab Chip 9:2123–2131
67. Liao W-Y, Weng C-H, Lee G-B, Chou T-C (2006) Development and characterization of an all-solid-state potentiometric biosensor array microfluidic device for multiple ion analysis. Lab Chip 6:1362–1368
68. Johnson RD, Gaualas VG, Daunert S, Bachas LG (2008) Microfluidic ion-sensing devices. Anal Chim Acta 613:20–30
69. Bange A, Halsall HB, Heineman WR (2005) Microfluidic immunosensor systems. Biosens Bioelectron 20:2488–2503

New Routes in the High-Throughput Screening of Toxic Proteins Using Immunochemical Tools

Mihaela Puiu and Camelia Bala

Abstract The chapter reviews several aspects related to the mechanism of action of toxic proteins used as biological warfare agents, together with the latest advances in immunosensor development. Emphasis will be put on the role played by the nanoparticle technology in the sensing and transduction design. The potential applications of the nanostructured immunosensors in point-of-care systems and the amenability of these devices for detection on-the-field will be critically commented.

Keywords Botulinum neurotoxin · Ricin · Immunosensor · Nanoparticles · Point-of-care-testing

1 Lethal Plant/Bacteria Proteins as Potential Warfare Agents: An Overview

Recent technological changes have increased the importance of protein toxins as biological warfare agents (BWA) [1]. These emerging threat agents are produced by living organisms: bacteria, plants or animals and differ significantly from replicating agents (viruses and bacteria) and from classic chemical agents. Chemical Weapons Convention (CWC) (1993) specifically includes protein toxins (abrin, botulinum toxins, *Clostridium perfringens* toxins, ricin, cholera toxin and tetanus toxin along with other highly toxic chemicals as chemical agents [2]. Since toxins are not volatile and with rare exceptions, do not directly affect the skin, an aggressor would have to present toxins to target populations in the form of aerosols, allowing contact with the more vulnerable inner surfaces of the lung. Aerosol attacks have two major

M. Puiu · C. Bala
R&D Center LaborQ, University of Bucharest, 4-12 Regina Elisabeta Blvd., 030018 Bucharest, Romania

C. Bala (✉)
Department of Analytical Chemistry, University of Bucharest, 4-12 Regina Elisabeta Blvd., 030018 Bucharest, Romania
e-mail: camelia.bala@g.unibuc.ro

D.P. Nikolelis and G.-P. Nikoleli (eds.), *Biosensors for Security and Bioterrorism Applications*, Advanced Sciences and Technologies for Security Applications,
DOI 10.1007/978-3-319-28926-7_3

advantages: (1) the bioagent particles are dispersed in the atmosphere and, driven by the wind, can drift over large areas, and (2) many diseases are more virulent when spread by the aerosol route. For maximum effect it is generally agreed that the bioagent particles should be in the size range of 1 to 10 μm. Larger particles will precipitate out of the air in too short a time; smaller particles will tend to be expelled from the lungs [3]. The acting mechanism of protein toxins, as a group, dictates both the manner they could be used as weapons and the manner they may be defended against. Protein toxins produced by bacteria and plants enter eukaryotic cells and inhibit the enzymatic synthesis of proteins. The majority of these toxins are referred to as AB-toxins because of their structural organization (Fig. 1) [4, 5]. The A moiety generally has enzymatic activity and modifies a cellular target upon entry into the cytosol which leads to cell death or other effects on cellular physiology. The B moiety, consisting of one or more subunits, binds the toxin to cell-surface receptors and can also play a role in the translocation of the A moiety to the cytosol [6]. Examples of intracellular targets are ribosomes, actin, small GTP-binding proteins like Rho, and heterotrimeric G-proteins [7–9] Commonly, an AB-toxin is synthesized in an inactive form that is activated by proteolytic processing (Fig. 1, [10]). Cholera toxin, ricin and clostridial neurotoxins are cleaved by the producing organism at a region between two cysteine residues [10]. *Clostridium botulinum* C2 toxin depends on proteolytic processing of the B moiety to expose a site that then binds the A moiety non-covalently [10]. The result of the proteolytic cleavage is a toxin with an enzymatically active part linked to the rest of the molecule by a disulfide bond.

The AB-toxins, bind to specific receptor molecules at the cell surface. For instance, the receptor for diphtheria toxin is the uncleaved precursor of the heparin binding EGF-like growth factor [11]; Shiga toxin binds to the glycolipid Gb3 [12] and cholera toxin binds the monosialotetrahexosylganglioside GM_1 [13]. Plant toxins, like ricin, usually bind to carbohydrates [14], regardless of whether they are attached to lipids or proteins.

1.1 Plant Protein Toxins

Plant toxic proteins belong to a group of phytotoxins, being glycoproteins with molecular weights of about 60 kDa, with two AB subunits linked into a dimer by a disulfide bond. The B chain is lectin with sites for carbohydrate binding; the A chain is a specific N-glycosidase which modifies the 28S rRNA-60S ribosomal subunit. The most toxic, ricin is a protein produced by the castor oil plant, *Ricinus communis*. The whole plant is poisonous, with ricin reaching the highest levels in seeds. Once in the target cell, a single ricin molecule inactivates around 1500 ribosomes per minute, ultimately resulting in cell death. A fatal ingested dose for human is about 1 μg/kg [8, 9, 15] but the levels of toxicity vary depending on the route of ricin exposure. The mean lethal dose (LD_{50}) in mice is approximately 1000–fold lower by injection or inhalation than by oral administration [9, 14]. The

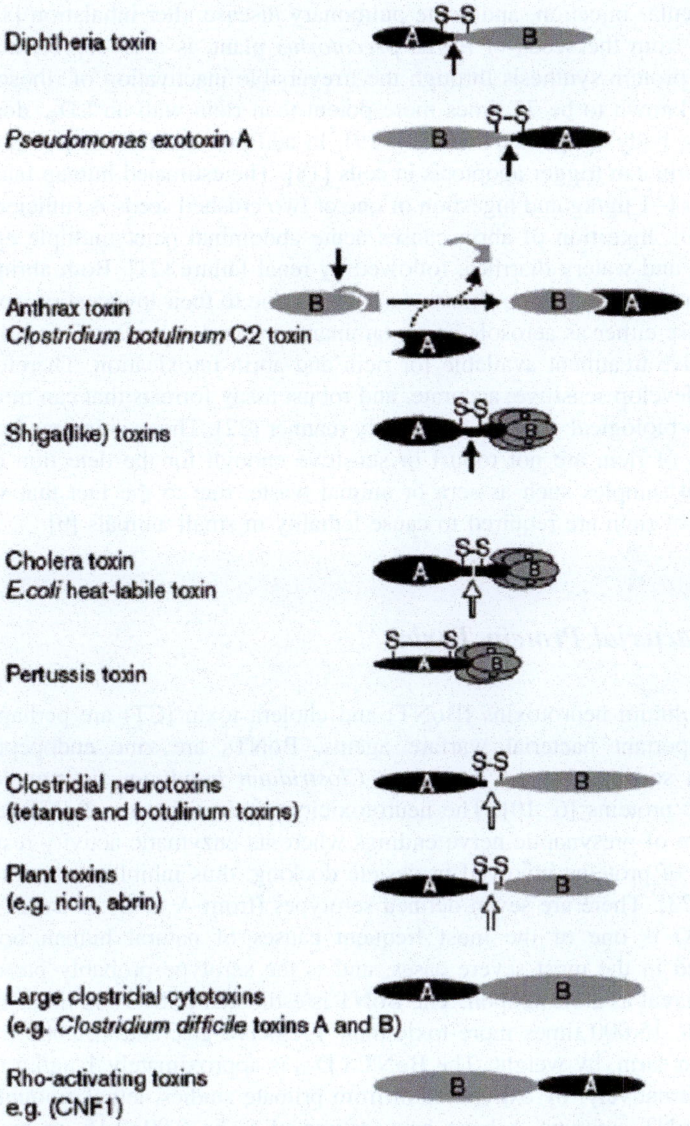

Fig. 1 Structural organization of AB-toxins reproduced from [7] with permission of Elsevier

lethal dose for an adult human is about 0.35–0.7 mg by inhalation, while the lethal oral dose has been estimated to be between 1 to 20 mg of ricin/kg body weight. This large discrepancy in oral and systemic toxicity is probably due to the harsh digestive conditions found in the stomach and epithelium, and innate immune barriers present in the intestinal tract [9, 16]. Ricin intoxication can manifest as gastrointestinal hemorrhage after ingestion, severe muscle necrosis after intramuscular injection, and acute pulmonary disease after inhalation [15]. Abrin, obtained from the seeds of *Abrus precatorius* plant, is a glycoprotein toxin that prevents protein synthesis through the irreversible inactivation of ribosomes [17]. Abrin is known to be 75 times more potent than ricin with an LD_{50} dose of only 2.8 μg/kg body weight of mice [17–19]. In addition to inhibition of protein synthesis, abrin can trigger apoptosis in cells [18]. The estimated human fatal dose for abrin is 0.1–1 μg/kg and ingestion of one or two crushed seeds is sufficient to cause death [20]. Ingestion of abrin causes acute abdominal pain, multiple episodes of vomiting and watery diarrhea, followed by renal failure [21]. Both abrin and ricin are of particular concern as biowarfare agents due to their low cost of isolation and ease of use either as aerosols or contaminants in food or water. Currently, there is no effective treatment available for ricin and abrin intoxication. Therefore, it is a need to develop sensitive, accurate, and robust assay formats that can monitor ricin in human biological samples in a timely manner [22]. The existing methods for the detection of ricin are not robust or sensitive enough for the detection of ricin in biological samples such as sera or animal waste, due to the fact that very small amounts of ricin are required to cause lethality in small animals [9].

1.2 Bacterial Protein Toxins

The Botulinum neurotoxins (BoNT) and cholera toxin (CT) are perhaps the two most important bacterial warfare agents. BoNTs are zinc endopeptidases of 150 kDa secreted by the bacterium *Clostridium botulinum* as complexes with non-toxic proteins [6, 10]. The neurotoxicity requires that the BoNT reaches the cytoplasm of presynaptic nerve endings where its enzymatic activity results in the cleavage of proteins involved in vesicle docking, thus inhibiting neurotransmitter release [23]. There are seven defined serotypes (from A to G) of BoNTs. Type A (BoNT/A) is one of the most frequent causes of natural human botulism, is implicated in the most severe cases, and is the serotype probably presenting the highest threat as a bioweapon. The BoNT is estimated to be 1000 times more toxic than ricin, 15,000 times more toxic than VX nerve gas, and 100,000 times more toxic than sarin, by weight. The BoNT LD_{50} is approximately 1 ng/kg in humans [24]. Alternatively, by extrapolation from primate studies, lethal amounts of crystalline BoNT serotype A have been estimated to be 0.09–0.15 μg by injection, 0.70–0.90 μg by inhalation, or 70 μg orally for a 70 kg human [25]. BoNTs must be detected rapidly to enable timely administration of neutralizing antibodies to the exposed persons and to prevent additional exposures. The risk of bioterrorism involving the release of BoNT into milk supplies was recently analysed [26]. Botulinum toxin produces a descending flaccid paralysis [27]. Usually, an identification of botulism is made through clinical manifestations and diagnosis, with subsequent confirmation by laboratory identification of clostridial spores or toxin in food, clinical or environmental samples [28]. The speed of recovery from botulism increases with the timely administration of antitoxin or medical interventions [29].

These findings emerged the need for assays that could be used in the field and that allow rapid detection of BoNTs. The reference method for detecting BoNTs continues to be the mouse lethality assay because it is a highly sensitive functional assay that directly measures the amount of toxic BoNT, despite of its well-known limitations. Foremost, this assay usually requires 24–48 h to yield results. However, antibody therapy is most efficient when given before the clinical signs of botulism, which can occur less than 12 h after exposure. Also, the need of maintaining the test mice makes it impractical outside reference laboratories and particularly in field conditions [26]. Cholera toxin (CT) is the major virulent factor of toxigenic strains of *V. cholerae* [30]. It has a common hetero-hexameric structure consisting of a single, enzymatically active A subunit (CT-A, 27 kDa), non-covalently linked to a pentameric core of five identical receptor-binding B subunits (CT-B, 58 kDa) [31, 32]. The CT's biological action is initiated by the binding of CT-B to the ganglioside GM_1 receptor on the intestinal cell membrane followed by internalization of CT-A into the cell where it activates adenylate cyclase [33]. This leads to increased intracellular levels of cyclic AMP, which in turn results in the secretion of large amounts of water and electrolytes into the bowels, causing severe diarrhea and death in the absence of a rapid treatment [32]. The lethal dose of CT in humans is relatively low (LD_{50} = 250 μg/kg [34]). CT has become a very important target in biological detection, and there has been increasing interest in the development of rapid and sensitive methods for the determination of CT. The majority of these techniques are based on recognition between the GM1 ganglioside and CT [35]. Tetanus neurotoxin (TeNT) is a 150 kDa tripartite toxin with a 50 kDa light chain which exhibits zinc endopeptidase activity and a 100 kDa heavy chain [36]. Along with BoNT, TeNT belongs to clostridial neurotoxins family Recently, TeNT was reported to enter central nervous system (CNS) neurons upon synaptic vesicle cycling that was mediated by the direct binding to synaptic vesicle protein 2(SV2); this implies that TeNT and BoNT utilize common mechanisms to enter CNS neurons [27]. Unlike BoNT, TeNT causes spastic paralysis [37]. Vaccination against the tetanus toxin is a routine procedure in Western countries. Thus, TeNT seems to be an appropriate proof-of-principle target in sensing human immune responses [38]. Antibodies able to neutralize TeNT are the main factors in protection against tetanus disease and preventive detection of this toxin. Antibody-based therapeutics for treatment of tetanus is currently available on the market but its production is tedious. Therefore, the preparation of tetanus-specific antibodies that could be easily produced in large scale in vitro is now in high demand [39].

The main issue concerning a biological attack is determining if and when the attack has occurred. Often is difficult to distinguish among the first symptoms of BWA infection and infection symptoms caused by more benign biological agents [40]. Therefore, molecular methods employing devices able to identify chemical markers from known biological agents are required. One can mention two classes of such devices which can be referred to as **detect-to-treat**, (allowing early identification of an infection), and **detect-to-protect** (providing a warning that a site is contaminated by BWAs to prevent infection) [41]. The criteria guiding these specific analytical methods are quite different. A **detect-to-treat** system must be able to identify a BWA, from biological samples, within a few hours of infection. Most of these methods are based on immunoassays using antigen/antibody reactions. ELISA (enzyme linked immunosorbent assay) and RIA (radio immunoassay) are widely used, although their application to biological samples is limited to urine or serum [42, 43]. These methods have important advantages such as rapidity, reduced sample manipulation and high sensitivity, but display significant disadvantages regarding the occurrence of false positives and negatives, the shortage of procedures with quantitative application and, sometimes, the lack of specificity [42, 44]. Therefore, the results require very often a further validation with other analytical methods based on chromatographic techniques in tandem with mass spectrometry. On the other hand, a **detect-to-protect** system must be able to provide a warning within a couple of minutes from a sample without any user intervention. Some analytical devices, which could satisfy both sets of conditions, are antibody-based sensors. They provide sufficiently low detection limits to detect a single BWA, being able to sample analytes from complex biological matrices, giving responses within a couple of minutes, giving rarely false positives and being portable devices which require no power to operate. Of course, no single device will match all these features in the near future; therefore the financial and scientific efforts should be focused on developing in situ adapted assays for the high throughput screening of BWAs [40].

2 Nanostructured Immunosensors for the on-Site Detection of Toxic Proteins in Food and Biological Samples

Antibodies (Abs) are probably the most used "affinity-recognition" elements in biosensing technologies, covering a wide range of applications from food safety and environmental monitoring to clinical analysis and medical diagnosis [45, 46]. The antibodies can be either polyclonal, if they are generated from different immune cells, or monoclonal, if they are generated from identical clones of a single parent cell. Polyclonal antibodies bind different regions from the antigen with different binding affinities while the monoclonal antibodies bind to the same target region (epitope), with equivalent affinity; they are far more selective than polyclonal antibodies [47]. Antibody-based sensors have been intensively used to quantify proteins, biological toxins, and biomarkers [48], based on the principle that antibody-antigen interactions are converted to a measurable signal by the transduction system (electrochemical, optical or piezoelectric). Antibodies can be immobilized onto different solid surfaces by physical or chemical adsorption, covalent coupling, using defined linkages such as glutaraldehyde, carbodiimide, proteins A and G, or gold nanoparticles (AuNP) [49]. Antibodies combined with nanomaterials are currently used to build-up powerful sensing systems [50, 51]

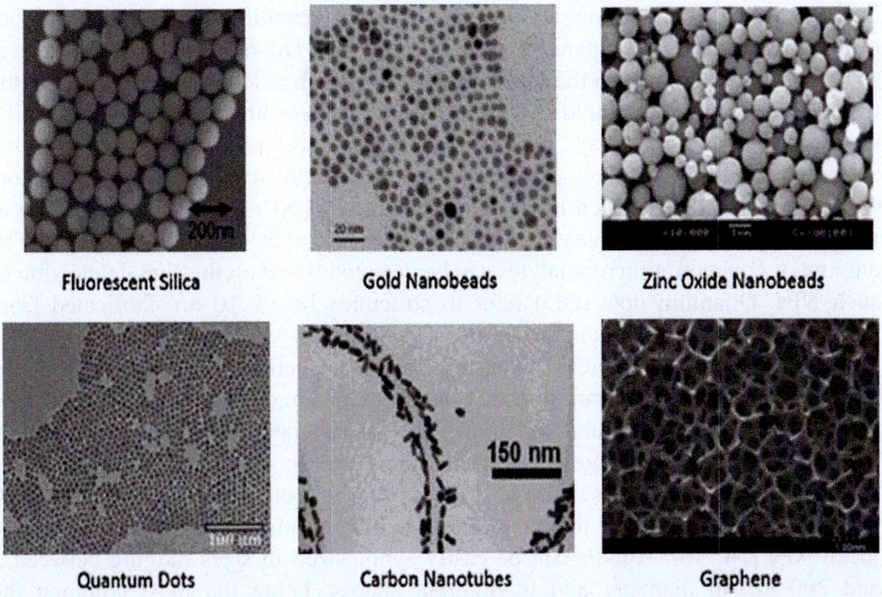

Fig. 2 Transmission electron microscopy monograph of NPs applied in immunosensor applications reproduced from [51] with permission of Elsevier

because nanomaterials can act either as electron transport mediator due to their good conductivity and electrocatalytic activity, or as support matrix displaying high specific surface area, strong adsorption capacity and the ability to be functionalized [51]. The metrics applied to immunosensors are devices that can detect the target(s) of interest in a fast and selective manner, with a minimal user input. The properties of nanoparticles (NPs) were exploited in the capture and concentration of analytical targets, in addition to transduction of the immune-analytical interaction [52]. Common NPs used in immunoassay design are depicted in Fig. 2. The properties of nanoparticles are adjustable due to the fact that the chemical and physicochemical characteristics are highly related to size on the nanoscale. Nanoparticles display a greater surface/volume ratio, therefore leading to an overall large working area to undertake chemistries, electrochemistry or immobilization of bioaffinity agents [53]. At nanoscale, the optical properties are significantly different compared to bulk materials. For example, AuNPs may appear red or black which is very different from the bulk material due to the influence of quantum effects (quantum confinement). Specifically, the electrons in NPs are densely packed having the energy difference between highest valence band and the low conduction band increased [51, 54].

Consequently, the energy to excite the particle increases, with higher energy being released when the electrons return to the ground state. This effect is visualized via a color shift in terms of optical characteristics and semi-conductor properties for phenomena such as surface plasmon resonance (SPR). Another property displayed by some NPs (e.g. metal oxides) is paramagnetism resulting from a lower number of unpaired electrons compared to the bulk material. The net result is a lower level of magnetism compared to the bulk material, but exhibits ferromagnetism under the influence of a magnetic field [51]. Macromolecule labeling such as metallic NPs conjugated to antibody plays a significant role in development of highly sensitive and specific bioassays, such as immunostrip and electro- chemical immunosensors assays. Biomolecules which have been labelled with NPs are, not only capable of maintaining their bioactivity but also interacting with their counterparts. The amount or concentration of analytes can be detected based on the signal detection of such NPs. Quantum dots (QD) refer to structures below 10 nm fabricated from materials taken from elements of groups II–VI and III–V [51]. QD absorb photons that generate exciton thereby releasing electromagnetic energy in the form of photoluminescence. QD emission wavelength depends on their size which can be controlled or tuned. Modulation of emission spectra opens the way for multiplex analysis with the advantage of intense fluorescence compared to standard fluorophores [55]. Metallic NPs could be employed in both of immunosensors and DNA sensors, where AuNPs are the most often used labels among all currently available metal NPs [54, 56]. AuNPs can be easily synthesized in sizes ranging between 3 and 200 nm in diameter and in different shapes, being the most common the quasi-spherical shape, mainly due to their surface energy that favors the formation of spherical particles. Generally, the method of choice to synthesize quasi-spherical gold NPs is the chemical reduction of Au(III) to Au(0) ions using sodium citrate as a reducing agent, a method first developed by Turkevich [57] and later optimized by Frens [56]. Besides gold and silver, inorganic NPs [58], platinum NPs [59], europium NPs [60, 61] were used to functionalize antibodies. It was reported that increasing the particle size increases the surface area to bind more Abs, which ultimately improves sensitivity [62]. Still, even if metallic NPs with higher diameter are expected to improve assay sensitivity, it was demonstrated that larger sized AuNPs are more prone to non-specificity than their smaller counterparts [63]. There are different approaches for the antibody functionalization of metal NPs, most of them being based on the following general protocols:

- Immobilization via noncovalent binding onto charged surfaces, without chemical coupling and retaining high activity of the antibody. Functionalized silica beads also provide a charged surface by which proteins can be adsorbed via ionic bonds [64]. Iron or zinc oxide beads can also be used for adsorbing bioaffinity agents by virtue of carrying a net negative charge at neutral pH [51, 65]. Covalent binding of affinity agents can also take place onto metal oxide nanobeads, although require to be functionalized with organic (e.g. lauric acid) or inorganic (silicon oxide) derivatives [66].
- Chemisorption (e.g. quasi-covalent binding of thiol functionalized antibody to gold NP) which allows oriented functionalization, but requires NPs with capping agents with weaker adsorption than the derivatization moiety. Due to thiol-modification, interferences by other chemical groups available for

Table 1 Nanoparticles commonly used in immunoassays for protein detection

Type of nanoparticle/material	Properties	Application in immunoassay	Reference
Transition metal/gold	Reacting with thiol groups of the immuno-affinity agents	Functionalization of the antibody or protein antigen Enzyme label	[17, 52, 68, 69] [24]
Transition metal/silver	Reacting with thiol groups of the immuno-affinity agents	Functionalization of the antibody or protein antigen	[70]
Metal oxides/iron oxide	Paramagnetic Reacting with organic/inorganic linker agents for protein conjugation Easy to use	Functionalization of the antibody	[71–73]
Non-metal oxide/silica	Encapsulating fluorescent dyes Biocompatibility	Fluorescent label of immuno-affinity agents Antibody immobilization	[64, 74]
Heavy metal salts/quantum dots	Semi-conductor Narrow emission spectra	Fluorescent label of immuno-affinity agents	[75, 76]
Carbon/nanotubes	Mechanical strength Electrical conductivity	Electrode Immobilization of antibodies	[77, 78]
Carbon/nanowires	Electrical conductivity	Electrode Immobilization of antibodies	[79]

adsorption to the gold surface, may occur; this approach may be prone to chemical degradation and surface oxidation of some NPs (e.g., silver) [67].

- Affinity-based (e.g. His-tag protein binding to nitrilotriacetic acid (Ni-NTA) derivatized gold NP); beside the oriented functionalization this approach provide very straightforward binding between affinity pairs. However, the drawback is that both NPs and biomolecules (antibodies or their antigen proteins) require modification with an affinity pair, which makes the approach limited to availability of suitable binding affinity pairs [56].

The main applications of nanoparticles in immunosensing techniques are summarized in Table 1.

Although nanoparticles usually enhance the analytical performance of the immunoassay techniques, it is still in demand to improve aspects such as the achievement of a higher specificity or the possibility of application to complex samples such as serum, gastric content and urine [42]. Meanwhile, it should be employed less complicated techniques with minimal number of steps that will allow the detection of the target toxin in the matrix, at the required concentration of

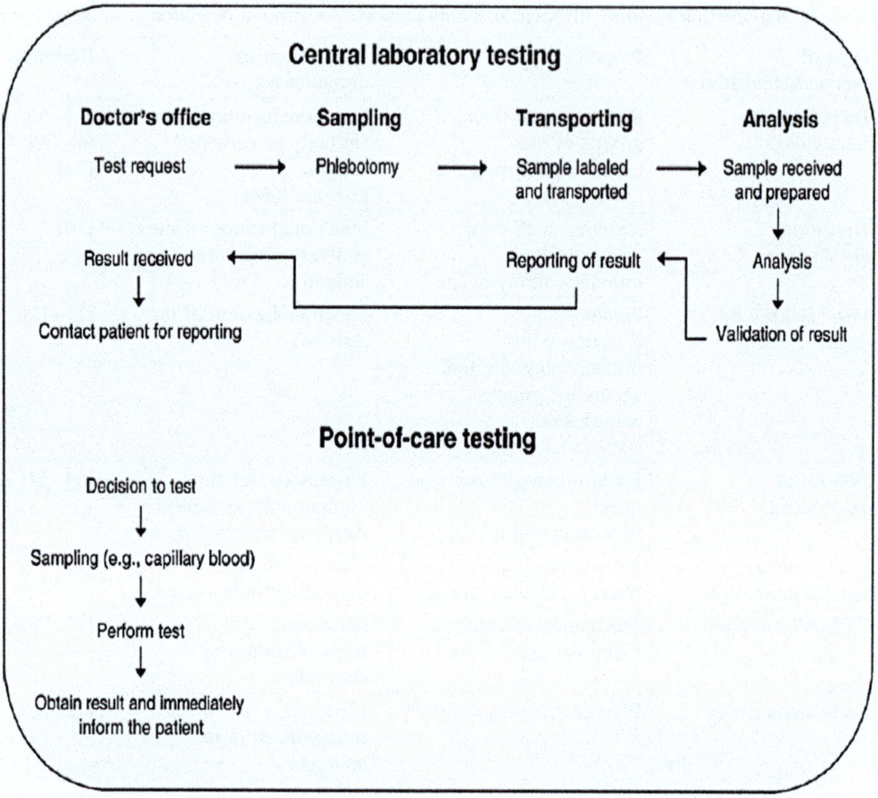

Fig. 3 General features of POCT platforms vs. compared to the current laboratory testing reproduced from [81] with permission of Elsevier

interest. For this reason, sample preparation needs to be tailored for the individual transducer and required detection level of the target [80].

Therefore, the immunosensors have to be used for the fast diagnosis in remote environments, where performing ELISA is not possible. The advantages of point-of-care-testing (POCT) platforms versus laboratory analysis are presented in the diagram of Lode (Fig. 3) [81].

Imunnosensing—based POCTs offer the possibility of carrying out "in situ" diagnosis with a minimum training using fast, inexpensive and multiple assays. The transduction techniques used immunoassay design can be described as electrochemical or optical based. The main advantage with electrochemical transduction is the simplified design, sensitivity and amenable to miniaturization [82]. Optical transduction is as sensitive or greater than electrochemical although has the disadvantage of needing camera/detectors for quantitative measurements [83]. The main concept in immunosenor design is to be operative outside the laboratory environment. Current commercial immunosensors for out-of-lab testing were until relatively recently limited to optical fibers, dip sticks, lateral flow immunoassay and amperometric devices [51] but the advances in nanotechnology enabled monitoring of the immunointeractions with portable platforms.

2.1 *Electrochemical Immunosensors*

The recently reported electrochemical immunosensors for BWA protein detection were mostly amperometric and impedimetric. In the past decade were reported redox sensors using voltammetric techniques such as alternating current differential pulse —and square wave voltammetry for the direct detection of an antibody [84, 85] or antigen protein [86]; In the first case the recognition element was a redox-tagged peptide epitope immobilized onto electrode surface. In the second case, the redox-tagged lipid microstructures containing the surface ligand GM_1 were immobilized onto gold surface for the direct Cholera toxin detection. The operating principle of the sensor was "signalling off", i.e. the binding of the target CT to GM_1 hindered the electron transfer from the redox tag to the electrode surface this causing a sharp decrease of the electrochemical signal. This approach provided a working range expanding from 1.0×10^{-8} to 5.0×10^{-7} M but was not available for real samples [40].

2.1.1 Amperometric Immunosensors with Highly Conductive Nanoparticle-Modified Transducers

Amperometry measures the current generated/consumed in redox reactions occurring at the surface of the working electrode. Most used are the three electrode systems which have a separate reference electrode that enables the applied potential to be controlled [51]. Amperometric immunosensors use antibodies or antigens immobilized onto electrode surface, but also require enzyme-labeling usually with horseradish peroxidase (HRP) or alkaline phosphatase (ALP) [87]. A sandwich ELISA system was used to detect ricin from water samples with multiwalled carbon nanotubes—modified paste electrode (MWCNTPE) providing a linear in the range 0.625–25 ng/ml [88]. The ricin capture antibody (raised in rat) was physically adsorbed on the modified electrode surface, while the ricin antigen and the revealing antibody (raised in rabbit), tagged with ALP were successively added. Each measurement requested around 20 min, far less time consuming compared with other ELISA-like assays. This portable ricin detection system using any hand held potentiostat can be used in field testing. An amperometric sensor for BoNT/A using gold nanoparticles as electronic bridges and signal amplifiers was recently reported [68]. Here AuNPs were first thetered with mixed PEG-aryl-diazonium salts and attached to the glassy carbon (GC) electrode by forming Au–C bonding, which displayed greater stability compared to Au and NH–Au-S bonds. Then, the thetered AuNPs were functionalized with the capture antibody. A similar procedure was used to functionalize thetered AuNPs with HRP-labeled anti BoNT/A antibody for the sandwich assay. Hydroquinone was used as mediator for HRP electron transfer (Fig. 4). The linear range was 4–35 pg/ml, and the detectable concentration of

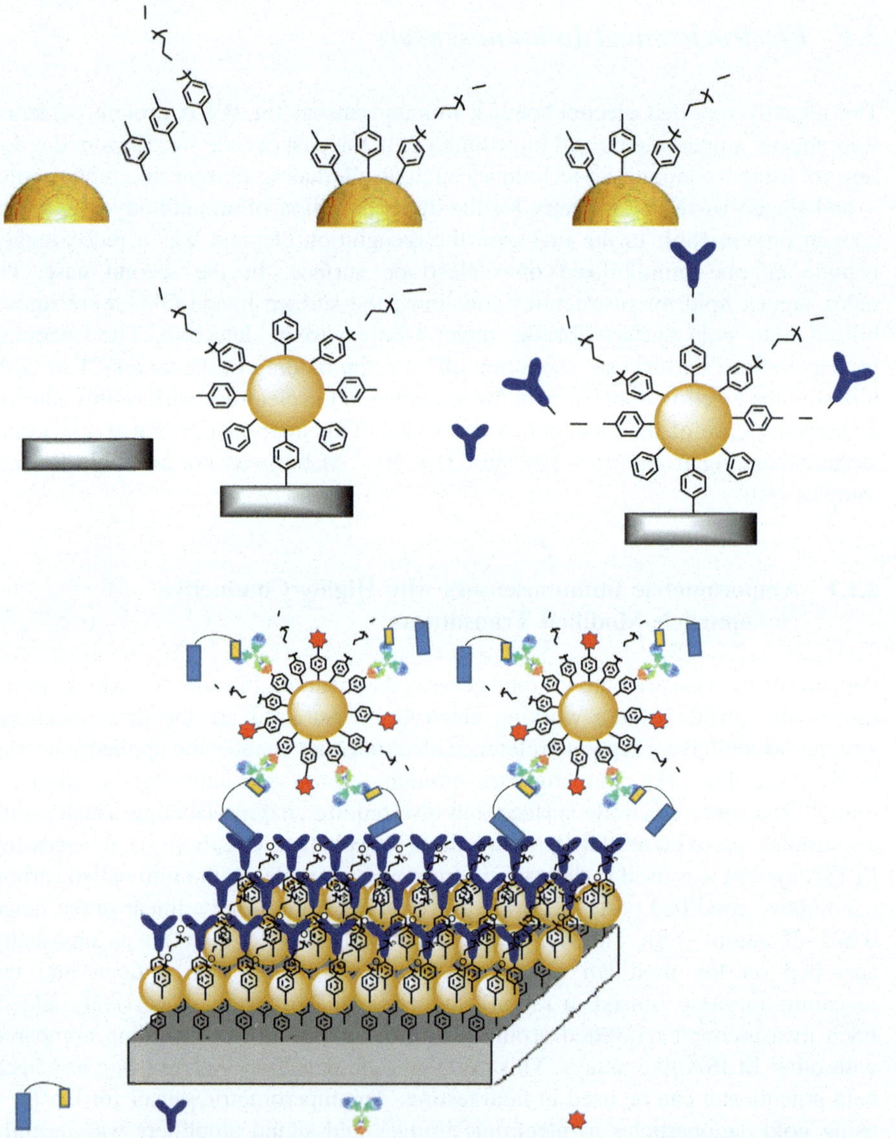

Fig. 4 Amperometric detection of BoNT/A: the AuNPs functionalized capture antibody (immobilized on GC electrode) and HRP labeled antibody bind each a different BoNT subunit. The enhanced electrical signal directly proportional with toxin concentration is due to the electronic bridge provided by the AuNPs (reproduced from [68] with permission of Elsevier)

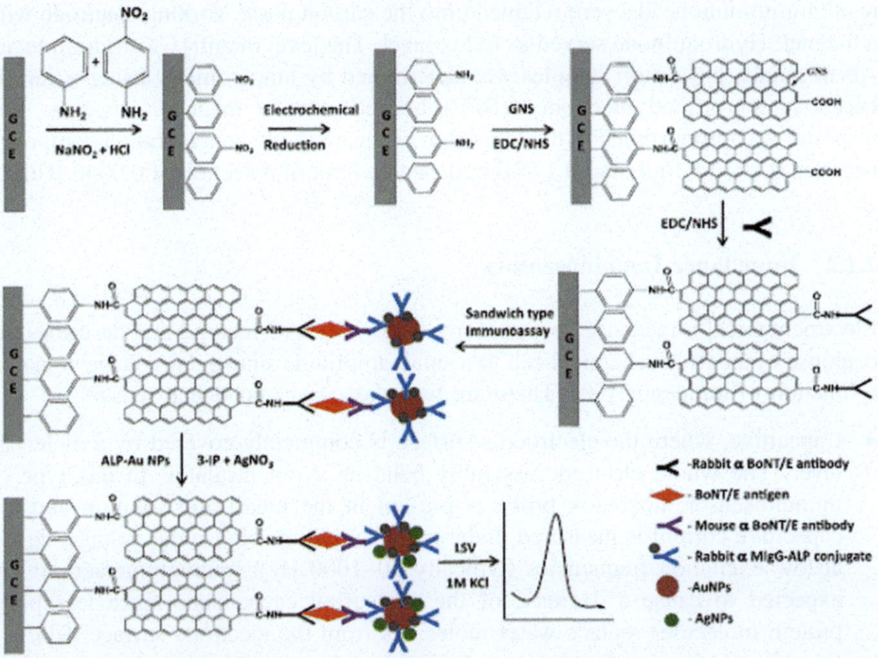

Fig. 5 Steps involved in GC electrode modification and sandwich immunoassay detection of BoNT/E based on the analysis of deposited silver nanoparticles (AgNPs) generated by mouse IgG-ALP/AuNPs catalysts and 3-indoxyl phosphate (3-IP) reaction (reproduced from [24] with permission of Elsevier)

BoNT/A was 1 pg/ml (6.67 fM) in Tris buffer and 5 ng/ml in spiked milk samples. The immunoassay displayed a short detection time (around 10 min) thus being amenable to miniaturized and portable devices for POCT.

A similar sandwich ELISA format was developed recently for BoNT serotype E. This method relied on graphene nanosheets—aryldiazonium salt modified glassy carbon electrodes (GCE) as sensing platform and enzyme induced silver nanoparticles (AgNPs) deposited on AuNPs as signal amplifier (Fig. 5) [24]. The developed electrochemical immunosensor could detect BoNT/E with linear range from10 pg/ml to10 ng/ml with LOD of 5 pg/ml and total analysis time of 65 min. The immunosensor was evaluated against food samples (orange juice and milk). The deposited AgNPs on electrode surface were determined by linear sweep voltammetry (LSV).

It is worth mentioning an electrochemical immunosensing assay for anti-*Clostridium tetani* Ab determination in serum [89] since no other immunosensor for the direct detection of tetanus toxin was reported so far. The antigen tetanus toxoid was immobilized on superparamagnetic nanobeads. The immunoreaction occurred in Eppendorf minitubes. The anti-TeNT Ab was incubated in the presence of the toxoid functionalized nanobeads, then reacted with HRP-labeled anti-IgG. The resulting immunobeads were retained onto the carbon paste working electrode with a magnet. Hydroquinone served as redox label. The level of anti- Clostridium tetani Ab in guinea pig serum samples was determined by amperometry using a carbon based screen-printed electrode (cSPE) housed onto a magnetic support. The response was proportional to the logarithm of the antibody concentration comprised between 0.0046 IU/mL and 0.175 IU/mL with a limit of detection of 0.0046 IU/mL.

2.1.2 Impedance Immunosensors

Electrochemical impedance spectroscopy (EIS) is an AC method that describes the response of an electrochemical cell to a small amplitude sinusoidal voltage signal as a function of frequency [90]. There are two classes of impedance sensors:

- Capacitive, where the electrode's surface is completely covered by a dielectric layer. The whole electrode assembly behaves as an insulator. In this type of immunosensor, no redox probe is present in the measuring solution and the capacitive current is measured under small amplitude sinusoidal voltage signal, at low excitation frequencies (typically 10–1000 Hz). Immunointeractions are expected to cause a decrease of the measuring capacitance since less polar protein molecules replace water molecules from the electrode surface [91].
- Faradaic, where the electrode's surface, which is partially or wholly covered by a non-insulating layer, or partially covered by an insulating layer is able to catalyze a redox probe, which exists in the measuring solution [90]. Here, the measured parameter is the charge transfer resistance (the real component of impedance at low frequency values, typically 0.1–1.0 Hz); antibody/antigen (Ab–Ag) interactions are expected to cause an increase of the charge transfer resistance, as the faradaic reaction becomes increasingly hindered. In general, faradaic impedance immunosensors display a higher sensitivity to Ab–Ag interactions [92].

A flow-injection capacitive immunosensor was developed for Cholera toxin using a monoclonal anti-CT which bound the B subunit of CT [32]. The immunosensor responded linearly to CT concentrations in the range from 1.0×10^{-13} to 1.0×10^{-10} M under optimized conditions. The limit of detection (LOD) was 1.0×10^{-14} M. Here the capture antibody was covalently attached to the alkylthiol self-assembled monolayer (SAM) on gold electrode. Capacitance measurements were performed by inserting the modified gold electrode as a working electrode in the flow cell. Another capacitive flow-injection system based on the adsorption of anti-CT antibodies on AuNPs incorporated on a poly-tyramine-modified gold electrode [69] with LOD limit of detection of 9×10^{-20} M (0.09 aM), and a dynamic range between 0.1 aM and 10 pM in water. While providing enhanced sensitivity in toxin detection, it seems that capacitive immunosensors are more time-consuming compared to amperometric ones, requiring longer periods to allow signal stabilization.

2.2 *Piezoelectric Imunosensors*

The piezoelectric transducer allows a binding event to be converted into a measurable signal, for example resonance frequency changes [87]. Quartz crystal microbalance (QCM) is a bulk acoustic wave (BAW) sensor that measures the change in frequency of a quartz crystal resonator induced by changes in mass and viscoelastic properties at the QCM surface (e.g. the binding of target to an immobilized bioaffinity agent) [51]. The resonance of the quartz crystal is monitored using AC impedance which measures the shift in frequency and can be related to the mass change [82]. A QCM assay for CT detection based on the CT-mediated surface-agglutination of ganglioside (GM_1)-functionalized liposomes was developed, in the absence of antibodies. The CT-specific agglutination at the surface of the electrode modified with a GM_1 supported lipid membrane was achieved via spontaneous spread of the liposomes on a SAM of a long-chain alkylthiol. In the presence of CT, the GM_1-incorporated liposomes in assay medium rapidly agglutinated at the electrode surface through the binding of CT to GM_1 on the electrode surface and the liposome interface. This results in a high mass loading on the piezoelelctric crystal as well as a significant increase of density and viscosity at the interface, thereby generating a decrease in frequency of the piezoelelctric crystal. The LOD was 25 ng/ml of CT in serum. For each assay the stabilization of the frequency shift needed around 40 min, again far more than in amperometric assay. Surface acoustic wave (SAW) sensing is a related technique to QCM and is classed in the broader group of acoustic wave devices. A surface acoustic wave represents a mechanical acoustic wave that propagates to a confined area of a cut piezoelectric crystal. The velocity and the amplitude of the wave are strongly dependent on the changes occurring in the media near the surface [93]. SAW sensors are sensitive to changes in the acoustic wave—modulation of surface (as opposed to the bulk) induced by density, mass, viscosity and temperature. No SAW sensors were reported as far as we know for BWA proteins but there are several promising works reporting SAW detection for BWA bacteria and viruses [94–96] and also for antibodies and other proteins in liquid samples [93, 97].

2.3 *Optical Immunosensors*

In optical biosensors, the immunoaffinity agent is immobilized onto the transducer's surface, thus responding to the interaction with the target either by generating an optical signal (fluorescence, phosphorescence), or by undergoing changes in optical properties, such as absorption, emission, reflectance and refractive index [51, 87]. A vast number of optical techniques (based on optical fiber- and planar waveguide transducers) are used to create optical changes, e.g., fluorescence, luminescence, total internal reflection fluorescence (TIRF), optical waveguide lightmode spectroscopy (OWLS), surface plasmon resonance (SPR), localized surface plasmon resonance (LSPR) and surface-enhanced Raman spectroscopy (SERS) [98, 99]. Among optical methods, SPR is the most used due its highly specific detection tools without any labeling of the reagents [51] but less used for BWA proteins since it cannot provide LODs below LD_{50}. The phenomenon of color change induced by the aggregation of AuNPs is widely used not only in colorimetric assays, but also in LSPR-based assays [100]. We will discuss further the techniques amenable for the fast screening of the BWA proteins.

2.3.1 Optical Fiber/Fluorescence-Combined Immunoassays

Fluorecence-based immunosensors sensors provide broad opportunities for multi-analysis of BWAs. For example, a multiplexing approach was described by Pauly et al. [101], for the simultaneous detection of BoNT/A,B ricin, abrin, and staphylococcal enterotoxin B in milk, yoghurt, iced coffee and carrot juice, using a suspension-bead-array-based optical (fluorescence) sensor. Toxin specific antibodies were immobilized to magnetic beads that were differentially fluorescently tagged. After mixing with homogenized samples, the beads containing toxin bound to the capture antibody were magnetically isolated, concentrated and washed before a secondary fluorescently labelled antibody was introduced. The assay required only 50 μl of sample. The matrix interference reduced the sensitivity to low nanogram per litre levels, but it was two to three orders of magnitude below the oral LD_{50}. BoNT serotypes A and B were detected in food and an animal sera using a sandwich ELISA with electrochemiluminescence (ECL) immunoassay platform developed by Meso Scale Discovery [102]. The reported LOD for BoNT/A was 3 pg/mL, and for BoNT/B 13 pg/mL. The ECL assay outperformed ELISA in detection sensitivity in most of the food matrices spiked with BoNT/A and in some foods spiked with BoNT/B. Another immunosensor exploiting the property of quantum dots to transfer Förster resonance energy was developed for BoNT/A detection Here, the generation of excitons (and consequently the emission light) was enhanced by closely located fluorescent molecules from the antibody label [103]. The LOD for BoNT/A was 5 pM in water. Less studies were dedicated to plant toxins other than ricin, but it worth mentioning an ECL immunoassay for abrin, based on a sandwich ELISA approach using magnetic microbeads as the carrier of ECL labels [104]. The assay exhibited a LOD of 2 ng/ml with a 2–200 ng/ml linear range.

One category of planar waveguide biosensors utilizes total internal reflection (TIR) for analysis of biomolecular events. Total internal reflection occurs when a propagating wave strikes a medium boundary at an angle larger than a particular critical angle with respect to the normal to the surface. If the refractive index (RI) is lower on the other side of the boundary and the incident angle is greater than the critical angle, the wave cannot pass through and is entirely reflected [98]. TIRF sensors use the evanescent wave to excite fluorophores at the interface between the waveguide and the lower RI medium (Fig. 6).

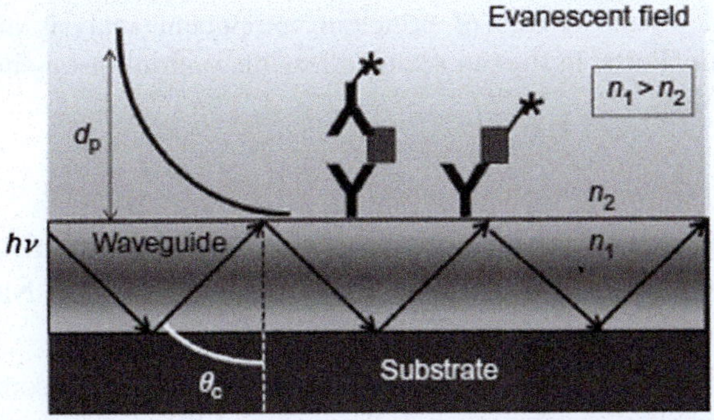

Fig. 6 Schematic illustration of evanescent field generation resulted from total internal reflection (TIR). TIR occurs when light (hv) is launched into a waveguide of RI n_1, that is placed into a lower RI (n_2) medium at an angle of incidence larger than the critical angle (ψc). Under TIR conditions, an evanescent field extends from the interface into the lower index medium (n_2). The limited penetration depth (d_p) of the evanescent field (the distance at which intensity drops to 1/e of its original value) can be used to interrogate minute changes in refractive index or excite fluorophores (shown with asterisks) in close proximity to the surface (reproduced from [98] with permission of Elsevier)

Once light is coupled into the waveguide, TIRF occurs through excitation of fluorophores within the evanescent wave, with little contribution or interference from components in the bulk solution. CT, ricin and botulinum toxoids were detected through TIRF planar waveguide arrays [105, 106] with LODs of 100 pg/ml and 1 ng/ml, respectively. It can be shown that SPR does not compete with TIRF in terms of sensitivity and LOD, since the SPR assay for BoNT/A provided a LOD of 1.0×10^{-11} M and a linear range from 1.0×10^{-9} to 1.0×10^{-6} M [32].

2.3.2 Surface-Enhanced Raman Scattering (SERS) Imunosensors

SERS is a technique that can be used to improve the Raman analysis. SERS can be employed as a way to potentially lower sample fluorescence and enhance Raman scattering coming from molecules which have been adsorbed on metal nanosized structures, in particular Au, Ag or Cu [99, 107]; both of these effects lead to a much improved limit of detection. A nanostructured metal substrate is used to enhance the signal from an analyte in contact with the substrate. The type of metal and the shape of the features of the nanostructure play an important role in the Raman signal enhancement [99, 108]. A rapid method that combined immunomagnetic separation (IMS) and surface-enhanced Raman spectroscopy (SERS) to detect ricin in whole milk was developed [108]. IMS was used to specifically bind the ricin from the milk. Then, SERS was applied to analyze the IMS eluate mixed with silver dendrite nanosubstrates. This approach allowed a LOD of 4 μg/mL ricin in milk within 20 min, based on the results of principal component analysis and partial least squares analysis. Portable Raman seems a feasible option for on-site detection in a processing facility.

3 Immunoassay Implementation: Labelled *Vs*. Label-Free Formats for the Detection of Proteins in Liquid Samples

The use of labelled immunoaffinity agents doesn't always overcome matrix interference. In unlabelled systems, such as those using SPR, it is sometimes possible to see the adsorption of the matrix components at the sensor surface as an unwanted signal that overlaps the specific binding of the target. In labelled systems, non-specific adsorption still occurs but, since the matrix component is not labelled, it would appear that there is no matrix effect [80]. In reality, the matrix binding to surface has not produced a signal but may prevent the specific interaction with labelled protein, thus reducing sensitivity [40]. There are advantages and disadvantages in the use of fluorescently labelled immunoafffinity agents (antibodies and receptors) over unlabelled equivalents. As an advantage is the possibility to increase the assay sensitivity by choosing a dye with an intense signal and/or changing the ratio of dye to binding protein. A larger ratio should increase the relative sensitivity since the signal from the binding of a smaller amount of antibody will be increased. The drawback include an additional labelling step that has to be well characterised to provide a reproducible result using different batches of labelled binding protein. Contamination with fluorescent biomolecules may also become an issue.

4 Solving Current Issues in High-Throughput Analysis: Decreasing the Detection Limit and the Response Time of the Sensor

A BWA sensor destined for POCT platforms will be required to have very low detection limits. This purpose can be achieved by using biorecognition molecules or surface chemistry which have higher equilibrium constants of binding, limiting the non-specific binding and finally using more sensitive transducers. Aptamers are molecules that display high affinity for proteins [40]. New aptamers can be produced through the SELEX method and can isolate new biorecognition molecules with affinities higher than natural antibodies. For labelled biosensors, which employ the sandwich ELISA the most common approach to increase the sensitivity of transduction is to amplify the label signal, so that fewer labels being required for the transducer to detect them. Enhancement of the label signal can be achieved through quantum dots [55], electrochemiluminescent molecules [102] and dye labelled liposomes [86]. The response time is strongly depended to ability of the target to diffuse to the sensing surface. Even in a small volume this could mean diffusing a long distance on the molecular scale [40]. One solution is to use highly porous transducers which allow the intimate mixing of the sample with the sensor. Highly porous transducers without using conventional labels are currently investigated [40, 109]. Porous silicon photonic crystals can be used as mirrors where binding events cause changes in reflectivity. The high porosity can provide rapid responses and the mirrors can have sharp reflectance peaks with subnanometer full width half maximum [40].

5 Future Applications in Point-of-Care Systems

The key requirements for POCT devices start with the needs of their users, but some features are common to virtually all users in all settings:

1. Simplicity of using
2. Reagents and consumables are robust in usage and storage
3. Results should be concordant with an established laboratory method
4. Devices, reagents and consumables are safe to use [110]

The main requirement for an immunosensor to be included on a POCT platform is portability. Portable biosensors have been already applied to the analysis of food contaminants, drug residues and airborne agents. We have shown that nanostructured electrochemical immunosensors are probably the most amenable to in situ analysis in terms of selectivity, stability, response time, and reproducibility even if are prone to matrix effects. On the other hand, optical sensors using labelled biomolecules overcome the effects of non-specific binding but are still difficult to handle by unskilled personnel. There is also a common limitation to the use of portable immunosensors: the power source. Real on-site testing would require either the use of a dedicated battery or power delivered via a laptop connection, both of which will limit the working time of the sensor depending on its power consumption [80]. Another limitation, specific for food contaminant analysis, is the necessity of sample preparation. Portable sensors are suitable for the analysis of liquid samples such as milk, water and fruit juices, but when more complicated sample preparation is needed (e.g. sample concentration), the sensor will be restricted by the pre-analysis sample preparation which must be performed in a laboratory [80]. Further research and financial efforts should be addressed to developing faster and simpler methods of sample preparation.

Acknowledgement The authors would like to acknowledge the financial support of: Romanian National Authority for Scientific Research grant project PN-II-ID-PCE-2011-3-0286 and the networking contribution by the COST Action CM1407 "Challenging organic syntheses inspired by nature—from natural products chemistry to drug discovery".

References

1. Millard C (2005) Medical defense against protein Toxin weapons. In: Lindler L, Lebeda F, Korch G (eds) Biological weapons defense. Humana Press, pp 255–283
2. Krutzsch W, Myjer E, Trapp R (2014) The chemical weapons convention: a commentary, OUP Oxford
3. U.S. Government (2011) Defense against Toxin weapons, General Books
4. Alouf JE, Ladant D, Popoff MR (2015) The comprehensive sourcebook of bacterial protein Toxins, Elsevier Science
5. Baldauf K, Royal J, Hamorsky K, Matoba N (2015) Cholera Toxin B: one subunit with many pharmaceutical applications. Toxins 7:974
6. Bigalke H, Rummel A (2005) Medical aspects of toxin weapons. Toxicology 214:210–220
7. Falnes PØ, Sandvig K (2000) Penetration of protein toxins into cells. Curr Opin Cell Biol 12:407–413
8. Sandvig K, Torgersen ML, Engedal N, Skotland T, Iversen T-G (2010) Protein toxins from plants and bacteria: probes for intracellular transport and tools in medicine. FEBS Lett 584:2626–2634
9. He X, McMahon S, Henderson TD, Griffey SM, Cheng LW (2010) Ricin Toxicokinetics and its sensitive detection in mouse sera or feces using immuno-PCR. PLoS ONE 5:e12858
10. Gordon VM, Leppla SH (1994) Proteolytic activation of bacterial toxins: role of bacterial and host cell proteases. Infect Immun 62:333–340
11. Naglich JG, Metherall JE, Russell DW, Eidels L (1992) Expression cloning of a diphtheria toxin receptor: Identity with a heparin-binding EGF-like growth factor precursor. Cell 69:1051–1061
12. Lindberg AA, Brown JE, Strömberg N, Westling-Ryd M, Schultz JE, Karlsson KA (1987) Identification of the carbohydrate receptor for Shiga toxin produced by Shigella dysenteriae type 1. J Biol Chem 262:1779–1785
13. Holmgren J, Lönnroth I, Månsson J, Svennerholm L (1975) Interaction of cholera toxin and membrane GM1 ganglioside of small intestine. Proc Natl Acad Sci USA 72:2520–2524
14. Sandvig K, van Deurs B (1996) Endocytosis, intracellular transport, and cytotoxic action of Shiga toxin and ricin. Physiol Rev 76:949–966
15. Jiří LS (2006) Patočka. Protein biotoxins of military significance, Acta Medica 49:3–11
16. Słomińska-Wojewódzka M, Sandvig K (2013) Ricin and ricin-containing immunotoxins: insights into intracellular transport and mechanism of action in vitro. Antibodies 2:236
17. Yang H, Deng M, Ga S, Chen S, Kang L, Wang J, Xin W, Zhang T, You Z, An Y, Wang J, Cui D (2014) Capillary-driven surface-enhanced Raman scattering (SERS)-based microfluidic chip for abrin detection. Nanoscale Res Lett 9:138
18. Bagaria S, Karande A (2014) Abrin and immunoneutralization: a review. In: Gopalakrishnakone P (ed) Toxinology. Springer, Netherlands, pp 1–21
19. Liu G, Li K (2015) Micro/nano optical fibers for label-free detection of abrin with high sensitivity. Sensor Actuat B Chem 215:146–151
20. Dickers K, Bradberry S, Rice P, Griffiths G, Vale JA (2003) Abrin Poisoning. Toxicol Rev 22:137–142
21. Subrahmanyan D, Dks JM, Raj M (2008) An unusual manifestation of Abrus precatorius poisoning: a report of two cases. Clin Toxicol 46:173–175
22. Ler SG, Lee FK, Gopalakrishnakone P (2006) Trends in detection of warfare agents: detection methods for ricin, staphylococcal enterotoxin B and T-2 toxin. J Chromatogr A 1133:1–12
23. Brunger AT, Breidenbach MA, Jin R, Fischer A, Santos JS, Montal M (2007) Botulinum neurotoxin heavy chain belt as an intramolecular chaperone for the light chain. PLoS Pathog 3:e113
24. Narayanan J, Sharma MK, Ponmariappan S (2015) Sarita, M. Shaik, S. Upadhyay, Electrochemical immunosensor for botulinum neurotoxin type-E using covalently ordered

graphene nanosheets modified electrodes and gold nanoparticles-enzyme conjugate. Biosens Bioelectron 69:249–256

25. Grate JW, Ozanich RM Jr, Warner MG, Bruckner-Lea CJ, Marks JD (2010) Advances in assays and analytical approaches for botulinum-toxin detection. TrAC Trend Anal Chem 29:1137–1156
26. Attrée O, Guglielmo-Viret V, Gros V, Thullier P (2007) Development and comparison of two immunoassay formats for rapid detection of botulinum neurotoxin type A. J Immunol Methods 325:78–87
27. Blum FC, Chen C, Kroken AR, Barbieri JT (2012) Tetanus toxin and botulinum toxin a utilize unique mechanisms to enter neurons of the central nervous system. Infect Immun 80:1662–1669
28. Lindström M, Korkeala H (2006) Laboratory diagnostics of botulism. Clin Microbiol Rev 19:298–314
29. Arnon SS, Schechter R, Inglesby TV et al (2001) Botulinum toxin as a biological weapon: medical and public health management. JAMA 285:1059–1070
30. Codeco C (2001) Endemic and epidemic dynamics of cholera: the role of the aquatic reservoir. BMC Infect Dis 1:1
31. Miller CE, Majewski J, Faller R, Satija S, Kuhl TL (2004) Cholera toxin assault on lipid monolayers containing ganglioside GM1. Biophysl Jl 86:3700–3708
32. Labib M, Hedström M, Amin M, Mattiasson B (2009) A capacitive immunosensor for detection of cholera toxin. Anal Chim Acta 634:255–261
33. Sack DA, Sack RB, Nair GB, Siddique AK (2004) Cholera. The Lancet 363:223–233
34. Gill DM (1982) Bacterial toxins: a table of lethal amounts. Microbiol Rev 46:86–94
35. Yu H-W, Wang Y-S (2011) Y. li, G.-L. Shen, H.-l. Wu, R.-Q. Yu, One Step Highly Sensitive Piezoelectric Agglutination Method for Cholera Toxin Detection Using GM1 Incorporated Liposome, Procedia. Environ Sci 8:248–256
36. Scott N, Qazi O, Wright MJ, Fairweather NF, Deonarain MP (2010) Characterisation of a panel of anti-tetanus toxin single-chain Fvs reveals cooperative binding. Mol Immunol 47:1931–1941
37. Calvo AC, Oliván S, Manzano R, Zaragoza P, Aguilera J, Osta R (2012) Fragment C of tetanus toxin: new insights into its neuronal signaling pathway. Int J Mol Sci 13:6883
38. Schirwitz C, Loeffler FF, Felgenhauer T, Stadler V, Breitling F, Bischoff FR (2012) Sensing immune responses with customized peptide microarrays. Biointerphases 7:47
39. Lukić I, Marinković E, Filipović A, Krnjaja O, Kosanović D, Inić-Kanada A, Stojanović M (2015) Key protection factors against tetanus: anti-tetanus toxin antibody affinity and its ability to prevent tetanus toxin—ganglioside interaction. Toxicon 103:135–144
40. Gooding JJ (2006) Biosensor technology for detecting biological warfare agents: recent progress and future trends. Anal Chim Acta 559:137–151
41. Brown K (2004) Up in the Air. Science 305:1228–1229
42. Yáñez-Sedeño P, Agüí L, Villalonga R, Pingarrón JM (2014) Biosensors in forensic analysis. A Rev Anal Chim Acta 823:1–19
43. Wang X, Lu X, Chen J (2014) Development of biosensor technologies for analysis of environmental contaminants. Trend Environ Anal Chem 2:25–32
44. Muzyka K (2014) Current trends in the development of the electrochemiluminescent immunosensors. Biosens Bioelectron 54:393–407
45. Justino CIL, Freitas AC, Amaral JP, Rocha-Santos TAP, Cardoso S, Duarte AC (2013) Disposable immunosensors for C-reactive protein based on carbon nanotubes field effect transistors. Talanta 108:165–170
46. Justino CIL, Rocha-Santos TA, Duarte AC, Rocha-Santos TA (2010) Review of analytical figures of merit of sensors and biosensors in clinical applications. TrAC Trend Anal Chem 29:1172–1183
47. Justino CIL, Freitas AC, Pereira R, Duarte AC, Rocha Santos TAP (2015) Recent developments in recognition elements for chemical sensors and biosensors. TrAC Trend, Anal Chem 68:2–17

48. Gan N, Du X, Cao Y, Hu F, Li T, Jiang Q (2013) An ultrasensitive electrochemical immunosensor for HIV p24 based on Fe3O4@SiO2 nanomagnetic probes and nanogold colloid-labeled enzyme-antibody copolymer as signal tag. Materials 6:1255
49. Baniukevic J, Kirlyte J, Ramanavicius A, Ramanaviciene A (2013) Application of oriented and random antibody immobilization methods in immunosensor design. Sensor Actuat B Chem 189:217–223
50. Willander M, Khun K, Ibupoto Z (2014) Metal oxide nanosensors using polymeric membranes, enzymes and antibody receptors as ion and molecular recognition elements. Sensors 14:8605
51. Warriner K, Reddy SM, Namvar A, Neethirajan S (2014) Developments in nanoparticles for use in biosensors to assess food safety and quality. Trends Food Sci Technol 40:183–199
52. Cao J, Sun T, Grattan KTV (2014) Gold nanorod-based localized surface plasmon resonance biosensors: a review. Sensor Actuat B Chem 195:332–351
53. Majdalawieh A, Kanan MC, El-Kadri O, Kanan SM (2014) Recent advances in gold and silver nanoparticles: synthesis and applications. J Nanosci Nanotechnol 14:4757–4780
54. Omidfar K, Khorsand F, Darziani Azizi M (2013) New analytical applications of gold nanoparticles as label in antibody based sensors. Biosens Bioelectron 43:336–347
55. Pisanic TR, Ii Y, Zhang TH (2014) Wang, Quantum dots in diagnostics and detection: principles and paradigms. Analyst 139:2968–2981
56. Doria G, Conde J, Veigas B, Giestas L, Almeida C, Assunção M, Rosa J, Baptista PV (2012) Noble metal nanoparticles for biosensing applications. Sensors 12:1657
57. Huang X, El-Sayed MA (2010) Gold nanoparticles: optical properties and implementations in cancer diagnosis and photothermal therapy. J Adv Res 1:13–28
58. Guo S, Dong S (2009) Biomolecule-nanoparticle hybrids for electrochemical biosensors. TrAC Trends Anal Chem 28:96–109
59. Zhang Y, Ge S, Wang S, Yan M, Yu J, Song X, Liu W (2012) Magnetic beads-based electrochemiluminescence immunosensor for determination of cancer markers using quantum dot functionalized PtRu alloys as labels. Analyst 137:2176–2182
60. Xia X, Xu Y, Ke R, Zhang H, Zou M, Yang W, Li Q (2013) A highly sensitive europium nanoparticle-based lateral flow immunoassay for detection of chloramphenicol residue. Anal Bioanal Chem 405:7541–7544
61. Gopinath SCB, Tang T-H, Citartan M, Chen Y, Lakshmipriya T (2014) Current aspects in immunosensors. Biosens Bioelectron 57:292–302
62. Lou S, Ye J-Y, Li K-Q, Wu A (2012) A gold nanoparticle-based immunochromatographic assay: the influence of nanoparticulate size. Analyst 137:1174–1181
63. Gopinath SCB, Awazu K, Fujimaki M, Shimizu K, Shima T (2013) Observations of immuno-gold conjugates on influenza viruses using waveguide-mode sensors. PLoS ONE 8: e69121
64. Kim D, Herr AE (2013) Protein immobilization techniques for microfluidic assays. Biomicrofluidics 7:041501
65. Gomez J, Tigli O (2013) Zinc oxide nanostructures: from growth to application. J Mater Sci 48:612–624
66. Frenzel H, Lajn A, Grundmann M (2013) One decade of fully transparent oxide thin-film transistors: fabrication, performance and stability, physica status solidi. (RRL)—Rapid Res Lett 7:605–615
67. Sperling RA, Parak WJ (2010) Surface modification, functionalization and bioconjugation of colloidal inorganic nanoparticles. Philos Trans Royal Soc London A: Math Phys Eng Sci 368:1333–1383
68. Liu G, Zhang Y, Guo W (2014) Covalent functionalization of gold nanoparticles as electronic bridges and signal amplifiers towards an electrochemical immunosensor for botulinum neurotoxin type A. Biosens Bioelectron 61:547–553
69. Loyprasert S, Hedström M, Thavarungkul P, Kanatharana P, Mattiasson B (2010) Sub-attomolar detection of cholera toxin using a label-free capacitive immunosensor. Biosens Bioelectron 25:1977–1983

70. Yuan Y, Zhang J, Zhang H, Yang X (2012) Silver nanoparticle based label-free colorimetric immunosensor for rapid detection of neurogenin 1. Analyst 137:496–501
71. Emami M, Shamsipur M, Saber R, Irajirad R (2014) An electrochemical immunosensor for detection of a breast cancer biomarker based on antiHER2-iron oxide nanoparticle bioconjugates. Analyst 139:2858–2866
72. Xu Y, Baiu DC, Sherwood JA, McElreath MR, Qin Y, Lackey KH, Otto M, Bao Y (2014) Linker-free conjugation and specific cell targeting of antibody functionalized iron-oxide nanoparticles. J Mat Chem B 2:6198–6206
73. Peterson RD, Chen W, Cunningham BT, Andrade JE (2015) Enhanced sandwich immunoassay using antibody-functionalized magnetic iron-oxide nanoparticles for extraction and detection of soluble transferrin receptor on a photonic crystal biosensor. Biosens Bioelectron 74:815–822
74. Hurley MT, Wang Z, Mahle A, Rabin D, Liu Q, English DS, Zachariah MR, Stein D, DeShong P (2013) Synthesis, characterization, and application of antibody functionalized fluorescent silica nanoparticles. Adv Funct Mater 23:3335–3343
75. Wang J, Han H, Jiang X, Huang L, Chen L, Li N (2012) Quantum dot-based near-infrared electrochemiluminescent immunosensor with gold nanoparticle-graphene nanosheet hybrids and silica nanospheres double-assisted signal amplification. Anal Chem 84:4893–4899
76. Esteve-Turrillas FA, Abad-Fuentes A (2013) Applications of quantum dots as probes in immunosensing of small-sized analytes. Biosens Bioelectron 41:12–29
77. Puertas S, de Gracia Villa M, Mendoza E, Jiménez-Jorquera C, de la Fuente JM, Fernández-Sánchez C, Grazú V (2013) Improving immunosensor performance through oriented immobilization of antibodies on carbon nanotube composite surfaces. Biosens Bioelectron 43:274–280
78. Venturelli E, Fabbro C, Chaloin O, Ménard-Moyon C, Smulski CR, Da Ros T, Kostarelos K, Prato M, Bianco A (2011) Antibody covalent immobilization on carbon nanotubes and assessment of antigen binding. Small 7:2179–2187
79. Lim SA, Ahmed MU (2015) A carbon nanofiber-based label free immunosensor for high sensitive detection of recombinant bovine somatotropin. Biosens Bioelectron 70:48–53
80. McGrath TF, Elliott CT, Fodey TL (2012) Biosensors for the analysis of microbiological and chemical contaminants in food. Anal Bioanal Chem 403:75–92
81. von Lode P (2005) Point-of-care immunotesting: approaching the analytical performance of central laboratory methods. Clin Biochem 38:591–606
82. Arugula AM, Simonian A (2014) Novel trends in affinity biosensors: current challenges and perspectives. Meas Sci Technol 25:032001
83. Thakur MS, Ragavan KV (2013) Biosensors in food processing. J Food Sci Technol 50:625–641
84. Gerasimov JY, Lai RY (2011) Design and characterization of an electrochemical peptide-based sensor fabricated via "click" chemistry. Chem Comm 47:8688–8690
85. Puiu M, Idili A, Moscone D, Ricci F, Bala C (2014) A modular electrochemical peptide-based sensor for antibody detection. Chem Comm 50:8962–8965
86. Cheng Q, Zhu S, Song J, Zhang N (2004) Functional lipid microstructures immobilized on a gold electrode for voltammetric biosensing of cholera toxin. Analyst 129:309–314
87. Puiu M, Bala C (2014) Jaffrezic-Renault. Progress in biosensor for mycotoxins assay, Chemical Sensors 1:4–11
88. Suresh S, Gupta AK, Rao VK, Om K, Vijayaraghavan R (2010) Amperometric immunosensor for ricin by using on graphite and carbon nanotube paste electrodes. Talanta 81:703–708
89. Patris S, De Vriese C, Prohoroff F, Calvo EB, Martínez JA, Kauffmann JM (2010) Anti-Clostridium tetani antibody determination in serum samples by amperometric immunosensing. Electroanal 22:41–48

90. Prodromidis MI (2010) Impedimetric immunosensors—A review. Electrochim Acta 55:4227–4233
91. K'Owino IO, Sadik OA (2005) Impedance spectroscopy: a powerful tool for rapid biomolecular screening and cell culture monitoring. Electroanal 17:2101–2113
92. Katz E, Willner I, Wang J (2004) Electroanalytical and bioelectroanalytical systems based on metal and semiconductor nanoparticles. Electroanal 16:19–44
93. Puiu M, Gurban A-M, Rotariu L, Brajnicov S, Viespe C, Bala C (2015) Enhanced sensitive love wave surface acoustic wave sensor designed for immunoassay formats. Sensors 15:10511
94. Moll N, Pascal E, Dinh DH, Lachaud JL, Vellutini L, Pillot JP, Rebière D, Moynet D, Pistré J, Mossalayi D, Mas Y, Bennetau B, Déjous C (2008) Multipurpose love acoustic wave immunosensor for bacteria, virus or proteins detection. IRBM 29:155–161
95. Matatagui D, Fontecha J, Fernández M, Gràcia I, Cané C, Santos J, Horrillo M (2014) Love-Wave sensors combined with microfluidics for fast detection of biological warfare agents. Sensors 14:12658
96. Rocha-Gaso M-I, March-Iborra C, Montoya-Baides Á, Arnau-Vives A (2009) Surface generated acoustic wave biosensors for the detection of pathogens: a review. Sensors 9:5740
97. Matatagui D, Fontecha J, Fernández MJ, Oliver MJ, Hernando-García J, Sánchez-Rojas JL, Gràcia I, Cané C, Santos JP, Horrillo MC (2013) Comparison of two types of acoustic biosensors to detect immunoreactions: love-wave sensor working in dynamic mode and QCM working in static mode. Sensor Actuat B Chem 189:123–129
98. Taitt CR, North SH (2015) 17—Total internal reflection fluorescence (TIRF) array biosensors for biothreat agents for food safety and food defense. In: Bhunia AK, Kim MS, Taitt CR (eds) High throughput screening for food safety assessment. Woodhead Publishing, pp 399–424
99. Zheng J, He L (2014) Surface-Enhanced Raman spectroscopy for the chemical analysis of food. Compr Rev Food Sci Food Saf 13:317–328
100. Shao Y, Xu S, Zheng X, Wang Y, Xu W (2010) Optical Fiber LSPR biosensor prepared by gold nanoparticle assembly on polyelectrolyte multilayer. Sensors 10:3585
101. Pauly D, Kirchner S, Stoermann B, Schreiber T, Kaulfuss S, Schade R, Zbinden R, Avondet M-A, Dorner MB, Dorner BG (2009) Simultaneous quantification of five bacterial and plant toxins from complex matrices using a multiplexed fluorescent magnetic suspension assay. Analyst 134:2028–2039
102. Cheng LW, Stanker LH (2013) Detection of botulinum neurotoxin serotypes A and B using a chemiluminescent versus electrochemiluminescent immunoassay in food and serum. J Agr Food Chem 61:755–760
103. Warner MG, Grate JW, Tyler A, Ozanich RM, Miller KD, Lou J, Marks JD, Bruckner-Lea CJ (2009) Quantum dot immunoassays in renewable surface column and 96-well plate formats for the fluorescence detection of botulinum neurotoxin using high-affinity antibodies. Biosens Bioelectron 25:179–184
104. Liu B, Tong Z-Y, Hao L-Q, Liu W, Mu X-H, Liu Z-W, Huang Q-B (2013) A new electrochemiluminescence immunoassay based on magnetic microbeads as carrier of labels. Chinese J Anal Chem 41:1807–1811
105. Rowe-Taitt CA, Hazzard JW, Hoffman KE, Cras JJ, Golden JP, Ligler FS (2000) Simultaneous detection of six biohazardous agents using a planar waveguide array biosensor. Biosens Bioelectron 15:579–589
106. Narang U, Anderson GP, Ligler FS, Burans J (1997) Fiber optic-based biosensor for ricin. Biosens Bioelectron 12:937–945
107. Das G, Mecarini F, Gentile F, De Angelis F, Mohan Kumar HG, Candeloro P, Liberale C, Cuda G, Di Fabrizio E (2009) Nano-patterned SERS substrate: application for protein analysis vs. temperature. Biosens Bioelectron 24:1693–1699

108. He L, Deen B, Rodda T, Ronningen I, Blasius T, Haynes C, Diez-Gonzalez F, Labuza TP (2011) Rapid detection of ricin in milk using immunomagnetic separation combined with surface-enhanced Raman spectroscopy. J Food Sci 76:N49–N53
109. Zhang H, Jia Z, Lv X, Zhou J, Chen L, Liu R, Ma J (2013) Porous silicon optical microcavity biosensor on silicon-on-insulator wafer for sensitive DNA detection. Biosens Bioelectron 44:89–94
110. St A, John CP (2014) Price, Existing and Emerging Technologies for Point-of-Care Testing, The. Clin Biochem Rev 35:155–167

Voltammetric Electronic Tongue for the Sensing of Explosives and Its Mixtures

Andreu González-Calabuig and Manel del Valle

Abstract This chapter presents recent work with electronic tongues, that is sensor analytical systems formed by an array of chemical sensors featuring low selectivity plus a chemometric tool to process the complex multivariate data that is generated. As the generic application covered is related to security, the described systems are those devised to identify and detect explosive compounds. These are characterized from their voltammetric features, whereas a particular fingerprint is used to identify particular compounds alone, or, in a more advanced application, to resolve mixtures of compounds, that is to quantify their presence in mixtures. Two are the main approaches shown, a first from the use of a voltammetric screen printed electrode, and a second one from an array of metallic electrodes. Detected compounds are different nitro-based energetic compounds, and later, also the identification of organic peroxide type compounds.

Keywords Electronic tongues · Voltammetry · Principal component analysis · Artificial neural networks · Explosives

1 Introduction

Concerns relating to homeland security have given rise to increased research into explosive detection as well as further developments for existing analytical techniques to enable faster, more sensitive, less expensive and simpler determinations to facilitate the trace identification of explosives.

Traditional security measures at airports include the use of metal detectors to identify concealed weapons in conjunction with X-ray machines to inspect baggage. One major problem is that explosive substances are not easily detectable using conventional approaches and that in this context terrorist groups may avoid

A. González-Calabuig · M. del Valle (✉)
Sensors and Biosensors Group, Chemistry Department, Universitat Autònoma de Barcelona, 08193 Bellaterra, Barcelona, Spain
e-mail: manel.delvalle@uab.es

D.P. Nikolelis and G.-P. Nikoleli (eds.), *Biosensors for Security and Bioterrorism Applications*, Advanced Sciences and Technologies for Security Applications,
DOI 10.1007/978-3-319-28926-7_4

the use of metallic objects. Approaches to detect volatile substances, such as ion mobility spectrometry in conjunction with swabbing, whilst in routine use, are largely only suitable for the screening of items of hand baggage.

Research in connection with security has received increased attention, through both the development of new, innovative detection approaches and through the improvement of existing techniques. The most commonly used techniques include Ion Mobility Spectrometry (IMS), Mass Spectrometry (MS), and Gas Chromatography (GC) followed by detection using a sensitive detector. Most of these devices are, however, rather bulky, expensive, and require time-consuming procedures. Over those, IMS is one of the most widely adopted detection techniques in routine use due to its ability to characterise the sample both qualitatively and quantitatively. Because of the above limitations, such systems are deployed only at strategic locations, e.g. airports or government buildings. Nevertheless, further complications arise when one considers not only these placements, where there is a reasonably controlled environment for sensing and detection, but also the virtually uncontrollable entry points to public places, transportation, etc. or in field use. Thus, to ensure security over those scenarios, mass deployment of miniature sensors that are sufficiently sensitive and selective, inexpensive, and amenable for mass production may be required.

Explosive compounds are widely used in warfare, mining industries, civil constructions as well as terrorist attacks. Those are categorized in four major classes: nitroaromatics, nitroamines, nitrate esters and peroxides according to their chemical structures, as schematized on Fig. 1 [1]; from there, the most widely used being 2,4,6-trinitrotoluene (TNT) or 1,3,5-trinitroperhydro-1,3,5-triazine (RDX). Among them, special attention must be paid to peroxide explosives since those compounds contain neither nitro groups nor aromatic functionalities, what makes them difficult to be detected with the more established analytical methods used to determine explosives [2]. That is, the challenge is that many current chemical identification techniques are based on the nitrogen and carbon content of a substance for its identification, and this practice is not suitable for peroxide explosive. Furthermore, the peroxide explosives are also not suitable for UV detection [3] because of their lack of chromophores and their instability under illumination of UV light—all necessary conditions for traditional detection procedures.

As an alternative, electrochemical devices are advantageous for addressing the growing need for detection of various explosives, satisfying previously described requirements [4]. In this direction, previous attempts were made to voltammetrically detect the aforementioned compounds employing different types of electrodes such as a bare screen printed electrode (SPCE), unmodified or modified gold electrodes and modified glassy carbon electrodes (GCE). Although the detection of such compounds can be achieved even at very low concentration levels, the main challenge now is that common real-life explosives are usually mixtures of two or more different explosive species. Therefore, it is interesting to discriminate between individual compounds and to resolve their mixtures.

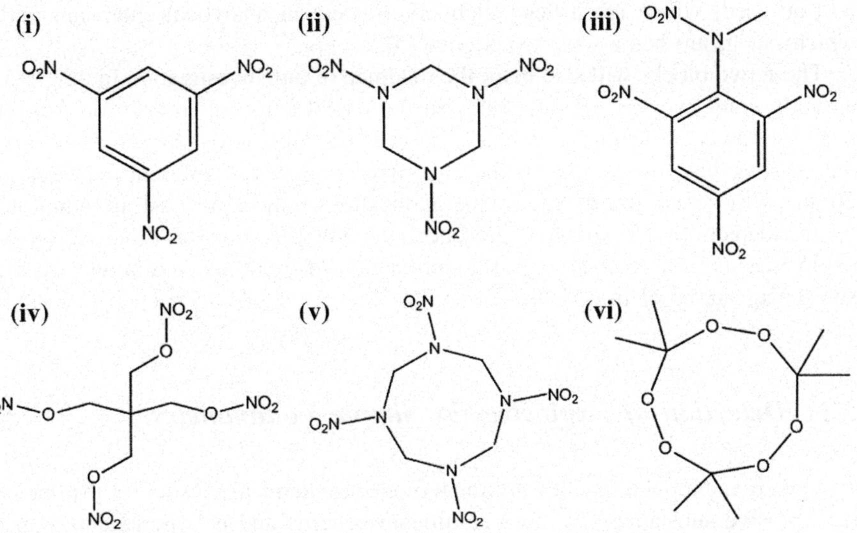

Fig. 1 Chemical structures of important explosive compounds commercially available or used in terrorist acts. The different nitro- compounds (*i*) 2,4,6-trinitrotoluene (TNT); (*ii*) 1,3,5-trinitroperhydro-1,3,5-triazine (RDX); (*iii*) N-methyl-N,2,4,6-tetranitroaniline (Tetryl); (*iv*) Pentaerythritol tetranitrate (PETN); (*v*) octahydro-1,3,5,7-tetranitro-1,3,5,7-tetrazocine (HMX) and the peroxo- type (*vi*) triacetone triperoxide (TATP)

On the other hand, voltammetric signals produced by these electrochemical methods correspond normally to a global overlapped, multiple peak voltammogram; i.e. there is a lack of specificity or identification of differentiated peaks for each of the compounds. Thus, to correctly discover the relationships among all variables and samples efficiently and to overcome the limitations found when analyzing the data using only one or two variables at a time, it would be desirable to process all of the data simultaneously, in this case with help of chemometrics. Chemometrics is the discipline for extracting information from multivariate chemical data using tools of statistics and mathematics. In our case, the used techniques will be divided into two classes depending on the nature of the extracted information, either qualitative methods such as principal component analysis (PCA), linear discriminant analysis (LDA) or quantitative ones such as partial least squares (PLS) or artificial neural networks (ANNs).

The methodology needed for this kind of work consists, then, of two main parts: the development of the (bio)sensors responding to the species sought and the use of the processing tools for the data treatment. In the case of sensors, different alternatives, which will be detailed later, have raised; however, in all cases, voltammetric sensors are the preferred technology if explosives are the target analytes. In the case of the processing tools, the concept and principles of electronic tongues will be used, which are a new relatively concept in analytical chemistry and on which our group has a wide experience [5].

These two blocks will also mark the roadmap of any investigation in this area. At an early stage we will mainly focus in the construction of the sensors first with response towards the nitro and later to the peroxo compounds. After the selection of the sensors that will be part of the electronic tongue, the different data treatment options will be evaluated; finally, the applications will be devised, in which it can be the detection of explosive presence, the identification of different types of explosives and the resolution and quantification of mixtures of different explosive substances, alone or in mixtures.

1.1 Detection of Explosives by Simple Voltammetry

As already mentioned, the development of electrochemical sensors for the detection of explosive substances provides significant benefits and is experiencing constant growth [6]. The main advantages of these types of sensors for on-field detection include its high sensitivity and selectivity, a wide linear range, minimal space and power requirements, and low-cost instrumentation. Moreover, both the sensor and the instrumentation can be readily miniaturized to yield compact and user-friendly hand-held meters for on-site (indoor and outdoor) testing [7].

In voltammetric sensors, there is a sweep in the potential applied to the working electrode to generate an electron transfer reaction (usually the reduction) of the explosive substance of interest, during which the current is measured. The inherent redox activity of commercial explosives [8], such as nitroaromatic or nitramine compounds, namely the presence of easily-reducible nitro groups, makes them ideal candidates for voltammetric monitoring. What is needed, is to obtain their particular voltammetric fingerprint, which is text used for its identification. Hence, electrochemical devices represent a promising solution for on-site explosives detection.

Upon selection of the measuring technique, the next step is the evaluation of the different strategies/technologies for the obtention of the sensor array that will allow the detection of the explosive compounds. In this direction, the most feasible options are the use of:

- Composites, in this case based on conductive phases dispersed in polymeric matrices. These materials combine the electrical properties of graphite with the ease of processing of plastics (epoxy, methacrylate, Teflon, etc.) and show attractive electrochemical, physical, mechanical and economical features compared to the classic conductors (gold, platinum, graphite, etc.) [9]. The main advantages derived from the use of composites include their ability to integrate different materials that allow to improve the sensitivity and selectivity, the flexibility in size and shape that they bring to their manufacture, the possibility of polishing its surface to obtain fresh material for the next measurements (obtaining reproducible results due to the composition of the "composite" being homogeneous), higher S/N ratio compared to pure conductors, which allows lower detection limits to be achieved, etc. The electrodes in the sensor array may incorporate catalysts, conducting polymers or nanoparticles in the composite formulation, according to existing knowledge, in order to display differentiated response.

- Screen Printed Electrodes (SPEs), which allow the miniaturization of the previous described system through the use of screen printed electrodes instead of bulk composite electrodes; increasing in this manner, the portability of the system, allowing disposable use, reducing the amount of sample required and the manufacturing cost [10].
- Electropolymerized conducting polymers [11]. The high application potential of conducting polymers in chemical and biological sensors is one of the main reasons for the intensive investigation and development of these materials. Although conducting polymers show almost no conductivity in the neutral (uncharged) state, their intrinsic conductivity results from the formation of charge carriers upon oxidizing (p-doping) or reducing (n-doping) their conjugated backbone. A process that can be done electrochemically, which additionally provides fine tuning of the doping level by adjusting the electrical potential. Moreover, sensitivity of conducting polymers to organic molecules (explosive compounds in our case) can be fine tuned based on the intrinsic affinity of the polymer backbone, on the affinity of side groups or on binding to immobilized receptors.
- Molecularly Imprinted Polymers (MIPs), also known as artificial antibodies, are polymers formed in the presence of the molecule that we aim to determine (template) and that at a later stage is removed, leaving a complementary cavity in the polymer with affinity to the chosen template molecule [12]. Such polymers present a high affinity towards the template molecule and can be used to manufacture sensors with a similar recognition mechanism to antibodies, therefore with a very high selectivity. Molecular imprinting is, in fact, making an artificial tiny lock for a specific molecule that serve as miniature key. Integration of MIPs with voltammetric sensors is feasible when the detected molecules are electroactive, which is the selected case. Adsorptive stripping techniques will be the ideal choice for improving detection limits to the sub-ppb level.
- Molecularly imprinted Au nanoparticle composites. The idea in the construction of these sensors is very similar to the previous one, but replacing the polymer with gold nanoparticles [13]. The imprinting process involves the electropolymerization of thioaniline-functionalized gold nanoparticles on a thioaniline monolayer-modified electrode in the presence of a carboxylic acid, acting as a template analogue for the respective explosive. Then, the high affinity of gold with thiol groups lead to the formation of an array of gold nanoparticles (similar to the polymer matrix) in which cavities might be also generated due to the presence of the template molecule.

1.2 Improvement of Voltammetric Results with Use of Chemometrics

After the data collection task, which typically involves several measurements made on many samples, the next step is data processing. Voltammetric multivariate data has traditionally been analyzed using one or two variables at a time. However, this approach fails to discover the relationships among all variables and samples efficiently. To overcome this, we must process all of the data simultaneously, in this case with the help of chemometrics [14]. Therefore, in order to extract useful information from what is not, and be able to interpret the data so they can be used in useful prediction models, the use of multivariate processing tools such as in the case of electronic tongues is required.

Electronic tongues are inspired by the sensory ability of taste in mammals, where a few receptors can respond to a large variety of substances [5]. This principle functions thanks to the complex data treatment applied in the brain, which allows the quantification or classification of a large amount of substances. These biomimetic systems, opposed to conventional sensing approaches, are directed towards the combination of low selectivity sensors array response (or with cross response features) in order to obtain some added value in the generation of analytical information. Moreover, once implemented and trained, the usage of such systems facilitates its implementation as a screening/monitoring device since it does not require of any technician presence. The pictorial concepts intervening in an electronic tongue are sketched on Fig. 2.

Multivariate analysis techniques allow disclosing the contribution of different factors in a result [15]. These factors are related to the explanatory variables of the system; in the case of electronic tongues these variables are usually determined

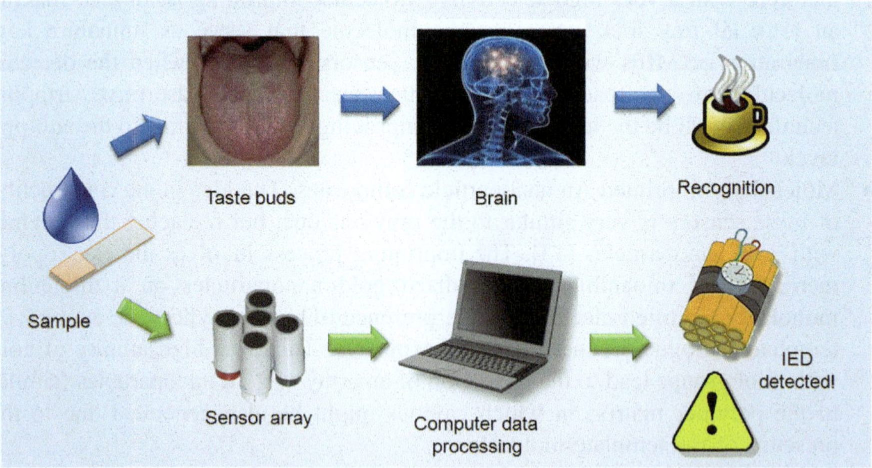

Fig. 2 The use of electronic tongue principles for the on-site detection of explosives and IEDs

experimentally and are related to each of the sensors' signals, while the response variable is related to the presence of a compound or the concentration value to be determined. The proposal and development of various methods based on mathematical, statistical and formal logic calculations are intended to establish procedures that allow to perform tasks such as the discrimination, quantification, classification and systems modelling trying to use the most relevant information from the analytical data available. On that account, data analysis and pattern recognition are a fundamental part of any electronic tongue system [5]. For the modelling, first, the data is preprocessed in order to make it independent from units, remove redundant information and to enhance signal-to-noise ratio. Following this, the model describing the relation between readings and outputs is then created. A known problem when voltammetric sensors are used is the large dimensionality of the generated data which hinders their treatment, that is, when a complete voltammogram is recorded for each sensor from the array. This is perhaps the main reason why this approach is not the most frequently used in the literature; especially if ANNs are to be used, in which case departure information needs to be preprocessed. Although signal preprocessing is not always strictly necessary (e.g. the whole data set may be employed in the case of PLS or PCA), it has been demonstrated that even in these cases, its use improves model predictive behaviour. In this sense, an attractive solution when dealing with a set of voltammograms is the use of a preprocessing stage for data reduction prior to modelling [5, 16]. The main objective of this step is to reduce the complexity of the input signal preserving the relevant information and making it compatible with ANN or other numeric modellers, which facilitates an advantageous reduction in training time, to avoid redundancy in input data and to obtain a model with better generalization ability. This compression stage may be achieved by the use of methods such as PCA [17], feature selection [18], "kernels" [19], discrete wavelet transform (DWT) [20] or even fast Fourier transform (FFT) [21].

After applying the desired preprocessing method, the next step is the modelling of the target variables using the obtained coefficients as inputs into the model stage. Among the various numeric procedures that can be implemented, PCA, PLS and ANNs are the most widely used for electronic tongue applications [5, 22]. Concretely, PCA is the most common one being used either as a qualitative visualization tool or as a preprocessing step; whereas most advanced qualitative modelling may be achieved with the use of PLS discriminant analysis (PLS DA), linear discriminant analysis (LDA) or support vector machines (SVM) [23]. While in the case of quantitative applications, easier models can be built either using multiple linear regression (MLR) or principal components regression (PCR), although better results can be achieved by using more powerful methods such as PLS or ANNs. Lastly, the new trends in data analysis are mainly related to the use of trilinear approaches such as PARAFAC for qualitative analysis, and multi-way PLS (nPLS) for quantitative models.

2 Systems Using a Single Sensor

A first attempt to perform the detection of explosive compounds involving the coupling of electrochemical measurements and advanced chemometric data processing was realized from voltammetric signals obtained at a disposable carbon electrode, and with use of proper chemometric tools. Three nitro-containing compounds found in the majority of explosive mixtures, namely hexahydro-1,3,5-trinitro-1,3,5-triazine (RDX), 2,4,6-trinitrotoluene (TNT) and pentaerythritol tetranitrate (PETN) were identified as individual energetic chemicals, and next, a quantitative application was also illustrated with the resolution of their trinary mixtures.

Screen printed electrodes (SPCE) were prepared following the conventional methodology previously described [8]. The electrochemical cell was formed by a carbon working electrode, a carbon counter electrode and a silver pseudo-reference electrode. Cyclic Voltammetry (CV) was the technique employed and a new electrode was used for each sample. Figure 3 shows some of the obtained voltammograms for the different mixtures of explosive compounds, as can be seen complex and highly overlapped signals were observed. This kind of signals, which exhibit different sensibility and selectivity, are an ideal departure point to be used in an ANN application.

In order to evaluate the capabilities of the proposed system to distinguish between different explosives, stock solutions of each of the pure compounds were analyzed, also mixtures based on usual commercial formulations were also prepared and measured. Hence, in this manner 10 different explosives mixtures were considered, the concentration of explosives was kept at 50 ppm for the case of pure compounds like TNT, RDX, HMX, Tetryl and PETN. For the commercial explosive mixtures like Semtex H (RDX:PETN at 1:1 ratio), Comp. B (RDX:TNT at 3:2 ratio), Comp. C-3 (RDX:TNT:Tetryl at 23:1:1 ratio), Pentolite (TNT:PETN at 1:1 ratio) and Tetrytol (TNT:Tetryl at 3:7 ratio) the total explosive concentration was

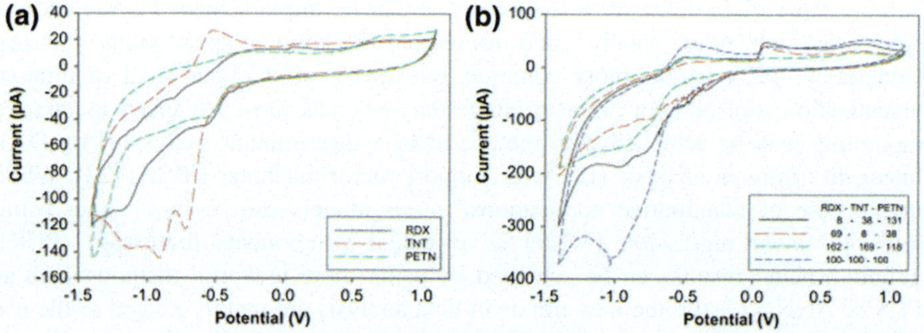

Fig. 3 Example of the voltammograms obtained for (**a**) 50 μg·mL^{-1} standard solutions of each of the three explosive compounds under study and (**b**) mixtures of the three explosive compounds (all concentrations are expressed in μg·mL^{-1}). Reprinted from [24], with permission from Elsevier

kept at 50 ppm.. All samples were prepared in triplicate and randomly measured employing a new sensor each time. Therefore, the set of samples under study was formed by 30 samples distributed in 10 different classes.

It is important to note that DWT was used as the feature extraction tool; PCA was used for qualitative analysis of the results, while quantitative analysis and classification was achieved by means of ANNs. DWT is a high performance signal processing technique developed inspired in the Fourier transform, with the key advantage over the latter of its temporal resolution: it captures both frequency and location information (location in time). DWT is used for signal decomposition onto a set of basic functions, obtained from dilations and translations of a unique function called mother wavelet, the most commonly used being Daubechies wavelet [25]. Transform is implemented using Mallat's pyramidal algorithm [26], which operates over a single discrete signal of length M by decomposing it into orthogonal subspaces of length ca. M/2 in each step [20]. In this way, by repeating this decomposition process n times, the signal compression ratio is increased at the expenses of the accuracy in the signal reconstruction.

Principal Component Analysis (PCA) allows the projection of the information contained in the original variables onto a smaller number of latent variables called principal components (PCs) with new coordinates called scores, obtained after data transformation. Plotting the PCs, one can view interrelationships between different variables, and detect and interpret sample patterns, groupings, similarities or differences [17]. As the PCA only is a visualization tool it has to be coupled with a modeling tool to be used as a classifier.

ANNs are excellent modellers, that consist of a number of simple processing units (also called neurons) linked by weighted modifiable interconnections [27], originally designed to mimic the function of the human brain. ANNs work by imitating the biological learning task, requiring a training process where the weights of those connections are adjusted, to build a model that will allow the prediction of the desired parameters. Such methods are known as supervised methods; the training data consists of a set of training examples (a fraction of the set cases) which are used to build the model plus an external set used to evaluate the performance of the model, the test subset. The main advantages of ANNs include a high modelling performance tool, particularly suited to non-linear sensor responses, and significant likeness to human pattern recognition [5].

The combination of CV and chemometric data treatment such as PCA or ANNs allowed the identification of each compound's voltammetric fingerprint and solved the problem of signal overlapping. To facilitate data treatment of information contained in each voltammogram, a compression step was selected [30]. In this specific case the reduction of the large data generated was achieved by means of DWT [20]. This allowed the reduction of signals from each voltammogram down to 82 coefficients without any loss of relevant information and achieving a compression ratio of 93.2 %. The obtained coefficients were used to build a model that predicts the desired parameters, either the discrimination of different samples or the quantitative resolution of the mixtures composition.

As was previously commented a total of 30 samples were analyzed, corresponding to triplicate measurements of 5 explosive compounds (HMX, PETN, RDX, Tetryl, TNT) and triplicate measurements of 5 typical commercial mixtures (C-3, Comp. B, Pentolite, Semtex, Tetrytol), with one complete voltammogram for each sample. Afterwards, responses were preprocessed employing DWT and the obtained coefficients were analyzed by means of PCA analysis and were grouped using cluster analysis tools. After the initial representation of data, an ANN model used with its binary output was used as classifier.

Upon completion of the PCA analysis, the accumulated explained variance was calculated with the three first PCs as ca. 94.3 %. This large value shows that nearly all the variance contained in the original data can be explained by just using the first new coordinates. Different clusters were obtained and plotted, outlined in Fig. 4; in this, observed patterns show evidence that samples are clearly grouped based on explosives mixtures composition.

Analyzing the plot more thoroughly, some expected trends could be observed; i.e. the fact that clusters corresponding to mixtures are located close to the pure compounds forming those mixtures. For example, Comp. B samples are clustered between RDX and TNT groups, or Tetrytol samples close to TNT and Tetryl groups, etc. Hence, despite some overlapping regions between the different pure compounds signals are observed, their fingerprints can be still distinguished due to the differentiated sensitivity shown by the electrode.

To confirm the discrimination of the samples shown in the PCA plot, a classifier based on a PCA-ANN model was built. The output of the ANN model was formed by binary predictors (1/0) for each of the classes. As usual, the ANN configuration first needed to be optimized. After some preliminary tests, the final ANN architecture model had 3 neurons (corresponding to the first three components of the

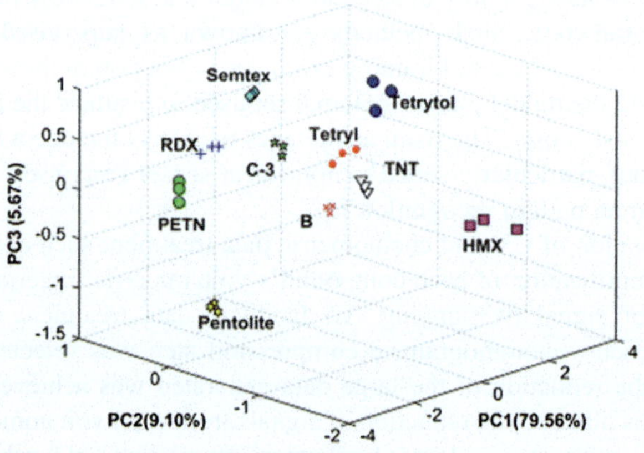

Fig. 4 Score plot of the first three components obtained after PCA analysis. A total of 30 samples were analyzed corresponding to triplicate determinations of: RDX, TNT, PETN, Tetryl, HMX, Semtex H, Comp. B, Comp. C-3, Pentolite and Tetrytol). Reprinted from [24], with permission from Elsevier

PCA) in the input layer, 6 neurons in the hidden layer and 11 binary (1/0) neurons in the output layer (one for each class plus an unknown class) with logsig transfer function connecting the three layers. The aim of including an unknown class in the classifier model was to somehow avoid that other explosive compounds could result in a misclassification [28].

The PCA–ANN model was trained with 67 % of the data (20 samples) and evaluated using the information of the testing set (remaining 33 % of the data; 10 samples) in order to characterize the accuracy of the identification model and obtain unbiased data. From the classification results, the corresponding confusion matrix was built. Correct classification for all the classes was obtained (i.e., a classification rate of 100 % for each of the groups), as indicated from the direct visualization of the PCA analysis. The percentage of correct classifications was estimated, from individual sample calculation in the test subset, as 100 %. The efficiency of the classification obtained was also evaluated according to its sensitivity, i.e., the percentage of objects of each class identified by the classifier model, and to its specificity, the percentage of objects from different classes correctly rejected by the classifier model. The value of sensitivity, averaged for the classes considered, was 100 %, and that of specificity was 100 %.

The quantitative resolution of mixtures of three explosive compounds was evaluated (i.e. RDX, TNT and PETN) to provide a tool that would allow the identification of the type of explosive, its quantification, and its specific composition. For this, a total set of 42 samples were manually prepared with a concentration range for the three species from 0 to 200 $\mu g \cdot mL^{-1}$ for each of the nitro-containing explosive compounds. The set of samples was divided into two data subsets: a training subset formed by 27 samples (64 %), which were distributed in a cubic design and used to establish the response model [18]; plus 15 additional samples (36 %) for the testing subset, randomly distributed along the experimental domain, and used to evaluate the model's predictive response.

In order to prove the capabilities of the approach to achieve the quantification of the explosive compounds, in addition to the qualitative identification of the commercial explosive, mixtures of the main constituents (i.e. RDX, TNT and PETN) were analyzed by means of CV. The voltammograms, as before, were compressed employing DWT and obtained coefficients were used as inputs to an ANN model [29].

The first step in building the ANN model is selecting the topology of the network used. This is a trial-and-error process where several parameters (number of neurons, transfer functions, etc.) are fine-tuned in order to find the best configuration that optimizes the performance of the model [27]. The final ANN architecture had 82 neurons (corresponding to the coefficients obtained from the DWT analysis) in the input layer, 5 neurons and tansig transfer function in the hidden layer and 3 neurons and purelin transfer function in the output layer (one for each compound).

The accuracy of the generated model was then evaluated towards samples on the external test subset by using it to predict the concentrations of explosives on those samples. Subsequently, comparison graphs of predicted vs. expected concentration for the three compounds were built, both for training and test subsets, to easily

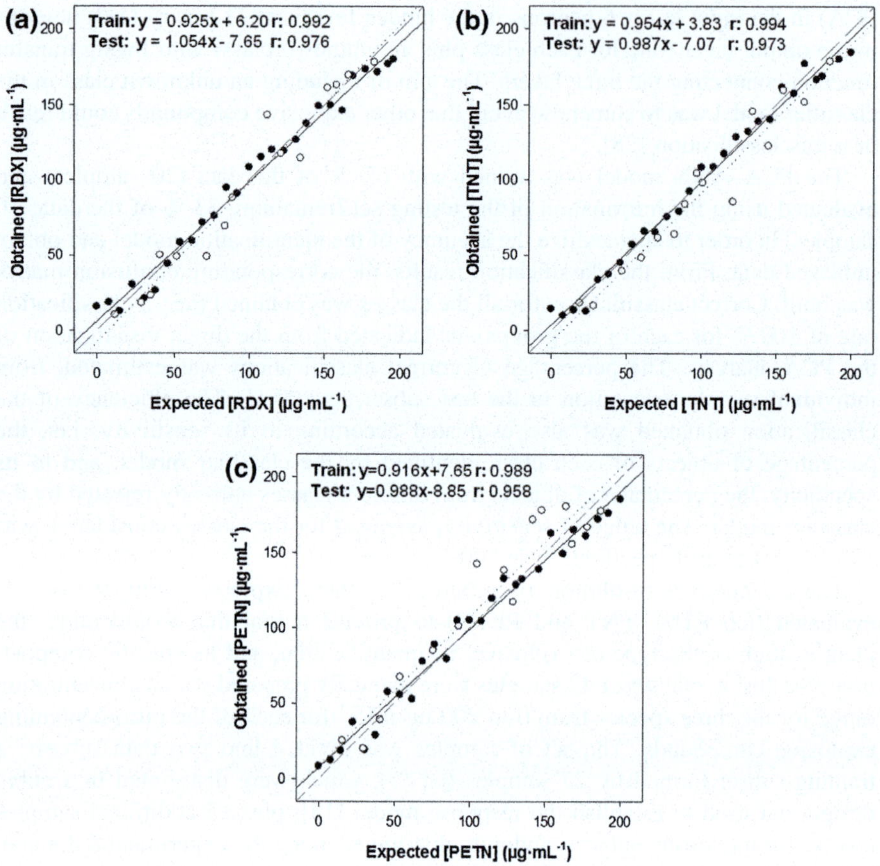

Fig. 5 Modeling ability of the optimized ANN. Sets adjustments of obtained versus expected concentrations for (**a**) RDX, (**b**) TNT and (**c**) PETN, both for training (●, *solid line*) and testing subsets (○, *dotted line*). *Dashed line* corresponds to theoretical diagonal line. Figure adapted from [24], with permission from Elsevier

check the prediction ability of the obtained ANN model (Fig. 5). A satisfactory trend is obtained for the three compounds observed in the figure, with regression lines almost indistinguishable from the theoretical ones. As is usual in ANN models, lower dispersion and uncertainties are obtained for the training subsets, as expected taking into account that the external test subset data is not employed at all for the modeling, so its goodness of fit is a measure of the model performance.

The figure also displays the resulting regression parameters of the obtained vs. expected comparison graphs. A good linear trend is attained for all the cases, but with improved correlation coefficients in the training subsets due to the lower dispersion. Despite this, the results obtained for both subsets are close to the ideal values, with intercepts ca. 0, and slopes and correlation coefficients ca. 1.

In this way, it has been demonstrated how the combination of CV with chemometric tools is an analytically powerful approach for the characterization and detection of individual explosive substances and its mixtures commercially available.

This is particularly useful since it does not only allow the identification of typical commercial mixtures, but also the quantification of the amount present and the ratio of the mixtures analyzed. In this manner home-made nitro-containing explosives and IEDs might be properly identified; this represents a major advantage compared to the qualitative approach.

The proposed sensor coupled with chemometrics represents a qualitative method to provide the identification of the voltammetric fingerprint of different explosive mixtures commercially or manufactured, such as IEDs. This represents a viable system with significant promise for in-field measurements given its simplicity, rapidity and portability. Nevertheless, to fully achieve the correct identification of almost all types of explosives, application of the proposed approach to the detection of peroxide-based explosives is still required. Its increasing use has led to considerable research into the detection of this group of improvised explosive substances; the challenge being that many current chemical identification techniques are based on the nitrogen and carbon content of a substance for identification and this practice is not suitable for peroxide explosive. In this context, electrochemical sensors offer an opportunity to detect peroxide-based explosives that would otherwise prove problematic.

3 Systems Using a Sensor Array

In a second application study, a voltammetric electronic tongue (ET) was developed towards the simultaneous determination of both nitro-containing and peroxide-based explosive compounds, two families that represent the vast majority of compounds employed either in commercial mixtures or in improvised explosive devices. The electronic tongue was formed by a multielectrode array constituted by graphite, gold and platinum electrodes, which exhibited marked mix responses towards the compounds examined; namely RDX, HMX, PETN, TNT, Tetryl and triacetone triperoxide (TATP).

In this particular study, our aim was to obtain a miniaturized sensor array. Thus the use of a quatrielectrode with inner counter and reference electrodes was proposed. Platinum, silver, gold and epoxy-graphite 1 mm diameter discs were used as electrodes. The metal electrodes were fabricated from its metal wires, the epoxy graphite electrode was done mixing epoxy and carbon; the electrodes were encased in inert epoxy resin using a PVC tube as the body [30].

To complete the electrochemical cell, a stainless steel tube was glued to the cylinder multi-electrode and used as a counter electrode while, the Ag electrode was converted into a Ag/AgCl pseudo-reference electrode. Figure 6 illustrates the design of the quatrielectrode used. Thus, with this adaptation all the electrodes are integrated in a small device suitable to be used in field applications, which in turn

Fig. 6 Quatrielectrode used during the measurements

allows using a smaller sample volume, also making possible to perform drop analysis.

As in the previous case, the set of samples was measured employing the multielectrode array to obtain a whole cyclic voltammogram for each of the sensors. The electrochemical measurements were carried out at room temperature (25 °C) under quiescent conditions. Some of those results are shown on Fig. 7. As can be seen, complex and overlapped signals are obtained along the whole voltammogram with differentiated signals obtained for the different kinds of sensors; this situation is ideal to implement the electronic tongue approach.

In a first experiment, discrimination of different explosive compounds by means of the electronic tongue was attempted. That is, to asses if the system presented herein was able to carry out the identification of the most common explosive compounds. For such purposes, we measured 18 samples, corresponding to 3 replicates of each explosive compound (RDX, HMX, TNT, TATP, PETN and tetryl).

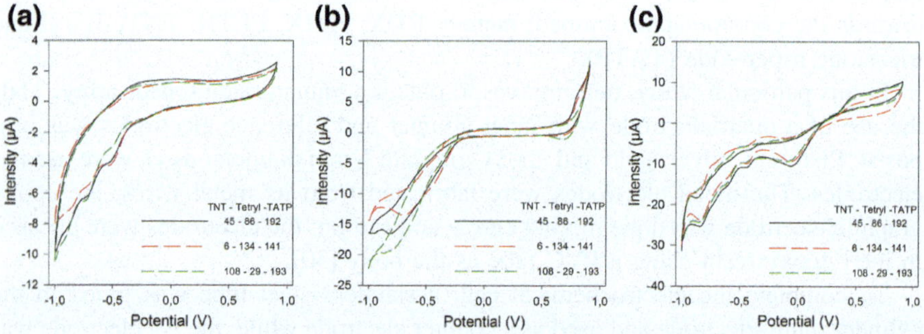

Fig. 7 Cyclic voltammograms for arbitrary mixtures of the three explosive compounds (all concentrations are expressed in $\mu g \cdot mL^{-1}$) for the graphite epoxy electrode (**a**), gold electrode (**b**) and platinum electrode (**c**)

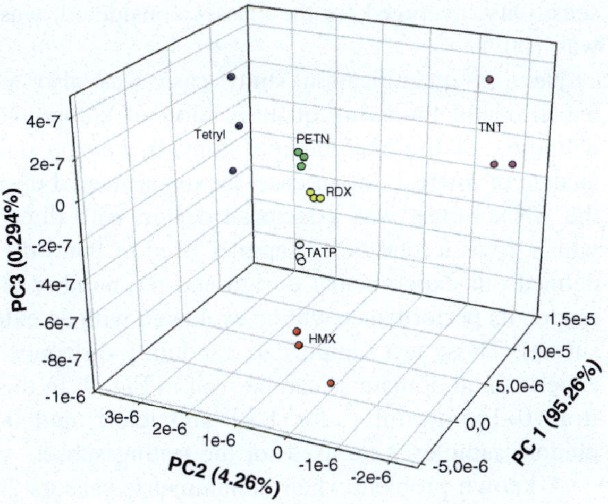

Fig. 8 Score plot of the first three components obtained after PCA analysis

To this aim, voltammetric responses obtained for the stock solutions of each of the pure compounds were analyzed by means of PCA and grouped using cluster analysis tools.

Upon the completion of the PCA analysis (Fig. 8), the accumulated explained variance was calculated with the three first PCs as ca. 99.81 %. This large value shows that nearly all the variance contained in the original data can be explained by just using the first new coordinates. In addition, patterns in the figure evidence that samples are clearly grouped based on each explosive compound, with replicas for the same class one to each other. Moreover, the low dispersion of PETN, TATP and RDX clusters shows a more reproducible behaviour. Nevertheless, despite the bigger dispersion obtained for the others compounds there is no overlap between clusters.

After the initial representation of the data, a fuzzy ARTMAP ANN model with binary inputs (1/0) was used as classifier, which allowed quantification of the classification performance system in contrast to PCA analysis which just provides a visualization of the grouping regions.

The PCA-ANN model was trained with the 67 % of the data (12 samples) and evaluated using the information of the testing set (remaining 33 % of the data; 6 samples) in order to characterize the accuracy of the identification model and obtain unbiased data. From the classification results, the corresponding confusion matrix was built. Correct classification for all the classes was obtained (i.e. a classification rate of 100 % for each of the groups), as indicated from the direct visualizations of the PCA. The percentage of correct classifications was estimated, from individual sample calculation in the test subset, as 100 %. The efficiency of the classification obtained was also evaluated according to its sensitivity, i.e. the percentage of each class identified by the classifier model, and to its specificity, the percentage of objects from different classes correctly rejected by the classifier model. The value of sensitivity, averaged for the classes considered, was 100 %, and that of specificity was 100 %.

Next, a quantification study case was also attempted. When designing an experimental involving quantification of compounds, a first consideration is the definition of the experimental domain, i.e. the range of concentrations for each analyte of interest. In our case, the experimental design used for the construction of the ANN model was a factorial design with three levels and three analytes (3^3) which gives a total of 27 samples used to build the model (training subset). Once defined our experimental design and the number of samples required to build the model, its performance will be evaluated with an external subset of samples (testing subset). Those test samples are randomly distributed along inside the limits of the experimental domain as can be seen in Fig. 9. In this case the concentration ranged from 0–165 $\mu g \cdot mL^{-1}$ for TNT and tetryl, and 0–300 $\mu g \cdot mL^{-1}$ for TATP; 10 random samples were used for the testing subset.

A known problem when voltammetric sensors are involved is the large dimensionality of the generated data (samples x sensors x polarization potentials) which hinders their treatment: especially if ANNs are to be used, in which case departure information has to be preprocessed. In this fashion, prior to building the quantification model the removal of less significant coefficients that barely contribute to the network was carried out by means of causal index (CI) pruning inputs [16, 31].

Briefly, CI is based on the usage of ANNs as feature selection tools, aimed to the selection of an optimal set of inputs that can successfully classify or predict the desired outputs. To this end, an ANN model is built employing the whole set of variables, followed by the determination of the contribution and relevance of each of the network inputs to the variance in the output layer. This can be achieved with the analysis of its connection weights, which allows to easily identifying the

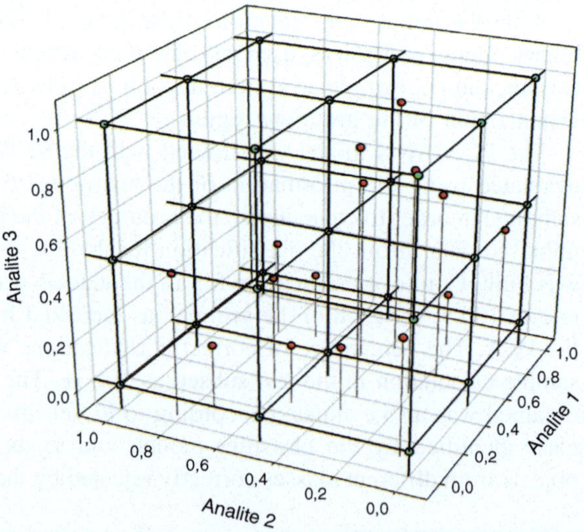

Fig. 9 Factorial design 33: training subset (*green*) and test subset (*red*)

important inputs since inputs that make relatively small contributions indicate that the input does not change significantly; and therefore, can be discarded, viz. pruned. Afterwards, selection of the most relevant inputs can proceed until a near-optimal, small, set of inputs is identified by repeating the training process of the ANN model with the reduced input set and selecting the most relevant ones each time. Finally, once the reduced set of inputs is identified, optimization of ANN architecture can proceed as usual [27].

The accuracy of the generated model was evaluated towards samples of the external test subset by using the built model to predict concentrations of the explosives of those samples. To evaluate the performance of the different models, its normalized root mean square error (NRMSE) for each combination of the transfer functions used in the hidden and output layers, and the number of neurons in the hidden layer.

$$NRMSE = \frac{\sqrt{\frac{\sum_{ij}(x_{expected} - x_{obtained})^2}{k \cdot n - 1}}}{x_{\max} - x_{\min}} \tag{1}$$

Thus, the optimum topology will be the one that also gives the lowest NRMSE value. As it can be seen in Fig. 10, by plotting the total NRMSE versus the number of inputs gives a clear view of the performance of the model.

Hence, once the predictors data matrix was obtained, the next step was the optimization of the ANN as previously done. In this case, the resulting ANN model has 48 neurons in the input layer (corresponding to the relevant data points previously selected), 8 neurons and *logsig* transfer function in the hidden layer, and 3 neurons and *purelin* transfer function in the output layer, providing the concentrations of the three species considered.

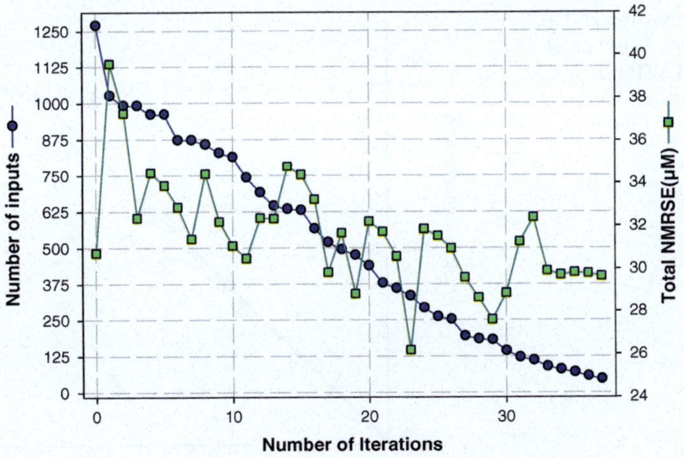

Fig. 10 Evolution of the total NRMSE as the number of inputs is decreased

Table 1 Results of the fitted regression lines for the obtained versus expected values for the samples of the training and testing subsets and the three considered explosive materials (intervals calculated at the 95 % confidence level)

Training subset					
Explosive	Correlation (r)	Slope	Intercept ($\mu g \cdot mL^{-1}$)	NRMSE	Total NRMSE
TNT	0.997	0.984 ± 0.030	1.4 ± 3.2	0.031	0.031
Tetryl	0.997	0.980 ± 0.030	1.7 ± 3.2	0.032	
TATP	0.998	0.981 ± 0.028	2.8 ± 5.6	0.030	
Testing subset					
Explosive	Correlation (r)	Slope	Intercept ($\mu g \cdot mL^{-1}$)	NRMSE	Total NRMSE
TNT	0.979	0.981 ± 0.166	2.2 ± 14.9	0.063	0.091
Tetryl	0.929	1.038 ± 0.337	5.4 ± 30.8	0.135	
TATP	0.945	0.959 ± 0.269	1.2 ± 44.7	0.080	

NRMSE Normalized root mean square error

Comparison graphs of predicted vs. expected concentration for the three compounds were built, both for training subset and testing subset. Table 1 shows the regression parameters for each compound in the training subset and the test subset. As can be seen in the plots (Fig. 11) a good linearity is achieved for all the cases, with better correlation in the training subset and less NRMSE. For the test subset, that is the one that really shows how the model is performing; the regression parameters are also close to the ideal values. Thus, in both subsets the intercepts are close to 0, and the slopes and the correlation coefficients close to 1.

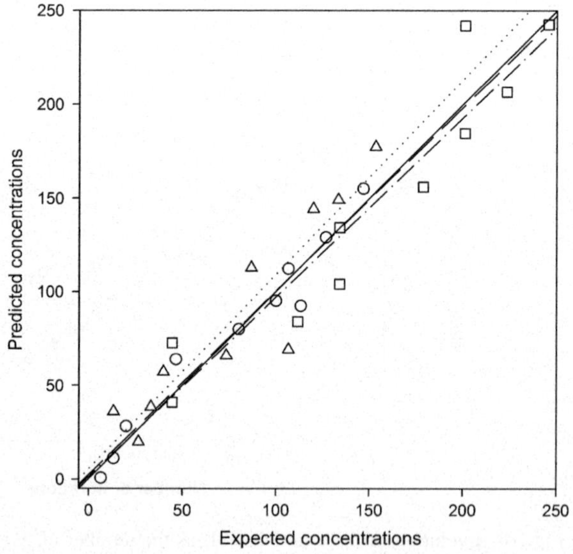

Fig. 11 Test plots of predicted versus expected concentrations for TNT (○, *dashed line*), Tetryl (Δ, *dotted line*) and TATP (□, *dashed-dotted line*)

Based on these results, we can confirm that the model is able to predict the concentration of the 3 individual compounds in a mixture sample with good performance. However, results show that the tetryl and the TATP show a bigger dispersion and that affects the regression parameters of these compounds.

The analysis of samples was based on the combination of cyclic voltammetry for the extraction of the fingerprints of the individual components and mixtures of these species, coupled with chemometric tools that allowed the resolution of signal overlapping and identification of the different compounds. The resolution and quantification of ternary mixtures was achieved employing an artificial neural network model. Obtained results suggest that voltammetric electronic tongues could be of application for the detection in real explosive formulation samples and a good candidate for homeland security applications.

4 Concluding Remarks

To summarize, these two application cases have shown how the voltammetric sensor systems used can be used as qualitative methods to provide the identification of the voltammetric fingerprint of different explosive mixtures commercially or manufactured, such as IEDs. The systems are also capable to detect peroxide compounds, due to their difficulty of detection its use has been increased and has led to considerable research into the detection of this group of improvised explosive substances; the challenge being that many current chemical identification techniques are based on the nitrogen and carbon content of a substance for identification and this practice is not suitable for peroxide explosives. In this context, electrochemical sensors offer an opportunity to detect peroxide-based explosives that would other-wise prove problematic. Alternatively, quantitative determination applications to determine explosives in mixtures can be also developed.

The results suggested that voltammetric electronic tongues could be of application for the detection of real explosive formulation samples and a good candidate for homeland security applications; leading to a new generation of on-site field deployable explosive detectors. These may be massively used, and with wireless communication allow for complete networked areas, even regions to prevent by their trace detection the approach of any of the considered substances. Alternatively, from environmental concerns, any affected area by these compounds may employ the described analytical systems a starting point for the monitoring of state and evolution of already polluted scenarios or the ones being remediated.

Acknowledgements Financial support from the Spanish Ministry of Economy and Innovation, MINECO (Madrid) through project CTQ2013-41577-P is gratefully acknowledged. M. del Valle thanks the support from program ICREA Academia. A. González-Calabuig thanks the Universitat Autònoma de Barcelona (UAB) for the award of PIF studentship.

References

1. Akhavan J (2004) The chemistry of explosives, 2nd edn. Royal Society of Chemistry, Cambridge
2. Schulte-Ladbeck R, Vogel M, Karst U (2006) Recent methods for the determination of peroxide-based explosives. Anal Bioanal Chem 386(3):559–565
3. Lu T, Yuan Y, He X, Li M, Pu X, Xu T, Wen Z (2015) Simultaneous determination of multiple components in explosives using ultraviolet spectrophotometry and a partial least square method. RSC Adv 5(17):13021–13027
4. Wang J (2007) Electrochemical Sensing of Explosives. Electroanalysis 19(4):415–423
5. del Valle M (2010) Electronic tongues employing electrochemical sensors. Electroanalysis 22 (14):1539–1555
6. Yinon J (2002) Field detection and monitoring of explosives. TrAC—Trends Anal Chem 21 (4):292–301
7. Wang J (2002) Portable electrochemical systems. TrAC Trends Anal Chem 21(4):226–232
8. Galik M, O'Mahony AM, Wang J (2011) Cyclic and square-wave voltammetric signatures of nitro-containing explosives. Electroanalysis 23(5):1193–1204
9. Alegret S, Alonso J, Bartroli J, Cespedes F, MartinezFabregas E, del Valle M (1996) Amperometric biosensors based on bulk-modified epoxy graphite biocomposites. Sens Mater 8(3):147–153
10. Caygill JS, Collyer SD, Holmes JL, Davis F, Higson SPJ (2013) Electrochemical detection of TNT at cobalt phthalocyanine mediated screen-printed electrodes and application to detection of airborne vapours. Electroanalysis 25(11):2445–2452
11. Lange U, Roznyatovskaya NV, Mirsky VM (2008) Conducting polymers in chemical sensors and arrays. Anal Chim Acta 614(1):1–26
12. Alizadeh T, Ganjali MR, Norouzi P, Zare M, Zeraatkar A (2009) A novel high selective and sensitive para-nitrophenol voltammetric sensor, based on a molecularly imprinted polymer–carbon paste electrode. Talanta 79(5):1197–1203
13. Riskin M, Ben-Amram Y, Tel-Vered R, Chegel V, Almog J, Willner I (2011) Molecularly imprinted Au nanoparticles composites on Au surfaces for the surface plasmon resonance detection of pentaerythritol tetranitrate, nitroglycerin, and ethylene glycol dinitrate. Anal Chem 83(8):3082–3088
14. Pravdová V, Pravda M, Guilbault GG (2002) Role of chemometrics for electrochemical sensors. Anal Lett 35(15):2389–2419
15. Ni Y, Kokot S (2008) Does chemometrics enhance the performance of electroanalysis? Anal Chim Acta 626(2):130–146
16. Cetó X, Céspedes F, del Valle M (2013) Comparison of methods for the processing of voltammetric electronic tongues data. Microchim Acta 180(5–6):319–330
17. Jolliffe I (2005) Principal component analysis. In: Encyclopedia of statistics in behavioral science, Wiley
18. Cetó X, Céspedes F, Pividori MI, Gutiérrez JM, del Valle M (2012) Resolution of phenolic antioxidant mixtures employing a voltammetric bio-electronic tongue. Analyst 137(2):349–356
19. Gutierrez-Osuna R, Nagle HT (1999) A method for evaluating data-preprocessing techniques for odour classification with an array of gas sensors. Syst, Man, Cybern, Part B: Cybern, IEEE Trans 29(5):626–632
20. Moreno-Barón L, Cartas R, Merkoçi A, Alegret S, del Valle M, Leija L, Hernandez PR, Muñoz R (2006) Application of the wavelet transform coupled with artificial neural networks for quantification purposes in a voltammetric electronic tongue. Sens Actuators B: Chem 113(1):487–499
21. Cetó X, Céspedes F, del Valle M (2012) BioElectronic Tongue for the quantification of total polyphenol content in wine. Talanta 99:544–551
22. Riul A Jr, Dantas CAR, Miyazaki CM, Oliveira ON Jr (2010) Recent advances in electronic tongues. Analyst 135(10):2481–2495

23. Richards E, Bessant C, Saini S (2002) Multivariate data analysis in electroanalytical chemistry. Electroanalysis 14:1533–1542
24. Cetó X, O'Mahony AM, Wang J, del Valle M (2013) Simultaneous identification and quantification of nitro-containing explosives by advanced chemometric data treatment of cyclic voltammetry at screen-printed electrodes. Talanta 107:270–276
25. Grossmann A, Morlet J (1984) Decomposition of Hardy functions into square integrable wavelets of constant shape. SIAM J Math Anal 15(4):723–736
26. Mallat SG (1989) A theory for multiresolution signal decomposition: the wavelet representation. Pattern Anal Mach Intell IEEE Trans 11(7):674–693
27. Despagne F, Massart DL (1998) Neural networks in multivariate calibration. Analyst 123 (11):157R–178R
28. Cetó X, Gutiérrez JM, Moreno-Barón L, Alegret S, del Valle M (2011) Voltammetric electronic tongue in the analysis of cava wines. Electroanalysis 23(1):72–78
29. Moreno-Barón L, Cartas R, Merkoçi A, Alegret S, Gutiérrez JM, Leija L, Hernandez PR, Muñoz R, del Valle M (2005) Data compression for a voltammetric electronic tongue modelled with artificial neural networks. Anal Lett 38(13):2189–2206
30. Gutés A, Calvo D, Céspedes F, del Valle M (2007) Automatic sequential injection analysis electronic tongue with integrated reference electrode for the determination of ascorbic acid, uric acid and paracetamol. Microchim Acta 157(1–2):1–6
31. Boger Z (2003) Selection of quasi-optimal inputs in chemometrics modeling by artificial neural network analysis. Anal Chim Acta 490(1–2):31–40

(Affordable; Sensitive; Specific; User-friendly; Rapid and Robust; Equipment-free; and, Delivered to those who need it).

The development of novel diagnostic platforms to meet the WHO requirements should therefore be focused on: (A) Cost-effective emerging technologies such as biosensors, lateral flow and agglutination tests, appropriate for application at community and primary-care level as well as in low resource settings; (SS) novel diagnostic targets and disease biomarker development, to achieve specificity (no false positives) and sensitivity (no false negatives); (U) analytical simplification in order to minimise pipetting, washing steps and manipulation of reagents to provide analytical tools requiring minimal training for final users; (R) robust tests, portable and stable at room temperature, able to provide rapid results and to enable taking actions immediately such as treatment at first visit; (E) bench-top and high cost instrumentation should be avoided and visual detection should be prioritised; (D) biomarkers should be selected from the major problems facing global health, including pathogenic bacteria occurring in food outbreaks, in order to ensure safety in food and water supplies in low resource settings, as well as major global infection disease such as AIDS and malaria, dengue, influenza and tuberculosis in low-income countries. To meet the demands, the dominant format currently in use is lateral-flow immunochromatographic immunoassay strip tests. Despite the poor sensitivity levels and the high reported LODs [14] generally observed for this technology, many methodological improvements has been done [15–18]. As complexity increases ranging from agglutination [19, 20], lateral flow [16], ELISA [21–23] and PCR [24], sensitivity and specificity also increase but so do cost and turnaround time, as well as the need of instrumentation [25]. The main challenge in bioanalysis is thus to provide low-cost yet simple methods wit any loss in the analytical performance and test accuracy [3].

As previously discussed, the amplification of target nucleic-acid sequences using techniques such as PCR can improve test sensitivity up to 100-fold over antigen detection tests, such as ELISA. However, the first generation of nucleic-acid amplification technologies requires instrumentation for temperature cycling [12]. Single-temperature or isothermal amplification has been developed and can be adapted to a point-of-care format [26, 27]. Improved efficiency of detection systems must also be achieved, using novel hybrid bionanomaterials or signal amplification strategies [28–30]. These bionanomaterials, including nanostructured carbon materials, inorganic nanoparticles (i.e., semiconducting, noble metal and magnetic nanoparticles), among others, appears to be keys in multiplex detection enhancing the biological reactions, providing high selectivity and improving the LODs [31–34].

A prominent development trend in recent years has been to miniaturize or integrate existing diagnostics into a biosensors devices or lab-on-a-chip [30, 35, 36]. These strategies potentially solve many issues by lowering test complexity in a platform that would be practical at the point of care. Cost is also reduced by using lower reagent volumes that would be housed and stored in a kit format. Microfluidic systems provide several advantages such as portability, lower reagent consumption, rapidity and possibility for automation. Usually, these systems are combined with agarose gel electrophoresis for DNA analysis, but they can also be coupled with other platforms [37]. The cost of production of microfabricated devices, requiring in most cases bench-top equipment for the readout, still constitutes a bottleneck and may put them out of range for end users in the developing world.

Biosensors are analytical devices, incorporating a bioreceptor in contact with a transducer. Despite the massive use of glucose biosensors with electrochemical transduction, examples of other commercial devices for applications including diagnosis of infection diseases are currently very limited. Although instrumentation may be required, it should be designed to be low maintenance, battery operated, and low cost [13].

Besides the progress in emerging technologies for diagnosis, the study of novel biomarkers for the early detection of infectious diseases is a worldwide challenge [38]. The WHO has defined a biomarker as 'almost any measurement reflecting an interaction between a biological system and a potential hazard, which may be chemical, physical, or biological' [39]. The measured response may be functional and physiological, biochemical at the cellular level, or a molecular interaction [40]. The identification of novel biomarkers represents a challenge not only for the improvement of early diagnostics, but also for patient monitoring and for evaluation of the efficiency of a therapeutic strategy. Biomarkers are also very promising candidates in achieving customisation of healthcare for personalised medicine.

To summarize, some of the new diagnostic tests that are needed are already on the market, but have not been adequately validated. In other cases, diagnostic targets and biomarkers have been identified, and it seems likely that an adequate test could be developed using existing technology. In these two cases, feasibility is high [3]. In other cases, the biomarkers have not yet been identified and, feasibility is thus low.

2 Magnetic Particles in ASSURED Diagnostic

Novel development in diagnosis that is needed involves preconcentration procedures on solid supports which can be easily integrated with emerging technologies. Microorganism and biomarkers in complex samples can be thus preconcentrated while the interfering matrix is removed at the same time, increasing the sensitivity and the specificity of the test. One of the most prominent materials to meet this challenge is magnetic particles (MPs) [41]. MPs can be tailored to specifically bind the biomarkers and concentrate them from the complex specimen under magnetic actuation, avoiding interference before testing [42–44].

Magnetic particles have been commercially available for many years. Nowadays several companies offer a wide range of products based on MPs, such as Adembeads® [45], Dynabeads® [46], BioMag® [47], SiMAG® [48], MACS® [49], among many others, which are widely used in laboratories to extract and preconcentrate desired biological components, such as cells, proteins, organelles or DNA, from a liquid. As shown in Fig. 1 they consist of an inorganic core of magnetic

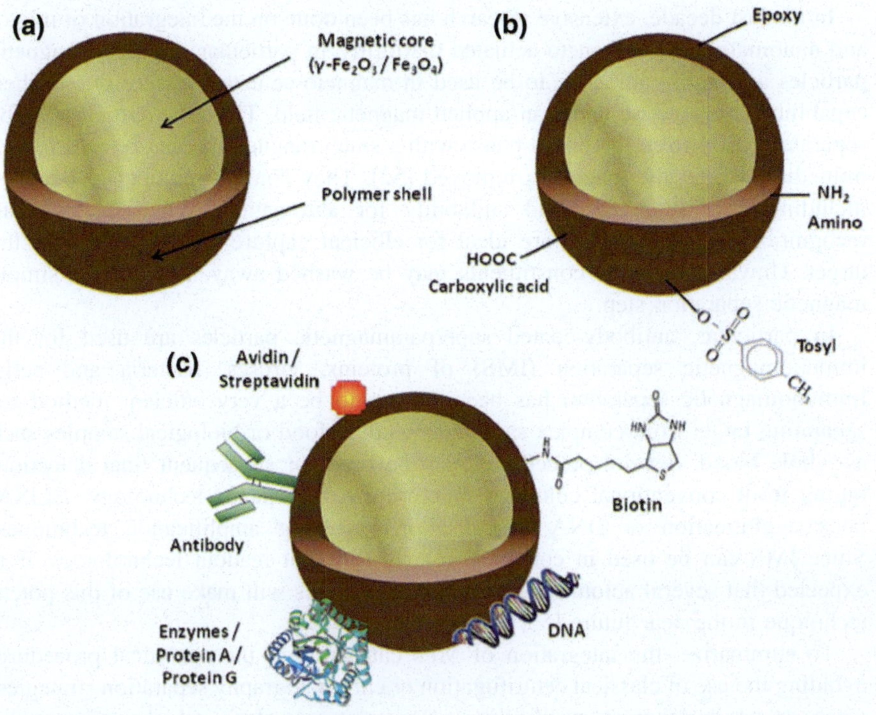

Fig. 1 Schematic representation of magnetic particles (**a**), activated with functional groups (**b**) and conjugated to biological molecules (**c**)

materials such as iron, nickel, cobalt, neodymium–iron–boron, samarium–cobalt or magnetite coated with polymer to confer stability (such as polystyrene, dextran, polyacrylic acid, or silica), which can be modified with functional groups to make subsequent conjugations easier. Although there are commercially available MPs already functionalised with biomolecules for a variety of bioanalytical and biotechnology applications, they can be modified with specific receptors, using surface chemical groups such as tosyl, amine, carboxyl or epoxy, for the identification/isolation of biomarkers including whole organisms, proteins and peptides, antibodies, DNA, among others [50]. Hence, magnetic particles can be tailored-modified with a whole range of ligands, including peptides, small molecules, proteins, antibodies, and nucleic acids.

Magnetic particles can have any size from a few nanometers up to a few micrometres. Nano-sized particles (5–50 nm) are usually composed of a single magnetic core with a polymer shell around it. Larger particles (30 nm–10 mm) are composed of multiple magnetic cores inside a polymer matrix. These particles can be used for efficient transport, faster assay kinetics, improved binding specificity and as labels for detection [51].

In the last decade, extensive research has been done on the integration of micro- and nanomaterials in magneto actuated platforms. In particular, superparamagnetic particles are highly attractive to be used in magneto-actuated devices due to their capability to magnetise under an applied magnetic field. Thus, the particles can be separated easily from the liquid phase with a small magnet, but can be redispersed immediately after the magnet is removed [52]. They confer a number of benefits, including easy separation and suitability for automation. When coated with recognition molecules, MPs are ideal for efficient capture and separation of the target. Unwanted sample constituents may be washed away, following a simple magnetic separation step.

In particular, antibody-coated superparamagnetic particles are used for the immunomagnetic separation (IMS) of proteins, viruses, bacteria and cells. Immunomagnetic separation has been proven to be a very efficient method for separating target from complex samples including food or biological samples such as whole blood. Several procedures may be used for subsequent final detection, raging from conventional culturing, microscopy, impedance technology, ELISA, latex agglutination or DNA hybridisation involving amplification techniques. Since IMS can be used in conjunction with different readout technologies, it is expected that several automated analytical procedures will make use of this potent technique in the near future [53, 54].

To summarize, the integration of MPs can simplify the analytical procedure, avoiding the use of classical centrifugation or chromatography separation strategies, since no pre-enrichment, purification or pretreatment steps, which are normally used in standard analytical methods, are required. Moreover, their use as solid support in bioassays has been shown to greatly improve the performance of the biological reactions.

3 Electrochemical Biosensors

The development of biosensors is a growing area, in response to the demand for rapid real-time, simple, selective and low cost techniques for analysis. Biosensors are compact analytical devices, incorporating a biological sensing element, either closely connected to, or integrated within, a transducer system. The combination of the biological receptor compounds (antibody, enzyme, nucleic acid) and the physical or physico-chemical transducer producing, in most cases, “real-time” observation of a specific biological event (e.g. antibody–antigen interaction) [55]. Depending on the method of signal transduction, biosensors can also be divided into different groups: electrochemical, optical, thermometric, piezoelectric or magnetic [56]. They allow the detection of a broad spectrum of analytes in complex sample matrices, and have shown great promise in areas such as clinical diagnostics, food analysis and environmental monitoring [57, 58]. The sensitivity of each of the sensor systems may

4 Magnetic Immobilisation Coupled with a Magneto-Actuated Electrode for Electrochemical Biosensing

As previously discussed, one of the most promising materials in bioanalysis is biologically modified magnetic particles, the use of which is based on the concept of magnetic bioseparation. Magnetic particles offer several novel attractive possibilities in biomedicine and bioanalysis since they can be coated with biological molecules and manipulated by an external magnetic field gradient. As such, the biomaterial, i.e., specific cells, proteins, or DNA, can be selectively bound to the magnetic particles and then separated from its biological matrix by applying an external magnetic field. The integration of MPs and electrochemical biosensing strategies improves analytical performance. Instead of direct modification of the electrode surface, both the biological reactions (immobilization, hybridization, enzymatic labeling, or affinity reactions) and the washing steps can be successfully performed on the MPs. After the modifications, the particles are easily captured by applying a magnetic field onto the surface of the GEC electrodes, which contain a small magnet (m-GEC) designed in our laboratories.

The preparation of the m-GEC electrode consists on different steps that are schematically explained in Fig. 2. To a female electric connector with a metal end of 2 mm diameter (i), a metal screw (6 mm d) is fitted (ii). After that, a copper disk with a diameter of 5.9 mm is placed on the female electric connector end (iii) with a welder using Sn wire. A cylindrical PVC tube (iv) (6 mm id, 8 mm od, 21.5 mm long) is then place over the female electric connector. A gap with a depth of 3 mm is thus obtained in the end of the body electrode, which is then filled with a thin layer of the graphite-epoxy composite (GEC) paste (v). The 3 mm diameter neodymium magnet is then placed in the center (vi). The electrode body gap is then completely filled with the soft GEC paste (vii). The electrode is then cured at 90 °C for 3 days until the paste become completely rigid. The electrodes can be stored in a dried place at room temperature.

This magneto-actuated electrode constitutes a versatile platform for electrochemical biosensing (both genosensors and immunosensors) for a broad range of application, including the detection of food contaminants such as pesticides, [75] antibiotic residues, [76] bacteria, [77, 78] food additives, [79] allergens, [80] or diseases biomarkers, such as malaria [81] or CD4 cells for AIDS monitoring [82]. In all cases, the electrochemical readout is achieved using horseradish peroxidase (HRP) as electrochemical reporter. Enzyme labelling has been transfer from non-isotopic classical methods to electrochemical biosensing. In electrochemical genosensing, the DNA duplex can be labeled with either strept(avidin)-HRP or antiDIG-HRP conjugates, depending on the tag of the DNA signaling probe (biotin or digoxigenin, respectively). Although a second incubation step is usually required for labelling, higher sensitivity and specificity have been reported for the enzyme

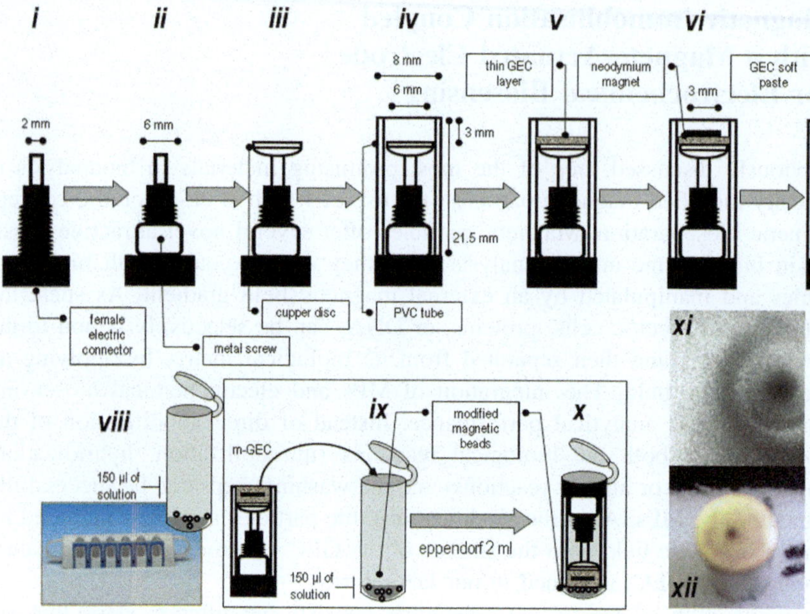

Fig. 2 Schematic representation of the construction of m-GEC (steps *i–vii*) and manipulation of the m-GEC electrodes, comprising the immobilization of DNA on magnetic beads (*viii*) following by capturing the modified beads on m-GEC electrode (*ix* and *x*). Scanning electron microphotographs showing the captured magnetic beads on the surface of m-GEC magneto sensor (500 µm, 15 kV) and a photograph showing the aspect of the sensor with the immobilized beads are also shown (*xi* and *xii*, respectively) (Number of magnetic beads: 6.2×10^6)

labelling method compared with the other reported methods [83, 84]. In electrochemical immunosensing, the enzymatic tag depends on the format of the immunoassay. In competitive immunoassays for small haptenic molecules, it is usually a conjugate obtained with HRP and the hapten. In other immunological formats, such as in sandwich assays or indirect approaches, the enzymatic label is typically a conjugate obtained with HRP covalently linked to the Fc part of the specific antibody. In all cases, amperometric determination is finally based on HRP activity following the addition of H_2O_2 and using hydroquinone as mediator. The modified electrode is immersed in the electrochemical cell containing hydroquinone and, under continuous magnetic stirring, a potential of –0.100 V vs. Ag/AgCl is applied. When a stable baseline is reached, H_2O_2 is added into the electrochemical cell (to a concentration able to saturate the total amount of enzyme employed in the labeling procedure) and the current is measured until steady state is reached (normally after 1 min of H_2O_2 addition).

5 Electrochemical Biosensors of Agents Affecting Food Safety

The World Health Organization (WHO) have considered Food Safety as the main topic of World Health´s Day in 2015 [85, 86]. It is estimated that two million deaths occur every year worldwide from contaminated food or drinking water, of each 200 are related to foodborne diseases, caused by harmful bacteria, viruses or parasites [87]. The emergence of foodborne infectious diseases in humans worldwide is attributed to several causes, such as the loss of biodiversity due to an intensive agriculture, food industry and land changes, together with the evolution of drug resistance [88]. Moreover, climatic factors related to the increase of the average global temperature and consequently to the raise of CO_2 concentrations, as well as precipitation changes have implications for food production, since they can potentiate a higher growth and survival of pathogenic microorganisms leading to the occurrence of food safety hazards [87, 89]. The socioeconomic status impact on food safety has also been studied, showing that low income individuals are more exposed to foodborne illness related to a poor hygiene and proper food handling practices and nutrition [90]. In Europe, over 320,000 human cases are reported each year being mostly related to the presence of pathogenic microorganisms in meat products, as well as fruits and vegetables, being *Salmonella* spp., *Escherichia coli* and *Listeria monocytogenes* the most common reported pathogens [91]. *S. enterica* sv. *Enteritidis* and *Typhimurium* are a frequent cause of foodborne outbreaks among *Salmonella* serovars [92]. One of the most recent cases was reported in August 2014, a multi-country outbreak reported in Austria, France and Luxembourg associated with eggs from Germany [86, 93].

One of the most serious outbreaks in Europe related to *E. coli* was reported in Germany, with a total of 3126 cases of diarrheal disease, including 17 deaths in Germany and additional extension to other countries as Norway, USA, Canada and Switzerland. It was caused by a Shiga toxin-producing E. coli (STEC) strain found in meat and fenugreek sprouts. This *E. coli* strain belongs to O104:H4 serotype, having the ability to excrete a shiga toxin which is lethal to humans [94, 95]. L. monocytogenes is a Gram positive pathogen that can survive in different environments, such as low temperatures and pH values or high salt concentrations. In the years of 2008 to 2012 an increase of number of listeriosis cases reported in the EU was registered with a total of 198 deaths in 2012 [96, 97].

"Make food safe" becomes a vital task worldwide, as a result, preventive approaches like Good Agricultural Practices (GAPs), Good Manufacturing Practices (GMPs), Hazard Analysis and Critical Control Point (HACCP) and the food code (Codex Alimentarius) have been implemented, which can considerably reduce the survival of pathogens during the process of handling, preparation and storage processing. These approaches include measures for the introduction of methods for decontamination, disinfection and cleaning, the implementation of analytical methods for screening feed and feed ingredients, to provide more inspection and control at industries and farms, as well as the implementation of communication strategies for consumers and the food industry [98, 99]. Examples of regulatory agencies or centers worldwide are the WHO, the US Centers for Disease Control and Prevention (CDC), the US Food and Drug Administration (FDA), the Public Health Agency of Canada (PHAC), the European Food Safety Authority (EFSA), the European Centre for Disease Prevention and Control (ECDC), OzFoodNet, PulseNet International (PNI), National Institute of Public Health, Japan, among many others. Identification and detection of foodborne bacteria is in general required for routine surveillance and monitoring, evaluation of the most common food sources responsible for specific foodborne, during regulatory actions or from investigation of a foodborne outbreak. A wide range of methods are available for foodborne bacteria identification and detection, in connection with these programs, for the prevention and identification of problems related to health and safety. The choice of the method is a key factor for the detection of foodborne pathogens and the intended use of the method, for instance whether for a qualitative or semi-quantitative screening, quantitative and/or confirmatory analysis, must be clearly defined [100]. Additionally, an ideal method should be rapid, providing results in a few hours, easy handling, accurate, applicable to several food matrices and foodborne bacteria [98–100]. Electrochemical biosensors are good candidates to meet these demands. The latest development that combines the use of MPs for preconcentration and electrochemical detection is of particular interest due to the considerable improvement achieved on the analytical features such as assay time and limit of detection. In this case, an immunomagnetic (IMS) or phagomagnetic (PMS) separation is integrated in electrochemical magneto genosensing and immunosensing approaches. Figure 3 summarized the scheme of three different strategies based on IMS/PMS separation coupled with electrochemical genosensing and immunosensing taking *Salmonella* as a model of food pathogen. In these approaches, magnetic particles have the dual function of (i) pre-concentrating bacteria from the complex matrix, using different biorecognition reactions (immunomagnetic (IMS) and phagomagnetic (PMS) separations) and (ii) improving the analytical features of both electrochemical genosensing and immunosensing of bacteria.

5.1 Immuno (IMS) and Phagomagnetic Separation (PMS) Coupled with Electrochemical Genosensing on Magneto Actuated Electrodes

The first approach, summarized as a "IMS/double-tagging PCR/m-GEC electrochemical genosensing" [101] was based on a double biorecognition of bacteria, in this case immunological followed by genetic biorecognition. The procedure consisted briefly of the following steps, as depicted in Fig. 3: (i) Immunomagnetic separation of the bacteria from food samples; (ii) Lysis of the bacteria and DNA separation; (iii) DNA amplification of Salmonella IS200 insertion sequence by double-tagging PCR; (iv) immobilization of the doubly-tagged amplicon in which the biotin tag of the dsDNA amplicon is immobilized on the streptavidin MPs;

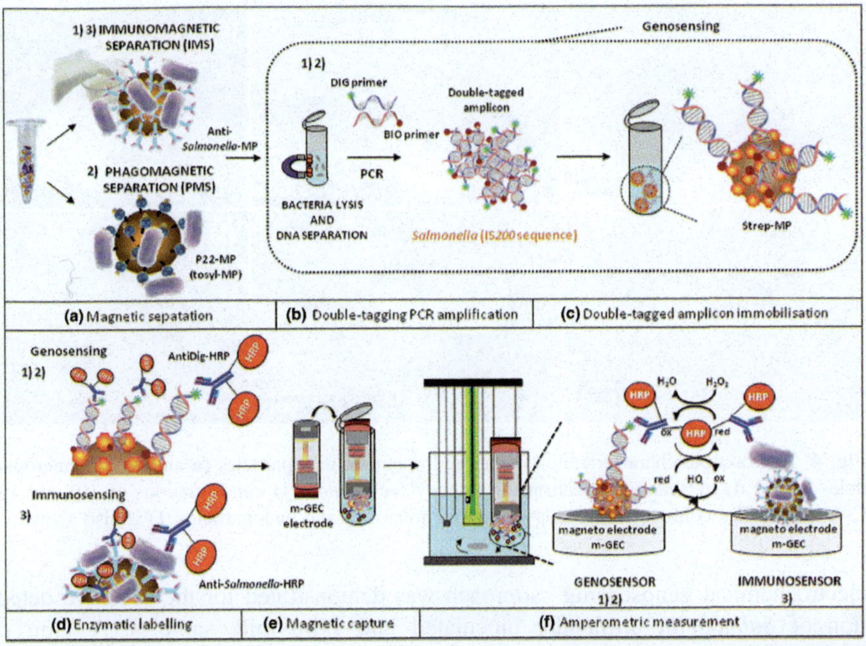

Fig. 3 Schematic representation of the electrochemical strategies for *Salmonella* spp. detection: (*1*) "IMS/double-tagging PCR/m-GEC electrochemical genosensing" [101] (*2*) "PMS/double-tagging PCR/m-GEC electrochemical genosensing" [103] and (*3*) "IMS/m-GEC electrochemical immunosensing" [109]

(v) enzymatic labelling using as enzyme label the antibody anti-DIG-HRP capable of reacting the other tag extreme of the dsDNA amplicon; (vi) magnetic capture of the modified magnetic particles; and (vii) amperometric determination [101].

In this approach, the bacteria can be captured and preconcentrated from food samples by IMS using both commercial magnetic microparticles [101] or tailored magnetic micro and nanoparticles with the specific antibody against Salmonella [102]. No significant differences were observed in the efficiency of the IMS using magnetic micro or nanoparticles, with the exception of the binding pattern, as shown in Fig. 4. After the lysis of the captured bacteria by IMS, further amplification of the genetic material by PCR with a double-tagging set of primers is performed to confirm the identity of the bacteria. Both steps (IMS and double-tagging PCR) are rapid alternatives to the time consuming classical selective enrichment and biochemical/serological tests.

The double-tagging PCR is performed with a set of two labeled PCR primers—one with biotin and the other with digoxigenin [77]. During PCR, not only amplification of the bacterial genome is achieved but also double-labeling of the amplicon ends with: (i) the biotinylated capture primer, to achieve immobilization on the streptavidin-modified MPs, and (ii) the digoxigenin signaling primer, to achieve electrochemical detection. The "IMS/double-tagging PCR/m-GEC

Fig. 4 Microscopic characterization of tailored magnetic nanoparticles (**a** and **b**) and microparticles (**c** and **d**) by scanning electron microscopy for *Salmonella* concentrations of 10^4 and 10^6 CFU mL^{-1}. The confocal microscopy study for micro (**e**) and nanoparticles (**f**) is also shown

electrochemical genosensing" approach was demonstrated for the sensitive detection of *Salmonella* artificially inoculated into skim milk samples. A limit of detection of 1 CFU mL^{-1} is obtained in 3.5 h without any pretreatment, in LB broth and in milk diluted 1/10 in LB. When the skim milk is pre-enriched for 6 h, the method is able to feasibly detect as low as 0.04 CFU mL^{-1} (1 CFU in 25 g of milk) with a signal-to-background ratio of 20 [101]. Interestedly, the specificity of this approach is conferred by both the antibody in the IMS and the set of primer during the double-tagging PCR, in this case for detecting *Salmonella* spp. The same approach could be also designed for detecting different *Salmonella* or *E. coli* serotypes by selecting a specific pair of primers or antibody.

The second strategy, "PMS/double-tagging PCR/m-GEC electrochemical genosensing" [103] was based on the use of bacteriophages, which offer several analytical advantages as biorecognition elements for the magnetic separation of pathogenic bacteria. The phage capabilities as a biorecognition element were explored by using the model phage nanoparticle P22 towards *Salmonella*. P22 bacteriophages were immobilised on tosyl-activated magnetic microparticles and carboxyl magnetic nanoparticles in an oriented way. The bacteria were then captured and pre-concentrated by the phage-modified magnetic particles through the phage–host interaction (Fig. 5). To confirm the identity of the bacteria, further double-tagging PCR amplification of the captured bacterial DNA and electrochemical magneto genosensing of the amplicon were performed, as schematically shown in Fig. 3.

In the detailed strategies, magnetic separation based on different affinity biorecognition principles was evaluated, i.e. immunomagnetic and phagomagnetic separation. Although similar analytical performance were obtained (LOD of 1 CFU mL^{-1} in 3 h assay time), the use of bacteriophages as a biorecognition element offers additional advantages. It must be highlighted that for the first time

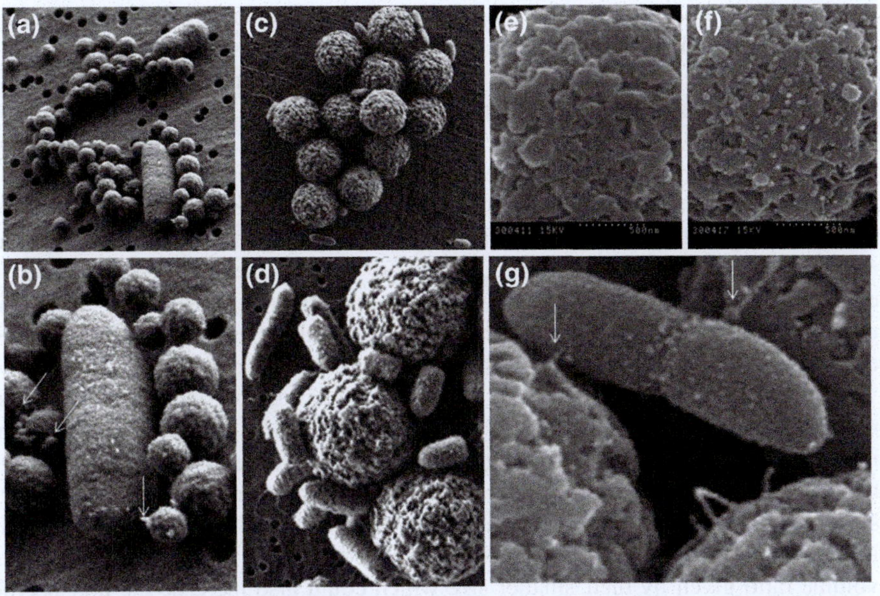

Fig. 5 Evaluation of the PMS by SEM at a Salmonella concentration of 3.2×10^6 CFU mL^{-1} using carboxyl magnetic nanoparticles (**a** and **b**) and tosylactivated magnetic microparticles (**c**–**g**). *Panels* **b** and **g** show the Salmonella cells attached to the magnetic nano and microparticles through tail spikes of the bacteriophages, signaled by arrows. *Panels* **f**shows the P22 bacteriophage immobilised on magnetic microparticles (2000 PFU/MP) while *panel* **e** shows the magnetic microparticle without any modification as a negative control. In all cases, identical acceleration voltage (15 kV) was used

non-modified bacteriophages were covalently coupled to magnetic particles, as shown in Fig. 5. Improved LODs (1 CFU mL^{-1}) were obtained in both cases when compared with the IMS and PMS followed by conventional gel electrophoresis (10^2 and 10^3 CFU mL^{-1}, respectively), as well as a significant reduction of the assay time when compared with IMS and PMS followed by the microbiological culture method (3 h vs. 18–24 h). The accuracy of the magnetic separation step coupled with microbiological culture is not measurable since agglomeration of particles often occurs and several target bacteria bound to the same particle give rise to only one colony forming unit (CFU) on the plating media, as shown in Fig. 5, panel A and C for magnetic nano and microparticles, respectively. Therefore, by coupling IMS or PMS with double-tagged PCR amplification and electrochemical magneto genosensing quantitative methods were achieved, due to the fact that a single cell is detected and these methods are not affected by the formation of aggregates. The double-tagging PCR also allows amplification of the analytical signal by amplifying the bacterial genome in a rapid way, instead of multiplication of the bacteria number by growing via traditional culturing methods. The magnetic separation and the double-tagging PCR provide specificity, as well as versatility to the assay, with the selection of different capture antibodies, bacteriophages or tagged primers.

Therefore, the models described can be widened to other bacterial targets affecting food safety and global health.

To summarize, bacteriophages are promising candidates to be used as a biorecognition element for the detection of pathogenic microorganisms. They provide many advantageous features such as outstanding selectivity, high sensitivity, and stability, which are three ideal attributes for any biorecognition probe that makes them suitable for in situ monitoring of food and environmental contaminants [104, 105]. Compared to antibodies, phages have distinct advantages as recognition receptors. On one hand, they are less fragile and less sensitive to environmental stress such as pH and temperature fluctuation reducing the environmental limitations, and on the other, their production besides being animal-free can be less complicated and less expensive than antibody production [106, 107].

The detection of *Salmonella* was demonstrated using both magnetic micro and nanoparticles modified with the bacteriophage P22 [108]. Although the covalent immobilization of P22 bacteriophages was successfully performed on both magnetic carriers achieving excellent coupling efficiencies, magnetic microparticles showed improved performance in terms of sensitivity and specificity, as well as lower matrix effect. These results could be related with the higher surface area per volume ratio given by their smaller size which could also increase the nonspecific adsorption, raising thus the influence of the matrix components during the assay.

5.2 Immunomagnetic Separation (IMS) Coupled with Electrochemical Immunosensing on Magneto Actuated Electrodes

In the third strategy, summarized as "IMS/m-GEC electrochemical immunosensing", a very simple and rapid method for the detection of *Salmonella* in milk is performed in which the detection of bacteria was achieved by a double immunological recognition (Fig. 3) [109]. In this approach, the bacteria were captured and pre-concentrated from milk samples with magnetic particles by immunological reaction with a specific antibody against *Salmonella*. A second polyclonal antibody labeled with peroxidase was used as serological confirmation, with electrochemical detection based on a magneto-electrode. Among the different procedures, better performances have been obtained using one-step immunological reactions. The "immunomagnetic separation step (IMS)/m-GEC electrochemical immunosensing" approach was employed, for the first time, in the detection of *Salmonella* artificially inoculated into skimmed-milk samples. A limit of detection of 7.5 10^3 CFU ml^{-1} in milk was obtained in 50 min without any pre-treatment. If the skimmed-milk is pre-enriched for 6 h, the method can detect as low as 1.4 CFU mL^{-1}, while following pre-enrichment for 8 h as few as 0.108 CFU mL^{-1} (2.7 CFU in 25 g of milk) are detected, thus complying with legislative criteria. IMS and detection with a second specific antibody can effectively replace "selective enrichment/differential plating" and "biochemical/serological testing" assays, respectively. Moreover, the assay time is considerably reduced, from 4 to 5 days to 50 min.

5.3 Simultaneous Electrochemical Magneto Genosensing of Foodborne Bacteria Based on Triple-Tagging Multiplex Amplification

Over the past years, a new challenge has been attracting researchers in this field, the design of novel biosensors with multiplexing capabilities, where the integration of nanomaterials plays an important role. These novel bionanomaterials appears to be keys in bacteria multiplex detection in biosensors [31–34].

The simultaneous detection of *Salmonella enterica*, *Listeria monocytogenes* and *Escherichia coli* based on triple-tagging multiplex PCR and electrochemical magneto genosensing on silica magnetic particles was also reported [110]. A set of tagging primers were selected for the specific amplification of yfiR (375 bp), hlyA (234 bp) and eaeA (151 bp), being one of the primers for each set labelled with fluorescein, biotin and digoxigenin coding for *S. enterica*, *L. monocytogenes* and *E. coli*, respectively. Afterwards, electrochemical magneto genosensing of the bacteria was achieved by using silica magnetic particles as a carrier and three different electrochemical reporters, specific for each pathogen. For the first time, silica magnetic particles were used as a platform for DNA immobilization followed by electrochemical genosensing of *S. enterica*, *L. monocytogenes* and *E.coli*, based on triple-tagged amplicons. Interestingly, the silica magnetic particles showed differential adsorption properties, based on the negative charge density, for longer dsDNA amplicon incorporating the tagged-primers over shorter ssDNA tagged-primers, showing to be not only a robust platform for the electrochemical detection of PCR products but also a promising magnetic carrier for fluorescence or other detection approaches. This method was able to clearly distinguish among the pathogenic bacteria tested within 50 min, with detection limits ranging from 12 to 46 pg μL^{-1}.

6 Electrochemical Biosensors of Infection Agents Affecting Global Health

6.1 Electrochemical Magneto-Actuated Biosensor for CD4 Count in AIDS Diagnosis and Monitoring

According to last WHO report in 2013, at the end of this year, around 35 million people were living with HIV. Developing countries in Africa and Asia are the most affected by this disease, for example, sub-Saharan Africa concentrates almost 70 %

of the global HIV-infected people, with 25 million people living with HIV in this region alone [111]. An additional problem that these regions face is that only 5.9 of 12.9 million people who need antiretroviral therapy have access to this treatment [112]. HIV is a retrovirus which infects primarily $CD4^+$ T lymphocytes. Progression to AIDS occurs as a result of chronic depletion of CD4 cells, when the count falls below 200 cells mL^{-1} of blood, at a functional level where opportunistic infections and malignancies cannot be controlled [113, 114]. HIV infection is commonly diagnosed through a blood test detecting antibodies against HIV, followed by a confirmatory assay [115]. The serological tests for detection of HIV antibodies are generally classified as screening and confirmatory, being ELISA and Western blot, respectively. Moreover, a variety of simple, instrument-free, rapid tests including agglutination, immunofiltration, immunochromatographic and dip-stick test, for example, OraQuick® Advance Rapid HIV-1/2, Reveal™ G-2 Rapid HIV-1 Antibody, Uni-Gold Recombigen® HIV, and Multispot HIV-1/HIV-2 Rapid Test are commercially available [116], all of the meeting the requirements of ASSURED given by the WHO.

However, after diagnosis, disease progression should be monitored through viral load based on viral nucleic acid detection or through the enumeration of CD4 cells by flow cytometry. Nucleic acid amplification test are laborious strategies, requiring dedicated equipment and trained technicians. In addition, flow cytometry requires complex and expensive equipment that requires regular maintenance and well trained personnel not only for data analysis, but also for the result interpretation. Currently, there are few cheaper alternative to the flow cytometer, mostly based on fluorescent labeling, requiring thus costly imaging equipment to achieve detection or manual counting by light microscopy [23, 117, 118].

Although RDTs are commercially available for diagnosis of HIV infection, CD4 cell counting is not available in the areas mostly affected [119–121], this control being imperative for assessing the progression towards AIDS [122]. As previously stated, the HIV virus infects the cells of the immune system, primarily $CD4^+$ T lymphocytes decreasing CD4 levels from the normal values (ranging from 500 to 1,200 cells μL^{-1}), which weakens the immune system and causes the progression to AIDS and death from cancer or opportunistic infections. When the number of the CD4 cells falls below 200 cells μL^{-1} of blood, it is considered to have progressed to AIDS [123]. AIDS it is also diagnosed with the emergence of one or more opportunistic illnesses regardless the CD4 count. Without treatment, people who progress to AIDS typically survive about 3 years. However, life-expectancy without treatment falls to about 1 year with the presence of opportunistic illness. Under antiretroviral treatment (ART) while maintaining a low viral load, a patient may enjoy a near normal life span without progression to AIDS. The $CD4^+$ T cell count is thus a critical parameter in monitoring HIV disease, since lower numbers of circulating $CD4^+$ T cells imply a more advanced stage of HIV disease and less competent defense mechanisms. In HIV infected patients, the $CD4^+$ T cell count is useful not only for assessing the degree of immune deterioration and speed of progression towards AIDS, but also for initiating ART, for deciding the timing for prophylaxis of opportunistic infections and, finally, for monitoring the efficacy of the treatment [119]. The new recommendations encourage all countries to initiate the treatment in HIV infected adults with CD4 cell count down to 500 cells μL^{-1} when their immune systems are still strong, regardless of the presence or absence of clinical symptoms. Unfortunately, the areas mostly affected by the HIV epidemic are resource-limited countries, wherein the CD4 count is not available due to laboratory requirements and cost of the assay [124].

Although there are many commercially available possibilities for Point-of-Care HIV diagnosis, there is still the need for novel affordable alternatives to flow cytometry for CD4 cell count in order to monitor the AIDS disease and the treatment in low resource settings.

Recent advances involve integration of MPs into bioassays for both diagnosis of HIV infection, as well as for the progression and follow-up of AIDS. For instance, a magneto-actuated electrochemical biosensor for CD4 count in whole blood was reported [82]. The $CD4^{+}$ T lymphocytes were isolated, preconcentrated and labeled from 100 μL of whole blood by immunomagnetic separation with magnetic particles modified with antiCD3 antibodies (Fig. 6).

The captured cells were labeled with a biotinylated antiCD4 antibody, followed by the reaction with the electrochemical reporter streptavidin-peroxidase conjugate, as schematically detailed in Fig. 7. The limit of detection for the CD4 counting magneto biosensor in whole blood was as low as 44 cells μL^{-1} while the logistic range was found to be from 89 to 912 cells μL^{-1}, which spans the whole medical interest range for CD4 counts in AIDS patients. The electrochemical detection together with the immunomagnetic separation confers high sensitivity, resulting in a rapid, inexpensive, robust, user-friendly method for CD4 counting. This approach is

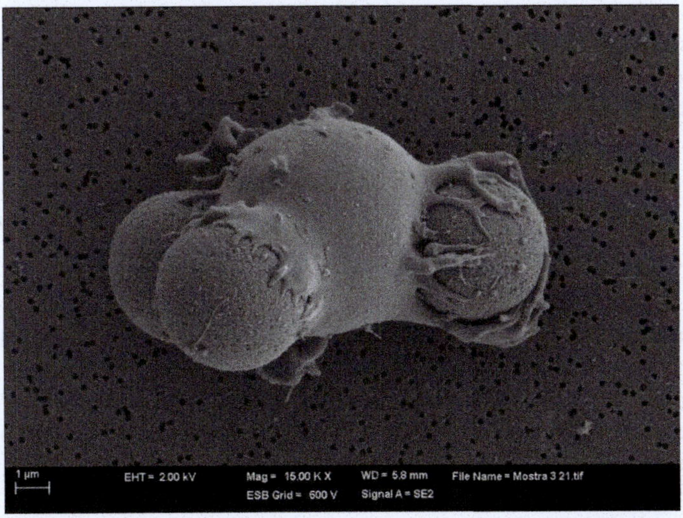

Fig. 6 Evaluation of the $CD4^{+}$ T lymphocytes immobilized on antiCD3 magnetic particles by SEM. Acceleration voltage (15 kV) was used

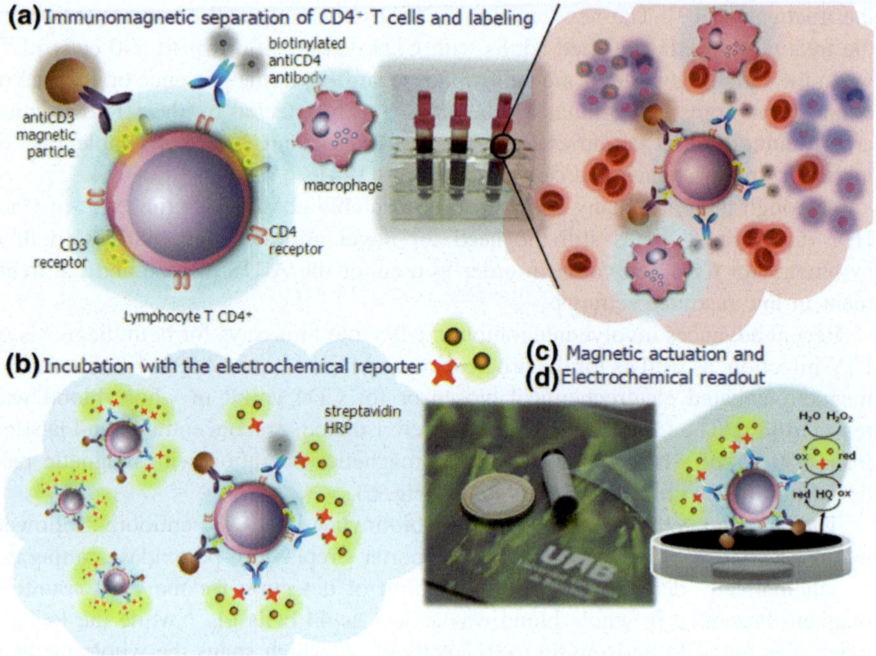

Fig. 7 Schematic representation of the CD4 counting magneto biosensor. **a** The $CD4^+$ T lymphocytes are captured from whole blood by the CD3-MPs and labeled in one step with antiCD4-biotin, **b** The incubation with the electrochemical reporter streptavidin-HRP is then performed. Finally, and after **c** the magnetic actuation, **d** the electrochemical readout is achieved the electrochemical readout is achieved

a promising alternative for costly standard flow cytometry and suitable as diagnostic tool at decentralized practitioner sites in low resource settings, especially in less developed countries.

6.2 Electrochemical Magneto-Actuated Biosensor for CD4 Count in AIDS Diagnosis and Monitoring

According to the WHO report, about 200 million people contracted malaria [125, 126] in 2012 and nearly 630,000 died of the disease [127]. Africa is the most affected area, counting one death every minute, most of them in children. Malaria is considered one of the major tropical parasitic disease and it is among the three most deadly communicable diseases [128].

In the absence of diagnostic tests, patients in low-resource settings are often treated based on clinical symptoms and local prevalence of disease. Whereas this approach captures most patients requiring treatment, it also unnecessarily treats patients who do not require treatment. Equally important, this latter group of patients is not being treated for their specific disease due to misdiagnosis. This syndromic management of disease may also increase drug resistance. Recently, strains of *Plasmodium falciparum* resistant to chloroquine have spread rapidly. Taking into account that other malarial treatments are significantly more expensive the correct diagnosis of infected individuals seems to be a cheaper strategy, rather than treating all those with similar symptoms [3]. Moreover, the most ethical policy is to ensure that the new generation of drugs are only used for true cases of malaria to avoid the appearance of resistant strains [129].

Light microscopy is considered the gold standard method for malaria diagnosis; a drop of blood from a finger prick is fixed with methanol on a glass slide and stained with dyes to visualise the parasite. Unfortunately, limited infrastructure in low-resources settings results in extremely poor performance of microscopy as a diagnostic tool for malaria, which shows an accuracy of only 70–75 %. Moreover, microscopic diagnosis requires highly trained and experienced staff, being thus in some instances not suitable for routine use at the community level. Modern methods for malaria diagnosis include fluorescent microscopy, flow cytometry, automated blood cell analyzers, antibody detection, molecular methods, and laser desorption mass spectrometry. The main disadvantage, in most of the cases, is their high cost [130].

Only a few examples of the integration of MPs for malaria diagnosis have been reported. For instance, MPs were used for the selective preconcentration of the protein biomarker HRPII (histidine-rich protein II) for *Plasmodium falciparum*. A rapid and simple magneto immunoassay, which can be coupled with both optical or electrochemical readout for the detection of HRPII, was reported (Fig. 8) [82]. The method involved covalent immobilisation of anti-HRP2 IgM monoclonal antibody on MNPs (Fig. 8, panel A), followed by reaction with an anti-HRP2 IgG antibody labeled with peroxidase (Fig. 8, panel B), which could be used as electrochemical or optical reporter (Fig. 8, panel C). In the magneto immunosensor, the MNPs were used to preconcentrate the biomarker from the clinical sample, to eliminate interference from the matrix and to immobilise the biomarker in close contact to the electrode surface, improving thus the limit of detection.

7 Final Remarks

In this chapter, recent advances in foodborne pathogens and infection diseases affecting global health have been discussed, with special focus on electrochemical biosensing devices with magneto actuated platforms. The most prominent format is the integration of a magnetic capture step prior to detection, to achieve the preconcentration of the biomarker from the complex interferences present in the samples. In general, the most common approach involves the integration of immunomagnetic separation prior the electrochemical readout, based on commercial or tailored-modified magnetic micro or nanoparticles, with a specific antibody.

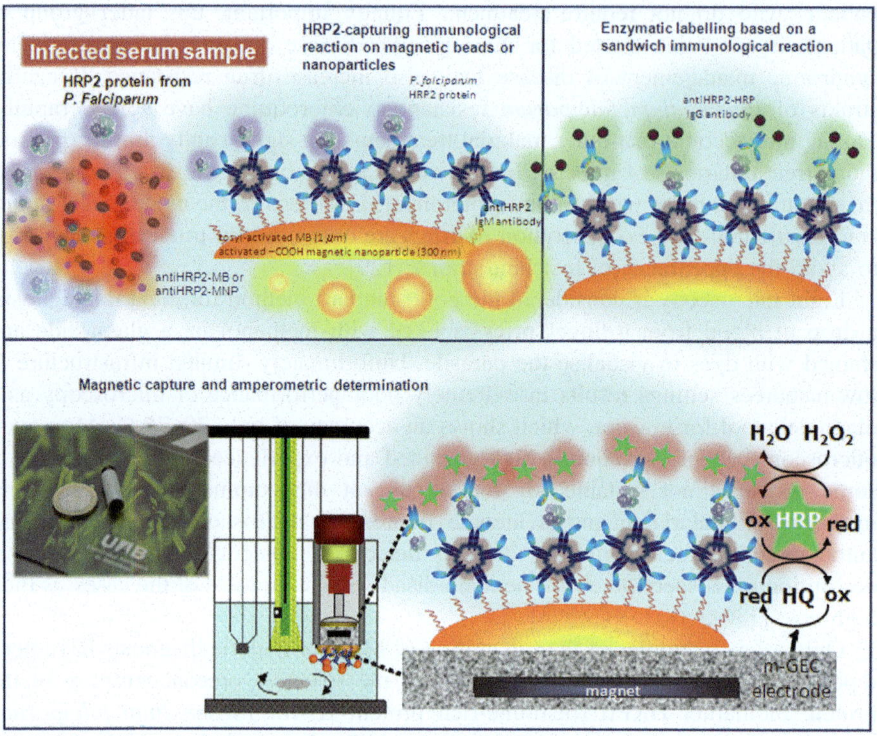

Fig. 8 Schematic representation of the sandwich immunoassay for Malaria detection performed on magnetic micro and nanoparticles with electrochemical readout

Magnetic capture was demonstrated to be compatible with different readout strategies, ranging from conventional methods such as culturing, microscopy and mass spectrometry, or emerging technologies such as lateral flow and biosensing devices, among others.

In some instances, the biomarker is preconcentrated and then released for further readout, although the detection of the biomarker is mostly performed while it is still attached to the MPs. In most formats, a label to achieve the readout is used, being a second antibody conjugated with enzymes, fluorophores, or nanomaterials (such as QDs or Au-NPs). Although the detection of biomarkers is in most instances performed by immunological recognition, there are some examples of the integration of MP in devices for DNA determination. Here, the MPs are used as a means of preconcentration of native DNA or RNA, or amplicons coming from PCR.

Although the MPs are mostly integrated in a preconcentration step prior to the readout, they can also be used as a carrier to achieve movement of the biomarkers along a microfluidic device through different reaction chambers, in order to achieve incubation and washing under magnetic actuation [131–133]. Unfortunately, there are still few examples of total integration into a chip from sample introduction to readout, requiring in all cases bench-top equipments to achieve detection. Other limitations of these chips are low sensitivity limited by the small sample volume, irreproducibility in microfabrication and high cost of scaling down. One of the most promising approaches for rapid diagnostic in low resource settings are lateral flow with qualitative visual detection.

Recently, the integration of MPs into lateral flow design was demonstrated, improving the sensitivity and providing quantitative results when required [134]. The equipment is inexpensive if compared with traditional ELISA readers. In these methods, the MPs are used not only for preconcentration, but also for the readout based on the magnetic moment of the superparamagnetic particles attached to the biomarker. Another approach based on the integration of MP is the magneto biosensor, mostly using electrochemical detection. In this format, MPs are used not only as a way to preconcentrate the sample, but also to immobilise the biomarker on the surface of the transducer, increasing thus the sensitivity of the assay when compared with conventional surface modification of the biosensor.

Comparing the performance of magnetic carriers, similar analytical performance was observed for the immunomagnetic separations of bacteria with both magnetic micro and nanoparticles. However, magnetic nanoparticles showed increased matrix effect together with aggregation and required longer time for magnetic actuation, as the actuation time and also the matrix effect are influenced by the size of the particles. In general, integration of magnetic particles with different approaches demonstrated improvement in analytical performance in terms of specificity and sensitivity. Their use as solid supports in bioassays has shown to greatly improve the performance of the biological reactions, due to several factors: (i) an increased surface area which improves the efficiency of the reactions, (ii) faster assay kinetics achieved because the particles are in suspension and the analytical target does not have to migrate very far, and (iii) a minimised matrix effect due to the improved washing and separation steps.

The integration of magnetic particles can thus simplify the analytical procedure, avoiding the use of classical centrifugation or chromatography separation strategies, since no pre-enrichment, purification or pretreatment steps, which are normally used in standard analytical methods, are required. In here, the preconcentration and purification is achieved by simply applying an external magnet. Biomarkers can be specifically isolated and preconcentrated from complex biological or food matrixes by magnetic actuation, increasing the specificity of the assay. MPs have been shown to be a robust and versatile material for the detection of a whole range of biomarkers including mammalian cells, whole viruses, bacteria, proteins, antibodies and DNA related with infectious diseases affecting global health. The integration of MPs into emerging technologies shows very promising features, although there is still a long way to achieve point of care devices following the ASSURED recommendations given by WHO (Affordable, Sensitive, Specific, User-friendly, Rapid and Robust, Equipment free, and Deliverable to those who need it).

References

1. Nordmann A (2004) Converging technologies—Shaping the Future of European Societies. Report. European Commission Research
2. EU Communication (2012) A European strategy for key enabling technologies—A bridge to growth and jobs
3. Mabey D, Peeling RW, Ustianowski A, Perkins MD (2004) Diagnostics for the developing world. Nat Rev Microbiol 2:231–240
4. Hauck TS, Giri S, Gao Y, Chan WCW (2010) Nanotechnology diagnostics for infectious diseases prevalent in developing countries. Adv Drug Deliv Rev 62:348–448
5. Koplan JP et al (2009) Towards a common definition of global health. Lancet 373:1993–1995
6. Plasència A (2012) Global health challenges and personalised medicine. Contrib Sci 8:175–180
7. Macfarlane SB et al (2008) In the name of global health: trends in academic institutions. J Public Health Policy 29:383–401
8. Velusamy V, Arshak K, Korostynska O, Oliwa K, Adley C (2010) An overview of foodborne pathogen detection: in the perspective of biosensors. Biotechnol Adv 28:232–254
9. Newell DG, Koopmans M, Verhoef L, Duizer E, Aidara-Kane A, Sprong H et al (2010) Food-borne diseases—the challenges of 20 years ago still persist while new ones continue to emerge. Int J Food Microbiol 139:S3–S15
10. Tomaso H, Kattar M, Eickhoff M, Wernery U, Al Dahouk S, Straube E et al (2010) Comparison of commercial DNA preparation kits for the detection of Brucellae in tissue using quantitative real-time PCR. BMC Infect Dis 10:100
11. Mandal PK, Biswas AK, Choi K, Pal UK (2011) Methods for the rapid detection of foodborne pathogens: an overview. Am J Food Technol 6:87–102
12. LaBarre P, Hawkins KR, Gerlach J, Wilmoth J, Beddoe A, Singleton J et al (2011) A simple, inexpensive device for nucleic acid amplification without electricity—toward instrument-free molecular diagnostics in low-resource settings. PLoS ONE 6:e19738
13. U.S. Department of Health and Human Services. Food and Drug Administration (2008) Clinical Laboratory Improvement Amendments of 1988 (CLIA) Waiver Applications for Manufacturers of In Vitro Diagnostic Devices
14. Seo K-H, Holt PS, Gast RK, Stone HD (2003) Simple and rapid methods for detecting Salmonella enteritidis in raw eggs. Int J Food Microbiol Chem 87:139–144
15. Gubala V, Harris LF, Ricco AJ, Tan MX, Williams DE (2012) Point of care diagnostics: status and future. Anal Chem 84:487–515
16. Posthuma-Trumpie GA, Korf J, van Amerongen A (2009) Lateral flow (immuno)assay: its strengths, weaknesses, opportunities and threats. A literature survey. Anal Bioanal Chem 393:569–582
17. Ngom B, Guo Y, Wang X, Bi D (2010) Development and application of lateral flow test strip technology for detection of infectious agents and chemical contaminants: a review. Anal Bioanal Chem 397:1113–1135
18. Anfossi L, Baggiani C, Giovannoli C, D'Arco G, Giraudi G (2013) Lateral-flow immunoassays for mycotoxins and phycotoxins: a review. Anal Bioanal Chem 405:467–480
19. Singer JM, Plotz CM (1956) The latex fixation test. I. Application to the serologic diagnosis of rheumatoid arthritis. Am J Med 1956(21):888–892
20. Park J, Kurosawa S, Watanabe J, Ishihara K (2004) Evaluation of 2-Methacryloyloxyethyl phosphorylcholine polymeric nanoparticle for immunoassay of C-Reactive protein detection. Anal Chem 2004(76):2649–2655
21. Marco MP, Gee S, Hammock BD (1995) Immunochemical techniques for environmental analysis I. Immunosensors. Trends Anal Chem 14:341–350

22. Banada PP, Bhunia AK (2008) Antibodies and Immunoassays for Detection of Bacterial Pathogens. In: Zourob M et al (eds) Principles of bacterial detection: biosensors, recognition receptors and microsystems. Springer, New York, pp 567–602
23. Carriere D, Vendrell JP, Fontaine C, Jansen A, Reynes J, Pages I et al (1999) Whole blood Capcellia CD4/CD8 immunoassay for enumeration of CD4+ and CD8 + peripheral T lymphocytes. Clin Chem 45:92–97
24. Niemz A, Ferguson TM, Boyle DS (2012) Point-of-care nucleic acid testing for infectious diseases. Expert Rev Mol Diagn 12:645–659
25. Lee HH, Allain JP (2004) Improving blood safety in resource-poor settings. Vox Sang 87 (Suppl 2):176–179
26. Weigl B, Domingo G, LaBarre P, Gerlach J (2008) Towards non- and minimally instrumented, microfluidics-based diagnostic devices. Lab Chip 8:1999–2014
27. Craw P, Balachandran W (2012) Isothermal nucleic acid amplification technologies for point-of-care diagnostics: a critical review. Lab Chip 12:2469–2486
28. Cháfer-Pericás C, Maquieira A, Puchades R (2012) Functionalized inorganic nanoparticles used as labels in solid-phase immunoassays. Trends Anal Chem 31:144–156
29. Johnson CJ, Zhukovsky N, Cass AE, Nagy JM (2008) Proteomics, nanotechnology and molecular diagnostics. Proteomics 8:715–730
30. Jokerst JV, Floriano PN, Christodoulides N, Simmons GW, McDevitt JT (2008) Integration of semiconductor quantum dots into nano-bio-chip systems for enumeration of CD4+ T cell counts at the point-of-need. Lab Chip 8:2079–2090
31. Mannoor MS, Tao H, Clayton JD, Sengupta A, Kaplan DL, Naik RR, Verma N, Omenetto FG, McAlpine MC (2012) Graphene-based wireless bacteria detection on tooth enamel. Nat Commun 3:763–771
32. Mao X, Baloda M, Gurung AS, Lin Y, Liu G (2008) Multiplex electrochemical immunoassay using gold nanoparticle probes and immunochromatographic strips. Electrochem Commun 10:1636–1640
33. Kell AJ, Somaskandan K, Stewart G, Bergeron MG, Simard B (2008) Superparamagnetic nanoparticle-polystyrene bead conjugates as pathogen capture mimics: a parametric study of factors affecting capture efficiency and specificity. Langmuir 24:3493–3502
34. Liu G, Lin Y (2007) Nanomaterial labels in electrochemical immunosensor and immunoassays. Talanta 74:308–317
35. Ramadan Q, Gijs MAM (2012) Microfluidic applications of functionalized magnetic particles for environmental analysis: focus on waterborne pathogen detection. Microfluid Nanofluid 13:529–542
36. Cheng X, Irimia D, Dixon M, Ziperstein JC, Demirci U, Zamir L, Tompkins RG, Toner M, Rodriguez WR (2007) A microchip approach for practical label-free CD4+ T-cell counting of HIV-infected subjects in resource-poor settings. J Acquir Immune Defic Syndr 45:257–261
37. Zhang C, Wang H, Xing D (2011) Multichannel oscillatory-flow multiplex PCR microfluidics for high-throughput and fast detection of foodborne bacterial pathogens. Biomed Microdevices 13:885–897
38. Ward JB, Henderson RE (1996) Identification of needs in biomarker research. Environ Health Perspect 104:895–900
39. Strimbu K, Tavel JA (2010) What are biomarkers? Curr Opin HIV AIDS 5(6):463–466
40. WHO International Programme on Chemical Safety. Biomarkers and Risk Assessment: Concepts and Principles. 1993. Available online at http://www.inchem.org/documents/ehc/ehc/ehc155.htm
41. Reddy LH, Arias JL, Nicolas J, Couvreur P (2012) Magnetic nanoparticles:design and characterization, toxicity and biocompatibility, pharmaceutical and biomedical applications. Chem Rev 112:5818–5878
42. Hsing IM, Xu Y, Zhao WT (2007) Micro- and nano-magnetic particles for applications in biosensing. Electroanalysis 19:755–768
43. Berensmeier S (2006) Magnetic particles for the separation and purification of nucleic acids. Appl Microbiol Biotechnol 73:495–504

44. Cao M, Li Z, Wang J, Ge W, Yue T, Li R (2012) Food related applications of magnetic iron oxide nanoparticles: enzyme immobilization, protein purification, and food analysis. Trends Food Sci Technol 27:47–56
45. Ademtech, consulted on 1st November 2015. http://www.ademtech.com/
46. Invitrogen, Life Technologies Corporation, consulted on 1st November 2015. http://www.invitrogen.com/
47. Bangs Laboratories, Inc. consulted on 1st November 2015. http://www.bangslabs.com/
48. Chemicell, consulted on 1st November 2015. http://www.chemicell.com/
49. Miltenyi Biotech, consulted on 1st November 2015. http://www.miltenyibiotec.com/en
50. Vikesland P, Wigginton A (2010) Nanomaterial enabled biosensors for pathogen monitoring—a review. Environ Sci Technol 44:3656–3669
51. Sharif E1, Kiely J, Luxton R (2013) Novel immunoassay technique for rapid measurement of intracellular proteins using paramagnetic particles. J Immunol Methods 388:78–85
52. Sole S, Merkoçi A, Alegret S (2001) New materials for electrochemical sensing III. Beads. Trends Anal Chem 20:102–110
53. Austin JW, Pagotto FJ (2003) Encyclopedia of food science and nutrition. In: Caballero B, Trugo LC, Finglas PM. Elsevier Science Ltd., Amsterdam, p 3886
54. Upmann M, Bonaparte C (1999) Encyclopedia of food microbiology. In: Robinson RK, Batt CA, Patel PD. Elsevier Science Ltd., Amsterdam, p 1887
55. Deisingh AK, Thompson M (2004) Biosensors for the detection of bacteria. Can J Microbiol 50:69–77
56. Terry LA, White SF, Tigwell LJ (2005) The application of biosensors to fresh produce and the wider food industry. J Agric Food Chem 53:1309–1316
57. Velasco-Garcia MN, Mottram T (2003) Biosensors for detection of pathogenic bacteria. Biosyst Eng 84:1–12
58. Patel PD (2002) (Bio)sensors for measurements of analytes implicated in food safety: a review. TRAC 21:96–115
59. Mellgren C, Sternesjo A, Hammer P, Suhren G, Bjorck L, Heeschen W (1996) Comparison of biosensor, microbiological, immunochemical and physical methods for detection of sulfamethazine residues in raw milk. J Food Prot 59:1223–1226
60. Sternesjo A, Mellgren C, Bjorck L (1995) Determination of sulphametazine residues in milk by a surface resonance based-biosensors assay. Anal Biochem 226:175–181
61. Barlen B, Mazumdar SD, Lezrich O, Kämpfer P, Keusgen M (2007) Detection of Salmonella by surface plasmon resonance. Sensors 7:1427–1446
62. Ivnitski D, Abdel-Hamid I, Atanasov P, Wilkins E, Stricker S (2000) Application of electrochemical biosensors for detection of food pathogenic bacteria. Electroanalysis 12: 317–325
63. Mehervar M, Abdi M (2004) Recent development, characteristics, and potential applications of electrochemcial biosensors. Anal Sci 20:1113–1126
64. Ricci F, Adornetto G, Palleschi G (2012) A review of experimental aspects of electrochemical immunosensors. Electrochim Acta 84:74–83
65. Skládal P, Kovář D, Krajíček V, Šišková P, Přibyl J, Švábenská E (2013) Electrochemical immunosensors for detection of microorganisms. Int J Electrochem Sci 8:1635–1649
66. Pemberton RM, Cox T, Tuffin R, Sage I, Drago GA, Biddle N, Griffiths J, Pittson R, Johnson G, Xu J, Jackson SK, Kenna G, Luxton R, Hart JP (2013) Microfabricated glucose biosensor for culture well operation. Biosens Bioelectron 42:668–677
67. Metters JP, Kadara RO, Banks CE (2011) New directions in screen printed electroanalytical sensors: an overview of recent developments. Analyst 136:1067–1076
68. Metters JP, Gomez-Mingot M, Iniesta J, Kadara RO, Banks CE (2013) The fabrication of novel screen printed single-walled carbon nanotube electrodes: electroanalytical applications. Sens Actuat B 177:1043–1052
69. Kong FY, Gu SX, Li WW, Chen TT, Xu Q, Wang W (2014) A paper disk equipped with graphene/polyaniline/Au nanoparticles/glucose oxidase biocomposite modified

Electrochemical Biosensors for Chemical Warfare Agents

Fabiana Arduini, Viviana Scognamiglio, Danila Moscone and Giuseppe Palleschi

Abstract Since the introduction of modern CWAs at the beginning of 20th century, there has been a continuous interest in the development of robust and reliable analytical tools for the detection of these agents, to provide early alarm in case of terroristic attacks, as well as to monitor their presence in the environment and prevent contamination. Nevertheless, some powerful analytical techniques, including chromatographic methods and mass spectrometry, may not be well suitable for field applications and fast early warning, due to the lack of portability, power requirements, long response time and expensive procedures. In this context, electrochemical (bio)sensors offer advantages in terms of high sensitivity, miniaturization, integration, low cost, and power requirements. The aim of this chapter is to highlight the important issues of electrochemical (bio)sensors for fast and cost-effective detection of CWAs in the field, considering the main advantages and limitations of this technology, and the last trends in nanotechnology, lab-on-chip, and functional materials.

Keywords Electrochemical (bio)sensors · Chemical warfare agents · Early warning

1 Introduction

Chemical warfare agents (CWAs) are highly toxic synthetic chemicals that can be dispersed in the environment as gas, liquid or aerosol or adsorbed to particles to become a powder [1].

F. Arduini (✉) · D. Moscone · G. Palleschi
Dipartimento di Scienze e Tecnologie Chimiche, Università di Roma Tor Vergata,
Via Della Ricerca Scientifica, 00133 Rome, Italy
e-mail: fabiana.arduini@uniroma2.it

V. Scognamiglio
Dipartimento Agroalimentare, IC-CNR Istituto di Cristallografia,
Via Salaria km 29.3, 00015 Rome, Italy

D.P. Nikolelis and G.-P. Nikoleli (eds.), *Biosensors for Security and Bioterrorism Applications*, Advanced Sciences and Technologies for Security Applications,
DOI 10.1007/978-3-319-28926-7_6

CWAs have been deliberately produced and employed in the battlefields during the 20th century with the purpose of killing or debilitating living organisms. The Chemical Weapons Convention (CWC) has classified them based on their volatility, chemical structure or the physiological effects produced on humans. Regarding the latter feature, CWAs are classified in nerve agents, blister agents or vesicants, blood agents, and chocking or pulmonary agents, as reported in Fig. 1.

In addition, a special attention is devoted to toxins, which have been largely exploited as potential chemical weapons, according to the Organisation for the Prohibition of Chemical Weapons (OPCW). In Table 1, different CWA groups, their persistence, and rate of action are reported, according to OPCW.

Thanks to the knowledge of synthetic chemistry, a considerable progression in the development of chemical compounds has occurred in the past century in human conflicts (Fig. 2). This common chemical and scientific knowledge allowed an increasing sophistication of CWAs, as demonstrated by the replacement of chlorine (Cl_2) by phosgene ($COCl_2$) and then by mustard gas.

The first use of chemicals as mass destruction weapons goes back to WW I (1915), when large amounts of chlorine-containing compounds and gas warfare were released by German military forces in Belgium, causing ~91 000 deaths (~1.3 million casualties) [2]. Warfare chemicals, including phosgene, hydrogen cyanide, arsenical compounds, sulphur mustard, and lewisite have been exploited until the WW II.

After WW II, chemical weapons were employed on a number of occasions: mustards were used during the soviet intervention into Afghan War (1978), and also used against rebels in Chad (1987); nerve agents and blister agents were used during Iran-Iraq War (1980–1988). Iraq used Sarin, hydrogen cyanide, and sulphur mustard against Iraqi Kurds. The use of chemical warfare agents, unfortunately,

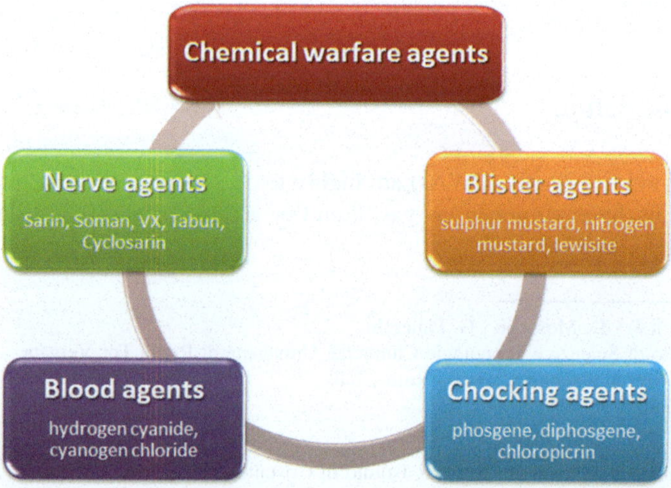

Fig. 1 CWAs classification

Table 1 CW Agent Group, persistency and rate of action

CW agent group	Persistency	Rate of action
Nerve agents		
Tabun (GA)	High	Very rapid
Sarin (GB)	Low	Very rapid
Soman (GD)	Moderate	Very rapid
Cyclosarin (GE, GF)	Moderate	Rapid
VX	Very high	Very rapid
Blister/vesicant agents		
Sulfur mustard (H, HD)	Very high	Delayed
Nitrogen mustard (HN)	High	Delayed
Phosgene oxime (CX)	Low	Immediate
Lewisite (L)	High	Rapid
Blood agents		
Hydrogen cyanide (AC)	Low	Rapid
Cyanogen chloride (CK)	Low	Rapid
Arsine (SA)	Low	Delayed
Choking/pulmonary agents		
Chlorine (Cl)	Low	Variable
Phosgene (CG)	Low	High
Diphosgene (DP)	Low	Delayed
Chloropicrin (PS)	Moderate	Rapid

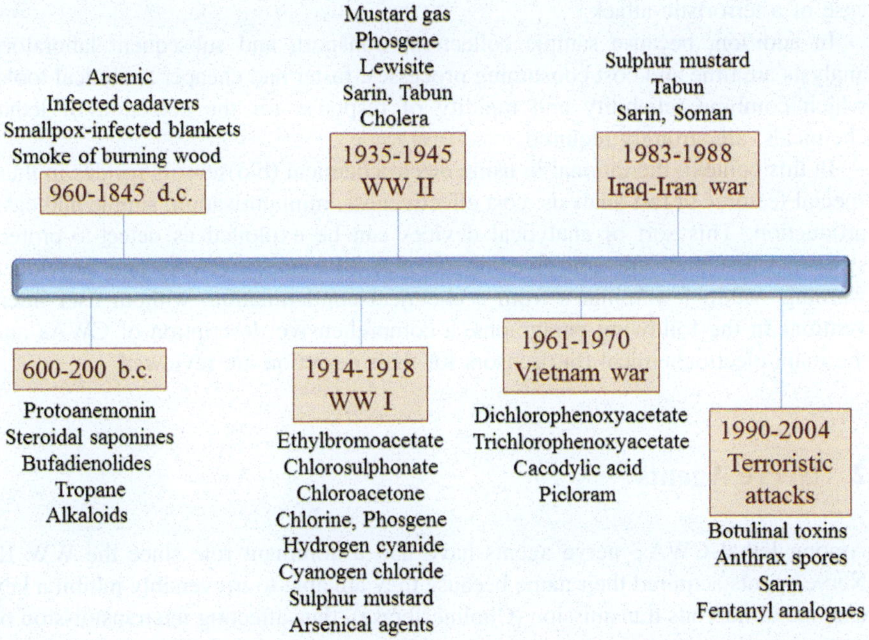

Fig. 2 Graphical timeline of CWAs use and related dates

was also extended to the terrorist attack: the Japanese cult group AUM Shinrikyo used the Sarin gas in Tokyo subway in 1995. After the tragic events at New York (2001) and at London (2005) the possibility of CWAs being used as means of terrorism is real.

In this overall scenario, due to the high number of exposures to CWAs through different sources (e.g. recent terrorist attacks or already contaminated sites) and to their chemical broad spectrum, there is an increasing interest in the development of highly sensitive, selective, contactless, and early detection systems for low concentrations below the median lethal dose (LD_{50}: dose required to kill half of the members of a tested population).

In this context, (bio)sensor technology has a great potential to address these challenges, providing the development of tailor-made small and portable instruments, with adequate sensitivity and selectivity, reasonable cost and fast time response, and powerful ability to identify CWAs. (Bio)sensors demonstrated to possess adequate detection methodologies to (i) detect the chemical species of weapons used in terroristic attacks; (ii) measure them in exposed people to early identify the chemical contamination; (iii) monitor their presence in the environment to provide warning on contaminated sites and thus prevent contamination. The detection of CWAs is generally carried out in dedicated centralized laboratories using large and expensive instrumentations, such as gas chromatographs coupled with mass spectrometers (GC-MS). Even if these laboratory set-up methodologies demonstrated to be useful and sensitive tools to evaluate the pollution of a contaminated site, their features do not match the requirements of an early detection in case of a terroristic attack.

In addition, because sample collection, transport, and subsequent laboratory analysis are time and cost consuming processes, faster and cheaper analytical tools, which combine reliability and rapidity of response for the detection of lethal chemicals, are strongly required.

In this context, the rational in using electrochemical (bio)sensors resides in their special features of fast analysis, cost effectiveness, miniaturisation, simple and easy production. This sort of analytical devices can be exploited as detect-to-protect system, since they are conceived to be able to identify CWAs and provide a warning within few minutes from a chemical contamination, without user intervention. In the following paragraphs, a comprehensive description of CWAs and the main electrochemical (bio)sensors for their detection are reviewed.

2 Nerve Agents

Among lethal CWAs, nerve agents have had a dominant role since the WW II. Nerve agents acquired their name because they are able to irreversibly inhibit a key enzyme of nervous transmission (Cholinesterase), thus affecting the transmission of nerve impulses in the nervous system. All nerve agents belong to the group of

2.2 *Biosensor Based on Organophosphate Hydrolase*

Biosensors based on the use of organophosphorus hydrolase (OPH) as bioreceptors are classified as substrate-type biosensors. Indeed, while AChE biosensor reveals OP by inhibition and the response is *inversely* proportional to the OP amount, the OPH biosensor detects OPs as a direct substrate and the response measured is *directly* proportional to the OP amount (Fig. 6).

The big advantage of this biosensor relies on the direct measurement of the analyte with a faster detection, but, on other hand, its sensitivity is lower in respect to ChE biosensor.

One of the first works on OPH biosensors was published in 1999 on *Analytical Chemistry* journal by Mulchandani's group. OPH was immobilised onto screen-printed carbon electrodes, and OP measurements were provided by the rapid anodic detection of the enzymatically generated *p*-nitrophenol product. This biosensor showed a linearity proportional to the concentration of the hydrolyzed paraoxon and methyl parathion substrates up to 40 and 5 μM, with detection limits of 9×10^{-8} and 7×10^{-8} M, respectively [13]. The same group developed also a microbial biosensor using genetically engineered PNP-degrader *Moraxella* sp. displaying OPH on the cell surface, immobilised onto oxygen electrode. OPH works in tandem with a PNP oxidation machinery to degrade PNP-substituted OPs with simultaneously oxygen consuming. The amount of oxygen consumed is proportional to the analyte concentration, allowing the detection of the analyte. Under optimised conditions, the biosensor showed a detection limit of 0.1 μM (27.5 ppb) using paraoxon as model, with an excellent selectivity against triazines, carbamates and OPs without PNP substituent [14].

A similar limit of detection (0.1 μM) was also reported by the OPH biosensor embedded in a flow system of Wang et al. [15] for the detection of paraoxon.

More recently, an interesting paper was published on *ACS nano* journal in 2010 by Chol et al. [16]. They reported the preparation of free-standing flexible conductive reduced graphene oxide/Nafion (RGON) hybrid films by a solution

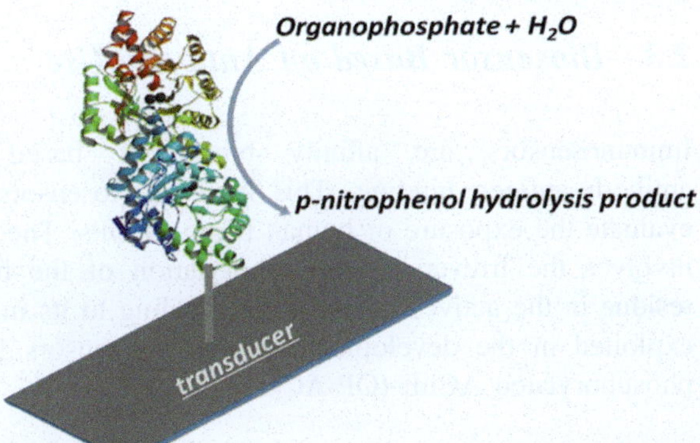

Fig. 6 Scheme of OPH biosensor

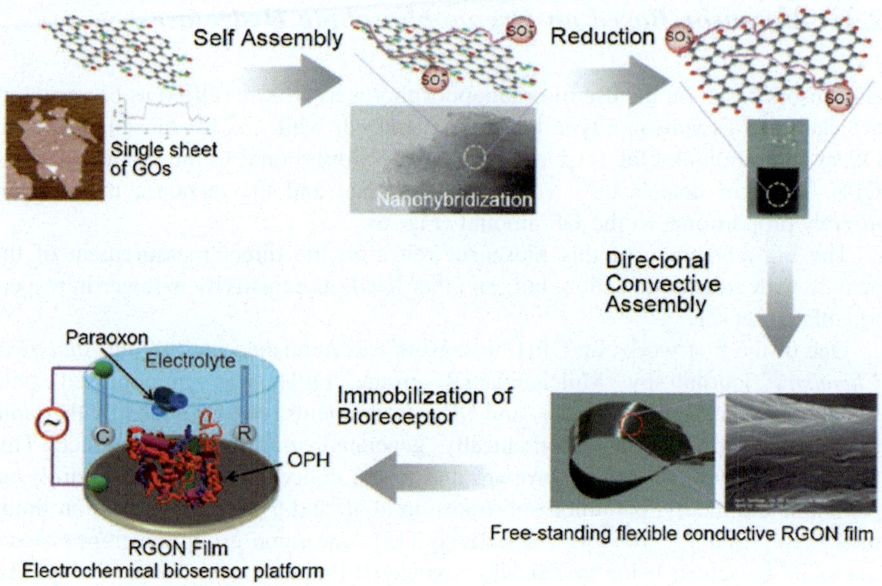

Fig. 7 Illustration of a procedure to design RGON hybrids and subsequently RGON platform for application in electrochemical biosensors. Reprinted (adapted) with permission from [16]. Copyright (2010) American Chemical Society

chemistry that utilizes both self-assembly and directional convective-assembly (Fig. 7). The performance characteristics of RGON were exploited immobilizing OPH and examining the amperometric response of the *p*-nitrophenol hydrolysis product at a potential of +0.85 V, reaching a detection limit of 1.37×10^{-7} M, and a response time of <3 s.

As illustrated in all examples reported above, the detection limit is around 0.1 μM, which is rather higher than the one obtained using ChE biosensor (nM/pM level), confirming the higher sensitivity of ChE biosensors.

2.3 Biosensor Based on Antibody Use

Immunosensors are affinity biosensors based on the measurement of antibody-antigen binding. This kind of biosensors is used in security field to evaluate the exposure of human being to OPs. The mechanism of OPs poisoning involves the irreversible phosphorylation of the hydroxyl group of the serine residue in the active site of AChE, leading to its inactivation. This mechanism is exploited in the development of immunosensors for evaluating the amount of phosphorylated AChE (OP-AChE).

A crucial example of immunosensors for OP detection was described by Du et al. [17]. They developed an integrated lateral flow test strip based on an electrochemical sensor (LFTSES) for biomonitoring of OP exposure, where anti-AChE antibody are immobilised on the immunochromatographic strip (Fig. 8). Serum samples of exposed people are added on the pad, and binding of both AChE and OP-AChE to anti-AChE antibody are electrochemically evaluated adding the enzymatic substrate acetylthiocholine. The proposed immunosensors were able to provide parallel measurements of post-exposure and baseline AChE enzyme activity, and reactivation of the phosphorylated AChE was exploited to measure the total amount of AChE (including inhibited and active) which serves as a baseline (control). Therefore, the quantification of phosphorylated adduct (OP-AChE) is realized by subtracting the active AChE from the total amount of AChE. These immunosensors represents a highly interesting and promising example of point-of-care tools for the diagnosis of OP pesticide poisoning and nerve agent exposure, useful to the first responders for the rapid identification of victims in a nerve agent attack and consequently for a prompt medical treatment.

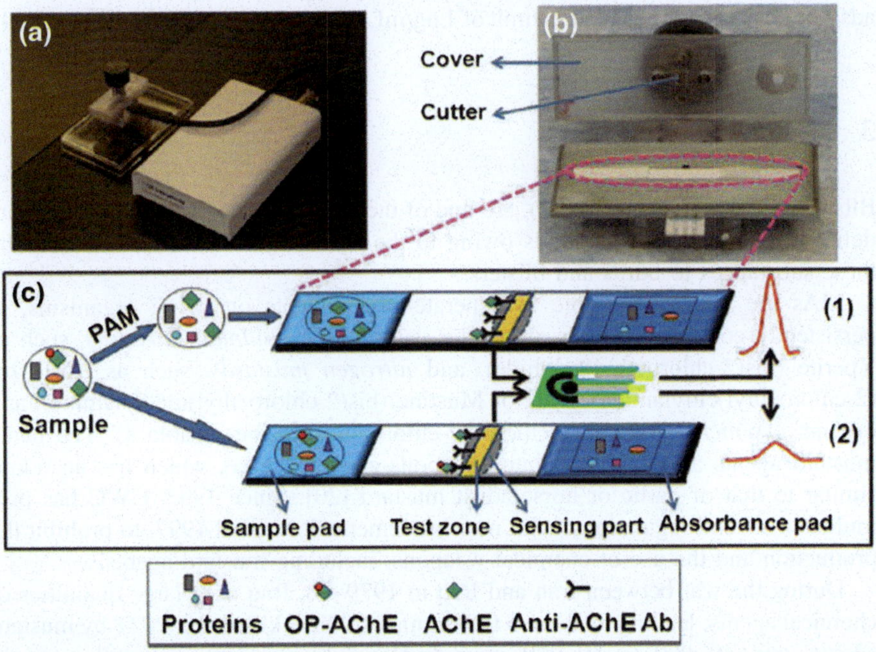

Fig. 8 **a** Entire portable analytical system, **b** LFTSES device, and **c** schematic illustration of the principle of LFTSES device. Reprinted (adapted) with permission from [17]. Copyright (2012) American Chemical Society

2.4 Sensors Based on Direct Electrochemical Detection

Electrochemical sensors can be an alternative to electrochemical biosensors, with the advantage to overcome the use of bioreceptors and the relative problem of storage stability, even if with the drawback of a lower selectivity due to the absence of the bioreceptor.

The use of electrochemical sensors for OP detection is largely demonstrated by Liu et al. [18], which exploited a carbon paste electrode to determine OP using the adsorptive stripping voltammetric technique. They exploited the ability of OPs to be strongly adsorbed on the surface of the carbon paste transducer in order to pre-concentrate them on the surface of the working electrode. As a result, a successive desorption step allowed to obtain a peak, which height is proportional to the amount of OP detected. This sensor showed a linear range of detection between 1 and 60 μM, using methyl parathion as organophosphate model (with 2 min of adsorption time), and a detection limit of 0.05 μmol/L with 10 min of adsorption time.

Based on the same principle, the authors developed a similar sensor using an electrode modified with zirconia nanoparticles, exploiting the strong affinity of zirconia for the phosphoric group. The improved sensor showed a linear range over the 5–100 ng/mL (ppb) using methyl parathion as organophosphate model with 2-min of adsorption time, and a detection limit of 1 ng/mL with 10-min as adsorption time [19].

3 Blister Agents

Blister or mustard agents (MAs), are one of the most common CW agents, and are defined as blistering compounds owing to the similarity of the lesions caused by these substances to burns and blisters.

MAs are compounds able to generate toxic effects on living organisms, to persistently contaminate soils and water, and embrace *sulphur mustards*, such as Yperite (bis(2-chloroethyl)sulphide), and *nitrogen mustards*, such as HN1 (bis (2-chloroethyl)ethylamine), HN2 or Mustine (bis(2-chloroethyl)methylamine), and organic *arsenical Lewisite* (dichloro (2-chlorovinyl)-arsine) (Table 4). The name mustard agents arises from impure weapons-grade material, which has an odour similar to that of garlic or horseradish mustard [20]. Since 1993, CWC has promulgated several regulations, with new implementations until 1997, to prohibit the production and the use of chemical weapons, including mustard agents.

During the war between Iran and Iraq in 1979–88, Iraq used large quantities of chemical agents, leading to about 5 000 Iranian soldiers killed, 10–20 % by mustard agents, and 40 000 to 50 000 injured. However, incidents are still annually occurring in the neighbourhood of Sweden, mainly involving fishermen exposed to mustard agents brought to the surface by fishing nets. Indeed, the background is found in the dumping of chemical weapons after the WW I in waters off the Danish and Swedish coasts.

class, *Phosgene* ($COCl_2$) is the most commonly used. It is a colourless highly toxic gas, causing severe lung irritation, pulmonary oedema, asphyxia and even death at a concentration as low as 2 ppm. The first recorded use of phosgene was in 1915, and it accounted for 80 % of all chemical fatalities during WW I.

Moreover, phosgene is also a valued industrial intermediate in organic synthesis. Therefore, developing a simple and sensitive method for phosgene detection is highly desirable not only for public safety against terrorist attack but also for industrial production [36].

An effective sensor for phosgene detection was realised by Virji et al. [37], based on a novel polyaniline nanofiber composite material. They constructed, using standard photolithography, a sensor array containing 18 sensors with each sensor having 35 pairs of fingers with 20 μm electrode gaps. The reaction mechanism was based on a nucleophilic substitution reaction of phosgene with the amine to generate an isocyanate and an acid. The acid dopes the polyaniline nanofibers increasing their conductivity by up to two orders of magnitude. With this sensor configuration, phosgene was detected well below the permissible exposure limit (0.1 ppm).

6 Toxins

According to OPCW, toxins are effective and specific poisons produced by living organisms, consisting of an amino acid chain which can vary in molecular weight between a couple of hundred (peptides) and one hundred thousand (proteins). Toxins are produced by numerous organisms, e.g., bacteria, fungi, algae and plants. Many of them are extremely poisonous, with a toxicity that is several orders of magnitude greater than the nerve agents. Because of the hybrid nature of toxins, they have sometimes been considered CW agents and sometimes biological agents (BW). Similarly, based on their mechanism of action, they are grouped as cardiotoxins, dermatotoxins, hepatotoxins, neurotoxins, etc. Being a heterogeneous group of compounds, they are able to interfere with biochemical processes, such as membrane function, ion transport, transmitter release and macromolecule synthesis. Human exposure to toxins can lead to serious health problems, including immunosuppression and carcinogenesis.

For these reasons, they started to attract military interest already during the first half of the last century. At that time, it was difficult to manufacture sufficiently large amounts of toxin, which caused interest to decrease. Many of the toxins considered at that time were sensitive to heat and light which made them unstable and unpractical to use. The U.S.A. ended its toxin programme in the late 1960s and destroyed its stockpile of, e.g., botulinum toxin. The Biological and Toxin Weapons Convention of 1972 prohibits the development, production and stockpiling of toxins as weapons.

The 1925 Geneva Protocol prohibition of use of chemical and bacteriological weapons also covers the use of weapons based on toxins. Since the definition of chemical weapons includes toxins, they are also covered by the Chemical Weapons Convention.

In the late 1970s, there was a rapid development of gene technology together with biotechnology. This led to again arise the threat from toxins as CW agents. Nowadays it became possible to more easily produce greater amounts of many toxins, in some cases even synthetically. Gene technology can be used to modify the toxin genes so that the end product shows new properties as, for example, to become less sensitive to sunlight.

The scientific and commercial development has together provided increased opportunities to incorrectly utilize biotechnology for military purposes. Recent researches, for example, made it possible to "target" toxins to different body organs or structures. This new knowledge mainly emanates from civilian research into, e.g., the treatment of cancer patients.

Toxins are still considered to be less suitable for dispersal on a large scale. Nonetheless, they could be used for sabotage or in especially designed inputs, e.g., against key persons. Since toxins have low volatility, they are dispersed as aerosols and then taken up foremost through inhalation. The new microencapsulation technology, which is easy to use, makes it possible to protect unstable toxins when dispersed.

A few examples of toxins which may be used as chemical warfare agents are listed below. The trichothecenes, mycotoxins obtained from, e.g., Fusarium genera, were alleged in the early 1980s to have been used as CW agents in Southeast Asia ("yellow rain"), but are of no military value today.

6.1 Mycotoxins

Botulinum toxin is the most poisonous substance known. It is also known as agent X. Arnon et al. estimated that 1 g of this toxin, when aerosolized, could kill more than one million people. The lethal dose for a 70-kg human is estimated to be approximately 0.7 μg if inhaled or 70 μg if ingested. Botulinum toxin is produced by the bacteria *Clostridium botulinum*, which grows on poorly preserved food and causes a severe form of food-poisoning (botulism). The incubation period is between one and three days after which the victim becomes ill with stomach pains, diarrhoea, disturbances to vision, giddiness and muscular weakness. The whole body including the respiratory musculature becomes paralyzed which leads to death by suffocation within a few days.

The toxin is a protein available in seven different forms, where the most poisonous is the type A (molecular weight 150 000 Da). It is possible to vaccinate against botulism, while once the victim has become poisoned, there is no antidote. Botulinum toxin is today commercially produced and is used in treating squinting and other muscular disorders.

The literature on sensing systems for the detection of botulinum neurotoxins, based on the modes of action of these toxins (which are well-established in literature), is very huge. Indeed, they are known to specifically cleave portions of the SNARE proteins SNAP-25 or VAMP. This interaction can be monitored by different transduction system. A central example is the study of Savage et al. [38], which reported the development of a SNAP-25 and a VAMP biosensors for detecting the activity of five botulinum neurotoxin serotypes (A–E) using electrochemical impedance spectroscopy, at concentrations as low as 25 fg/mL, in a short time-frame compared with the current standard methods of detection.

Nevertheless, the use of biosensors to measure botulinum finds many application in different fields, being the detection of botulinum essential for identifying its presence in potential cases of terrorism and food contamination, as well as in diagnosis of botulism. As an example, Chan et al. [39] described a reduced graphene oxide(rGO)/Au electrode based electrochemical biosensor for ultrasensitive detection of BoNT serotype A light chain (BoNT-LcA) protease activity in milk samples. They fabricated rGO/Au electrodes to realise a robust and biocompatible platform with enhanced electron transfer capability and large surface area for SNAP-25-GFP peptide immobilization. Once immobilised, SNAP-25-GFP peptide is specifically cut at the cleavage sites upon the addition of BoNT-LcA, consequently releasing the cut section from the electrode surface. This enzymatic activity of BoNT-LcA on SNAP-25-GFP peptide substrate was detected by monitoring the enhanced redox probe transfer rate by differential pulse voltammetry with a linear detection range up to 1 ng/mL and LOD around 8.6 pg/mL. The specificity of the biosensor reported by Chan et al. was demonstrated, as well as its feasibility in complex matrices.

A different application of botulinum biosensors is presented in the pharmaceutical field. Halliwell et al. [40] reported the development of two electrochemical bioassays for the detection of active botulinum neurotoxin in pharmaceutical samples based on gold electrodes modified with self-assembled monolayers of the SNARE protein SNAP-25, which is selectively cleaved by active botulinum neurotoxin A. The addition of the toxin to the layer was measured by cyclic voltammetry and electrochemical impedance spectroscopy, performed on the modified working electrodes to observe changes of toxin concentrations as low as 25 fg/ml, with results being obtained in less than an hour, outperforming the mouse bioassay. The presented sensing assay demonstrated to be able to replace currently standard methods based on the mouse bioassay, actually considered as the most reliable method for the detection of the active form of this toxin, but also time-consuming and expensive.

Among mycotoxins, *ricin* is a very potent toxin of plant origin, isolated from the seeds of castor oil, *Ricinus communis*. It inhibits ribosome proteins, and the toxic dose for humans is about 0.1–1.0 μg/kg, depending on the mode of administration. It was used in the famous “umbrella tip” assassination of the Bulgarian journalist. Iraq produced and weaponized several highly potent toxins such as botulinum toxin and aflatoxin.

A small number of electrochemical sensing systems have been reported in literature in the last years. A recent example is represented by the paper-based assay platform of Cunningham et al. [41] for detection of an immunological ricin chain.

The group assembled a paper platform, by simple origami paper folding, to measure the toxin by means of quantitative, electrochemical detection of silver nanoparticle labels linked to a magnetic microbead support via a ricin immunosandwich. The reported sensor demonstrated to be highly advantageous thanks to the high sensitivity (34 pM), robustness even in the presence of 100-fold excess hoax materials, cost-effectiveness ($0.30 per assay), and speed response (9.5 min per assay).

A similar sandwich immunoassay format was described by Suresh et al. [42] for the immunosensing of ricin. In particular, they developed an amperometric immunosensor for the specific detection of *Ricinus communis* based on screen-printed electrodes modified with gold nanoparticles loaded with a multi-walled carbon nanotubes- chitosan film. Specific antibodies tagged with the enzyme alkaline phosphatase were used to convert the substrate 1-naphthyl phosphate into 1-naphthol, consequently determined with the amperometric technique and correlated with the concentration of ricin. The reported method showed high stability due to the chitosan film, short response time with good reproducibility and increased shelf life of the electrodes with immobilised antibodies. Furthermore, the optimization of composition of carbon nanotubes and gold nanoparticles was able to improve the electrochemical activity of the electrodes, which showed under optimal conditions, a wide linear response to the concentration of ricin in the range of 2.5–25 ng mL^{-1}, a sensitive limit of detection of 2.1 ng mL^{-1}, a relative standard deviation of 5.1 % and storage life of 32 days.

6.2 Phycotoxins

Many toxins are produced by marine organisms. Likewise, phycotoxins may produce undesirable effects also at low concentrations, causing intoxication syndromes.

One example is saxitoxin, which is synthesized by a type of blue-green algae (cyanobacteria). These algae provide food for different shellfish, e.g., mussels. The mussels themselves are not influenced by the poison, but human beings who later eat the mussels may become seriously ill. *Saxitoxin* attacks the nervous system and has a paralyzing effect, but causes no symptoms in the gastro-intestinal tract. The development of the illness is extremely rapid and at high doses death may occur within less than 15 min. The LD50 for man is at about 1 mg. Saxitoxin is a small molecule with a molecular weight of 370 D. It is not sensitive to heat but is destroyed by oxygen.

Several analytical methodologies have been developed for the detection of the main marine toxins, including biosensors for domoic acid, okadaic acid, and tetrodotoxin [43–50]. For instance, Micheli et al. have developed a competitive indirect test immunosensor coupled to differential pulse voltammetry for the detection of domoic acid using disposable screen-printed electrodes [44]. Domoic acid was conjugated to bovine serum albumin and was coated onto the working electrode of the SPE, followed by incubation with sample (or standard toxin) and anti- domoic antibody. An anti-goat IgG-alkaline phosphatase conjugate was used for signal measurement. The developed immunosensor is able to detect domoic with LOD equal to 5 ng/mL. For tetrodotoxin detection, a direct competitive electrochemical immunosensor was developed by Neagu et al. [46]. In the construction of this immunosensor, antigen labelled with alkaline phosphatase was prepared in order to evaluate the amount of tetrodotoxin in the sample analysed, reaching a dynamic range comprised between 2 and 50 ng/mL and LOD of 1 ng/mL. An electrochemical multi-enzymatic system based on phosphatase-2A inhibition for okadaic acid was assembled by Volpe at a [50]. The amount of okadaic acid was estimated evaluating the degree of inhibition of protein phosphatase-2A, which is proportional of the okadaic amount. The basic reaction of this system is the one catalyzed by the Phosphorylase *a* that converts glycogen and phosphate (P) into glucose-1-P, which in turn produces glucose through the use of alkaline phosphatase. The glucose is then converted by glucose oxidase into H_2O_2 which is electrochemically oxidized at the platinum electrode inserted into a flow injection analysis, allowing for a working range of 30–250 pg/mL.

7 Electrochemical (Bio)Sensors for in Situ Measurement Versus Laboratory Set-up Methodologies

Having an analytical device easily used by unskilled personnel for in situ application, has led scientists to integrate the electrochemical (bio)sensors in portable and embedded apparatuses. In this context, significative examples were reported for nerve agent detection. For instance, Du et al. [51] have developed a portable set-up based on the use of screen-printed electrode modified with carbon nanotubes combined with a microflow-injection device for OP detection. This analytical system was able to assess the presence of inhibited AChE in saliva via regeneration of AChE, demonstrating the suitability for subclinical organophosphate exposure evaluation.

In the direction of automatisable systems, an interesting work was reported by the Hart's group; in this case, an array on six *engineered* acetylcholinesterase screen-printed electrodes was integrated in a novel automated instrument coupled with a neural network program. The system was successfully applied in several samples like water, food and vegetable extracts, demonstrating the possibility to cover the detection of several pesticides in a number of matrices by using an automatic and portable system [52].

The first example of lab-on-a-chip based on an electrochemical cholinesterase biosensor was developed by us using butyrylcholinesterase immobilised on screen-printed electrodes modified with Prussian Blue. In this case, the probe can be able to detect inhibitors like Sarin in gas phase using a micro fun for air sampling. The integrated miniaturized circuit is able to apply the potential, to register the current, to turn on-off the fan and eventually give an alarm switching on a led, demonstrating the readiness of the electrochemical biosensor for a rapid alarm in the case of nerve agent pollution in air [53].

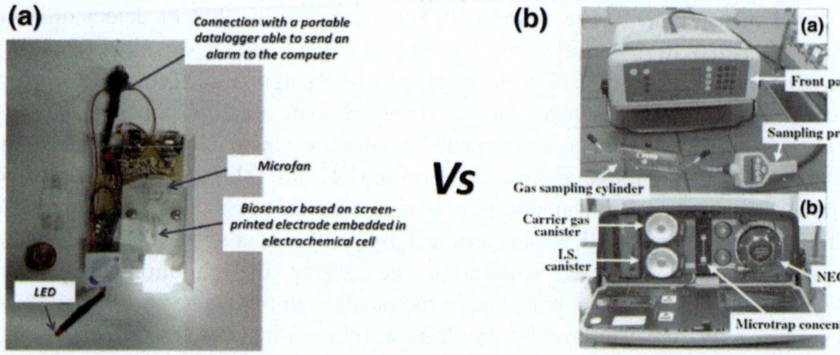

Fig. 9 **a** ChE biosensor integrated into a lab-on-a-chip. The prototype is composed of a cell in which is inserted the biosensor, a little fan able to sampling air and an electronic circuit. The circuit is able to apply the potential, to register the current, to turn on-off the fan and eventually to give an alarm by led or wireless at a personal computer thorught the datalogger [53] versus **b** Hapsite instrument [57]. Reprinted (adapted) with permission from [53, 57]

In order to understand the feasibility of the electrochemical biosensor use, a comparison with commercial available portable instruments is mandatory.

Various kinds of on-site nerve agent detection equipment are used by the military and police mobile teams. The portable devices are based on ion mobility [54], surface acoustic wave [55], gas chromatographs coupled with mass spectrometer (Hapsite® instrument) [56]. The commercial available systems are characterized by high selectivity and sensitivity (i.e. Hapsite), but at the same time, the instruments are difficult to further miniaturize and rather expensive. On the contrary, the lab on a chip based on cholinesterase biosensor is characterized by cost-effectiveness, easiness to use, can be easily automatized (Fig. 9) and applied for security monitoring in airports, undergrounds, etc.

In this direction, the lab on a chip based on electrochemical (bio)sensors is an useful first alarm system able to give a rapid alarm, sustaining the management of chemical risk in the case of terroristic attacks.

References

1. Ganesan K, Raza SK, Vijyaraghavan R (2010) Chemical warfare agents. J Pharm Bioallied Sci 2:166–178
2. Szinicz L (2005) History of chemical and biological warfare agents. Toxicology 214:167–181
3. Guilbault GG, Kramer DN, Jr PL (1962) Cannon, electrical determination of organophosphorous compounds. Anal Chem 34:1437–1439
4. Fennouh S, Casimiri V, Burstein C (1997) Increased paraoxon detection with solvents using acetylcholinesterase inactivation measured with choline oxidase biosensor. Biosens Bioelectron 12:97–104
5. Cremisini C, Di Sario S, Mela J, Pilloton R, Palleschi G (1995) Evaluation of the use of free and immobilised acetylcholinesterase for paraoxon detection with an amperometric choline oxidase biosensor. Anal Chim Acta 311:273–280

6. Ciucu AA, Negulescu C, Baldwin RP (2003) Detection of pesticides using an amperometric biosensor based on ferophthalocyanine modified carbon paste electrode immobilised bienzymatic system. Biosens Bioelectron 18:303–310
7. Ricci F, Amine A, Palleschi G, Moscone D (2003) Prussian Blue based screen printed biosensors with improved characteristics of long-term lifetime and pH stability. Biosens Bioelectron 18:165–174
8. Upadhyay S, Rama Rao G, Sharma MK, Bhattacharya BK, Rao VK, Vijayaraghavan R (2009) Immobilization of acetylcholinesterase–choline oxidase on a gold–platinum bimetallic nanoparticles modified glassy carbon electrode for the sensitive detection of organophosphate pesticides, carbamates and nerve agents. Biosens Bioelectron 25:832–838
9. Lee WE, Thompson HG, Hall JG, Bader DE (2000) Rapid detection and identification of biological and chemical agents by immunoassay, gene probe assay and enzyme inhibition using a silicon-based biosensor Biosens. Bioelectron. 14:795–804
10. Pohanka M, Adam V, Kizek R (2013) An Acetylcholinesterase-based chronoamperometric biosensor for fast and reliable assay of nerve agents. Sensors 13:11498–11506
11. Pohanka M, Dobes P, Drtinova L, Kuca K (2009) Nerve agents assay using cholinesterase based biosensor. Electroanalysis 21:1177–1182
12. Arduini F, Ricci F, Amine A, Moscone D, Palleschi G (2007) Fast, sensitive and cost-effective detection of nerve agents in the gas phase using a portable instrument and an electrochemical biosensor. Anal Bioanal Chem 388:1049–1057
13. Mulchandani A, Mulchandani P, Chen W, Wang J, Chen L (1999) Amperometric thick-film strip electrodes for monitoring organophosphate nerve agents based on immobilized organophosphorus hydrolase. Anal Chem 71:246–2249
14. Mulchandani P, Chen W, Mulchandani A (2006) Microbial biosensor for direct determination of nitrophenyl-substituted organophosphate nerve agents using genetically engineered *Moraxella* sp. Anal Chim Acta 568:217–221
15. Wang J, Krause R, Block K, Musameh M, Mulchandani A, Schoning MJ (2003) Flow injection amperometric detection of OP nerve agents based on an organophosphorus/hydrolase biosensor detector. Biosens Bioelectron 18:255–260
16. Choi BG, Park H, Park TJ, Yang MH, Kim JS, Jang SY, Heo NS, Yup Lee S, Kong J, Hong W (2010) Solution chemistry of self-assembled graphene nanohybrids for high-performance flexible biosensors. ACS Nano 4:2910–2918
17. Du D, Wang J, Wang L, Lu D, Lin Y (2012) Integrated lateral flow test strip with electrochemical sensor for quantification of phosphorylated cholinesterase: biomarker of exposure to organophosphorus agents. Anal Chem 84:1380–1385
18. Liu G, Lin Y (2005) Electrochemical stripping analysis of organophosphate pesticides and nerve agents. Electrochem Commun 7:339–343
19. Liu G, Lin Y (2005) Electrochemical sensor for organophosphate pesticides and nerve agents using zirconia nanoparticles as selective sorbents. Anal Chem 77:5894–5901
20. Mansour Razavi S, Salamati P, Saghafinia M, Abdollahi M (2012) A review on delayed toxic effects of sulfur mustard in Iranian veterans. J. Pharm Sci 20:51–59
21. Haines DD, Fox SC (2014) Acute and long-term impact of chemical weapons: lessons from the Iran-Iraq war. Forensic Sci Rev 26:97–114
22. Fu D, Calvo JA, Samson LD (2012) Balancing repair and tolerance of DNA damage caused by alkylating agents. Nat Rev Cancer 12:104–120
23. Roshan R, Rahnama P, Ghazanfari Z, Montazeri A, Soroush MR, Naghizadeh MM, Melyani M, Tavoli A, Ghazanfari T (2013) Long-term effects of sulfur mustard on civilians' mental health 20 years after exposure (The Sardasht-Iran Cohort Study). Health Qual Life Outcomes 11:69–76
24. Zhang SW, Swager TM (2003) Fluorescent detection of chemical warfare agents: Functional group specific ratiometric chemosensors. JACS 125:3420–3421
25. Burnworth M, Rowan SJ, Weder C (2007) Fluorescent sensors for the detection of chemical warfare agents. Chemistry: A Eur J 13:7828–7836

26. McGill RA, Nguyen VK, Chung R, Shaffer RE, DiLella D, Stepnowski D, Melsna TE, Venezky DL, Dominguez D (2000) The "NRL-SAWRHINO": a nose for toxic gases. Sens Actuat B 65:10–13
27. Singh VV, Nigam AK, Boopathi M, Pandey P, Singh B, Vijayaraghavan R (2012) In situ electrochemical sensing of 2-chloroethyl ethyl sulphide a CWWA simulant using CuPc/RTIL composite gold electrode. Sens Actuat B 161:1000–1009
28. Singh VV, Nigam AK, Yadav SS, Tripathi BK, Srivastava A, Boopathi M, Singh B (2013) Graphene oxide as carboelectrocatalyst for in situ electrochemical oxidation and sensing of chemical warfare agent simulant. Sens Actuat B 188:1218–1224
29. Arduini F, Scognamiglio V, Covaia C, Amine A, Moscone D, Palleschi G (2015) A choline oxidase amperometric bioassay for the detection of mustard agents based on screen-printed electrodes modified with Prussian blue nanoparticles. Sensors 15:4353–4367
30. Lindsay AE, Greenbaum AR, O'Hare D (2004), Analytical techniques for cyanide in blood and published blood cyanide concentrations from healthy subjects and fire victims. Anal Chim Acta 511:185–195
31. http://www.atsdr.cdc.gov/toxprofiles/tp.asp?id=72&tid=19. Accessed 31 Oct 2015
32. Abbaspour A, Asadi M, Ghaffarinejad A, Safaei E (2005) A selective modified carbon paste electrode for determination of cyanide using tetra-3, 4-pyridinoporphyrazinatocobalt (II). Talanta 66:931–936
33. Taheri A, Noroozifar M, Khorasani-Motlagh M (2009) Investigation of a new electrochemical cyanide sensor based on Ag nanoparticles embedded in a three-dimensional sol–gel. J Electroanal Chem 628:48–54
34. Zacharis CK, Tzanavaras PD, Voulgaropoulos AN, Karlberg B (2009) Amperometric determination of cyanides at the low ppb level by automated preconcentration based on gas diffusion coupled to sequential injection analysis. Talanta 77:1620–1626
35. http://www.globalsecurity.org/wmd/intro/cw-choking.htm. Accessed 31 Oct 2015
36. Feng D, Zhang Y, Shi W, Li X, Ma H (2010) A simple and sensitive method for visual detection of phosgene based on the aggregation of gold nanoparticles. Chem Commun 46:9203–9205
37. Virji S, Kojima R, Fowler JD, Villanueva JG, Kaner RB, Weiller BH (2009) Polyaniline nanofiber composites with amines: Novel materials for phosgene detection. Nano Res 2:135–142
38. Savage AC, Buckley N, Halliwell J, Gwenin C (2015) Botulinum neurotoxin serotypes detected by electrochemical impedance spectroscopy. Toxins 7:1544–1555
39. Chan CY, Guo J, Sun C, Tsang MK, Tian F, Hao J, Chen S, Yang M (2015) A reduced graphene oxide-Au based electrochemical biosensor for ultrasensitive detection of enzymatic activity of botulinum neurotoxin A. Sens Actuat B 220:131–137
40. Halliwell J, Savage AC, Buckley N, Gwenin C (2014) Electrochemical impedance spectroscopy biosensor for detection of active botulinum neurotoxin. Sens Bio-Sens Res 2:12–15
41. Cunningham JC, Scida K, Kogan MR, Wang B, Ellington AD, Crooks RM (2015) Paper diagnostic device for quantitative electrochemical detection of ricin at picomolar levels. Lab Chip 15:3707–3715
42. Suresh S, Gupta M, Kumar GA, Rao VK, Kumar O, Ghosal P (2012) Synergic effect of multi-walled carbon nanotubes and gold nanoparticles towards immunosensing of ricin with carbon nanotube–gold nanoparticles–chitosan modified screen printed electrode. Analyst 137:4086–4092
43. Stevens RC, Soelberg SD, Eberhart B-TL, Spencer S, Wekell JC, Chinowsky TM et al (2007) Detection of the toxin domoic acid from clam extracts using a portable surface plasmon resonance biosensor. Harmful Algae 6:166–174
44. Micheli L, Radoi A, Guarrina R, Massaud R, Bala C, Moscone D, Palleschi G (2004) Disposable immunosensor for the determination of domoic acid in shellfish. Biosens Bioelectron 20:190–196

with other species of bacteria, e.g., *V. cholera*, these microorganisms are sources of serious waterborne diseases.

Different analytical methods are available for continuous monitoring of microorganism contaminants in water and food products [1], however, most of them use procedures involving culturing of these microorganisms. Unfortunately, this culturing is time consuming, usually taking 24–48 h and rather expensive.

Importantly, short determination time and easy operation with minimum interference of close analogs of analytes are two of many important requirements for successful application of chemosensors in a real world analysis of contaminants. For that, chemosensors using electrochemical [2] and electrical [3] transductions offer easy operation towards determination of target analytes. Additionally, possibility of miniaturization, ease of use, and a low cost make these determinations promising. Other than these, several transductions are used to develop sensing platforms for on-line monitoring of microorganisms. These include piezoelectric microgravimetry (PM) [4, 5] and surface plasmon resonance (SPR) [6, 7]. Unfortunately, selectivity of chemosensors using these transductions without any recognition is low.

To incur desired selectivity in chemosensing, surface of transducers is modified mostly with polymer based recognition units [8, 9]. Typically, these units, fabricated in the form of thin films, are deposited directly on the transducer surface. Detectability and selectivity of these polymer-based chemosensors are appreciable.

Selective receptors using target analytes as templates have become more and more popular [9–13]. These receptors often reveal stability constants of formation of complexes with analytes similar to those of biological receptors [14]. Generally, the procedure of molecular imprinting involves polymerization of functional monomers with cross-linking monomers in the presence of a template, which is most often the analyte itself (Scheme 1). Subsequent removal of the template leaves in the resulting molecularly imprinted polymer (MIP) molecular cavities complementary in their

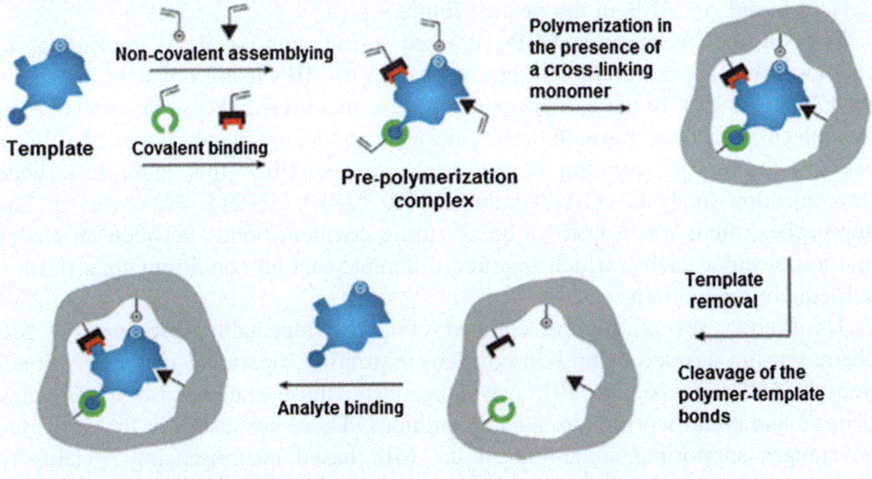

Scheme 1 Consecutive steps of molecular imprinting (Adapted from [13].)

size, shape, and orientation of recognition sites to those of the template molecule. This imprinting generates shape-selective molecular cavities, which allow discriminating between close structural analogues of the analytes.

Three main strategies were developed to design molecular cavities in MIPs, namely, (i) covalent, (ii) non-covalent, and (iii) semi-covalent imprinting [15].

In covalent imprinting, templates are covalently bound to functional monomers during MIP preparation [16]. After polymerization, the template is removed by bond cleavage and the recognition sites of the molecular cavities left are capable of binding the target analyte molecule by re-establishing the same covalent bonds. Unfortunately, this imprinting involves derivatization of the template molecule, which adds an additional preparation step.

In non-covalent imprinting [17–19], interactions between functional monomers and the template during polymerization, including hydrogen bonding, ion-pairing as well as dipole-dipole and hydrophobic interactions, are reproduced in subsequent analyte binding by the template-free molecularly imprinted cavities of an MIP. Due to its simplicity, this strategy is most commonly used.

Semi-covalent imprinting combines advantages of both covalent and non-covalent imprinting [20–22]. In this strategy, initially, a template is trapped in the polymer matrix by covalent bonds. Then, after template removal, the analyte is bound in a molecular cavity through non-covalent interactions.

Application of molecular imprinting for devising recognition units of chemical sensors [10] and stationary phase materials for solid-phase extraction (SPE) [23–25] appears promising for MIP commercialization. Now, MIP based SPE sorption materials for extraction of traces of toxins, such as patulin, aminoglycosides, and bisphenol A, are commercially available. Unfortunately, despite variety of selective chemosensors devised in last decades [9, 10, 15], no chemosensor based on MIP is available on the market yet. Although sensing results presented in all reports cited above are very promising, the chemosensors described seek much improvement before commercialization. Undoubtedly, we will see first commercial chemosensing devices based on MIPs in the nearest future.

For on-line measuring, MIPs devised using non-covalent imprinting for chemosensing appeared efficient in comparison to MIPs using covalent imprinting [8, 17–19, 26–29]. In the earlier approaches, as mentioned above, an analyte binds in molecular cavities through weak and easy to disrupt interactions. Mostly, an excess of a carrier solution is sufficient to accomplish this latter task under flow-injection analysis (FIA) conditions [8, 17–19, 26–29]. However, in later approaches, there was a need to break strong covalent bonds between an analyte and a molecular cavity, which required different reaction conditions than those of subsequent recognition.

Until now, several reports claimed obtaining repeatable responses of MIP chemosensors devised using non-covalent imprinting, operating under FIA conditions [10, 17, 18, 26, 28, 29]. Advantageously, these chemosensors were inexpensive and could work under harsh conditions. These are just only few important advantages supporting suitability of the MIP based chemosensing systems for on-line analyte detection.

Table 1 Analytical parameters of MIPs devised for determination of bacteria

Bacteria species	Functional monomer/cross-linker	Method of imprinting	Template extraction with	Determination or recognition technique	Comment	Ref.
Eduscho grande	Acrylic acid/divinylbenzene	Stamping	Hot water	PM	LOD = 10^4 cells mL^{-1}	[40]
S. cerevisiae	Polyurethane	Stamping	Hot water	PM		[40]
S. cerevisiae	Polyurethane	Stamping	Hot water	PM	–	[41]
S. cerevisiae	Polyurethane Titanium ethylate/(3-aminopropyl) methyldiethoxysilane	Stamping Stamping	Hot water Hot water	PM PM	LOQ = 10^4–10^9 cells mL^{-1}	[42]
E. coli MRE-600	Acrylamide/*N*,*N′*-methylenebis acrylamide	Bulk imprinting	Lysozyme enzyme/50 mM SDS/50 mM Tris-Cl pH = 8.5	Electrophoresis	Difference in migration time was observed for templated bacteria and other bacterial strains.	[36]
Bacillus thuringiensis kurstaki	Polyamide	Lithography	12 M HCl: methanol (50: 50, *v*: *v*)	Fluorescence	MIP and NIP captured 39 % and 13 % of spores, respectively.	[43]
Bacillus subtilis	Pyrrole	Electropolymerization	DMSO	EIS	LOQ = 10^4–10^7 cfu mL^{-1}	[44]
Deinococcus radiodurans *E. coli* *B. subtilis* *S. natans*	Tetraethoxysilane	Bulk imprinting	Ethanol (96 %)	Fluorescence	LOQ = 10^7–10^9 cfu mL^{-1}	[39]
Pseudomonas aeruginosa	Pyrrole	Electropolymerization	–	Fluorescence	–	[45]
E. coli	Tetraethoxysilane	Bulk imprinting	Ethanol	PM	LOD = 10^2 cfu mL^{-1}	[46]
Sulfate reducing bacteria	Chitosan-reduced graphene	Electropolymerization	Acetone	EIS	LOQ = 10^4–10^8 cfu mL^{-1}	[47]
E. coli *P. aeruginosa* *B. subtilis* *S. cerevisiae*	Pyrrole	Electropolymerization	Electrochemical overoxidation	PM	–	[48]

cfu, colony forming unit; DMSO, dimethylsulfoxide; EIS, electrochemical impedance spectroscopy; LOD, limit of detection; LOQ, limit of quantification; PM, piezoelectric microgravimetry; SDS, sodium dodecyl sulfate.

Table 2 Analytical parameters of MIPs devised for determination of viruses

Virus species	Functional monomer/cross-linker	Method of imprinting	Template extraction with	Determination or recognition technique	Comment	Ref.
TMV	Methacrylic acid/styrene, divinylbenzene	Stamping	Hot water/ 2 % SDS	PM	Only one concentration of virus was measured	[49]
	Polyallylamine hydrochloride	Bulk imprinting	Ethanol (70 %, 24 h)/ 1 mM NaCl (1 h, 100 °C)/ 1 M NaCl (100 °C)/ 3 days in water	UV-vis spectroscopy	MIP binding = 8.8 mg/g, NIP binding = 4.2 mg/g	[34]
	Polyallylamine hydrochloride	Bulk imprinting	1 M NaOH	Fluorescence spectroscopy	MIP binding = 2.73 mg/g, NIP binding = 1.18 mg/g	[30]
	Methacrylic acid/*N*,*N*′-(1,2-dihydroxyethylene) bis-acrylamide/1-vinyl-2-pyrrolidinone	Stamping	–	EIS	–	[50]
	Azobenzene bearing acrylate polymer	–	1 % SDS (24 h)	Chemiluminescence	–	[51]
Dengue	Acrylic acid, acrylamide/*N*-benzylacrylamide	Bulk imprinting	20 mM phosphate buffer (pH = 4)	PM	–	[52]
Influenza A H1N3	Acrylamide, methacrylic acid, methylmethacrylate/*N*,*N*′-(1,2-dihydroxyethylene) bisacrylamide	Stamping	10 % HCl (45 °C, 3 h)	PM	up to 2.5×10^7 particles/mL	[53]
Influenza H5N1	Acrylamide, methacrylic acid/methylmethacrylate	Bulk imprinting	10 % HCl (45 °C, 3 h)	PM	–	[54]

(continued)

Table 2 (continued)

Virus species	Functional monomer/cross-linker	Method of imprinting	Template extraction with	Determination or recognition technique	Comment	Ref.
Human norovirus VLPs	Tetraethylorthosilicate/aminopropyltriethoxysilane	Surface imprinting	0.1 M HCl (with 0.01 % Triton-X)	Indirect ELISA	–	[55]
Apple stem pitting	Acrylamide, *N*-isopropyacrylamide/ *N*,*N*′-methylene bisacrylamide	Bulk imprinting	Ethanol/NaOH	Diffraction grating	LOD = 10 ng mL^{-1}	[38]
Bacteriophage MS2	*N*-isopropylacrylamide, acrylic acid, *N*,*N*′-methylene-bis-(acrylamide), *N*-tetrabutyl acrylamide, *N*-(3-aminopropyl) methacrylamide hydrochloride	Surface imprinting	Water (60 °C)	SPR	LOQ = 0.33–27 pmol	[56]

EIS, electrochemical impedance spectroscopy; ELISA, enzyme-linked immunosorbent assay; PM, piezoelectric microgravimetry; SPR, surface plasmon resonance; TMV, tobacco mosaic virus; VLP, virus-like particle.

The present chapter critically summarizes the available information on the MIP based materials devised so far for microorganism determination in waters and food products (Tables 1 and 2). Moreover, we highlight here with some representative examples of imprinting of microorganisms using different strategies (Tables 1 and 2).

2 Microorganism Imprinting

In comparison to relatively easy imprinting of small molecules, molecular imprinting of whole microorganisms is challenging. Major difficulty in the latter imprinting is a huge size of microorganisms. Therefore, this size restricts their removal from the molecular cavities formed. Then, during recognition, this huge size makes the microorganism diffusion within the polymer matrix slow resulting in a long response time of chemosensors. To cope with these challenges, several strategies have so far been developed. They aim at optimizing MIP preparation procedures and MIP matrix composition to improve the imprinting. For instance, MIPs were prepared with different degree of cross-linking [30] or in a form of nanoparticles with the well-defined surface [31] for easy diffusion of the microorganism analyte to and from molecular cavities. Moreover, on the large surface of a microorganism, density of possibly accessible recognition sites is high. This high density requires either high number of mono-functional monomer molecules or functional monomers with many functionalities for establishing non-covalent interaction equilibria.

The earliest successful attempt of bioanalyte imprinting engaged emulsion polymerization resulting in polyamide beads templated with bacteria on the interface [32, 33] (Scheme 2). On the surface of the resulted beads, there were anisotropic patches of addressable functionalities of exactly the same size and shape as those of the microorganism template. To decrease non-specific binding, the exposed surface was derivatized with diisocyanato-terminated perfluoropolyether. Interestingly, this lithographic procedure did not use any specialized reagents or instrumentation. The SEM imaging of these beads after template removal showed the presence of 100–200 nm deep indentations (Scheme 2). The MIP material obtained based on this procedure was suitable for selective removal of bacteria cells from a test solution.

3 Imprinting of Microorganisms in Bulk Polymer

Another early approach described microorganism imprinting in hydrogel monoliths by bulk polymerization of monomer mixtures containing the microorganism template and a small amount of solvent acting as the porogen [30, 34–36]. Hydrogel is a three-dimensional insoluble polymer network, which can be homogeneously dispersed in water. High hydrophilicity of a hydrogel allows stronger biomolecule accommodating [37]. Moreover, hydrogels are biocompatible. For their preparation,

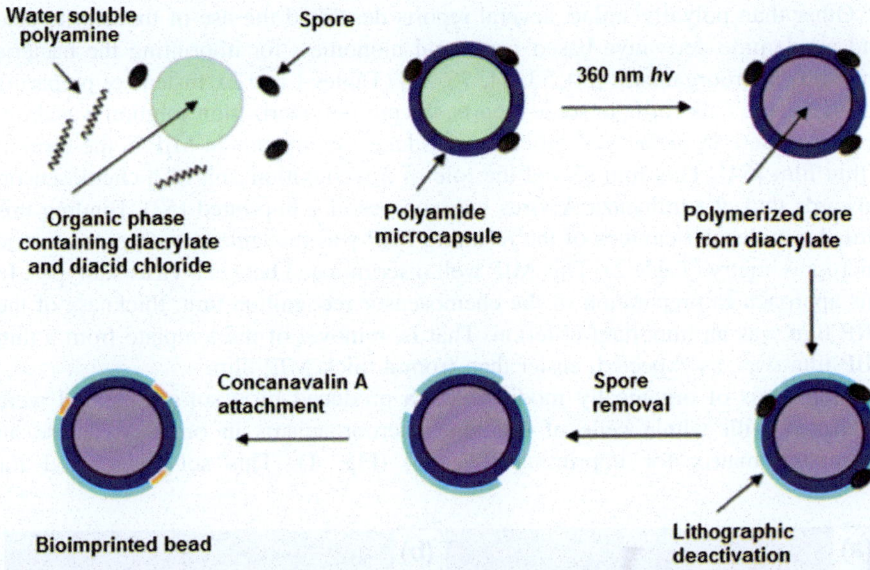

Scheme 2 Basic steps of the lithographic strategy of preparation of bacteria imprinted polyamide beads (Adapted from [32].)

polyallylamine was mostly used as the functional monomer [34, 35] (Table 2). The resulting imprinted polymeric hydrogel monolith was cut in order to produce sufficiently small-size particles with a reasonable fraction of accessible molecularly imprinted sites. Hydrogels of this type were designed to work as "sponges" capable of selective removal of pathogenic microorganisms from blood [34, 35].

For preparation of homogeneous molecularly imprinted cavities, virus aggregation in a pre-polymerization solution was prevented by using an optimized amount of functional monomers. In one example, an MIP hydrogel prepared for recognition of the tobacco mosaic virus (TMV) showed 2.1-fold higher affinity to TMV than the non-imprinted polymeric hydrogel did [34] (Table 2). This MIP hydrogel was suitable for separation purposes but it was not suitable for more advanced applications, such as that of serving as the recognition unit of a chemosensor.

Recently, a hydrogel MIP based chemosensor was fabricated for apple stem pitting virus (ASPV) [38] (Table 2). To provide additional selectivity to the molecular cavities imprinted, a polymerizable aptamer was devised and used as the functional monomer for ASPV imprinting. In the presence of the ASPV template, this monomer was copolymerized with additional functional monomers (*N*-isopropylacrylamide, acrylamide, and *N*,*N*′-methylene bisacrylamide) as well as the cross-linking monomer (*N*,*N*,*N*′,*N*′-tetramethylethylendiamine) to form imprinted cavities featuring multiple recognition sites. This hydrogel shrank in the presence of ASPV. This shrinking linearly depended on the virus concentration in the range of 1.0×10^{-2} to 1.0 μg mL^{-1}.

Other than polyallylamine, several reports described the use of the acrylic acid and acrylamide derivative based functional monomers for imprinting the bacteria and virus microorganisms [50, 53, 54, 56, 57] (Tables 1 and 2). Instead of preparing an MIP in the bulk form, in some reports, the pre-polymerization solution was drop- or spin-coated on surfaces of different transducers to prepare an MIP in the form of a thin film [54]. This film served the role of a recognition unit of a chemosensor. Towards that, the influenza A virus was successfully imprinted [53]. Binding this virus by molecular cavities of the resulting MIP was recognized with piezoelectric microgravimetry (Table 2). This MIP well discriminated between virus subtypes. In this approach to preparation of the chemosensor recognition unit, thickness of the MIP film was an important criterion. That is, removal of the template from a thin MIP film was, as expected, easier than from a thick MIP film.

Thin films of organically modified silica produced by a sol-gel method were imprinted with whole cells of different microorganisms in order to devise an alternative matrix for imprinting [39, 46] (Fig. 1). This sol-gel method for

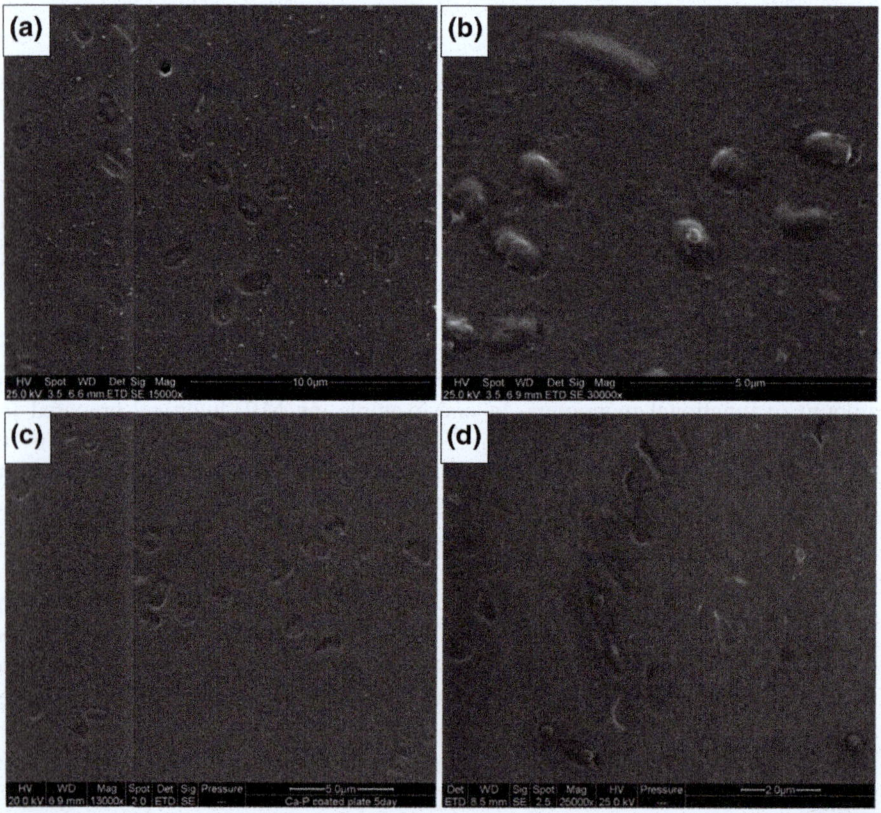

Fig. 1 SEM images of the bacteria-imprinted sol-gel films after removal of bacteria templates (**a**) *B. subtilis*, (**c**) *E. coli* as well as the same films (**b**) and (**d**), respectively, after 30-min exposure to the respective bacteria suspensions (Adapted from [39].)

polycondensation of silanes, used ambient conditions in comparison to those mentioned above of acrylate polymerization. Worth mentioning, an acrylate based functional monomer requires UV light or heat (~60 °C) for polymerization. However, these conditions are destructive for microorganisms. Therefore, several microorganisms including *D. radiodurans*, *E. coli*, *S. natans*, and *B. subtilis* [39, 46] were imprinted in sol-gel matrices (Fig. 1). After removal of the entrapped bacteria template, membrane components of these bacteria were left entrapped in the molecular cavities of the film [39]. These components provided selectivity to the imprinted matrix.

4 Microorganism Imprinting on a Polymer Surface

An interesting strategy involved "stamping technique" to form a polymer film deposited on a transducer surface with well-organized molecular cavities for bulky macromolecules [40–42, 49, 53, 57, 58]. For that, microorganisms were first drop- or spin-coated on glass microscope slides to prepare molecular stamps (Scheme 3). Then, theses stamps were pressed against the pre-polymerized solution coated transducer surfaces. Constant pressing force applied to the resulting sandwich of the stamp-template-sensor film transducer provided good mechanical contact between the polymer and the cell walls of the microorganisms (Scheme 3) [59]. Most often, polyurethane matrix was used for this preparation [40, 41, 53]. Typically, polyurethanes cure overnight under ambient conditions. The microorganism stamp was reasonably easy to remove with hot water leaving behind well-arranged molecular

Scheme 3 Steps of the stamping technique for microorganism imprinting (Adapted from [53].)

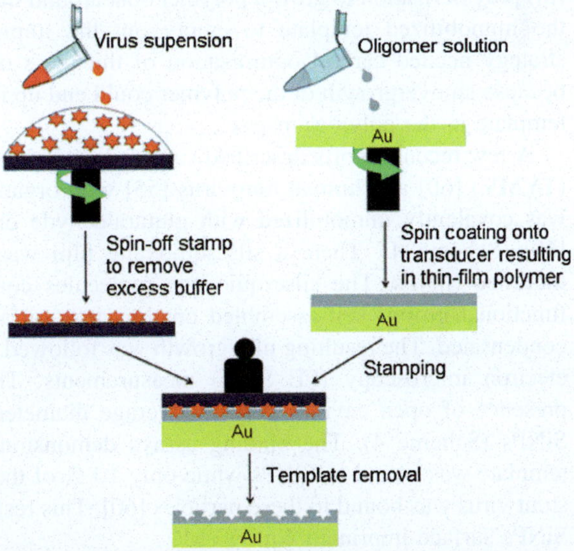

cavities. For improving operation performance, the polymer composition was optimized to increase reversible interactions between the imprinted cavities and the microorganisms. For that, a phenolic functionality was added to the highly cross-linked polymer matrix. Besides, the phenolic group afforded wetting to the polymer surface.

Similarly, microorganism stamps were prepared with silanes. They were produced by the sol-gel method [42]. The resulting relatively rigid matrix was suitable for fabrication of materials to be used in an on-line measuring system. The sol-gel method provided robust MIP films efficiently operating under harsh conditions. Importantly, high affinity towards microorganisms of the imprinted sol-gel films was governed not only by morphology of the imprinted cavity (rod, coccus, or tetrad) but also by binding functional groups present on the outer surface of the bacteria cell. These groups guided orientation of recognizing functionalities in the course of formation of these cavities. In effect, the resulting MIP film was able to discriminate between different species of the bacteria tested despite their similar shape.

The stamping technique was used to form imprints of TMV and human Rhinovirus serotype 2 in a microfluidic device [50] (Fig. 2). This devise contained microinterdigitated capacitor (μIDC) strips coated by MIP films for TMV detection by contact-less dielectric measurements. The virus stamps were prepared on a polyvinylpyrrolidone, PVP, and poly(methacrylic acid), PMA, copolymer.

The results obtained with the surface stamping technique were promising, therefore prompting the necessity of formation of homogenous molecularly imprinted cavities on the polymer surface. For that, an alternative strategy was then developed. This strategy consisted of three distinct steps, which included (i) immobilization of the microorganism template on the polymer surface, (ii) equilibration of the template-modified surface with a mixture of functional monomers, (iii) polymerization to grow a polymer coat around the template, and (iv) removal of the immobilized template to empty out the imprinted cavities. However, this strategy needed careful optimization of thickness of the polymer coat. This was because an overgrowth of the polymer could end up in permanent entrapment of the template in the polymer matrix.

A few recent reports described surface imprinting of turnip yellow mosaic virus (TYMV) [60] and human norovirus [55] with organosilanes. For that, the TYMV was covalently immobilized with glutaraldehyde on silica nanoparticles (SiNPs) [60] (Scheme 4). Then, a silsesquioxane film was grown on top of this virus modified SiNPs. The silsesquioxane molecules containing the −OH and $-NH_2$ functional groups self-assembled on the virus surface, and then they were polycondensated. The resulting film growth was followed by the field emission scanning electron microscopy (FE SEM) measurements. The SEM image revealed the presence of open cavities with an average diameter of 20 nm on the surface of SiNPs (Scheme 4). The binding assays demonstrated that 84 % of the TYMV template was bound to SiNPs while only 10 % of the non-templated tomato bushy stunt virus was bound to these particles [60]. This result confirmed selectivity of the SiNPs surface imprinted with TYMV.

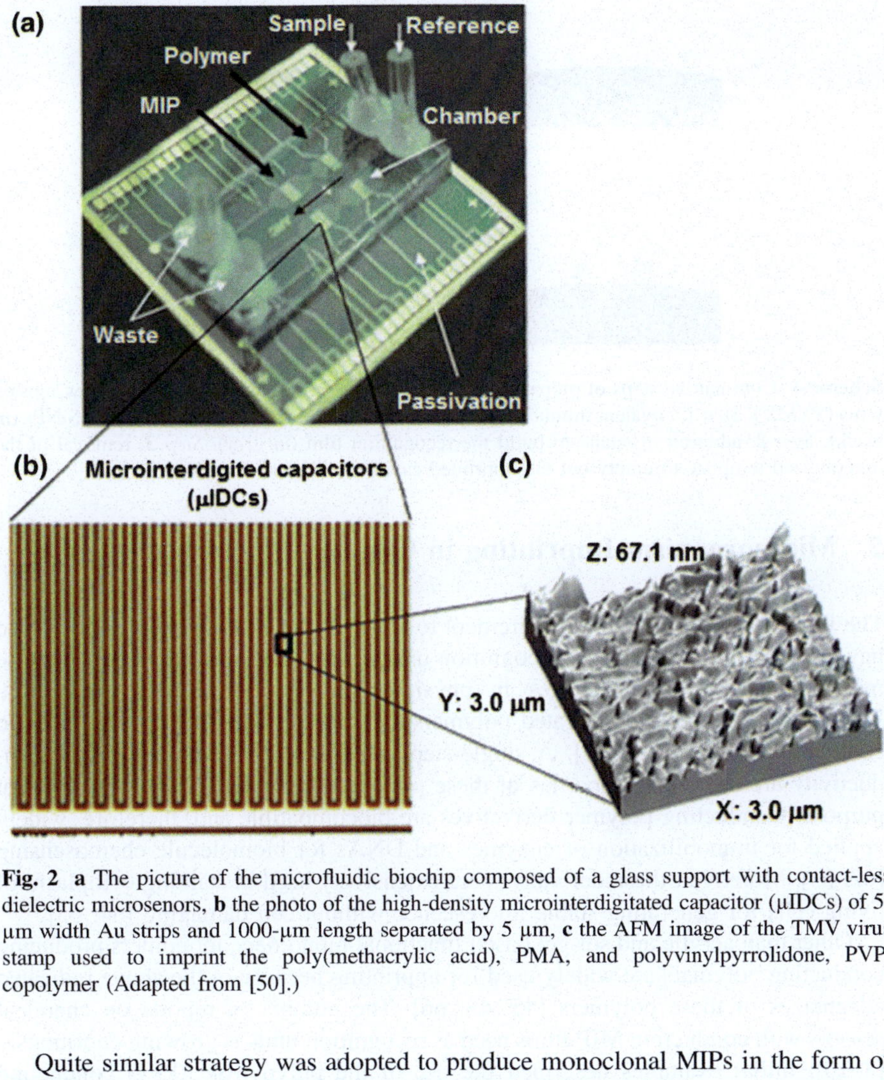

Fig. 2 **a** The picture of the microfluidic biochip composed of a glass support with contact-less dielectric microsenors, **b** the photo of the high-density microinterdigitated capacitor (μIDCs) of 5-μm width Au strips and 1000-μm length separated by 5 μm, **c** the AFM image of the TMV virus stamp used to imprint the poly(methacrylic acid), PMA, and polyvinylpyrrolidone, PVP, copolymer (Adapted from [50].)

Quite similar strategy was adopted to produce monoclonal MIPs in the form of nanoparticles [56]. For synthesis of these MIPs, MS2 bacteriophage was employed as the template. Nanoparticles of these MIPs were prepared using a new automated solid-phase method. That is, the MS2 template was immobilized on the surface of glass beads for imprinting. Then, acrylamide based functional monomers were used to imprint this surface immobilized bacteriophage. A dynamic light scattering (DLS) measurement confirmed the success of the synthesis of nanoparticles with the size of 200–230 nm. These nanoparticles were then immobilized on SPR chips for devising a chemosensor for the MS2 bacteriophage (Table 2). The MIP nanoparticles fabricated that way were able to bind the MS2 bacteriophage reversibly. The mean affinity of the MIP to the MS2 bacteriophage was at the nM level. The limit of detection of the resulting chemosensor was 5×10^6 plaque forming unit mL^{-1}.

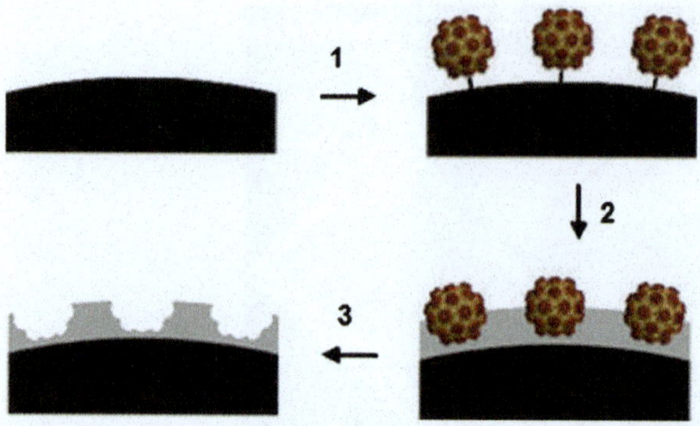

Scheme 4 Consecutive steps of preparation of MIP surface imprinted with turnip yellow mosaic virus (TYMV). *Step 1*, covalent immobilization of the TYMV template on the surface of SiNPs (in black). *Step 2*, addition of silane to build the recognition film (*in grey*). *Step 3*, removal of the immobilized template to empty out the imprinted cavities (Adapted from [60].)

5 Microorganism Imprinting in Conducting Polymers

The first and most important requirement to construct a sensitive chemosensor is to have a close proximity of the recognition unit to the transducer surface for precise measurement of small changes in physico-chemical parameters of this unit. Therefore, conducting conjugated polymers are still more frequently used ones to construct these units [8, 61]. A single-step preparation procedure and high conductivity are desirable properties of these polymers making them suitable for that purpose. Conducting polymer derivatives are biocompatible and, therefore, widely applied for immobilization of enzymes and DNAs for biomolecule chemosensing [62, 63]. Moreover, these derivatives are extensively used in staining cytoplasm of living cells for generating stable fluorescence without cell damaging [64, 65].

Other than acrylic and sol-gel based functional monomers, monomers producing conducting polymers are widely used for imprinting because of the above indicated advantages of these polymers [45, 48, 66]. The number of reports on chemical sensors with conducting MIP films used as recognition units is growing enormously [8–10]. Initial examples described the use of the polypyrrole (PPy) conducting polymer for imprinting of ions [67, 68]. Those studies concluded that ion dopants were incorporated in this conducting polymer to compensate for a cationic charge incurred by the polymer backbone [67, 68]. This feature was then utilized for selective recognition of some inorganic anions [67, 68].

Similarly, efforts were undertaken to imprint microorganisms in conducting polymers [44, 45, 48, 66, 69]. Several reports revealed successful entrapment of bacteria in PPy films without losing their viability [45, 48, 66]. The pK_a values of the *B. Subtilis* bacteria species used as the template for imprinting were 4.8, 6.9, and 9.4, corresponding to the carboxy, phosphate, and hydroxy functional surface groups, respectively. These values confirmed that anionic groups dominate on the

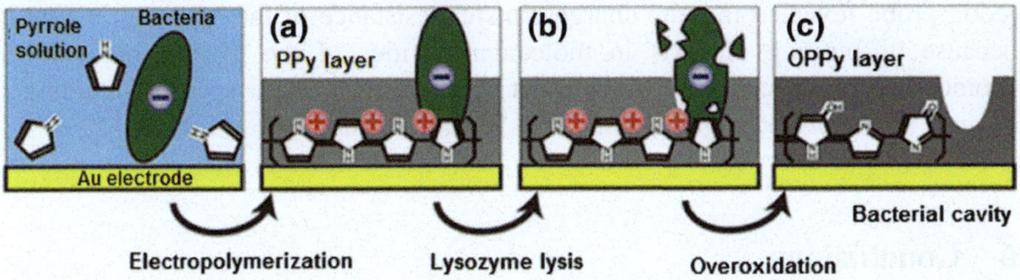

Scheme 5 Consecutive steps of entrapment of the bacteria template in a conducting polymer during polymerization; OPPy stands for overoxidized polypyrrole (Adapted from [48].)

microorganism surface in neutral solutions. Therefore, direct insertion of bacteria in a positively charged conducting polymer was possible (Scheme 5). Further fluorescence microscopy imaging proved entrapment of living bacteria in the polymer matrix [66].

Application of conducting polymers for microorganism imprinting solved one of the most important MIP issues, i.e., the problem of incomplete removal of templates from molecularly imprinted cavities. That is, the microorganism templates were electrostatically ejected from the conducting polymer matrix simply by electrochemical overoxidation [48]. Moreover, verification of template removal in many instances was quite easy [8, 10]. Importantly, this overoxidation treatment preserved the cavity shape in the polymer matrix. Toward that, the PM [48] and CV [44] based chemosensors were prepared for determination of the *E. coli* [48] and *Bacillus subtilis* [44] spores, respectively.

Now, deposition of conducting MIP films by electropolymerization is the most common procedure suitable for these films directly growing on the surface of transducers [10]. Therefore, this procedure was further extended to deposit biocompatible chitosan for imprinting of the sulfate-reducing bacteria, SRB, (Fig. 3) [47] (Table 1). Faradic impedance measurements in the presence of the $Fe(CN)_6^{4-/3-}$

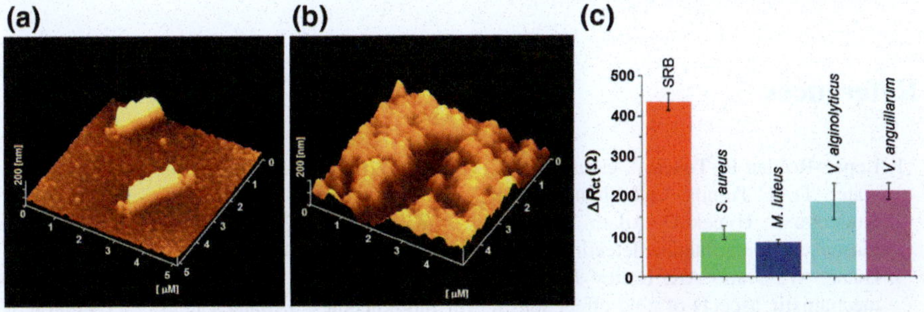

Fig. 3 AFM images of the polymer film imprinted with sulfate-reducing bacteria, SRB **a** before and **b** after the SRB template removal, **c** the change in the charge transfer resistance of the impedimetric chemosensor before and after immobilization of different bacteria species (Adapted from [47].)

redox probe revealed that the charge transfer resistance of the MIP film increased because of binding of SRB in molecular cavities of the MIP. Moreover, the chemosensor was selective with respect to different bacteria species including *S. aureus*, *M. luteus*, *V. alginolyticus*, and *V. anguillarum*.

6 Conclusions

After decades of improvement, several contemporary analytical devices have revealed the capability of on-line detection of the microorganism contamination of foods, waters, and environmental samples. To a large extent, application of synthetic polymers as recognition units in these devices has provided the much needed selectivity with respect to microorganism determination. Selective cavities generated in these polymers can distinguish different species of bacteria and viruses. Additionally, development of different procedures of MIP synthesis has allowed improving the MIP performance. Tables 1 and 2 summarize figures of merit of majority of the procedures developed to-date in this active area of research.

Undoubtedly, microorganisms are entrapped in the imprinted polymer matrices as templates. However, subsequent removal of these templates is still challenging, especially from the polymer bulk. This removal is slow and requires application of rather drastic conditions (Tables 1 and 2). Application of these conditions limits the lifetime of MIPs. In contrast, template removal from a surface imprinted polymer is easier and much faster (Table 1). Therefore, application of surface imprinting for devising chemosensors for determination of microorganisms is increasing. These chemosensors appeared to be more robust than the relevant biosensors and chemosensing less time consuming than the ELISA test.

Acknowledgments We thank the Polish National Science Centre, NCN (Grant Nos. 2013/11/N/ST5/01907 to Z.I., and 2014/15/B/NZ7/01011 to W.K.), and the Ministry of Science and Higher Education of Poland (Grant No. 0005/E-64/9/2014, and IP 2014-041473 to P.S.S.) and the U.S. National Science Foundation (Grant No. CHE-1401188 to F.D.).

References

1. Lopez-Roldan R, Tusell P, Courtois S, Cortina JL (2013) On-line bacteriological detection in water. TrAC Trends Anal Chem 44:46–57
2. Krejcova L, Hynek D, Adam V, Hubalek J, Kizek R (2012) Electrochemical sensors and biosensors for influenza detection. Int J Electrochem Sci 7:10779–10801
3. Huang W, Diallo AK, Dailey JL, Besar K, Katz HE (2015) Electrochemical processes and mechanistic aspects of field-effect sensors for biomolecules. J Mater Chem C 3:6445–6470
4. Farka Z, Kovář D, Skládal P (2015) Rapid detection of microorganisms based on active and passive modes of QCM. Sensors 15:79–92
5. Szalontai H, Adanyi N, Kiss A (2014) Comparative determination of two probiotics by QCM and OWLS-based immunosensors. New Biotechnol 31:395–401

6. Brogioni B, Berti F (2014) Surface plasmon resonance for the characterization of bacterial polysaccharide antigens: a review. Med Chem Commun 5:1058–1066
7. Dudak FC, Boyacı IH (2009) Rapid and label-free bacteria detection by surface plasmon resonance (SPR) biosensors. Biotechnol J 4:1003–1011
8. Huynh T-P, Sharma PS, Sosnowska M, D'Souza F, Kutner W (2015) Functionalized polythiophenes: recognition materials forchemosensors and biosensors of superior sensitivity, selectivity, and detectability. Prog Polym Sci 47:1–25
9. Sharma PS, Iskierko Z, Pietrzyk-Le A, D'Souza F, Kutner W (2015) Bioinspired intelligent molecularly imprinted polymers for chemosensing: a mini review. Electrochem Commun 50:81–87
10. Sharma PS, Pietrzyk-Le A, D'Souza F, Kutner W (2012) Electrochemically synthesized polymers in molecular imprinting for chemical sensing. Anal Bioanal Chem 402:3177–3204
11. Schirhagl R (2014) Bioapplications for molecularly imprinted polymers. Anal Chem 86:250–261
12. Huang D-L, Wang R-Z, Liu Y-G, Zeng G-M, Lai C, Xu P, Lu B-A, Xu J-J, Wang C, Huang C (2015) Application of molecularly imprinted polymers in wastewater treatment: a review. Environ Sci Pollut Res 22:963–977
13. Alexander C, Andersson HS, Andersson LI, Ansell RJ, Kirsch N, Nicholls IA, O'Mahony J, Whitcombe MJ (2006) Molecular imprinting science and technology: a survey of the literature for the years up to and including 2003. J Mol Recognit 19:106–180
14. Zhang H (2014) Water-compatible molecularly imprinted polymers: promising synthetic substitutes for biological receptors. Polymer 55:699–714
15. Whitcombe MJ, Kirsch N, Nicholls IA (2014) Molecular imprinting science and technology: a survey of the literature for the years 2004–2011. J Mol Recognit 27:297–401
16. Wulff G (1995) Molecular imprinting in cross-linked materials with the aid of molecular templates—a way towards artificial antibodies. Angew Chem Int Ed 34:1812–1832
17. Huynh T-P, Pietrzyk-Le A, KC CB, Noworyta KR, Sobczak JW, Sharma PS, D'Souza F, Kutner W (2013) Electrochemically synthesized molecularly imprinted polymer of thiophene derivatives for flow-injection analysis determination of adenosine-5′-triphosphate (ATP). Biosens Bioelectron 41:634–641
18. Pietrzyk A, Kutner W, Chitta R, Zandler ME, D'Souza F, Sannicolo F, Mussini PR (2009) Melamine acoustic chemosensor based on molecularly imprinted polymer film. Anal Chem 81:10061–10070
19. Sharma PS, Dabrowski M, Noworyta K, Huynh T-P, KC CB, Sobczak JW, Pieta P, D'Souza F, Kutner W (2014) Fullerene derived molecularly imprinted polymer for chemosensing of adenosine-50-triphosphate (ATP). Anal Chim Acta 844:61–69
20. Tang Y-W, Fang G-Z, Wang S, Li J-L (2011) Covalent imprinted polymer for selective and rapid enrichment of ractopamine by a noncovalent approach. Anal Bioanal Chem 401:2275–2282
21. Hashim SNNS, Boysen RI, Schwarz LJ, Danylec B, Hearn MTW (2014) A comparison of covalent and non-covalent imprinting strategies for the synthesis of stigmasterol imprinted polymers. J Chromatogr A 1359:35–43
22. Qi P, Wang J, Wang L, Li Y, Jin J, Su F, Tian Y, Chen J (2010) Molecularly imprinted polymers synthesized via semi-covalent imprinting with sacrificial spacer for imprinting phenols. Polymer 51:5417–5423
23. Cheong WJ, Yang SH, Ali F (2013) Molecular imprinted polymers for separation science: a review of reviews. J Sep Sci 36:609–628
24. Dias ACB, Figueiredo EC, Grassi V, Zagatto EAG, Arruda MAZ (2008) Molecularly imprinted polymer as a solid phase extractor in flow analysis. Talanta 76:988–996
25. Toth B, Horvai G (2012) Chromatography, solid-phase extraction, and capillary electrochromatography with MIPs. Top Curr Chem 325:267–306
26. Huynh T-P, Pieta P, D'Souza F, Kutner W (2013) Molecularly imprinted polymer for recognition of 5-fluorouracil by RNA-type nucleobase pairing. Anal Chem 85:8304–8312

27. Pietrzyk A, Suriyanarayanan S, Kutner W, Chitta R, Zandler ME, D'Souza F (2010) Molecularly imprinted polymer (MIP) based piezoelectric microgravimetry chemosensor for selective determination of adenine. Biosens Bioelectron 25:2522–2529
28. Pietrzyk A, Suriyanarayanan S, Kutner W, Chitta R, D'Souza F (2009) Selective histamine piezoelectric chemosensor using a recognition film of the molecularly imprinted polymer of bis(bithiophene) derivatives. Anal Chem 81:2633–2643
29. Pietrzyk A, Suriyanarayanan S, Kutner W, Maligaspe E, Zandler ME, D'Souza F (2010) Molecularly imprinted poly[bis(2,2′-bithienyl)methane] film with built-in molecular recognition sites for a piezoelectric microgravimetry chemosensor for selective determination of dopamine. Bioelectrochemistry 80:62–72
30. Bolisay LD, Culver JN, Kofinas P (2007) Optimization of virus imprinting methods to improve selectivity and reduce nonspecific binding. Biomacromolecules 8:3893–3899
31. Sankarakumar N, Tong YW (2013) Preventing viral infections with polymeric virus catchers: a novel nanotechnological approach to anti-viral therapy. J Mater Chem B 1:2031–2037
32. Aherne A, Alexander C, Payne MJ, Perez N, Vulfson EN (1996) Bacteria-mediated lithography of polymer surfaces. J Am Chem Soc 118:8771–8772
33. Alexander C, Vulfson EN (1997) Spatially functionalized polymer surfaces produced via cell-mediated lithography. Adv Mater 9:751–755
34. Bolisay LD, Culver JN, Kofinas P (2006) Molecularly imprinted polymers for tobacco mosaic virus recognition. Biomaterials 27:4165–4168
35. Bolisay LD, Kofinas P (2010) Imprinted polymer hydrogels for the separation of viruses. Macromol Symp 291–292:302–306
36. Bacskay I, Takátsy A, Végvári Á, Elfwing A, Ballagi-Pordány A, Kilár F, Hjertén S (2006) Universal method for synthesis of artificial gel antibodies by the imprinting approach combined with a unique electrophoresis technique for detection of minute structural differences of proteins, viruses, and cells (bacteria). III: gel antibodies against cells (bacteria). Electrophoresis 27:4682–4687
37. Wang R, Li Y (2013) Hydrogel based QCM aptasensor for detection of avianin fluenza virus. Biosens Bioelectron 42:148–155
38. Bai W, Spivak DA (2014) A double-imprinted diffraction-grating sensor based on a virus-responsive super-aptamer hydrogel derived from an impure extract. Angew Chem Int Ed 53:2095–2098
39. Cohen T, Starosvetsky J, Cheruti U, Armon R (2010) Whole cell imprinting in sol-gel thin films for bacterial recognition in liquids: macromolecular fingerprinting. Int J Mol Sci 11:1236–1252
40. Dickert FL, Hayden O, Halikias KP (2001) Synthetic receptors as sensor coatings for molecules and living cells. Analyst 126:766–771
41. Hayden O, Dickert FL (2001) Selective microorganism detection with cell surface imprinted polymers. Adv Mater 13:1480–1483
42. Dickert FL, Hayden O (2002) Bioimprinting of polymers and sol-gel phases. Selective detection of yeasts with imprinted polymers. Anal Chem 74:1302–1306
43. Harvey SD, Mong GM, Ozanich RM, Mclean JS, Goodwin SM, Valentine NB, Fredrickson JK (2006) Preparation and evaluation of spore-specific affinity-augmented bio-imprinted beads. Anal Bioanal Chem 386:211–219
44. Namvar A, Warriner K (2007) Microbial imprinted polypyrrole/poly(3-methylthiophene) composite films for the detection of Bacillus endospores. Biosens Bioelectron 22:2018–2024
45. Tokonami S, Saimatsu K, Nakadoi Y, Furuta M, Shiigi H, Nagaoka T (2012) Vertical immobilization of viable bacilliform bacteria into polypyrrole films. Anal Sci 28:319–321
46. Starosvetsky J, Cohen T, Cheruti U, Bilanović D, Armon R (2012) Effects of physical parameters on bacterial cell adsorption onto pre-imprinted sol-gel films. J Biomater Nanobiotechnol 3:499–507

47. Qi P, Wan Y, Zhang D (2013) Impedimetric biosensor based on cell-mediated bioimprinted films for bacterial detection. Biosens Bioelectron 39:282–288
48. Tokonami S, Nakadoi Y, Nakata H, Takami S, Kadoma T, Shiigi H, Nagaoka T (2014) Recognition of gram-negative and gram-positive bacteria with a functionalized conducting polymer film. Res Chem Intermed 40:2327–2335
49. Hayden O, Bindeus R, Haderspock C, Mann K-J, Wirl B, Dickert FL (2003) Mass-sensitive detection of cells, viruses, and enzymes with artificial receptors. Sens Actuator B 91:316–319
50. Birnbaumer GM, Lieberzeit PA, Richter L, Schirhagl R, Milnera M, Dickert FL, Bailey A, Ertl P (2009) Detection of viruses with molecularly imprinted polymers integrated on a microfluidic biochip using contact-less dielectric microsensors. Lab Chip 9:3549–3556
51. Ikawa T, Kato Y, Yamada T, Shiozawa M, Narita M, Mouri M, Hoshino F, Watanabe O, Tawata M, Shimoyama H (2010) Virus-templated photoimprint on the surface of an azobenzene-containing polymer. Langmuir 26:12673–12679
52. Tai D-F, Lin C-Y, Wu T-Z, Huang J-H, Shu P-Y (2006) Artificial receptors in serologic tests for the early diagnosis of dengue virus infection. Clin Chem 52:1486–1491
53. Wangchareansak T, Thitithanyanont A, Chuakheaw D, Gleeson MP, Lieberzeit PA, Sangma C (2013) Influenza A virus molecularly imprinted polymers and their application in virus sub-type classification. J Mater Chem B 1:2190–2197
54. Wangchareansak T, Thitithanyanont A, Chuakheaw D, Gleeson MP, Lieberzeit PA, Sangma C (2014) A novel approach to identify molecular binding to the influenza virus H5N1: screening using molecularly imprinted polymers (MIPs). Med Chem Commun 5:617–621
55. Sykora S, Cumbo A, Belliot G, Pothier P, Arnal C, Dudal Y, Corvinia PF-X, Shahgaldian P (2015) Virus-like particles as virus substitutes to design artificial virus-recognition nanomaterials. Chem Commun 51:2256–2258
56. Altintas Z, Gittens M, Guerreiro A, Thompson K-A, Walker J, Piletsky S, Tothill IE (2015) Detection of waterborne viruses using high affinity molecularly imprinted polymers. Anal Chem 87:6801–6807
57. Dickert FL, Hayden O, Bindeus R, Mann K-J, Blaas D, Waigmann E (2004) Bioimprinted QCM sensors for virus detection–screening of plant sap. Anal Bioanal Chem 378:1929–1934
58. Dickert FL, Lieberzeit P, Miarecka SG, Mann KJ, Hayden O, Palfinger C (2004) Synthetic receptors for chemical sensors-subnano- and micrometre patterning by imprinting techniques. Biosens Bioelectron 20:1040–1044
59. Hayden O, Lieberzeit PA, Blaas D, Dickert FL (2006) Artificial antibodies for bioanalyte detection-sensing viruses and proteins. Adv Funct Mater 16:1269–1278
60. Cumbo A, Lorber B, Corvini PF-X, Meier W, Shahgaldian P (2013) A synthetic nanomaterial for virus recognition produced by surface imprinting. Nat Commun 4:1503
61. Ates M (2013) A review study of (bio) sensor systems based on conducting polymers. Mater Sci Eng C 33:1853–1859
62. Gerard M, Chaubey A, Malhotra BD (2002) Application of conducting polymers to biosensors. Biosens Bioelectron 17:345–359
63. Guimard NK, Gomez N, Schmidt CE (2007) Conducting polymers in biomedical engineering. Prog Polym Sci 32:876–921
64. Guimard NKE, Sessler JL, Schmidt CE (2009) Towards a biocompatible, biodegradable copolymer incorporating electroactive oligothiophene units. Macromolecules 42:502–511
65. Palamà I, Maria FD, Viola I, Fabiano E, Gigli G, Bettini C, Barbarella G (2011) Live-cell-permeant thiophene fluorophores and cell-mediated formation of fluorescent fibrils. J Am Chem Soc 133:17777–17785
66. Nagaoka T, Shiigi H, Tokonami S, Saimatsu K (2012) Entrapment of whole cell bacteria into conducting polymers. J Flow Injection Anal 29:7–10

67. Chen Z, Takei Y, Deore BA, Nagaoka T (2000) Enantioselective uptake of amino acid with overoxidized polypyrrole colloid templated with L-lactate. Analyst 125:2249–2254
68. Okuno H, Kitano T, Yakabe H, Kishimoto M, Deore BA, Siigi H, Nagaoka T (2002) Characterization of overoxidized polypyrrole colloids imprinted with L-lactate and their application to enantioseparation of amino acids. Anal Chem 74:4184–4190
69. Ramanaviciene A, Ramanavicius A (2004) Molecularly imprinted polypyrrole-based synthetic receptor for direct detection of bovine leukemia virus glycoproteins. Biosens Bioelectron 20:1076–1082

Electrochemical DNA Biosensors for Bioterrorism Prevention

Hafsa Korri-Youssoufi, Anna Miodek and Wadih Ghattas

Abstract In the wake of letters containing anthrax spores terrifying the USA and other letters containing unidentified white powders circulating all over the world, the threat of bioterrorism attracts the attention of the general public as well as scientist. Therefore, it is urgent to develop rapid, sensitive, and high-throughput diagnostic methods able to counter attacks of bioterrorism by elucidating the suitable actions that should be implemented to prevent serious epidemic diseases. Numerous such methods are in development but Nucleic Acid Detection is the standard employed for identifying most biological agents that are used in bioterrorism. This method is based on PCR assays *via* the classical techniques of amplification and fluorescent detection. On the other hand, electrochemical biosensors are promising platforms that could achieve rapid highly sensitive and selective onsite detection of such agents. This chapter will present the recent developments in electrochemical biosensors for preparing DNA detection platforms that could be used to prevent attacks of bioterrorism.

Keywords Bioterrorism · Bio-agents · DNA · Biosensors · Electrochemical

1 Introduction

Biological agents (bio-agents) were used as weapons of war for many centuries but more recently the threat of bioterrorism has attracted great attention because of the letters containing anthrax spores, which terrified the USA and the other letters containing unidentified white powders, which circulated all over the world [1]. The Center for Disease Control and Prevention (CDC) classifies bio-agents based on their potential risk and those that can be used in attacks of bioterrorism are found mainly in three classes (A, B and C). More than 160 species of microorganisms

H. Korri-Youssoufi (✉) · A. Miodek · W. Ghattas
ICMMO, UMR-CNRS 8182, Université Paris Saclay, ECBB, Bat 420,
91405 Orsay, France
e-mail: hafsa.korri-Youssoufi@u-psud.fr

D.P. Nikolelis and G.-P. Nikoleli (eds.), *Biosensors for Security and Bioterrorism Applications*, Advanced Sciences and Technologies for Security Applications,
DOI 10.1007/978-3-319-28926-7_8

have been recognized as pathogenic of which, thirty could be used in bioweapons. Examples of such bioagents include *Bacillus anthracis*, *Yersinia pestis*, *Brucella spp.*, *Francisella tularensis*, *Burkholderia pseudomallei* and *Clostridium botulinum*. More importantly, some of these possess the characteristics that make them ideal candidates for preparing attracts of bioterrorism. These characteristics are the eased availability, production, storage and dissemination as well as the high virulence, infectivity and lethality. Additionally, bio-agents that can infect *via* the respiratory route by inhalation of aerosols are favored for bioterrorism but other possible route of infection, such as digestive contaminations (ingestion of contaminated water or food) and percutaneous contaminations could also be exploited.

The rapid detection and identification of the threatening bio-agents are crucial to counter attacks of bioterrorism by elucidating the suitable actions that should be implemented to disinfect pollutants and cure infected individuals. Therefore, it is urgent to develop rapid, sensitive, and high-throughput diagnostic methods able to tackle bioterrorism and prevent serious epidemic diseases. Furthermore, the upmost advantage is to develop a portable and user-friendly instrument capable of onsite simultaneous identification of multiple bio-agents. Numerous methods have been used for the detection and identification of bio-agents but DNA detection is the hallmark method of accurate identification of most specimens. It is based on PCR assays through classical techniques of amplification and fluorescence detection. New advanced DNA sensing technologies are developed by using a recognition system paired with a transducer that transforms the recognition into an analyzable signal. Electrochemical biosensors are promising platforms that achieve highly sensitive and selective detection of bio-agent. This chapter will present the recent developments on electrochemical biosensors for DNA detection platforms that combine a biological recognition system with artificial transducers.

2 Bioterrorism Agents

2.1 Brief History of Bioterrorism

The use of biological agents as weapons of war has marked the history of international conflicts. Historians agree that in 1346 the agent of plague *Yersinia pestis* has been unintentionally employed by the Genoese Tatars during the siege of the city of Caffa. The Tatars smuggled the bodies of their contaminated dead comrades inside the city walls, hopping to further grime the life of the besieged citizens and accelerate their surrender. Their act resulted in a plague-ridden city, which was abandoned by survivors most of whom traveled across Europe spreading plague and causing one of the origins of Black Death that killed 20 to 30 million Europeans [2]. In the 20th century, the field of biological weapons was significantly advanced by modern microbiology and by the knowledge gathered from multiple applications that took place in international and civil wars. For example, during the First World

War, German agents used anthrax to infect the animals of the allied powers and in the Second World War as well as in the Cold War several countries had secret programs for biological weapons development.

In 1972, the Biological Weapons Convention prohibited the development, production and stockpiling of biological weapons. 173 countries have ratified this convention but unfortunately this did not stop isolated attacks of bioterrorism. Examples such as the intentional contamination by *Salmonella typhimurium* of salad bars in Oregon restaurants by some of the Rajneeshee cult followers in 1984 [3] and the release of nerve gas sarin in a Tokyo subway by some of the Aum Shinrikyo cult followers in 1995 [4] illustrate the feasibility of using bio-agents by terrorists either for political, religious, or other purposes. Biohazards may also be inflicted accidentally similarly to the incident that took place in April 1979 in Sverdlovsk (USSR). Information about the incident remains classified but it is suspected that aerosolized spores of *Bacillus anthracis* were accidentally leaked from a military laboratory infecting 79 people of whom 68 died [5]. Such examples demonstrate the potential high lethality of bio-agents and emphasize the need for developing efficient counter measures.

2.2 Description of Bioterrorism Agents

The NATO glossary defines a biological agent as "A microorganism that causes disease in personnel, plants, or animals or causes the deterioration of materiel". This includes toxins which are substances naturally produced by bio-agents. In this case, the limit between biological warfare agents and chemical agents is blurred. Bio-agents possess unique properties that enhance their attractiveness to individuals or groups wanting to inflict high morbidity and mortality on a human population. As mentioned earlier they are classified in three categories. Category A agents pose the greatest threat because of their relative ease of transmission, infliction of high rate of mortality as well as their ease of production, transport and dissemination. Category B agents are moderately transmissible and inflict morbidity with low rate of mortality. Finally, Category C agents refer to emerging pathogens and are potential risks for the future. Table 1 sites some of the main biological agents that may be used for bioterrorism along with their most notable characteristics [6].

2.3 Approach of Detection

There are currently several methods in use and in development for the identification of biological agents [7]. Some detection systems are based on metabolomics by following for example the consumption patterns of characteristic substrates or by detecting characteristic fatty acid profiles. Other more specific methods include immunological detection and protein imprint identification using proteomics.

Table 1 Characteristic of the main biological agents that could be used in bioterrorism

Agents	Diseases	Biological threat level	Lethality
Bacillus anthracis	Anthrax	A	+++
Yersinia pestis	Plague	A	+++
Clostridium botulinum	Botulism	A	+++
Francisella tularensis	Tularemia	A	+++
Variola Major	Small pox	A	+++
Listeria monocytogenes	Listeriosis	B	++
Brucella melitensis	Ovine Brucellosis	B	+
Escherichia coli O157: H7	Hemolytic Uremic Syndrome	B	+
Salmonella genus	Typhoid fever, paratyphoid fever	**B**	+

Nonetheless, the most commonly used technic is still nucleic acid detection via quantitative PCR. It is based on PCR assays through classical techniques of amplification and fluorescent detection. This method is reliable and specific able to detect variants and genetically modified strains. Nevertheless quantitative PCR is relatively time consuming and cannot be miniaturized to portable devices that are essential for taking fast actions in case of attacks. On the other hand, electrochemical biosensors are promising platforms that could achieve rapid highly sensitive and selective onsite detection of such agents.

3 Electrochemical DNA (E-DNA) Biosensor

3.1 Definition of Biosensor

According to The IUPAC definition, a biosensor is a device that uses specific biochemical reactions mediated by isolated enzymes, immune-systems, tissues, organelles, DNA or whole cells to recognize targeted chemical or biological compounds. The effect of the recognition process is transformed by a composite called transducer into an observable and measurable electrical signal. The modification of the signal is directly proportional to the concentration of the target in the sample and can be measured by different techniques such as electrochemical [8], optical [9], piezoelectric [10], conductimertic [11], spectrophotometric [12] or calorimetric [13]. Biosensors are classified according either to the nature of the biomolecules immobilized on their surface and are responsible for the recognition or to the type of interaction they engage with the target. DNA biosensors are devices in which oligonucleotides are attached to the transducers and in which the detection is owed to the formation of double-strand (ds) DNA via the hybridization between the single-strand (ss) DNA acting as a probe linked to the transducer and the target ssDNA comprising the complementary sequence of nucleotides specific to the probe.

The important parameters that characterize biosensors are the dynamic range and the linear range of detection, the limit of detection, the sensitivity and the selectivity. The dynamic range represents the concentration range of target up to the highest that provides an observable response signal. The linear range represents the concentration range of target that provides a linear response signal. The slope of the linear range corresponds to the sensitivity of the biosensor. The lowest concentration that can be measured corresponds to the limit of detection (LOD) and by extrapolation of the dynamic range curve the detection limit (DL) can be calculated according to various methods described in analytical techniques. The most commonly used methods in biosensor devices are those that take into account signal to noise ratio of 3, where DL is obtained by the equation:

$$x_{DL} = \frac{\alpha_0 + 3s_{bl}}{\alpha_1}$$

Where s_{bl} is the standard deviation, α_o the result of the measurement obtained with a blank test and α_1 is the sensitivity.

Electrochemical biosensors are attractive devices for the identification of biomolecules due to the possibility of their miniaturization, their low manufacturing cost and their ability to directly measure the electrical signal derived from the detection of targets. The choice of the transducer has a significant impact on the characteristics of the biosensor. Transducers that are commonly reported for the construction of electrochemical biosensors are conductive organic polymers (polypyrrole [14], polythiophene [15], polyaniline [16]), carbon nanotubes [17], graphene [18], metal nanoparticles (gold nanoparticles [19]), and gold electrodes modified with self-assembled monolayers [20]. To enhance the electrochemical signal, transducers can be additionally associated with a redox marker as for example ferrocene [21], quinone [22] and metalloporphyrins [23]. Different methods are described for the attachment of biomolecules to transducers such as, physical adsorption, electrostatic interactions, chemical cross-linking, covalent grafting, immobilization through affinity systems like biotin/streptavidin [24] or adamantane/β-cyclodextrin [25], entrapment in polymers [26] or sol-gels [27].

The detection by DNA biosensors has been achieved using both indirect and direct methods. Indirect methods require further steps after DNA hybridization to accomplish the measurement, while direct electrochemical DNA sensing approaches are capable of directly measuring DNA hybridization without any further step.

3.2 DNA Detection Based on Indirect Strategy

Indirect methods of detection rely on sensors in which two different DNA probes are used. The primary probe DNA is attached to the transducer and the secondary probe DNA is labeled either with an enzyme, a redox marker or nanoparticles. The hybridization reaction is then monitored either *via* the redox signal of the product of the enzymatic reaction, the redox signal of the electroactive markers or the enhancement of redox signal by the presence of conducting nanoparticles, respectively. Other indirect strategies rely on using redox DNA intercalators that possess an affinity for dsDNA. In this case, the detection of DNA hybridization is observed *via* the increase in the intensity of the redox signal of the intercalator, which is proportional to the formed dsDNA. Some examples of indirect methods often used for DNA detection are presented in this section.

3.2.1 DNA Labelling with Enzymes

DNA sensing approaches based on enzymatic reactions are often based on sandwich structures for signal amplification. After the hybridization reaction between the primary probe DNA and the target DNA, an additional step is added. This step involves the hybridization of a different section of the target DNA with labeled secondary probe DNA. The secondary DNA is comprised of a redox enzyme such as horseradish peroxidase (HRP) [28, 29] alkaline phosphatase (ALP) [30, 31] or glucose oxidase [32]. The hybridization reaction may be followed through the catalytic properties of the enzymes in the presence of their substrates. The immobilized enzymes on hybridized DNA lead to the production near the electrode surface of electroactive products, which produce a current related to the amount of hybridized target DNA (Fig. 1).The detection of various bio-agent such as *Salmonella* and *Listeria* has been demonstrated by using this method [33].

3.2.2 DNA Labelling with Nanoparticles

Probe DNA can be labeled by tagging with nanoparticles like gold nanoparticles (AuNPs), [34] carbon nanotubes (CNTs) or dendrimers [35]. AuNPs and CNTs can greatly intensify the electrochemical responses signal because of their high conductivities. This fact is demonstrated in assays such as those based on the sandwich structure used in combination with CNTs conjugates encapsulating Cadmium Sulfide (CdS) nanoparticles [36]. The conjugates could be attached to a secondary DNA probe via the biotin/streptavidin system that served as tags. When compared with the conventional single-particle stripping hybridization assays a substantial (~500-fold)

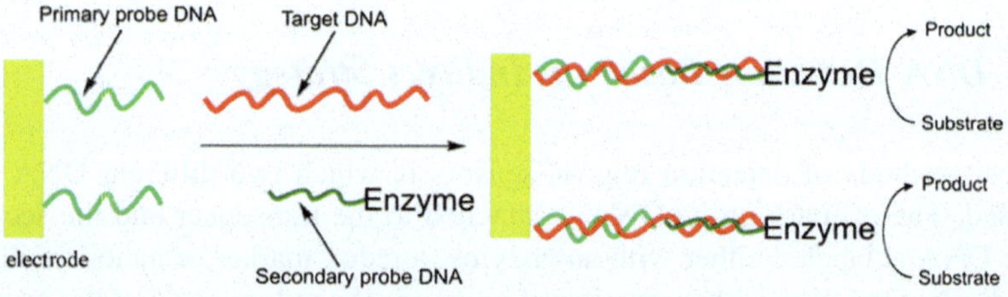

Fig. 1 Schematic representation of the sandwich structure for the detection by enzyme labeled probe DNA

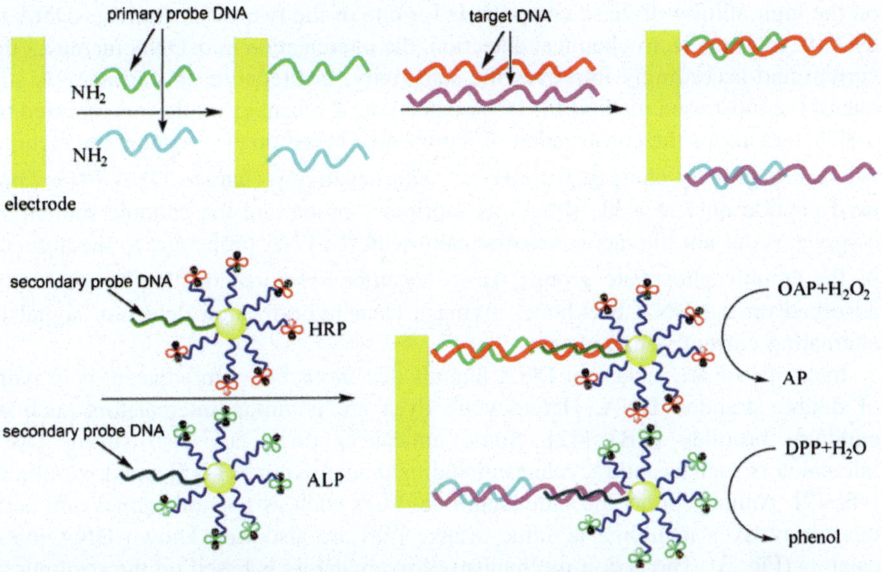

Fig. 2 Specific and highly sensitive dual target biosensor designed by Li et al. [40]

lowering of the detection limit was observed. AuNPs were also conjugated either with thiolated ferrocene, [37, 38] thionine [39] or enzymes. The specific electrochemical detection of two DNA targets sequences in one sample was performed by Li et al. using enzyme functionalized AuNPs as catalytic labels (Fig. 2) [40]. This DNA sensor was constructed by using the sandwich-assay detection strategy in which, two different primary probes DNA were immobilized on the surface of the transducer. Each primary probe DNA hybridized specifically with its complimentary DNA target. The two secondary DNA probes were also each specific for one of the targets but both were associated with AuNPs. Finally, one secondary DNA probe was linked with HRP and the other with ALP. Consequently, the electrochemical signal was generated from the products of the enzymatic catalysis of phenol by HRP and/or of alkaline phosphate by ALP. In addition, enhanced detection sensitivity was obtained because the AuNPs carriers increased the amount of enzyme molecules per hybridization.

3.2.3 Redox Intercalators

The monitoring of hybridization reactions can be performed using redox molecules that possess affinity for some DNA bases or some dsDNA structures by insertion into double helix DNA structures or interactions with minor or major DNA helix grooves. This affinity takes place because of interactions in well-defined binding sites via intercalation and/or electrostatic binding. Both organic compounds and cationic metal complexes can be used as specific DNA binders. The detection relies on the high affinity of these compounds for one of the two DNA forms: ssDNA or dsDNA. During electrochemical detection, the intercalation in dsDNA increases the current and accordingly improves the sensitivity. In literature, this strategy is still one of the most used method for DNA detection. A recent example was reported by Steichen et al. for the construction of a biosensor based on electrostatic interactions between positively charged $Ru(NH_3)_6^{2+}$ and negatively charged DNA [41]. They used peptide nucleic acids (PNA) as a primary probe and the cationic ruthenium complexes did not interact electrostatically with the PNA probe due to the absence of the anionic phosphate groups. However after hybridization, $Ru(NH_3)_6^{2+}$ was adsorbed on the DNA backbone, giving a clear hybridization detection signal in alternating current voltammetry.

Intercalators are a class of DNA ligands that insert between adjacent base pairs of double-stranded DNA. Heterocyclic dyes are common intercalators such as ethidium bromide (EB) [42]. Some anticancer drugs are also strong DNA intercalotors such as anthracyclines including daunomycin [43–45] and doxorubicin [46, 47]. Antipsychotic and antihistaminic drugs such as phenothiazines and acridine derivatives including acridine orange [48] are also well known DNA intercalators (Fig. 3). The action mechanism of intercalators is based on the stacking of planar, aromatic groups between nucleic base pairs in an approximately perpendicular position to the double-helix axis. These interactions could be selective as for example daunomycin, which was found to inserts into the DNA duplex preferentially between GC base pairs.

Threading intercalators are those that carry substituents on the periphery of the intercalating moiety. When intercalated into dsDNA, these substituents rest in the major and the minor grooves simultaneously. Examples of threading intercalators include the naphthalene diimide derivative carrying ferrocenyl groups (Fig. 4), which has been demonstrated a detect limit of 10 zmol of DNA.

Other DNA binders rely on the strong electrostatic binding of the negatively charged sugar phosphate backbone of DNA. Examples include most metallic DNA stains like $Ru(NH_3)_6^{2+}$, $Ru(bpy)_3^{2+}$, $Os(bpy)_3^{3+}$, $Co(bpy)_3^{3+}$, $Co(phen)_3^{3+}$ and manganese complex of rutin MnR_2. Such stains can also show enantiomeric selectivity, for examples $Co(bpy)_3^{3+}$, $Co(phen)_3^{3+}$ [50, 51], and $Ru(bpy)_3^{2+}$ [52] are minor groove helix DNA binders while $Ru(phen)_3^{2+}$ [53] is a major grove helix DNA binder.

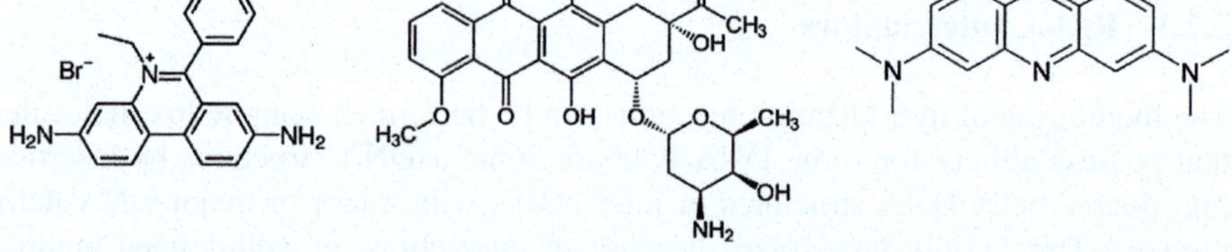

Fig. 3 Structure of different intercalators. From *left to right* ethidium bromide, doxorubicin and acridine orange

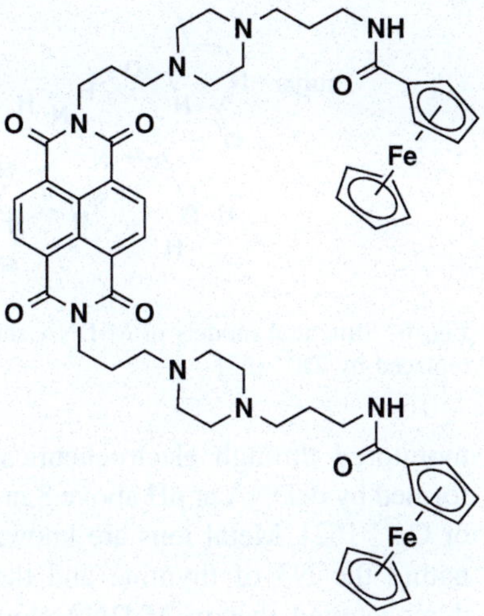

Fig. 4 Structure of naphthalene diimide carrying ferrocene [49]

DNA binders can combine an affinity for GC-rich or for AT-rich sequences within a preference for one helix structure [54]. For example, Hoechst 33258 specifically binds to dsDNA in a solution by recognizing AT-rich sequences within minor grove helixes (Fig. 5). Other DNA binders such as $Ru(bpy)_3^{2+}$ rely on electrostatic affinity and shows preference for G-rich DNA sequences [55, 56], while Methylene Blue (MB) is a DNA intercalator that has affinity for GC-rich sequences [57–59].

The use of intercalators in E-DNA biosensors is now a very promising approach. For example, intercalators are used in PCR as markers for electrochemical detection instead of optical detection [60]. In the same way, detection in microsystems including PCR have been achieved with redox intercalators and lead to a detection limit lower than 10 aM [61].

3.2.4 Metal Ions

Another strategy is based on detecting DNA hybridization through the metalation of dsDNA to form metal complexes of DNA (M-DNA). The detection is based on the modulation of the conductivity of DNA since the formation of M-DNA decreases the resistance of the electronic transport of the electrode interface, which can be

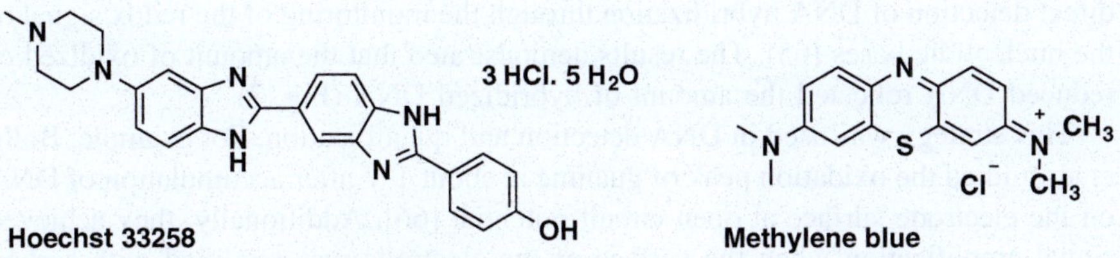

Fig. 5 Structures of Hoechst 33258 and methylene blue

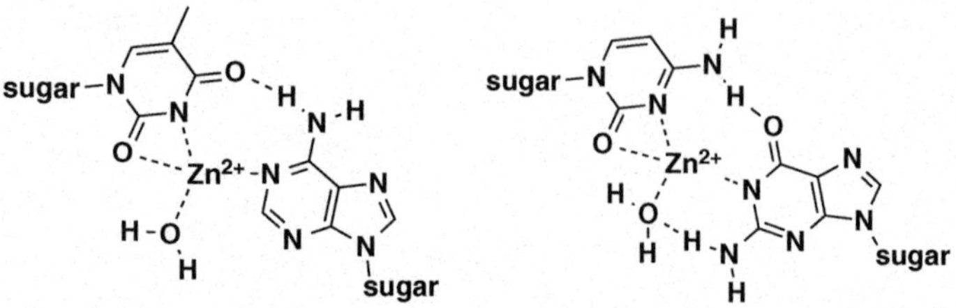

Fig. 6 Structural models of M-DNA showing that the imino protons of the T and G bases are replaced by Zn^{2+} [63]

monitored through electrochemical impedance spectroscopy. M-DNA is usually formed by dsDNA at pH above 8 in the presence of Zn^{2+}, Co^{2+} or Ni^{2+} but not Mg^{2+} or Ca^{2+} [62]. Metal ions are known to bind in the center of DNA helixes, coordinating the N3 of thymine and the N1 of guanine in every base-pair. Xu et al. demonstrated that in M-DNA double-stranded chains, Zn^{2+} is a better electrons transporter than both Co^{2+} and Ni^{2+} [63] (Fig. 6).

3.3 DNA Detection Based on Direct Strategy

The second strategy for monitoring DNA hybridization is based on direct detection. The main advantage of this method is that not require the use of labels or indicators. There are various possibilities for direct DNA detection: (1) monitoring of the redox signal of DNA bases, (2) monitoring of the conformational changes of DNA tagged with some redox marker after the hybridization with the target and (3) monitoring of the changes in the electrochemical properties of transducers due to the hybridization, which could be obtained by monitoring the decrease or the increase of the redox signal of the transducers.

3.3.1 Detection Based on Redox Properties of Guanine

The observation of the redox peaks of DNA bases due to reduction and oxidation reactions, leads to the monitoring of DNA hybridization. Guanine and adenine are the most electroactive DNA bases. Palecek et al. [64] reported the first example of direct detection of DNA hybridization through the monitoring of the redox signal of the nucleotidic bases [65]. The results demonstrated that the amount of oxidized or reduced DNA reflected the amount of hybridized DNA (Fig. 7).

This strategy was used in DNA detection and quantification. For example, Bollo et al. studied the oxidation peak of guanine at about 1 V after accumulation of DNA on the electrode surface at open circuit potential [66]. Additionally, they achieved signal amplification when the surface of the electrode was modified with carbon nanotubes entrapped in chitosan. For up to 90.0 ppm of dsDNA, a linear

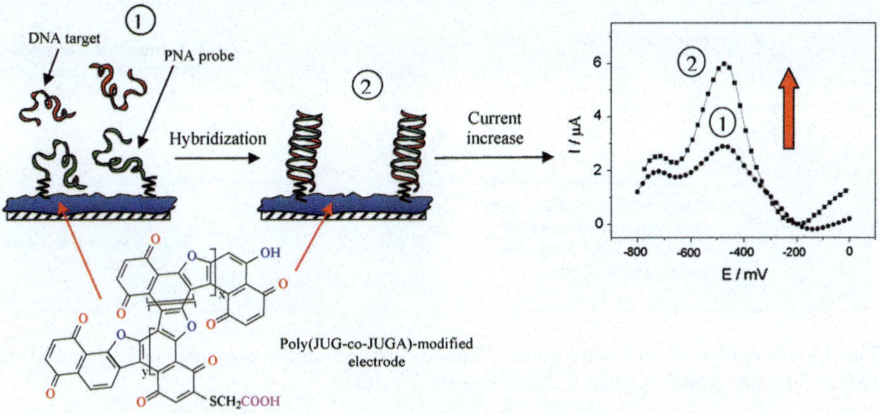

Fig. 12 Schematic illustration of "signal on" biosensor based on JUG polymer [85]

molecule comes from quinone groups, which can provide an intense redox signal. These polymers are able to form hydrogen bonds with single-strand DNA resulting in a decrease in the electroactivity of JUG. When hybridization occurred, the dissociation of hydrogen bonds and the release of dsDNA were observed. This restored the redox activity of the polymer based on juglone and consequently the signal on was obtained (Fig. 12).

3.4 Electrochemical Detection Without PCR Amplification

PCR-less target DNA amplification methods are based on the combination of novel two-component oligonucleotide-modified gold nanoparticles (NPs) and single-component oligonucleotide-modified magnetic microparticles (MMPs) followed by the detection of amplified target DNA in the form of bar-code DNA using a chip-based method. Two components oligonucleotide-modified nanoparticle probes have been designed and used in the bio-bar-code assay, which showed a sensitivity limit of 500 zmolar target DNA. Because the DNA Bar-Code Amplification (BCA) approach is a pseudo-homogeneous system with both MMPs and NPs in solution, a large concentration of the probe DNA can be used to achieve very efficiently binding to target DNA, thereby reducing the time of experiments required for highly sensitive detection (Fig. 13). Indeed, an advantage of the DNA-BCA approach over conventional microarray sandwich assays is that the entire assay can be carried out in 3–4 h, regardless of target concentration. Additionally, the system has an excellent dynamic range and is ideally set up for multiplexing [88].

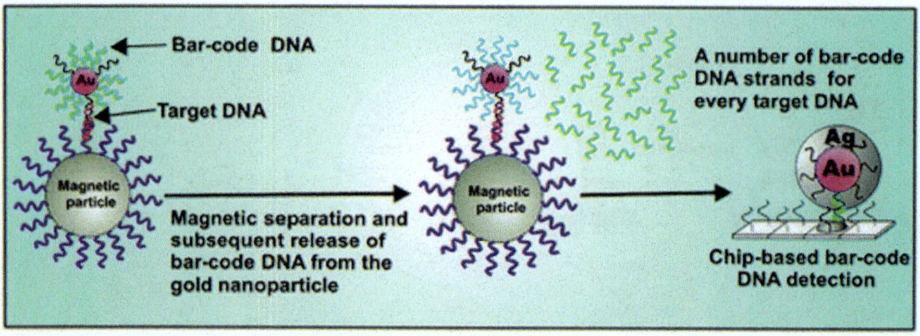

Fig. 13 Illustration of an electrochemical biosensor based on the combination NPs and MMPs followed by the detection using a chip-based BCA method

4 Conclusion

The main electrochemical methods in development for the detection of DNA are discussed and some of the most notable examples are highlighted above. The detection of pathogenic DNA is of upmost priority for tackling bioterrorism. Some E-DNA biosensors have been already reported in literature. The concern mainly class B and class C pathogens particularly *E. Coli*, *Salmonella* and *Listeria*. There are various DNA probes in literature databases that can be used for the development of E-DNA biosensors for the detection of pathogens and biothreats. While better performing platforms should be developed, many system presented in this chapter provided rapid and accurate identification of biothearts agent.

E-DNA detection is a promising application due to sensitive detection and eased implementation into miniaturized and automated devices suitable for rapid screening of multiple unprocessed samples. Because of the advantages of electrochemical biosensors, these are looming as efficient DNA identifiers to replace conventional methods for the detection of biothreats.

References

1. Hughes JM, Gerberding JL (2002) Anthrax bioterrosim: lessons learned and future directions. Emerg Infect Dis 8:1013–1014
2. Slack P (1989) The black death: past and present. Trans R Soc Trop Med Hyg 83:461–463
3. Török TJ, Tauxe RV, Wise RP et al (1997) A large community outbreak of salmonellosis caused by intentional contamination of restaurant salad bars. JAMA 278:389–395
4. Okumura T, Hisaoka T, Yamada A et al (2005) The Tokyo subway sarin attack-lessons learned. Toxicol Appl Pharmacol 207:S471–S476
5. Meselson M, Guillemin J, Hugh-Jones M et al (1994) The Sverdlovsk anthrax outbreak of 1979. Science 266:1202–1208

6. Broussard LA (2001) biological agents: weapons of warfare and bioterrorism. Molecular Diagnosis 6:323–333
7. Lim DV, Simpson JM, Kearns EA et al (2005) Current and developing technologies for monitoring agents of bioterrorism and biowarfare. Clinic Microbiol Rev 18:583–607
8. Heller A (1996) Amperometric biosensors. Curr Opin Biotechnol 7:50–54
9. Dostálek J, Ladd J, Jiang S, Homola J (2006) SPR biosensors for detection of biological and chemical analytes. Chem Sens Biosens 4:177–190
10. Bunde RL, Jarvi EJ, Rosentreter JJ (1998) Piezoelectric quartz crystal biosensors. Talanta 46:1223–1236
11. Muhammad-Tahir Z, Alocilja EC (2003) A conductometric biosensor for biosecurity. Biosens Bioelectron 18:813–819
12. Sattarahmady N, Tondro GH, Gholchin M, Heli H (2015) Gold nanoparticles biosensor of *Brucella spp. genomic* DNA: visual and spectrophotometric detections. Biochem Eng J 97:1–7
13. Zhang Y, Tadigadapa S (2004) Calorimetric biosensors with integrated microfluidic channels. Biosens Bioelectron 19:1733–1743
14. Ramanavicius A, Ramanaviciene A, Malinauskas A (2006) Electrochemical sensors based on conducting polymer-polypyrrole. Electrochim Acta 51:6025–6037
15. Uygun A (2009) DNA hybridization electrochemical biosensor using a functionalized Polythiophene. Talanta 79:194–198
16. Chang H, Yuan Y, Shi N et al (2007) Electrochemical DNA biosensor based on conducting polyaniline nanotube array. Anal Chem 79:5111–5115
17. Wang J, Lin Y (2008) Functionalized carbon nanotubes and nanofibers for biosensing applications. Trends Anal Chem 27:619–626
18. Pumera M, Ambrosi A, Bonanni A et al (2010) Graphene for electrochemical sensing and biosensing. Trends Anal Chem 29:954–965
19. Pingarrón JM, Yanez-Sedeno P, González-Cortés A (2008) Gold nanoparticle-based electrochemical biosensors. Electrochim Acta 53:5848–5866
20. Li C-Z, Long Y-T, Sutherland T et al (2006) Electronic biosensors based on dna self-assembled monolayer on gold electrodes. In: Xing WL, Cheng J (eds) The Frontiers in Biochip Technology. Springer, China, pp 274–291
21. Miodek A, Castillo G, Hianik Korri-Youssoufi H (2014) Electrochemical aptasensor of cellular prion protein based on modified polypyrrole with redox dendrimers. Biosens Bioelectron 56:104–111
22. Piro B, Reisberg S, Noel V, Pham MC (2007) Investigations of the steric effect on electrochemical transduction in a quinone-based DNA sensor. Biosens Bioelectron 22:3126–3131
23. Cannone F, Perrée-Fauvet M, Mahy JP et al (2008) Electrochemical detection of DNA sequences based on metalloporphyrins-polypyrrole towards a multi-detection analysis. Sens Lett 6:570–576
24. Cass T, Ligler FS (eds) (1998) Immobilized biomolecules in analysis a practical approach. Oxford University Press, New York
25. Holzinger M, Bouffier L, Villalonga R, Cosnier S (2009) Adamantane/beta-cyclodextrin affinity biosensors based on single-walled carbon nanotubes. Biosens Bioelectron 24:1128–1134
26. Cosnier S (1999) Biomolecule immobilization on electrode surfaces by entrapment or attachment to electrochemically polymerized films. Biosens Bioelectron 14:443–456
27. Kandimalla VB, Tripathi VS, Ju H (2006) Immobilization of biomolecules in sol–gels: biological and analytical applications. Crit Rev Anal Chem 36:73–106
28. Campbell CN, Gal D, Cristler N et al (2002) Enzyme-amplified amperometric sandwich test for RNA and DNA. Anal Chem 74:158–162
29. Marchand G, Delattre C, Campagnolo R et al (2005) Electrical detection of DNA hybridization based on enzymatic accumulation confined in nanodroplets. Anal Chem 77:5189–5195
30. Carpini G, Lucarelli F, Marrazza G, Mascini M (2004) Oligonucleotide-modified screen-printed gold electrodes for enzyme-amplified sensing of nucleic acids. Biosens Bioelectron 20:167–175

31. Hernández-Santos D, Díaz-González M, González-García MB et al (2004) Enzymatic genosensor on streptavidin-modified screen-printed carbon electrodes. Anal Chem 76:6887–6893
32. Kavanagh P, Leech D (2006) Redox polymer and probe DNA tethered to gold electrodes for enzyme-amplified amperometric detection of DNA hybridization. Anal Chem 78:2710–2716
33. Farabullini F, Lucarelli F, Palchetti I et al (2007) Disposable electrochemical genosensor for the simultaneous analysis of different bacterial food contaminants. Biosens Bioelectron 22:1544–1549
34. Wang J (2003) Nanoparticle-based electrochemical DNA detection. Anal Chim Acta 500:247–257
35. Gao H, Jiang X, Dong YJ et al (2013) Dendrimer-encapsulated copper as a novel oligonucleotides label for sensitive electrochemical stripping detection of DNA hybridization. Biosens Bioelectron 48:210–215
36. Wang J, Liu G, Jan MM, Zhu Q (2003) Electrochemical detection of DNA hybridization based on carbon-nanotubes loaded with CdS tags. Electrochem Commun 5:1000–1004
37. Wang J, Li J, Baca AJ et al (2003) Amplified voltammetric detection of DNA hybridization via oxidation of ferrocene caps on gold nanoparticle/streptavidin conjugates. Anal Chem 75:3941–3945
38. Jin Y, Lu W, Hu J et al (2007) Site-specific DNA cleavage of EcoRI endonuclease probed by electrochemical analysis using ferrocene capped gold nanoparticles as reporter. Electrochem Commun 9:1086–1090
39. Wang W, Song L, Gao Q et al (2013) Highly sensitive detection of DNA using an electrochemical DNA sensor with thionine-capped DNA/gold nanoparticle conjugates as signal tags. Electrochem Commun 34:18–21
40. Li XM, Fu PY, Liu JM, Zhang SS (2010) Biosensor for multiplex detection of two DNA target sequences using enzyme-functionalized Au nanoparticles as signal amplification. Anal Chim Acta 673:133–138
41. Steichen M, Decrem Y, Godfroid E et al (2007) Electrochemical DNA hybridization detection using peptide nucleic acids and $[Ru(NH3)6]^{3+}$ on gold electrodes. Biosens Bioelectron 22:2237–2243
42. Castro H, Ana C, Erick GF (2014) Preparation of genosensor for detection of specific DNA sequence of the hepatitis B virus. Appl Surf Sci 314:273–279
43. Mascini M, Palchetti I, Marrazza G (2001) DNA electrochemical biosensors. J Anal Chem 369 (1):15–22
44. Yang Y, Wang Z, Yang M et al (2007) Electrical detection of deoxyribonucleic acid hybridization based on carbon-nanotubes/nano zirconium dioxide/chitosan-modified electrodes. Anal Chim Acta 584:268–274
45. Cai H, Cao X, Jiang Y et al (2003) Carbon nanotube-enhanced electrochemical DNA biosensor for DNA hybridization detection. Anal Bioanal Chem 375:287–293
46. Ting BP, Zhang J, Gao Z, Ying JY (2009) A DNA biosensor based on the detection of doxorubicin-conjugated Ag nanoparticle labels using solid-state voltammetry. Biosens Bioelectron 25:282–287
47. Zhanga Y, Wang J, Xu M (2010) A sensitive DNA biosensor fabricated with gold nanoparticles/poly(p-aminobenzoic acid)/carbon nanotubes modified electrode. Colloids Surf B 75:179–185
48. Kapuscinski J, Darzynkiewicz Z, Melamed MR (1983) Interactions of acridine orange with nucleic acids properties of complexes of acridine orange with single-stranded ribonucleic acid. Biochem Pharmacol 32:3679–3694
49. Takenaka S, Yamashita K, Takagi M et al (2000) DNA sensing on a DNA probe-modified electrode using ferrocenyl naphthalene diimide as the electrochemically active ligand. Ana. Chem 72:1334–1341
50. Millan KM, Mikkelsen SR (1993) Sequence-selective biosensor for DNA based on electroactive hybridization indicators. Anal Chem 65:2317–2323

51. Millan KM, Saraullo A, Mikkelse SR (1994) Voltammetric DNA biosensor for cystic fibrosis based on a modified carbon paste electrode. Anal Chem 66:2943–2948
52. Pyle AM, Rehmann JP, Meshoyrer R et al (1989) Mixed-ligand complexes of ruthenium(II): factors governing binding to DNA. J Am Chem Soc 111:3051–3058
53. Barton JK, Goldberg JM, Kumar CV et al (1986) Binding modes and base specificity of tris (phenanthroline)ruthenium(II) enantiomers with nucleic acids: tuning the stereoselectivity. J Am Chem Soc 108:2081–2088
54. Hashimoto K, Ito K, Ishimori Y (1994) Sequence-specific gene detection with a gold electrode modified with DNA probes and an electrochemically active dye. Anal Chem 66:3830–3833
55. Yang IV, Thorp HH (2001) Modification of indium tin oxide electrodes with repeat polynucleotides: electrochemical detection of trinucleotide repeat expansion. Anal Chem 73:5316–5322
56. Lim SH, Wei J, Lin J (2004) Electrochemical genosensing properties of gold nanoparticle-carbon nanotube hybrid. Chem Phys Lett 400:578–582
57. Erdem A, Kerman K, Meric B et al (2000) Novel hybridization indicator methylene blue for the electrochemical detection of short DNA sequences related to the hepatitis B virus. Anal Chim Acta 422:139–149
58. Sun W, Qi X, Zhang Y et al (2012) Electrochemical DNA biosensor for the detection of Listeria monocytogenes with dendritic nanogold and electrochemical reduced graphene modified carbon ionic liquid electrode. Electrochim Acta 85:145–151
59. Li J, Liu Q, Liu Y et al (2005) DNA biosensor based on chitosan film doped with carbon nanotubes. Anal Biochem 346:107–114
60. Miranda-Castro R, Marchal D, Limoges B et al (2012) Homogeneous electrochemical monitoring of exonuclease III activity and its application to nucleic acid testing by target recycling. Chem Commun 48:8772–8774
61. Ferguson BS, Buchsbaum SF, Swensen JS et al (2009) Integrated microfluidic electrochemical DNA sensor. Anal Chem 81:6503–6508
62. Lee JS, Latimer LJP, Reid RS (1993) A cooperative conformational change in duplex DNA induced by zinc and other divalent metal ions. Biochem Cell Biol 71:162–168
63. Xu Y, Jiang Y, Cai H et al (2004) Electrochemical impedance detection of DNA hybridization based on the formation of M-DNA on polypyrrole/carbon nanotube modified electrode. Anal Chim Acta 516:19–27
64. Palecek E (1960) Oscillographic polarography of highly polymerized deoxyribonucleic acid. Nature 188:656–657
65. Palecek E (1988) Adsorptive transfer stripping voltammetry: determination of nanogram quantities of DNA immobilized at the electrode surface. Anal Biochem 170:421–431
66. Bollo S, Ferreyra NF, Rivas GA (2007) Electrooxidation of DNA at glassy carbon electrodes modified with multiwall carbon nanotubes dispersed in chitosan. Electroanalysis 19:833–840
67. Kang D, Zuo X, Yang R et al (2009) Comparing the properties of electrochemical-based DNA sensors employing different redox tags. Anal Chem 81:9109–9113
68. Ricci F, Zari N, Caprio F et al (2009) Surface chemistry effects on the performance of an electrochemical DNA sensor. Bioelectrochemistry 76:208–213
69. Xiao Y, Qu X, Plaxco KW, Heeger AJ (2007) Label-free electrochemical detection of DNA in blood serum via target-induced resolution of an electrode-bound DNA pseudoknot. J Am Chem Soc 129:11896–11897
70. Fan C, Plaxco KW, Heeger AJ (2003) Electrochemical interrogation of conformational changes as a reagentless method for the sequence-specific detection of DNA. Proc Natl Acad Sci USA 100:9134–9137
71. Anne A, Bouchardon A, Moiroux J (2003) 3'-Ferrocene-labeled oligonucleotide chains end-tethered to gold electrode surfaces: Novel model systems for exploring flexibility of short DNA using cyclic voltammetry. J Am Chem Soc 125:1112–1113
72. Immoos CE, Lee SJ, Grinstaff MW (2004) DNA-PEG-DNA triblock macromolecules for reagentless DNA detection. J Am Chem Soc 126:10814–10815

73. Li D, Song S, Fan C (2010) Target-responsive structural switching for nucleic acid-based sensors. Acc Chem Res 43:631–641
74. Lai RY, Lagally ET, Lee S-H et al (2006) Rapid, sequence-specific detection of unpurified PCR amplicons via a reusable electrochemical sensor. PNAS 103:4017–4021
75. Yang W, Lai RY (2011) Langmuir. Comparison of the stem-loop and linear probe-based electrochemical DNA sensors by alternating current voltammetry and cyclic voltammetry 27:14669–14677
76. Li X-M, Fu P-Y, Liu J-M, Zhang Shu-Sheng (2010) Biosensor for multiplex detection of two DNA target sequences using enzyme-functionalized Au nanoparticules as signal amplification. Analytica Chimica Acta 673:133–138
77. Yang J, Yang T, Feng Y, Jiao K (2007) A DNA electrochemical sensor based on nanogold-modified poly-2,6-pyridinedicarboxylic acid film and detection of PAT gene fragment. Anal Biochem 365:24–30
78. Korri-Youssoufi H, Garnier F, Srivastava P et al (1997) Toward bioelectronics: specific DNA recognition based on an oligonucleotide-functionalized polypyrrole. J Am Chem Soc 119:7388–7389
79. Tlili C, Korri-Youssoufi H, Ponsonnet L et al (2004) Electrochemical impedance probing of DNA hybridisation on oligonucleotide-functionalised polypyrrole. Talanta 68:131–137
80. Lê HQA, Chebil S, Makrouf B et al (2010) Effect of the size of electrode on electrochemical properties of ferrocene-functionalized polypyrrole towards DNA sensing. Talanta 81:1250–1257
81. Bouchet A, Chaix C, Marquette CA (2007) Cylinder-shaped conducting polypyrrole for labelless electrochemical multidetection of DNA. Biosens Bioelectron 23:735–740
82. Miodek A, Mejri N, Gomgnimbou M et al (2015) E-DNA sensor of *Mycobacterium tuberculosis* based on electrochemical assembly of nanomaterials (MWCNTs/PPy/PAMAM). Anal Chem 87:9257–9264
83. Lien TTN, Lam TD, An VTH et al (2010) Multi-wall carbon nanotubes (MWCNTs)-doped polypyrrole DNA biosensor for label-free detection of genetically modified organisms by QCM and EIS. Talanta 80:1164–1169
84. Reisberg S, Piro B, Noel V, Nguyen TD, Nielsen PE, Pham MC (2008) Investigation of the charge effect on the electrochemical transduction in a quinone-based DNA sensor. Electrochim Acta 54:346–351
85. Reisberg S, Dang LA, Nguyen QA et al (2008) Label-free DNA electrochemical sensor based on a PNA-functionalized conductive polymer. Talanta 76:206–210
86. Reisberg S, Piro B, Noel V, Pham MC (2006) Selectivity and sensitivity of a reagentless electrochemical DNA sensor studied by square wave voltammetry and fluorescence. Bioelectrochemistry 69:172–179
87. Zhang QD, Piro B, Noël V et al (2012) An electroactive conjugated oligomer for a direct electrochemical DNA sensor. Synthetic Met 162:1496–1502
88. Nam J-M, Stoeva SI, Mirkin CA (2004) Bio-bar-code-based DNA detection with PCR-like sensitivity. J Am Chem Soc 126:5932–5933

Biosensors for the Express Evaluation of the Level of Genotoxicity of Chemical Substances

Nickolaj F. Starodub

Abstract The special attention is given to methods for control of genotoxicity. Among them, the detailed analysis is given that approaches which are based on the traditional molecular genetics tests, common instrumental systems and a modern biosensor devices. It is provided as general characteristics of the developed widely dispersed, most applicable in the practice approaches and detailed description of the basic principles of their functioning. The last a specially concerns instrumental analytical methods and, in particular, cell biosensors considering the possible type of the transducers, types of cells as sensitive structures, their integration in sensor elements and way of specific signal registration. In general it is analysed sensitivity and field of application of the existed approaches for the control of total toxicity and genotoxicity.

Keywords Genetoxicity · Molecular genetics · Control · Instrumental approaches · Biosensors

1 Introduction

The ecological toxicity on the living organisms may be revealed as affect on the cells in respect of their metabolic changes, or full deeds, or some reconstruction of some carriers genetic information, which are presented by DNA or RNA in separate organisms. That is way, the specific effects may have a different implications for cells: (a) the repaired damage without any further consequences; (b) that are remained unrepaired and leads to death, as well as (c) that induce an error-prone repair pathways realizing in mutagenesis or in cancerogenesis. Last two effects are as basis for the development of the approaches for the testing of genotoxicity of environmental factors, in generally, and with involving modern instrumental methods including based on the principles of biosensorics.

N.F. Starodub (✉)
Laboratory of Biosensorics, National University of Life and Environmental Sciences of Ukraine, 15, Geroyiv Oborony, Kyiv 03041, Ukraine
e-mail: nikstarodub@yahoo.com

D.P. Nikolelis and G.-P. Nikoleli (eds.), *Biosensors for Security and Bioterrorism Applications*, Advanced Sciences and Technologies for Security Applications,
DOI 10.1007/978-3-319-28926-7_9

2 Traditional Approaches

Today are more than 100 different methods to assess genotoxicity but really no more than 20 test systems are practically used [31]. The most common method in this respect was proposed Bruce Ames and it is based on the application of His-mutants of *Salmonella typhimurium* which do not synthesize histidine and survive on non-histidine media only when they have mutation to wild-type His+. Revertants wild type form colonies on medium without histidine as an indicator of gene mutations (http). Construction of test strains which are the most sensitive to the action of a mutagen is achieved by the inactivation of the excision repair system in their cells. Furthermore, the cells used in the Ames test strains have and other features that increase their sensitivity to mutagenic. In recent years test Ames has been greatly improved: automated testing procedure, increased sensitivity to certain types of mutagens.

In most cases the mutation of single genes in higher organisms are not determined since they are very rare. As a rule, it is restricted by estimation of the level of mutations in the chromosome as a whole.

The first such method for the detection and determination of the frequency of mutations in *Drosophila* was proposed by Muller [21, 29]. This method allowed to distinguish newly arisen and existed mutations which plays the role of "closers" for crossingover (C) and has a recessive lethal effect (l) in the genotype. This chromosome was labeled by a dominant gene (Bar) which is reducing facet eye. And as the result of it the normal spherical eyes of female heterozygotes acquire bean-shape, and the males are slotted. Females of the test SlV line was crossed with irradiated males (as a mutagen may serve and chemical compound). From the first generation of females it was chosen SlV/+ persons for statement of the individual crosses. Since males with genotype ClB/Y die the splitting on the gender in the second generation will be 2:1. Lack of male in the second testified about the lethal mutation in the X chromosome. Its frequency is expressed as the ratio of the number of X-chromosomes in the population. It is needed to remember that the first generation of females between the sex chromosomes can sometimes be a double crossing-over, resulting in reduction of the true frequency of lethal mutations.

Currently C1B method has lost its practical value. Instead of it was proposed Muller-5 method. Females-line analyzers, both X chromosomes contain two inversions, non-lethal effect: sc8 (reduced bristles) captures a large portion of the X chromosome 49—inversion in the middle of the X-chromosome. As result of these reversals is substantially complete exclusion of crossing between chromosomes. Additionally both the X chromosome gene labeled female body and yellow color bristles yellow (y). Males in this line viable. If taken for the study of male wild-type no mutation on the X chromosome, then, after crossing it with the female line of the analyzer, the second generation we get to 2 phenotypic classes of females and males. If in the analyzed X-chromosome of male lethal mutation arose in the second generation of all males will belong to the same phenotypic class (scsy d49)—yellow with reduced bristles. Moreover, each individual culture of the second generation, which is the

of the SOS repair system are the next tests: SOS-Chromo (Quillardet and Hofnung [53], Umu [46], Lux-Fluoro [5], VitoTOX® [74] and some biosensors variants [50]. The Lux-Fluoro test is a unique combination of two bioassays [56], which coincidentally measure genotoxicity (SOS-Lux test) and cytotoxicity (Lac-Fluoro test) of substances and mixtures of substances. The SOS-Lux assay, like the SOS-Chromo test or the Umu test, is based on the measurement of DNA damage-dependent induction of the bacterial SOS system in genetically modified *Salmonella typhimurium* TA1535 bacteria [51], which have been transformed with the plasmid pPLS-1 carrying the promoter less lux genes of *Photobacterium leiognathi* as reporter element under the control of a DNA damage-dependent SOS promoter from ColD as sensing element [6]. This system reacts to agents, which induce DNA damages inside these bacterial cells with the dose-dependent production of bioluminescence. The bioluminescence as a signal for DNA damage is an enzymatic reaction of a photolyase with its specific substrate, both encoded by the luxCDABFE genes of *Photobacterium leiognathi*, in presence of oxygen. Since the bioluminescent light can be registered by an appropriate detector like a photomultiplier without destroying the cells, the kinetics of the processing of the DNA damage by the SOS system can be followed in living cells. The SOS-Lux test as a bioassay for genotoxicity can be used partly or fully automatically for routine measurements and can be employed for high throughput screening. The analogue Lac-Fluoro-test detects the cellular responses to cytotoxins [54]. It is based on the constitutive expression of green fluorescent protein (GFP) mediated by the bacterial protein expression vector pGFPuv as GFPuv expression is not under regulatory constraints in *Salmonella typhimurium*, due to the lack of a functional lacI repressor in this species. In response to cytotoxic agents, this system reacts with a dose-dependent reduction of GFP-fluorescence. The recombinant *S. typhimurium* strains carrying either the SOS-Lux plasmid or the lac-GFPuv plasmid are used to determine in parallel in one well of a microplate the genotoxic and the cytotoxic potential of the test compounds. Light and fluorescence emission as well as absorption of cells in the test samples and in the controls are measured in the microplate luminometer-fluorometer-photometer combination Victor2 and the calculated luminescence induction as well as fluorescence reduction is used to determine the genotoxic and/or cytotoxic potential of the applied compound. Victor2 device is a complete platform for quantitative detection of fluorescence, flash or glow luminescence, absorbance and photometry using specific filters. The instrument was equipped with the following filters: absorbance: at 490 nm (20 nm band width), fluorescence: excitation at 405 nm (15 nm band width), emission at 510 nm (10 nm band width), luminescence: open.

Already it was shown that the above mentioned reporter system reacts with a high level of light production to different classes of DNA damaging agents already at concentrations that have nearly no effect on cell survival in *S. typhimurium* TA1535 cells [29]. For higher concentrations, when tested agents induce cytotoxic effects, the determination of SOS induction is influenced by the proportion of dying cells of the exposed population. As simultaneous measurements of luminescence and fluorescence allow for discrimination between genotoxic and cytotoxic potency of the tested compound it is possible to correct light output yields for the proportion of surviving cells. The resulting induction factor Fi can be used to identify genotoxicity: A test compound is considered to be cytotoxic, if fluorescence and/or bioluminescence of exposed cells are decreased; a test compound is considered to be genotoxic if bioluminescence is increased and induction factors Fi exceed double the amount of control levels. It was proved the test's special value in respect to its advantages over other test systems: (i) the in vivo measurement itself is non-disruptive and can be repeated several times with 96 samples in parallel; (ii) the whole kinetics of the SOS induction by a DNA-damaging substance can be followed up for several hours in the same sample, thereby stating possible growth delays which cannot be seen in other tests and which may falsify comparisons of substances when only one measurement is performed after a fixed period of time; and (iii) first indications on a substance's genotoxic potential can already be derived after 2–3 h of incubation [3, 6, 54, 74]. Genotoxic chemical and environmental samples were successfully identified by the SOS-Lux test, simultaneously the Lac-Fluoro test confirmed the absence of cytotoxic components interfering with the results of the SOS-Lux test. The lower detection limit of 4-nitroquinoline 1-oxide was 8.1×10^{-3} μg L^{-1}. No matrix effect was observed with the Lux-Fluoro test. The 2-aminoanthracene was identified to showthe highest genotoxic response of all tested substances with detection limit after metabolic activation with S9 of 4 μg L^{-1}. The lowest concentration of N-methyl-N'-nitro-N-nitrosoguanidine detected as genotoxicwas 0.216 μg L^{-1}. The not filtered surface water showed a limited genotoxic reaction only after incubation with S9 fraction in a 1 % dilution and no concentration dependency. The Lux-Fluoro test showed a strong positive signal for the effluent water of a textile industry in demonstrating the genotoxicity of this sample.

Mammalian μ-FADU assay which is based on alkaline DNA unwinding have recently been presented [44]. This assay is operated in a 96-well format, thus greatly increasing throughput. The number of cells required has been reduced to less than 10,000 per data point. The threshold for detection of X-ray-induced DNA strand breaks is 0.13 Gy. The total assay time required for a typical experiment to assess DNA strand break repair is 4–5 h [43]. It has established a robust and convenient method measuring of formation and repair of DNA single-strand breaks in live cells.

4 Biosensor Tests

The start in the development of these approaches for the determination of genotoxicity was done not long ago [66]. Their appearance was stimulated not wishing to appreciate medium toxicity only and to have information about gene toxic effect in regime on line.

The yeast-based biosensors consist of two components: the RNR3 gene in lacZ system wich serve as sensor since it induces during the DNA damage only and reporter. In spite that a series of genetic manipulations allowed to make the RNR3-lacZ system highly sensitive but for application in biosensors it should be improved [17]. So, the lacZ reporter based on a colorimetric determination of the β-galactosidase activity, which requires cell disruption. To achieve high efficiency and simplicity operating genotoxic testing the system lacZ reporter was replaced by a yEGFP gene encoding yeast-enhanced green fluorescent protein which was optimized for the expression in *Saccharomyces cerevisiae* early [72, 76]. Recombinant yeast (*S. cerevisiae*) containing fluorescent markers such as green or red fluorescent protein (GFP or RFP) are ideal candidates for microscreening. GreenScreenTM has been employed for screening a different genotoxic industrial products and environmental contaminants [7].

Due to the number genetic investigations it was proposed a yeast-cell based HUG1-GFP biosensor as a sensitive genotoxic testing system to detect multiple genotoxins. HUG1 promotor (hydro-xyurea and UV, gamma radiation induced), which is regulated by the Mec1check point pathway [8, 14]. At the comparison of two biosensors it was stated [7] that maximum induction and linear regression of the HUG1-yEGFP biosensor is about twice as sensitive as RNR3-yEGFP one. Perhaps the most significant improvement in such types of biosensors concerns manipulation with seven genes from among dozens of candidate for the disarming all these systems that play roles in the protection of yeast cells from effects of environmental factors. As results of it there is possibility to created a hypersensitive host strain that enables reporters like HUG1-yEGFP and RNR3-yEGFP to detect extremely low doses of genotoxins with a more than 300-fold increase in sensitivity and in certain aspects surpasses the current industrial gold standards like the Ames test and SOS chromotest. It should be noted that the septuple mutant strain can be utilized by other yeast genotoxicity testing systems including those based on cell survival or mutagenesis and can also be further improved in combination with either in vivo or in vitro metabolic activation of certain chemicals. The yEGFP based reporter in combination with appropriate mutant strains can also be utilized to detect other non-genotoxic environmental chemicals.

In spite of existed improving biosensors based on yeast transcriptional response to genotoxicity there is necessary in investigation of many agents which involved in effect on the metabolic activation in mammals to become genotoxic and carcinogenic. Unfortunately such responsible activation systems are largely lacking in yeast cells. That is why, the future research should be directed towards humanizing yeast cells for the metabolic activation of pre-genotoxic/pre-carcinogenic compounds and application of new type system in biosensors for express screening of environmental factors.

It was constructed a bacterial biosensor on *E. coli* strain with a transformed egfp gene as a reporter one under the control of the promoter of recA gene and developed an SOS-EGFP test. By this test, the biosensor cells treated by chemicals can produce brighter fluorescence than the untreated control if the chemicals can induce substantial DNA damage [38]. The constructed biosensor is probably useful to simultaneously evaluate the genotoxicity and cytotoxicity.

The products of a number of SOS-dependent genes are involved in DNA-repair-mechanisms which are activated at the occurrence of DNA single strands. Such DNA lesions are induced by compounds that form the DNA-adducts. The expression of the SOS genes is regulated by the LexA protein that specifically binds to SOS-responsive promoter sequences. In order to detect the SOS-response the SOS-sensitive promoters like the recA or umuDC genes are fused to lacZ or phoA encoding the enzymes β-galactosidase or alkaline phosphatase. For the bacterial test systems it is of crucial importance to mimic the metabolism of xenobiotics that takes place in the liver of vertebrates and which can lead to a formation of bio-activated and thus genotoxic intermediates. Usually, this is done by the addition of the S9-fraction that is prepared from the liver of induced rodents. It is composed of a complex mixture of enzymes involved in the metabolism of xenobiotics, in particular the microsomal bound cytochrome-P450-dependent monooxygenases. They catalyze the oxidation of organic compounds by molecular oxygen. The cytochrome-P450-dependent monooxygenases are activated by a reduction step concomitantly with the consumption of NADPH [42]. The electrochemical signal can be detected via para-aminophenyl β-d-galactopyranoside (pAPG) since the reporter enzyme, β-galacosidase cleaves the glycosidic bond in this substance. The reaction product p-aminophenol (pAP) can be oxidized electrochemically to p-iminoquinone even it is possible without cell-lysis [12]. Direct electrochemical signal detection is preferable in the comparison with others ones since the use of a simple set of electrodes would greatly reduce the complexity, size and costs that are typically associated with the optical detection. However, it has yet to be proven that electrochemistry can also compete in terms of sensitivity with the colorimetric signal detection. This is of special interest because of the mandatory presence of the uncharacterized mixture of potentially electro-active enzymes and metabolites (S9-fraction of liver homogenate), as well as several cofactors (e.g. NADP) in the standard assay reaction, that are added to metabolically activate pre-genotoxic substances. These compounds might interfere with the electrochemical signal detection and decrease its sensitivity. It was reported [37] about the using electrochemiluminescent arrays for the genotoxicity testing of metabolites of benzo[a]pyrene that are generated in situ by various immobilized cytochrome (cyt) P450 or imbedded microsomes as their source. In contrast of these investigation it was described the aforementioned electrochemiluminescent arrays for the detection of the DNA-damage without any cellular context, i.e. the formation of the adducts with the purified DNA and not a cellular response. The chromo-amperometric electrochemical signal was characterized by the detection following the induction of the bacterial SOS-response in the presence of S9-mix. It was demonstrated that the unique substrate mediated electrochemical detection is simple to use, can be integrated on a miniaturized whole cell bio chip and yield satisfactory results in comparison to the respective ISO standard (ISO 13829: [34]. Chromo-amperometry based on the screen printed electrodes was compared with a standardized colorimetric assay for the detection of genotoxic samples by reporter gene induction (lacZ) via the bacterial SOS-system. The amperometric method was optimized in terms of substrate concentration for the reporter gene β-galactosidase that cleaves pAPG to pAP which in turn is oxidized to p-iminoquinone at the electrode. It was found that a final concentration of 6 mM of pAPG is suitable to guarantee its cleavage by pseudo zero-order kinetics even if the reporter enzyme is strongly induced. By means of linear sweep voltammetry it was shown that a potential range of 300–400 mV is most suitable for the detection of pAP in a potential whole cell-based biosensor even in the presence of a large excess of pAPG. A comparison of the colorimetric and electrochemical detection methods shows a high correlation of the determined SOS-induction factors indicating the usability of the amperometric signal detection in principle. But the noise level of the electrochemical detection at 300 mV is substantially increased compared to the colorimetric assay limiting its potential for the assessment of environmental samples because of a decrease in sensitivity. In contrast, the noise level of the amperometric detection of pAP at 400 mV is very similar to the colorimetric standard method. Such biosensor will contain bacterial reporter strains and all necessary compounds for the metabolic activation of xenobiotics (S9-fraction and cofactors) which are lyophilized on top of the electrode in a small reaction chamber. The freeze-dried biological compounds could be dissolved by the sample before the eventual induction of the SOS-response. The electrodes will be exchanged after the measurement [9].

Today it was proposed many mammalian cell-based gene mutation assays but only four cell lines of Chinese hamster V97 and CHO cells, human lymphoblastoid TK6 cells, mouse lymphoma L5178Y cells as well as three genetic loci of HPRT (hypoxanthine-guanine phosphoribosyltransferase), TK (thymidine kinase) and the cell membrane Na^+/K^+ ATPase genes are well validated and widely used. But they have a low sensitivity that is still a problem in these mammalian cell-based gene mutation assays [26].

In mammalian cells the transcription factor p53 works as a guard keeper of the genome by inducing DNA damage repair, cell cycle arrest and apoptosis in response to cellular stresses leading to DNA damage, thus it is also called tumor suppressor. The DNA repair gene P53R2 which encodes a subunit of ribonucleotide reductase is a p53-target gene activated in response to cellular DNA damage. The p53R2-mediated luciferase reporter gene were used in the bioassay system for genotoxicity detection using human cells with wild-type p53 [47, 48]. Validation of this assay system indicated that it could be a rapid and reliable tool in the screening of genotoxic chemicals. The GADD45a-mediated GFP reporter gene were applied in the bioassay system for the genotoxicity detection in human TK6 cells. It was found that this assay system had both high specificity and high sensitivity in genotoxicity detection of different genotoxicants [28, 36]. The cyclin-dependent kinase 1A inhibitor of p21CIP1/WAF1 is the major downstream target gene of activated p53 and is responsible for causing cell cycle arrest following DNA damage. These p21-mediated eGFP reporter gene were used in the bioassay system for genotoxicity detection in human hepatoma HepG2 cells [113]. A fish cell biosensor system for genotoxicity detection was created by the integration of three plasmids of pGL3-p21-luc (p21 promoter linked to firefly luciferase gene), pRL-CMV (CMV promoter linked to *Renilla luciferase* gene) and pcDNA3.1 into FG cells [26]. In that biosensor system two reporter genes were introduced and they were simultaneously expressed and measured sequentially within a single test system. The expression of firefly luciferase is correlated with the DNA damage response to genotoxicants. The expression of *Renilla luciferase* serves as an internal control normalizing the experimental variability caused by differences in cell viability or extraneous influences in dual-reporter assays including pipetting volumes, cell lysis and assay efficiency. It was obtained more reliable data by this fish cell biosensor system in comparison with the single luciferase reporter systems and was concluded that the fish cell biosensor system may become a specific and sensitive tool for genotoxicity detection of new chemicals and drugs. Moreover, that the FG cell line has been established and widely used to study the toxic effects and mechanisms of environmental pollutants on fish species [69, 78]. Unlike mammalian cells, FG cells can be easily maintained in a wide range of temperatures from 15–30 °C. This will provide an extraordinary merit in the shelf life and transportation once this fish cell biosensor system is marketed.

The one of very important problem which arouses at the creation of any biosensors is the optimization of the integration of the biological selective structures with the transducer surface. Especially it is appeared at the application of the different types of cells. As a rule for this purpose the number of organic and polymeric materials [61–63]. The recombinant bacteria were incorporated in soft gels such as agarose, polyacrylamide or calcium and strontium alginates and sol-gel [10, 55]. The main problems at the immobilization of genetic engineering of bacteria for the expression of the reporting enzymes in response to physiological stress conditions are connected with the soft hydrogel supports, biodegradation susceptibility, diffusion limitation due to the thick films involved, low physical deformation resistance and the instability of the alginates in calcium-poor solutions and in the presence of calcium chelates. It was used the encapsulation of cells by a dialysis membrane [33] and that based on a glycerol–acryl vinyl acetate copolymer latex [40]. A very good results were obtained with the application of sol-gel when all the immobilized bacteria maintained viability and luminescence activity for several months [10]. The bacteria–silicate hybrids can be used either as disposable sensors or in multiple use sensing test-kits and they can be also integrated in early warning devices operated in continuous flow conditions.

4.1 Fiber Optic SOS-Type Biosensor for the Control of the Genotoxicity of Some Environmental Objects

We early [60, 64] have developed the fiber optic immune biosensor based on the principle of enhanced chemiluminescence for the medical diagnostics, namely for the simultaneous determination of the content of the lidocaine and phenatoine. Late [65] it was proposed the fiber optic biosensor directed on the control of the level of the luminescence level of *Daphnia's* living medium at the control of the toxicity some chemical substances, in particular, mycotoxins.

Next on the basis of such type device it was created biosensor based on the fiber optics at the determination of the genotoxicity effects of the number of the chemical toxic agents [67, 68]. This biosensor was tested at the determination of the genotoxicity of the number substances as: ethanol, dimethylsulfate and mitomycin C. The sensitivity of the proposed biosensor corresponds to the approaches based on the application based on the traditional, complicate and expensive devices. The developed biosensor may be used for the express analysis, namely during 20 min if the optrodes with the appropriate immobilised cells will be prepared in advance. It was informed that according to the preliminary results the functional activity of such optrodes may be served up to one day. It was concluded that the biosensor may have perspective in future for the using in field conditions.

5 Conclusion

The control of the genotoxicity level of the different objects has a special and very important significance since we have now a increasing loading environment by the different chemical substances. Some of them may have not only general toxicity and can generate mutagenic or different genetic effects too. Moreover, there may even be a situation that genotoxicity appears at the low concentrations of the active agent when the overall its toxicity is still quite difficult to detect. At present, there are many approaches that have already convincingly being used in practice. Significant progress in this direction achieved with the development of instrumental methods, but much progress in the control of genotoxicity contributed to the development of biosensor approaches that are able to meet all requirements of the practice, not only in terms of sensitivity analysis but its simplicity, fulfillment in on-line regime and in field condition. There are high hopes not only on progress in the development of biosensor methods in further, but also on the intensification of their practical application. The main directions of both ways of the development of the instrumental methods including biosensors is outlined in this article.

Acknowledgements This work was partially supported by a Science for peace and security NATO programme "Nanostructured materials for the catalytic decontamination of chemical warfare agents", project SPS Ref: 984481 as well as National University of Life and Environmental Sciences of Ukraine, projects "Development of new generation of the instrumental analytical approaches based on the principles of nano-biotechnology and biosensorics for providing systems of biosafety" and "Creation of effective methods for the construction of artificial selective sites for the instrumental analytical devices of new generation based on the principles of biosensorics", N 110/476-Appl. and № 110/71-fund., appropriate.

References

1. Ames BN, McCann J, Yamasaki E, Choi E (1975) Detection of carcinogens and mutagens in *Salmonella*/microsome test. Mutant Res 31:347–364
2. Bartha B, Huber C, Harpaintner R, Schroder P (2010) Effects of acetaminophen in Brassica juncea *L. Czern.*:investigation ofup-take, translocation, detoxification, and the induced defense pathways. Environ Sci Pollut Res 17:1553–1562

3. Baumstark-Khan C, Horneck G (2007) Results from the "Technical Workshop on Genotoxicity Biosensing" on the microscale fluorometric assay of deoxyribonucleic acid unwinding. Anal Chim Acta 593:75–81
4. Baumstark-Khan Ch, Rode A, Rettberg P, Horneck G (2001) Application of the Lux-Fluoro test as bioassay for combined genotoxicity and cytotoxicity measurements by means of recombinant *Salmonella typhimurium* TA1535 cells. Anal Chim Acta 437:23–30
5. Baumstark-Khan C, Khan RA, Rettberg P, Horneck G (2003) Bacterial Lux-Fluoro test for biological assessment of pollutants in water samples from urban and rural origin. Anal Chim Acta 487:51–60
6. Baumstark-Khan Ch, Rabbow E, Rettberg P, Horneck G (2007) The combined bacterial Lux-Fluoro test for the detection and quantification of genotoxic and cytotoxic agents in surface water: Results from the "Technical Workshop on Genotoxicity Biosensing". Aquatic Toxicol. 85:209–218
7. Benton MG, Glasser NR, Palecek SP (2007) The utilization of a *Saccharomyces cerevisiae* HUG1P-GFP promoter-reporter construct for the selective detection of DNA damage. Mut. Res. 633:21–34
8. Benton MG, Glasser NR, Palecek SP (2008) Deletion of MAG1 and MRE11 enhances the sensitivity of the saccharomyces cerevisiae HUG1P-GFP promoter-reporter construct to genotoxicity. Biosens Bioelectron 24(4):736–741. doi:10.1016/j.bios
9. Ben-Yoav H, Biran A, Pedahzur R et al (2009) Whole cell electrochemical biosensor for water genotoxicity bio-detection. Electrochim Acta 54(25):6113–6118
10. Bettaieba F, Ponsonnet L, Lejeunec P et al (2007) Immobilization of *E. coli* bacteria in three-dimensional matrices for ISFET biosensor design. Bioelectrochem 71(2):118–125
11. Bohmdorfer G, Schleiffer A, Brunmeir R et al (2011) GMI1, a structural-maintenance-of-chromosomes-hinge domain-containing protein, is involved insomatic homologous recombination in Arabidopsis. Plant J 67:420–433
12. Buchingera S, Grill P, Morosow V et al (2010) Evaluation of chrono-amperometric signal detection for the analysis of genotoxicity by a whole cell biosensor. Anal Chim Acta 659:122–128
13. Chalfie M, Tu Y, Euskirchen G, Ward WW, Prasher DC (1994) Green fluorescent protein as a marker for gene expression. Science 263:802–805
14. Chen Z, Lu M, Zou D, Wang H (2012) An E. Coli SOS-EGFP biosensor for fast and sensitive detection of DNA damaging agents. J Environ Sci 24(3):541–549
15. Collins AR (2004) The comet assay for DNA damage and repair:principles, applications and limitations. Mol Biotechnol 26:249–261
16. Cormack BP, Valdivia RH, Falkow S (1996) FACS-optimized mutants of the green fluorescent protein (GFP). Gene 173(1):33–38
17. Cormack BP, Bertram G, Egerton M (1997) Yeast-enhanced green fluorescent protein (yEGFP): a reporter of gene expression in Candida albicans. Microbiology 143(Pt2):303–311
18. Crameri A, Whitehorn EA, Tate E, Stemmer WP (1996) Improved green fluorescent protein by molecular evolution using DNA shuffling. Nat Biotechnol 14:315–319
19. Dammeyer T, Tinnefeld P (2012) Engineered fluorescent proteins illuminate the bacterial periplasm. Comput Struct Biotech J 3(4):e201210013
20. De las Heras A, Carreño CA, Martínez-García E, de Lorenzo V (2010) Engineering input/output nodes in prokaryotic regulatory circuits. FEMS Microbiol Rev 34(5):842–865
21. Demerec M (1948) Introduction of mutations in drosophila by dibenzanthracene. Genetics 33:337–348
22. Elad T, Belkin S (2013) Broad spectrum detection and "barcoding" of water pollutants by a genome-wide bacterial sensor array. Water Res. doi:10.1016/j.watres.04.011
23. Fiskesjo G (1988) The allium test—an alternative in environmental studies: the relative toxicity of metal ions. Mutat Res 197:243–260
24. Fu Y, Pastushok L, Xiao W (2008) DNA damage-induced gene expression in *Saccharomyces cerevisiae*. FEMS Microbiol Rev 32(6):908–926

25. García-Alonso J, Greenway GM, Hardege JD et al (2009) A prototype microfluidic chip using fluorescent yeast for detection of toxic compounds. Biosens Bioelectron 24:1508–1511
26. Geng D, Zhang Z, Guo H (2012) Development of a fish cell biosensor system for genotoxicity detection based on DNA damage-induced trans-activation of p21 gene expression. Biosensors 2:318–340. doi:10.3390/bios2030318
27. Hastwell PW, Chai LL, Roberts KJ et al (2006) High-specificity and high-sensitivity genotoxicity assessment in a human cell line: validation of the GreenScreen HC GADD45a-GFP genotoxicity assay. Mutata Res 607(2):160–175. doi:10.1016/j.mrgentox.04.11. eScholarID:1c6091
28. Hastwell PW, Webster TW, Tate M et al (2009) Analysis of 75 marketed pharmaceuticals using the GADD45a-GFP 'GreenScreen HC' genotoxicity assay. Mutagenesis 24:455–463
29. Herskowitz IM (1950) The differential induction of lethal mutations by formalin in the two sexes of drosophila. Science 112(2907):302–303. doi:10.1126/science.112.2907.302
30. Ho YL, Ho SK (1981) Screening of carcinogens with the prophage clts857 induction test. Cancer Res 41:532–536
31. MedicalPlanet http://medicalplanet.su/genetica/111.html
32. Ichikawa K, Eki T (2006) A novel yeast-based reporter assay system for the sensitive detection of genotoxic agents mediated by a DNA damage-inducible LexA-GAL4 Protein. J Biochem 139:105–112
33. Ikariyama Y, Nishiguchi S, Koyama T et al (1997) Fiber-optic-based biomonitoring of benzene derivatives by recombinant E. coli bearing luciferase gene-fused TOL-plasmid immobilized on the fiber-optic end. Anal Chem 69:2600
34. ISO 13829 (2000) Water quality—Determination of the genotoxicity of water and waste water using the umu-test
35. Kamisugi Y, Schaefer DG, Kozak J et al (2012) MRE11 and RAD50, but not NBS1 are essential for gene targeting inthe moss Physcomitrella patens. Nucleic Acids Res 40:3496–3510
36. Knight AW, Birrell L, Walmsley RM (2009) Development and validation of a higher throughput screening approach to genotoxicity testing using the GADD45a-GFP GreenScreen HC assay. J Biomol Screen 14:16–30
37. Krishnan S, Hvastkovs EG, Bajrami B et al (2008) synergistic metabolic toxicity screening using microsome/DNA electrochemiluminescent arrays and nanoreactors. Anal Chem 80:5279–5285
38. Ku WW, Bigger A, Brambilla G et al (2007) Strategy for genotoxicity testing-metabolic considerations. Mutat Res 627(1):59–77
39. Kumaravel TS, Vilhar B, Faux SP, Jha AN (2009) Comet assay measurements: a perspective. Cell Biol Toxicol 25:53–64
40. Lyngberg OK, Stemke DJ, Schottel JL, Flickinger MC (1999) A single-use luciferase-based mercury biosensor using Escherichia coli HB101 immobilized in a latex copolymer film. J Ind Microbiol Biotechnol 23:668–676
41. McArt DG, McKerr G, Saetzler K et al (2010) Comet sensitivity in assessing DNA damage and repair indifferent cell cycle stages. Mutagenesis 25:299–303
42. Mie Y, Kowata K, Hirano Y et al (2008) Comparison of enzymatic recycling electrodes for measuring aminophenol: development of a highly sensitive natriuretic peptide assay system. Anal Sci 24(5):577–582
43. Zhang M, Zhang C, Li J et al (2011) Inactivation of YAP1 enhances sensitivity of the yeast RNR3-lacZ genotoxicity testing system to a broad range of DNA-damaging agents. Toxicol Sci 120(2):310–321. doi:10.1093/toxsci/kfq391
44. Moreno-Villanueva M, Pfeiffer R, Sindlinger T et al (2009) A modified and automated version of the 'fluorimetric detection of alkaline DNA unwinding' method to quantify formation and repair of DNA strand breaks. BMC Biotechnol 9:39
45. Nelson DL, Cox MM (2005) Lehninger: Principles of Biochemistry, 4th edn. W.H. Freeman and Company, New York, 1098 pp
46. Oda Y, Nakamura S, Oki I et al (1985) Evaluation of the new system (umu-test) for the detection of environmental mutagens and carcinogens. Mutat Res 147:219–229

47. Ohno K, Tanaka-Azuma Y, Yoneda Y, Yamada TG (2005) Genotoxicity test system based on p53R2 gene expression in human cells: examination with 80 chemicals. Mutat Res 588:47–57
48. Ohno K, Ishihata K, Ohno K et al (2008) A genotoxicity test system based on p53R2 gene expression in human cells: assessment of its reactivity to various classes of genotoxic chemicals. Mutat Res 656:27–35
49. Ostling O, Johansons KJ (1984) Microelectrophoretic study of radiation-induced DNA damage inindividual mammalian cells. BBRC 123:291–298
50. Polyak B, Bassis E, Novodvorets A et al (2000) Optical fiber bioluminescent whole-cell microbial biosensors to genotoxicants. Water Sci Technol 42:305–311
51. Ptitsyn LR, Horneck G, Komova O et al (1997) A biosensor for environmental genotoxin screening based on an SOS lux assay in recombinant *Escherichia coli* cells. Appl Environ Microbiol 63:4377–4384
52. Quillardet P, Huisman O, D'Ari R, Hofnung M. (1982) SOS chromotest, a direct assay of induction of an SOS function in *Escherichia coli* K-12 to measure genotoxicity. Proc Natl Acad Sci USA 79(19):5971–5975
53. Quillardet P, Hofnung M (1993) The SOS chromotest: a review. Mutat Res 297:235–279
54. Rabbow E, Rettberg P, Baumstark-Khan C et al (2003) The SOSLux-lac-fluoro toxicity test on the International Space Station (ISS). Adv Space Res 31:1513–1524
55. Premkumar JR, Rosen R, Belkin S, Lev O (2002) Sol–gel luminescence biosensors: Encapsulation of recombinant E. coli reporters in thick silicate films. Anal Chim Acta 462: 11–23
56. Rettberg P, Bandel K, Baumstark-Khan C, Horneck G (2001) Increased sensitivity of the SOS-LUX-Test for the detection of hydrophobic genotoxic substances with *Salmonella typhimurium* TA1535 as host strain. Anal Chim Acta 426:167–173
57. Rhee SW, Taylor AM, Tu CH et al (2004) Patterned cell culture inside microfluidic devices. Lab Chip 4(5):102–107
58. Schroder P, Collins CJ (2002) Conjugating enzymes involved in xenobiotic metabolism of organic xenobiotics in plants. Int J Phytorem 4:247–265
59. Shimomura O, Johnson FH, Saiga Y (1962) Extraction, purification and properties of aequorin, a bioluminescent protein from the luminous hydromedusan, Aequorea. J Cell Comp Physiol 59:223–239
60. Starodub NF, Arenkov PJ, Starodub AN, Berezin VA (1994) Fiber optic immunosensors based on enhanced chemiluminescence and their application to determine different antigens. Sens Actuators, B 18–19:161–165
61. Starodub NF, Samodumova IM, Starodub VN (1995) Usage of organosilanes for integration of enzymes and immunocomponents with electrochemical and optical transducers. Sens Actuators, B 24–25:173–176
62. Starodub NF, Rebriev AV (2007) Liquid photopolymerizable compositions as immobilized matrix of biosensors. Bioelectrochem 71(1):29–32
63. Starodub NF (2011) Photopolymerizable materials in biosensorics. In: Ekundayo E (ed.) Environmental Monitoring, pp. 299–326. InTech. ISBN:978-953-307-724-6
64. Starodub NF (2013a) Biosensors for the control of biochemical parameters in the diagnostics of diseases. In: Nikolelis DP, Varzakas T, Erdem A, Nikoleli G-P (eds) Portable Biosensing of Food Toxicants and Environmental Pollutants. Book of Series in Sensors, pp. 743–775. CRC Press, Taylor & Francis, Group Boca Raton, London, NewYork
65. Starodub NF (2013b) Efficiency of biosensors in environmental monitoring. In: Nikolelis DP, Varzakas T, Erdem A, Nikoleli G-P (eds) Portable Biosensing of Food Toxicants and Environmental Pollutants. Portable Biosensing of Food Toxicants and Environmental. CRC Press, Taylor & Francis Group Boca Raton London, New York, pp. 515–560
66. Starodub NF (2015) Genotoxiticy: modern instrumental approaches for its control in environmental objects. J Biosens Bioelectron 6(169):2. http://dx.doi.org/10.4172/2155-6210.1000169

67. Starodub NF, Ruban J, Taran MV et al (2015a) Development of instrumental analytical approaches based on the principles of biosensorics for control of quality of milk and milk products. In: Abstract of International Science-Technology Conference State and Perspective of Food Science and Industry, Ternopil (Ukraine), 8–9 Oct 2015, pp. 176–177
68. Starodub NF, Shavanova KE, Taran MV, Shpirka NF (2015b) Modern instrumental analytical approaches on the basis of principles of biosensorics for the estimation of general toxicity and genotoxicity of environmental objects. In: Abstract Conference "Modern Achievements of Biotechnology and Nanopharmacology", 22–23 Oct 2015, Kiev, pp. 112–113
69. Su F, Zhang SC, Li HY, Guo HR (2007) In vitro acute cytotoxicity of neonicotinoid insecticide imidacloprid to gill cell line of flounder Paralichthy olivaceus. Chin J Oceanol Limnol 25:209–214
70. Taran MV, Shavanova KE, Marchenko OA (2011) Investigtion of mutagenic effect of synthetic flavors. In: Abstract of International Science Conference of Young Scientists, 19–22 Sept 2011, Kiev, p. 9
71. Taran MV, Shavanova KE (2011) Effect of food additives on mitotic activity Allium sepa. In: Abstract of International Conference "Youth and Progress of Biology", 5–8 Apr 2011, Lviv, Ukraine, pp. 171–172
72. Tsien RY (1998) The green fluorescent protein. Ann Rev Biochem 67:509–544
73. Van der Meer JR, Belkin S (2010) Where microbiology meets microengineering: design and applications of reporter bacteria. Nat Rev Microbiol 8(7):511–522
74. Verschaeve L, Van Gompel J, Thilemans L et al (1999) VITOTOX bacterial genotoxicity and toxicity test for the rapid screening of chemicals. Environ Mol Mutagen 33:240–248
75. Vollmer AC, Belkin S, Smulski DR et al (1997) Detection of DNA damage by use of *Escherichia coli* carrying recA':lux, uvrA':lux, or alkA':lux reporter plasmids. Appl Environ Microbiol 63(7):2566–2571
76. Ting W, Zhang C, Xin X et al (2013) Construction and evaluation of two biosensors based on yeast transcriptional response to genotoxic chemicals. Biosens Bioelectron 44:138–145
77. Wentzel JF, Gouws C, Huysamen C et al (2010) Assessing the DNA methylation status ofsingle cells with comet assay. Anal Biochem 400:190–194
78. Zager V, Cemazar M, Hreljac I et al (2010) Development of human cell biosensor system for genotoxicity detection based on DNA damage-induced gene expression. Radiol Oncol 44: 42–51
79. Zeringue HC, Rutledge JJ, Beebe DJ (2004) Early mammalian embryo development depends on cumulus removal technique. Lab Chip 5(1):86–90

Efficiency of Instrumental Analytical Approaches at the Control of Bacterial Infections in Water, Foods and Feeds

Nickolaj F. Starodub, Yulia O. Ogorodniichuk and Oleksandra O. Novgorodova

Abstract Bacterial agents are those that are pathogenic to humans, plants and animals either by causing infectious diseases or by producing toxins. The review provides the description of some pathogenic bacteria, which are a threat to human health and life and must be strictly controlled in the food, water and the environment. With increasing reports on bioterrorism and other bio-threats, rapid and real time detection methods for various pathogens are warranted. Nowadays is important to develop strategies for early detection and monitoring bacterial agents under any conditions that warrant their recognition, including clinical-based diagnostics and biological warfare applications. The review is devoted to the microorganism's indication methods. A study of methods of exposure and authentication of biological agents is important design of biosensors and automatic microorganism's indication. In this study the authors compare the characteristics of the immune biosensors based on the SPR, TIRE, quartz crystal acoustic wave, amperometry, chemiluminescence and on the ISFETs with CeO_x gate surface and conclude that they have similar sensitivity. Special attention is paid to biosensors and last tendencies of their creation based on nanostructures such as deposited nanorods, quantum dots and graphene nanostructures. The achievements of the authors in this field and other researching groups all over the world are described.

Keywords Microroganisms · Identification · Control level · Biosensors · Bioterrorism

N.F. Starodub (✉) · Y.O. Ogorodniichuk · O.O. Novgorodova
Laboratory of Biosensorics, National University of Life and Environmental Sciences of Ukraine, 15, Geroyiv Oborony, Kyiv 03041, Ukraine
e-mail: nikstarodub@yahoo.com

Y.O. Ogorodniichuk
e-mail: ogorodniichuk@mail.ru

O.O. Novgorodova
e-mail: oleksandra_n@yahoo.com

D.P. Nikolelis and G.-P. Nikoleli (eds.), *Biosensors for Security and Bioterrorism Applications*, Advanced Sciences and Technologies for Security Applications,
DOI 10.1007/978-3-319-28926-7_10

1 Introduction

The world we live in has been dealt with key security challenges for years and it seems that nowadays, human security as well as national security level interchangeable. Terrorism by itself is a shadow in almost every part of our lives; however, it is also a reason for all of us to work on preventing and preempting possible consequences of its strike. That could be done through addressing the threat and forming responsible policy regarding it on a timely basis.

Foodborne infections and intoxications are a global problem. Changed agricultural production methods and consumption habits as well as modern harvesting, processing and packaging technologies promote the development of new and emerging pathogens. Travel and border crossing trade facilitate their global expansion. Diseases considered beaten are on the rise again. Bioterrorism emerges as a new threat to food safety [1].

"Food security is a huge issue for businesses, governments and society. As populations grow and climate change and competition for land use are taken into consideration, the problem of securing future food supplies is only going to get worse" says Neil Smith, Lloyd's Emerging Risks & Research Manager. He continues, "Insurance is likely to play a key role in mitigating some of the risks relating to food security, including agroterrorism" [4].

Agro-terrorism is a new term for one kind of ecological terrorism, which means terrorists attempts to destroy or damage agriculture in a country by misusing plant or animal pathogens to cause disruption in affected field. However, the term is relatively new, while the practice is not.

Biological warfare agents can be classified as microorganisms and toxins, these can be extremely toxic to humans and animals and the substances work to kill or incapacitate a population. Biological weapons may be used to target living organisms such as humans, animals or vegetation as well to contaminate essential materials such as air, water and soil. Massive epidemic rooted by pathogenic bacteria have been reported throughout human history. Thus, the increased threat of bioterrorism attacks to the public has stimulated growing demand for rapid detection of biological warfare agents (BWAs) in order to take effective countermeasures [98].

Bacterial agents are those that are pathogenic to humans, plants and animals either by causing infectious diseases or by producing toxins. Antibiotic medications have been successful in controlling some bacterial infections; however, bacteria can gain antibiotic resistance, either naturally or through genetic modification [69].

Nowadays is important to develop strategies for early detection of pathogens responsible for emerging foodborne diseases, and re-evaluate pathogens that are gaining importance for food safety. The rapid detection of foodborne pathogens is of vital importance to keep the food supply rid of contamination. The review provides the description of some pathogenic bacteria, which are a threat to human health and life and must be strictly controlled in the food, water and the environment. One of the global challenges is to provide both rapid and highly sensitive determination of pathogens. High-specific and rapid methods of bacteria detection are urgently needed in all segments such as medical, environmental, food, and military. Furthermore, over the past few years, there has been a shift toward more pathogen detection on-site e.g. in food-processing plants and this demands portability and non-laboratory technics. We believe that the problem can be solved using new methods of analysis based on principles of biosensors. In the review, we provide overview of the most common pathogenic bacteria that pose a threat for human health and describe the most frequently used conventional and rapid methods. Special attention is paid to biosensors and last tendencies of their creation based on nanostructures such as deposited nanorods, quantum dots and graphene nanostructures. The achievements of the authors in this field and other researching groups all over the world are described.

2 Pathogens

The presence of microorganisms in a commodity is not necessarily an indicator of hazard to a consumer or of inferior quality. Moulds, yeasts, and bacteria are almost always found in food and water unless they are sterilized. When these are contaminants, microorganisms may be innocuous, others may cause spoilage, and still others may cause disease. The possibility of commodities becoming hazardous to consumers increases significantly if sanitation or hygiene is compromised. As a result, many international organizations such as the International Commission on Microbiological Specifications for Foods (ICMSF) and the Joint FAO/WHO Codex Alimentarius Commission as well as regional/country jurisdictions (e.g., European Commission Regulation [EC] No. 2073/2005) have established hygienic practices, sampling plans and microbiological specifications as well as other composite programs such as the Hazard Analysis Critical Control Point System (HACCP) to help prevent food- and water-borne diseases [31].

In recent years public concern about the safety of foods of animal origin has heightened due to problems that have arisen with bovine spongiform encephalopathy (BSE), dioxin contamination, outbreaks of foodborne bacterial infections, as well as growing concern about veterinary drug residues and microbial resistance to antibiotics. These problems have drawn attention to feeding practices within the livestock industry and have prompted health professionals and the feed industry to closely scrutinize food quality and safety problems that can arise in foods of animal origin as a result of animal feeding systems [104].

While the World Organization for Animal Health (OIE) has attempted to include animal diseases that present public health risks in the compilation of these lists (e.g., anthrax, brucellosis), it is clear that the primary focus has been, until recently, on animal diseases of economic importance. Zoonotic pathogens associated with waterborne and foodborne diseases such as Salmonella enterica, *E. coli* O157:H7 and other enterohaemorrhagic *E. coli*, *Campylobacter jejuni*, *Giardia duodenalis*, and *Cryptosporidium parvum* are not included in these lists. Leptospirosis is a list B 386 Waterborne Zoonoses disease. The lack of representation of many zoonotic waterborne and foodborne pathogens in these lists is explained by the fact that they are not frequently associated with severe animal disease. However, the OIE has recently recognized the increasing importance of food safety programmes to member countries in satisfying national and international trade requirements and the fundamental importance of controlling zoonotic pathogens in food animal populations as a cornerstone of these efforts [101].

The gathering, identification of pathogens and bacteriological defense measures in advance in order to be effective, is very difficult. The environment (water, air, soil) reside in different amounts pathogenic microbes and various organic compounds, so when identifying bacteriological pathogens difficulties arise.

2.1 Foodborne Bacterial Infections

2.1.1 *Salmonella* Species

Salmonella spp. is one of the most frequently occurring food borne pathogens affecting the microbial safety of food and causes great concern in the food industry. Representatives of this species can cause different pathogenic diseases which provide a harmful influence to all body systems and are hardly treated. Majority of *Salmonella* is pathogenic for human and animals but in epidemiological meaning only a few of them have especially significant part. Infection spreads very fast and can afflict large quantity of people or animals. For example, it is estimated that about 1.4 million *Salmonella* infections occur each year in the United States, and cause more than 500 deaths annually [91]. Another big concern is *Salmonella spp.* antibiotic resistance. Among 180 raw food samples, including chicken, beef, pork, and shellfish samples about half (50.5 %) of the isolates is resistant to at least one antibiotic. Furthermore, from all food types are isolated multiresistant *Salmonella* isolates resistant to at least three different classes of antibiotics [93]. This fact complicates treatment of diseases caused by *Salmonella* representatives, and makes it possible to use resistant *Salmonella spp.* as a biological weapon [78].

Many species of *Salmonella* have been implicated in diseases of farm animals. Of these, *S. typhimurium* is universally distributed while *S. enteriditis* has emerged as a regular pathogen of poultry and contaminant of eggs and chicken meat. Animal feeds are thought to be an important source of these bacteria. Meat and bone meal and fishmeal are frequently contaminated with *Salmonella*. Intensive pasture utilization provides an additional source through contamination of faeces from infected animals. Furthermore, the practice of spreading cattle slurry on to pastures in conventional and organic farms is another potentially significant source of infection.

In many parts of the world poultry manure is used as a feed for ruminants. For example in the USA, two poultry waste products are available for such use: dried poultry waste and dried poultry litter [45]. Dried poultry waste represents undiluted excreta generally derived from caged layer flocks, whereas dried poultry litter is a mixture of excreta and litter. These products are heated to reduce bacterial contamination but are, nevertheless, not sterile. It is reassuring to note that in the trail of [45] that all samples of processed poultry litter collected from 13 dairy farms were totally free of *Salmonella* contamination even though virtually all samples contained Enterobacteriacea, non-glucose fermenting Gram-negative and Gram-positive bacteria.

Numerous analytical methods were developed for the detection of Salmonella antibodies including Widal agglutination test, enzyme linked immunosorbent assay [12, 68], antiglobulin haemagglutination test [19], indirect haemagglutination test [65], latexagglutination test [67] and counter immunoelectrophoresis [40]. However, these methods require complex sample pretreatment procedures and expert personnel and also they are time consuming. Among the serological tests, Widal test is commonly employed for reactive diagnosis of typhoid fever but it is not very reliable [52, 73].

2.1.2 *Escherichia Coli* Including *E. Coli* O157

It is widely recognised that cattle feeds contain *E. coli* through contamination with faeces. There is particular concern over the occurrence of *E. coli* O157 since this form has been definitively linked with specific outbreaks of illness in humans.

The application of slurry on to pastures means that there is potential for the transfer of faecal *E. coli* to grazing animals, a practice that has caused some disquiet among those concerned with food safety. It is important to distinguish the notion of `reservoir' from that of an incidental host as regards the source of *E. coli* O157. They were not able to support the thesis that cattle are the reservoir species of this form of the bacterium. The impact of spreading slurry on land still requires evaluation but proceeded to recommend the introduction of precautionary measures with this practice. Thus, the contamination of pasture grass with *E. coli* O157 from slurry remains a contentious issue.

Other forms of *E. coli* occur extensively in cattle feeds. Thus, it was reported that just over 30 % of cattle feed samples collected from 13 dairies and four feed mills in USA were found to be contaminated with *E. coli*, although *E. coli* O157 was not detected in any of these samples. In five dairies, concentrations of *E. coli* exceeded 1000 colony forming units/g feed. These authors suggested that attention should be focused on the replication of *E. coli* in moist feeds and duration of storage in feed bunks. Other studies in the USA also demonstrated the occurrence of *E. coli* (non-O157) in cattle feeds. Jeffrey et al. [45] reported that 13 non-O157 *E. coli* bacteria were isolated from 52 samples of dried poultry litter destined for dairy feeds [17].

2.2 *Waterborne Bacterial Infections*

The environmental burden of excreta from domestic animals, wildlife, and human beings will increase in coming decades, and excreta are likely the largest source of pathogens for the environment. The potential for transfer of these pathogens to surface water and groundwater is evident. Current water treatment systems have been designed to address some of the well known waterborne diseases (e.g., cholera). However, pathogens that are important causes of waterborne illness today and those that represent potential emerging threats present significant challenges for current strategies to prevent waterborne illness [26].

Management of water quality has been applied mainly to receiving waters contaminated by point sources of human pollution. Waters affected by point sources of pollution are usually subject to regulation, because the pollution frequently impacts bathing beaches or shellfish harvesting waters, as well as drinking-water supplies. Water resources contaminated by dispersed, unidentified sources of pollution, the type usually associated with animals, have not been given special attention until fairly recently. This is the result of a greater awareness of emerging waterborne pathogenic zoonotic microorganisms and improved technical methods to measure water quality. This, in turn, has led to some technical problems that were not anticipated when microbial indicators of faecal contamination were first proposed as a means of monitoring water quality. These problems include the methodology and microbes traditionally.

The monitoring of water quality and the recognition of emerging zoonotic pathogens as a waterborne risk, resulting, in some instances, from changing host population behaviour patterns. At the beginning of the 20th century, Escherich noted that harmless, easily cultivated bacteria occurred in faeces. The organism was suggested for use as an indicator of the presence of faeces from warm-blooded animals in water. The practice of using *Bacillus coli* (later named *Escherichia coli*) as a measure of water quality was adopted, and, over the years, it was used in many forms, dependent on the practical methodology available. Coliforms, faecal coliforms, and *E. coli* were measured, respectively, over time as new specific methods became available. The measurement of *E. coli* in water served as evidence of the presence of faeces and even as a guide to how much faecal contamination was present. This approach appeared to work, because it was applied to waters near relatively small urban centres with poor infrastructure, little water treatment, and populations with little mobility. However, this approach to measuring water quality has been shown to have many inadequacies in identifying public health risk, no doubt related to extending its practical use to unintended applications that have never been validated. Urban centres have expanded beyond their central confines, to the extent that suburban spread is encroaching on feral animal habitat. Another pressure not envisioned when microbial indicators, such as *E. coli*, were proposed is the commercialization of aquatic food resources harvested from waters that were frequently contaminated with human and animal faecal wastes. Similarly, the practice of confining large numbers of animals to feedlots has created a situation where very large volumes of faecal waste frequently reach waterways, leading to pollution of aquatic food sources and water resources used for recreation. In addition, much of the increased leisure time available to humans is spent on recreational water activities far from urban centres and nearer to natural animal habitats. This population movement has led to regulations attempting to govern the safety of all surface waters, including those with non-anthropogenic sources of faecal contamination. However, the means by which we monitor surface waters to provide evidence that they are free of faeces—and, therefore, zoonotic enteric pathogens—has changed little in the last 100 years. The microbial indicator of faecal contamination has not changed, and the methodology for measuring these faecal indicator organisms has changed very little. The methods still rely on bacterial growth in culture media to enumerate their presence in water samples. This usually requires a 24-h period of growth before results are obtained, presenting a situation where the potential risk associated with a water body is detected long after a risk activity has occurred [59].

2.2.1 *Pseudomonas aeruginosa*

Pseudomonas aeruginosa is a common environmental organism and can be found in faeces, soil, water and sewage. It can multiply in water environments and also on the surface of suitable organic materials in contact with water. *Pseudomonas aeruginosa* is a recognized cause of hospital-acquired infections with potentially serious complications. It has been isolated from a range of moist environments such as sinks, water baths, hot water systems, showers and spa pools [59].

The widespread habitat of *P. aeruginosa* in nature, which includes soil, water, food and the surfaces of plants and animals, makes very difficult controlling the organism in hospital settings, and prevention of contamination is almost impossible. The list of materials from which this bacterium can be isolated is almost endless and, despite the recent development of different detection methods with no need for target amplification and enrichment, no sufficiently quick and reliable diagnostic systems for *P. aeruginosa* detection are available to date [22].

The risk of illness associated with exposure to foods harvested from animal contaminated waters or the risk of illness due to direct exposure to these waters is real, but largely unquantified. Although we are fully aware of waterborne zoonotic illnesses through individual cases and outbreaks of illness, we have no way of predicting illness associated with animal-contaminated waters. Our current system for monitoring water quality is designed to protect human health from human-derived pathogens. Unfortunately, the source of faecal indicator bacteria used to monitor water quality is not specific to humans. Currently used indicator methodology cannot distinguish an *E. coli* from animals from an *E. coli* whose source is human.

This shortcoming has led many countries to treat animal-contaminated water as if it were water contaminated by humans and therefore of equal risk. Some solutions to these problems are presented in the chapters of this section. The means to define risk associated with waterborne zoonoses, techniques for identifying sources of animal pollution, and techniques for measuring the safety of animal-contaminated water are discussed with respect to providing tools and information for managing water resources [26].

Detection of contaminated water by pathogenic microorganism is an important concern for ensuring water safety, security and public health. A clean and treated water supply to each house may be the norm in Europe and North America, but in developing countries, access to both clean water and sanitation are not in the prime focus thus waterborne infections are common. Two and a half billion people have no access to improved sanitation, and more than 1.5 million children die each year from water borne diarrheal diseases [59].

In regard to waterborne pathogen detection, there are several important issues to consider. These chips are designed to detect minute quantities of target molecules. The target molecules must be free from contaminants before being applied to the chips. In order to utilize these chips, water samples must go through extensive cell amplification (growth in culture) or sample concentration by filtration, separation, absorption, or centrifugation. Particles in a water sample will easily block the channels used in the microfluidic biochips. The necessity of preparing a water sample prior to analysis will not allow the biochips to provide "real-time" detection of pathogens in water at this time. Another concern is viability of the pathogens to be detected by biochips. Monitoring the presence of some target molecule will provide evidence only for the presence or absence of the target pathogen, not for the viability of the pathogen in question. Some form of culture enrichment to ensure growth is still needed in order to obtain meaningful results. The potential of biochips and microarrays for waterborne pathogen detection is great; at this time, however, much more research is needed to make this technology a reality in applied water microbiology [26, 59].

2.3 Animal Feed Pathogens

Potential bacterial pathogens found in the feed are *Cl. Perfringens, Cl. Botulinum, Listeria* s. (*L. monocytogenes*), *Escherichia coli, Salmonella* spp. Feed, therefore, may contain a diverse microflora originating from soil that is characterized by its ability to survive under conditions of desiccation and a wide range of microenvironments. The dissemination of bacteria may be assisted by boring insects that invade seeds. Low water activity limits bacterial grown in stored grain. When animal feed is stored under moist, anaerobic conditions such as silage or haylage, bacterial diversity is inhibited by low pH. However, bacteria in feed have received relatively little attention compared to moulds and fungi [60].

2.4 Approaches for Microbial Diversity Characterization

Researchers have investigated microbial population behavior in complex environments by examining DNA sequences that are conserved across pathogen species, plasmids, or by using restriction enzymes to characterize genomes based on sequences vulnerable to digestion. [16] used small subunit (ss) rRNA genes to examine microbial population diversity in pasture soil. [42] used a combination of

16S rDNA analysis and BOX-PCR (targeting repetitive intergenic sequence elements of *Streptococcus* spp.) to investigate nitrogen-fixing *Azoarcus* spp. in soil. Restriction fragment length polymorphism (RFLP) analysis, in which population diversity is measured by variations in the length of an amplified 16S rDNA gene, has been used in environmental samples such as bioreactor sludge, termite guts, and seagrass (*Halophila stipulacea*), and to determine bacterial population differences in cow versus human fecal populations [13, 57, 103]. If specific microorganisms could be tracked throughout the environment, a clearer understanding of transmission cycles may lead to new advances in slowing disease spread.

Molecular analysis has also been applied as a surveillance method to detect microbial contamination in animal feed samples [62]. These techniques have an advantage over accepted methods involving the isolation of bacteria by enrichment, selective culturing, and biochemical or serological identification is that molecular assays are more rapid. Traditional culturing methods can take up to 5–7 days to produce results versus 24–36 h for molecular analysis. Thus, a feed mill that produces several tons of feed per hour can monitor its production in a more timely manner. There are numerous commercial PCR kits available. A sampling of PCR kits for the detection of *Salmonella* spp. include: BAXTM (Qualicon, Wilmington, DE.), ProbeliaTM (Sanofi Diagnosic Pasteur, Marnes La Coquette, France), and TaqManTM (PE Applied Biosystems, Foster City, CA). All of these assays have reported greater than 95 % accuracy in detecting the bacterium in various matrices. Part of the difficulty with molecular assays lies with the problem of extracting and recovering representative samples from feeds for molecular analyses. Several methods for the extraction of microbial nucleic acids have been tested to create a sampling strategy that may be used to isolate and amplify microbial DNA from animal feeds using polymerase chain reaction (PCR). A detergent-based initial detachment of microorganisms from feeds followed by the modified procedure of Widmer et al. [21, 103] was found to be the most robust extraction technique for removing inhibitory compounds [61]. However, the extraction requires an overnight incubation step, which complicates its use as a rapid method. Incubation at lower temperatures (−70 °C instead of −20 °C) may be able to shorten this incubation. As more commercial assays become available standard protocols for extraction, enrichment and PCR can be more systematically evaluated to determine the efficacy, detection limits, and the minimum enrichment time required for routine analyses of animal feeds [60].

3 Traditional Methods for Detection of Waterborne and Foodborne Pathogens

The effective testing of bacteria requires methods of analysis that meet a number of challenging criteria. Analysis time and sensitivity are the most important limitations related to the usefulness of bacterial testing. An extremely selective detection methodology is also required, because low numbers of pathogenic bacteria are often present in a complex biological environment along with many other non-pathogenic bacteria [27].

Many advances have been made in rapid methods and automation in microbiology in the past 20 years. Fung comprehensively reviewed the subject in the inaugural issue of the journal Comprehensive Review in Food Science and Food Safety, published by the Institute of Food Technologists. The history and key developments in the field were described. In addition, advances in sample preparation and treatments, total viable cell count methodologies, miniaturization and diagnostic kits, immunological testing, instrumentation and biomass measurements, genetic testing, and biosensors were described. Finally, US and world market and testing trends and predictions of future developments were included. Of particular interest are the developments in genetic testing and biosensors, which can be used in water microbiology. Traditional microbiological methods rely on growth characteristics of organisms at various temperatures, colony morphology in various cultural media, biochemical reactions in various carbohydrates, and immunological antibody–antigen reactions. These characteristics are influenced by environmental factors and growth conditions of cells [21, 26, 84, 88, 105].

Existing and prospective microbiological control methods may be inherent physical, chemical and biological principles detection of biological agents [98]. Biological methods to control bacterial contamination lies in the fact that the studied samples (water, air, soil, food, swabs techniques, etc.). secrete microorganisms, sow them in culture media and after the incubation temperature count the number of colonies that grew [32]. During testing samples for the presence of microorganisms prefer membrane filtration method. This method air, water or suspension of the sample is passed through a special membrane bacterial filter. More filter placed in a nutrient medium, incubated with the appropriate temperature and count the number of colonies that grew [29]. This method makes it possible to confirm the presence of viable microorganisms in the sample with the highest possible probability. The main factors affecting the efficiency of determining the degree of bacterial contamination, is the object of the sample for analysis, engineering planting, nutrient composition, time and temperature crops, etc. [29, 102].

Microscopic method is used for the detection of the morphologically similar bacteria in the selected material to reveal the suspected agents of disease. This method provides a basis for the approximate diagnosis and usually requires further confirmation with the help of the bacteriological and serological methods. The last ones include the following reactions: agglutination, precipitation, bakteriolysis, haemolysis, complement fixation and others. Bacteriological method is particularly important in the diagnostics of most infectious diseases and it demands the obtaining a pure culture of the causative agent of disease and the further investigations. This method is a key in the diagnosis of bacterial diseases [82].

Existing biological methods of bacterial analysis, despite the high efficiency, usually quite labor-intensive, requiring a lot of time to get results and expensive media for culturing microorganisms [27]. All this prevents use these methods for permanent monitoring and rapid analysis.

In pathogenic bacteria more dangerous are products of their decomposition or life. For food characterized by the presence of toxins and for injectable solutions—lipopolysaccharide (pyrogens). Pyrogens—substances of different origin (mostly biological) that cause a feverish condition of the body with a sharp increase in body temperature [27]. It is possible that with pyrogenic reactions and endotoxins [102]. In this case, pyrogenicity test can be used to determine the presence of the samples. Wide application acquired control method for pyrogenicity by raising body temperature in response to an investigational intravenous drugs tested on rabbits. This method is recorded only in the State Pharmacopoeia of our country but also in the pharmacopoeias of other countries [64].

However, years of experience defining pirogenicity of rabbits revealed shortcomings of this method: the high cost of purchasing and operating the equipment, keeping animals and care for them, the dependence of the results on the individual animal, the need to test the sample for three or more animals to obtain reliable results [64].

At this time in the US Pharmacopoeia and Canada introduced a new method for determining the biological pyrogenicity which is to measure the optical density of the gel amebocyte lysate crab Limulus polyphemus influenced by Gram-negative bacteria endotoxin. This method has been tested on different types of bacteria *Escherichia coli*, *Proteus vulgaris*, *Klebsiella aerogenes*, *Salmonella thyphimurium*, *Pseudomonas aeruginosa*. Its sensitivity is in the range: *E. coli*—750, *P. vulgaris*—200, *P. aeruginosa*—12,000 cells. By A. M. Greek was developed the method of pyrogenicity were asked to determine water for injections largest ratio of fluorescence intensity excimer to monomer pyrene. Also the possibility of determining liposacharyds using photoacoustic spectroscopy, thermal-lens spectrometry.

If microbial samples before testing process solution ethylenediaminetetraacetic acid or methyl thymol blue [74], the reaction threshold is increased and the test gives a positive result for the presence of *E. coli* 125 cells, 50 cells of *P. vulgaris* and *P. aeruginosa* 400 cells per unit volume [25]. This method, despite its advantages, has some significant disadvantages: difficulty obtaining amebocyte lysate, the need for special storage conditions, instability lysate, complexity catching, transportation, holding crabs, lobsters and Xiphosura. That lack of classic biological analyzes the driving force behind the development of methods specific indication and diagnosis of infectious diseases.

Methods for determining the presence of microorganisms in the samples using probes—dyes. The essence of it is in the studied sample injected rhodamine 6G, record its fluorescence intensity on the wavelength of 555 nm in the presence of bacteria that is compared with the intensity of fluorescence standard solution indicator on the same wavelength. The results of measurements are concluding the presence of microorganisms. For research took 3 types of microorganisms—*E. coli*, *Bacillus subtilis*, *P. aeruginosa*. The magnitude of changes in fluorescence intensity depended on the type of indicator bacteria and their strength. The resazurin as indicator, example you can use to determine the total microbial number during the microbiological control of foodstuffs. Experiments were performed cultures: *Bacillus cereus*, *B. subtilis*, *E. coli*, *Staphylococcus aureus*, *Candida utilis* [27].

Most of the microorganisms have the property of absorbing light and intrinsic fluorescence in the ultraviolet area by tryptophan and tyrosine residues of protein toxins [34]. On this basis, it was proposed to use the said property to identify microorganisms in aqueous solutions and in dry soil samples [95], and the fluorescence and light scattering to determine the presence of pathogens in the air [46].

Microorganisms have the ability to not only the absorption in the ultraviolet (UV), but also in the infrared (IR) section. With the shift of the maximum absorption of infrared-spectral range in 1600–1700 cm^{-1} can determine the degree of cell viability [49].

In connection with the development of sensitive equipment to register superweak luminescence (chemiluminescence, bioluminescence and electrochemiluminescence) of various biological objects the ability to use this feature to test the presence of microorganisms in different environments. Some types of microorganism's bioluminescence observed in the presence of ions Mg [30].

The glow is due to the addition of oxygen to the complex luciferase luciferin adenosine monophosphate. The source adenosine diphosphate serves as adenosine triphosphate (ATP) microbes. Sensitivity of 10^3 microbial cells in 1 ml.

Bioluminescence method can be used to control the presence of microorganisms in different environments [18]. In addition, microorganisms can decompose hydrogen peroxide using catalysis in the presence of chemiluminescent compound. It was used during the development of the method of determination of the total concentration and the presence of live microorganism's cells. Applying as oxidizing agent of potassium dihydrogen phosphate, can increase the sensitivity chemiluminescence method [49].

The most sensitive radiometric methods are that allow the definition based on the number $C^{14}O_2$ that emerged in the course of microorganisms with C^{14} labeled carbohydrate nutrient medium, to identify individual cells *E. coli* (even one in the water) [8]. The calculation was performed on a liquid scintillation spectrometer. The presence of the negatively charged bacterial membrane allows to determine their presence in different environments. Technically method is carried out as follows: in the cooking chamber with liquid nutrient and studied a model containing two electrodes—measuring and standard. Standard covered electrode material, permeable to microorganisms.

When the camera reached some cell concentration, measured potential difference proportional to their number [105]. Possible areas of application: determine the intensity of contamination of samples machinery, food products, evaluation of the effectiveness of antibacterial agents.

We also consider a method that allows the change in the electrical conductivity of the sample slurry to conclude that the degree of growth of microorganisms in the medium and their number. The sensitivity of this method is low. Changing the voltage or current to the electrode is recorded only when the concentration of microorganisms 10^5–10^6 cells per 1 ml [89].

Create a constant electric potential on the graphite electrode leads to inhibition of respiratory activity of microbial cells of *Saccharomyces cerevisiae*, *B. subtilis*, *E. coli*, placed directly on this electrode.

The degree of respiratory activity depends on the electrode potential and the type of bacteria. Fall respiratory activity correlated with a decrease in the number of viable cells. All this suggests the possibility of creating an electrochemical method indicating the presence of microorganisms.

Some interest electric method of identifying and counting the bacteria in suspensions based on determining the change in the polarization of the laser beam that occurs after its passage through the bacteria suspension, oriented in an electric field [58].

There is a way to identify *Enterobacteriaceae* based on tryptychnomu solubilization and decomposition of proteins by their laser mass spectrometry desorption ionization with matrix [75].

One of the methods of high definition and identification of bacteria is to compare the outer membrane protein with a database full set of proteins. For a complete set of tabulations masses used proteins laser mass spectrometry method with liquid chromatography [94, 100]. Identification of proteins on the surface of the outer membrane of bacterial cells is carried out by ligands that fluorescing and have a greater propensity to bind to proteins [28]. This method can be used to indicate and to differentiate between pathogenic and non-pathogenic strains [53].

The surface-enhanced Raman spectroscopy [23, 37] of electromagnetic waves in the optical range sorted on metals chemicals and biological agents (viruses, bacteria, toxins) can also be used to determine small concentrations of highly toxic substances. Two-dimensional spectral resonance scattering in the ultraviolet region allowing identification of bacteria and identify similar genetic samples [10].

In 1985 Karl Myullis with team developed a method for cloning DNA sequences in vitro, known as polymerase chain reaction (PCR). This method of amplification in vitro, by which for a few hours can be identified and multiplying a sequence of DNA in excess of 108 times the output in [10]. PCR can be made fast and highly sensitive determination of Mycobacterium tuberculosis [36, 63, 97], human immunodeficiency virus (HIV), papilloma virus [35] and so on. It is not necessary to work with radioactive isotopes, as amplified viral DNA segment appears immediately after electrophoretic separation of DNA and dye coloring it accordingly [35].

Genotypic characteristics of a cell are far more stable than phenotypic expressions. Genetic methods have been developing rapidly in the past two decades. First, DNA and RNA hybridization techniques were perfected and widely used to rapidly detect organisms such as *Salmonella*, *Listeria*, *Campylobacter*, etc. More recently, PCR has been widely used for the detection of a great variety of microorganisms in clinical, food, industrial, and water microbiology. Basically, a DNA molecule (double helix) of a target pathogen (e.g., *Salmonella*) is first denatured at about 95 °C to form single strands, then the temperature is lowered to about 55 °C for two primers (small oligonucleotides specific for *Salmonella*) to anneal to specific regions of the single-stranded DNA. The temperature is then increased to about 70 °C for a special heat-stable polymerase, the TAQ enzyme from *Thermus aquaticus*, to add complementary bases (A, T, G, or C) to the single-stranded DNA and complete the extension to form a new double strand of DNA. This is called a thermal cycle. After this cycle, the tube is heated to 95 °C again for the next cycle. After one thermal cycle, one copy of DNA will become two copies. After about 21 cycles and 31 cycles, 1 million and 1 billion copies of the DNA will be formed, respectively. This entire process can be accomplished in less than an hour in an automatic thermal cycler. After PCR reactions, one still needs to detect the presence of the PCR products to indicate the presence of the pathogen to be detected. In the original PCR procedure, PCR products were detected by electrophoresis, which is time consuming and laborious. Several new PCR protocols have recently been developed to efficiently report successful PCR reactions. It was described an automatic system for screening a family of PCR assays for pathogens, which combines DNA amplification and automated homogeneous detection to determine the presence or absence of specific targets. All primers, polymerase, and deoxyribonucleotide bases necessary for PCR as well as a positive control and an intercalating dye are incorporated into the single tablet. The system works directly from an overnight enrichment of the target organisms. No DNA extraction is required. Assays are available for *Salmonella*, *E. coli* O157:H7, *Listeria*, and *L. monocytogenes*. The system uses an array of 96 blue light emission diodes as the excitation sources and a photomultiplier tube to detect the emitted fluorescent signal indicating successful PCR reactions. This integrated system improves the ease of use of the assay. The inclusivity and exclusivity of this system reach almost 100 %, meaning that false-positive and false-negative rates are almost zero. Additionally, this automated system can now be used with assays for the detection of *Cryptosporidium parvum* and *Campylobacter jejuni/coli* and for the 372 Waterborne Zoonoses quantitative and qualitative detection of genetically modified organisms in soy and corn described a self-contained PCR system that can report the successful PCR reaction by measuring fluorescence reaction in the experimental chambers. A special molecule is annealed to the single-stranded DNA to report the linear PCR amplification. The molecule has the appropriate sequence for the target DNA. It also has two attached particles. One is a fluorescent particle, and the other is a quencher particle. When the two particles are close to each other, no fluorescence occurs. However, when the TAQ polymerase is adding bases to the linear single strand of DNA, it will break this molecule away from the strand. As this occurs, the two particles will separate from each other, and fluorescence will occur. By measuring fluorescence in the tube, a successful PCR reaction can be determined. A new system using molecular beacon technology has been developed. In this technique, all of the reactions are in the same tube. A molecular beacon is a tailor-made hairpin-shaped hybridization probe. The probe is used to attach to target PCR products. On one end of the probe, a fluorophore is attached; on the other end is a quencher of the fluorophore. In the absence of the target PCR product, the beacon is in a hairpin shape and there is no fluorescence. However, during PCR reactions and the generation of target PCR products, the beacons will attach to the PCR products and cause the hairpin molecule to unfold. As the quencher moves away from the fluorophore, fluorescence will occur, and this can be measured. By using molecular beacons containing different fluorophores, one can detect different PCR products in the same reaction tube and thus perform "multiplex" tests of several target pathogens (Bio-Rad 2003). One of the major problems of PCR systems is that of contamination of PCR products from one test to another. A system developed by the Pasteur Institute attempts to eliminate PCR product contamination by substituting the base uracil for thymine in the entire PCR protocol. Thus, in the reaction tube, there are adenine, uracil, guanine, and cytosine, but not thymine. During the PCR reaction, the resultant PCR products will be AUGC pairing and not the natural ATGC pairing. After one experiment is completed, a new sample is added to another tube for the next experiment. That tube contains an enzyme called uracil-D-glycosylase (UDG), which will hydrolyse any DNA molecules that contain a uracil base. Therefore, if there is contamination from a previous run, the AUGC-DNA will be destroyed before the beginning of the new run. Before a new PCR reaction, the tube with all reagents is heated to 56 °C for 15 min for UDG to hydrolyse any contaminants. During the DNA denaturation step, the UDG will be inactivated and will not act on the new AUGC-PCR products. This Detection/enumeration methods 373 system can detect *Salmonella, Listeria monocytogenes*, and other. The nucleic acid sequence-based amplification (NASBA) technique has been perfected in recent years for water microbiology. It has advantages over the PCR technologies. Since the target is RNA, it can be used to detect RNA viruses and functional mRNA targets. It is isothermal and thus does not require a thermal cycler for the reaction. It is rapid and sensitive for detection of target molecules. PCR, NASBA, and related genetic technologies can be powerful tools for water microbiology once the existing problems are solved and the systems are automated. Also, analysts have to be convinced to invest time and money to convert to these technologies for the routine analysis of water microbiology [17].

4 Biosensors for Detection of Waterborne and Foodborne Pathogens

There are now a large number of biosensors available for detection of target microorganisms in a variety of food, water, clinical, and industrial samples. [43] provided a comprehensive overview of different physicochemical instrumental techniques for direct and indirect identification of bacteria, including infrared and fluorescence spectroscopy, flow cytometry, chromatography, and chemiluminescence techniques, as a basis for biosensor construction [43]. Biosensor development and application are exciting fields in applied microbiology. The basic idea is simple, but the actual operation is quite complex and involves much instrumentation. Basically, a biosensor is a molecule or a group of molecules of biological origin attached to a signal recognition material. When an analyte comes in contact with the biosensor, the interaction will initiate a recognition signal that can be reported in an instrument. Many types of biosensors have been developed, including a large variety of enzymes, polyclonal and monoclonal antibodies, nucleic acids, and cellular materials. In some applications, whole cells can also be used as a biosensor. Analytes detected include toxins (e.g., staphylococcal enterotoxins, tetrodotoxins, saxitoxin, and botulinum toxin); specific pathogens (e.g., *Salmonella*, *Staphylococcus*, and *Escherichia coli* O157:H7); carbohydrates (e.g., fructose, lactose, and galactose); insecticides and herbicides; ATP; antibiotics (e.g., penicillins); and others. The recognition signals used include electrochemical (e.g., potentiometry, voltage changes, conductance and impedance, and light addressable); optical (e.g., ultraviolet, bioluminescence, chemiluminescence, fluorescence, laser scattering, reflection and refraction of light, surface plasmon resonance, and polarized light); and miscellaneous transducers (e.g., piezoelectric crystals, thermistor, acoustic waves, and quartz crystal). Recently, much attention has been directed to the development of "biochips" and "microchips," which may detect a great variety of molecules associated with waterborne and foodborne pathogens.

Due to the advancements in miniaturization technology, as many as 50 000 individual spots (e.g., DNA microarrays), each spot containing millions of copies of a specific DNA probe, can be immobilized on a specialized microscope slide. Fluorescent labelled targets can be hybridized to these spots and detected. Biochips can also be designed to detect all kinds of waterborne bacteria by imprinting a variety of antibodies, or DNA molecules, against specific pathogens on the chip for the simultaneous detection of pathogens such as *Salmonella*, *Listeria*, *E. coli*, and *Staphylococcus aureus* on the same chip. Biochips are an exceedingly important technology in life sciences, and the market value is estimated to reach $5 billion by the middle of this decade. This technology is especially important in the rapidly developing field of proteomics, which requires massive amounts of data to generate valuable information. The development of these biochips and microchips provides an impressive method for obtaining a large amount of information for biological sciences [17, 26].

4.1 SPR Based Biosensors

Optical biosensors are powerful analytical tools for detection of microorganisms, toxins and other substances and represent the largest group among sensor devises. Due to their high sensitivity, ease of use and short time of analysis some of them were successfully commercialized and have been applied in numerous important fields such as medical diagnostics, environmental monitoring, food safety and security. Phenomenon of surface plasmon resonance (SPR) in different modifications is widely used for biosensors creating. SPR phenomenon was first described by Kretschmann in 1971 and then the proposed scheme for SPR was widely used for detection of organic mono- and multi-layers on metal surfaces. After [56] used it in the purposes of biosensing SPR was deeply studied and became a powerful tool for label-free studying of interactions between biological sensing elements and target molecules. By far, the most common SPR sensor platforms are based on prism scheme [41] and angular modulation, but recently, a lot of attention have been paid to the study of waveguides with optical phase detection technique since it demonstrates high sensitivity for detection of bio-reactions [66]. As a biological sensing elements the proteins (e.g., antibodies) and peptides are the most frequently used. In addition, it was shown that immobilization of biomolecules to the bare transducer surfaces has negative impact to their reactivity therefore different methods of previous surface preparation are used. The main aim of surface modification is to provide maximum interaction between biomolecule (ligand) and analyte. Alternative to physical adsorption is covalent binding between amino groups of the protein and activated carboxyls on self-assembled monolayers (SAM) of alkanethiolates or within a dextran matrix. Avidin (or streptavidin)-biotin interactions are widely used for providing effective immobilization of biomolecules, such as oligonucleotides or for site-specific binding of antibodies. In case of antibodies immobilization to get a biomolecular layer with high level of site-oriented molecules intermediate binding molecules are used such as Fc-related proteins A [81, 83], G [72] and hybrid immunoglobulin highly-specific binding proteins LG and LA [47, 90]. Covalent binding of antibodies with SAM-molecules or Langmuir-Blodgett films via thiol groups increases the activity of antibodies up to 70 % [5]. The sensitivity of analysis strongly depends on the algorithms of detection fulfillment. There are next variants: (1) the "direct" way when the specific antibodies, which were immobilized on the transducer surface interact with solution contented different concentration of antigen; (2) the "competitive" way-1 when antigen to be analyzed compete with its conjugate (covalent linked complex antigen with enzyme or some protein) for the specific antibodies which were immobilized on the transducer surface; (3) the "competitive" way-2 when antigen in form of conjugate (covalent linked complex antigen with enzyme or some protein) are immobilized on the transducer surface and it competes with free antigen to be analyzed for the specific antibodies in solution; (4) the "to saturate" way when the specific antibodies which were immobilized on the transducer surface interact at first with free antigen to be analyzed and then with antigen in form of conjugate (covalent linked complex antigen with the enzyme or some protein); (5) the "sandwich method" when antigen interacts at first with specific antibodies which were immobilized on the transducer surface and then with second portion of specific labeled antibodies.

In recent years, the interest and application of biosensors based on SPR for detection and control of pathogens has highly increased. For this purpose immune reaction and direct way of analysis is commonly used. As it was shown before by the numerous studies antibodies immobilization via physical adsorption is ineffective for antigen binding, therefore previous modification of transducer surfaces for providing oriented binding of antibodies has wide application. Koubova´ et al. demonstrated *Salmonella enteritidis* detection with SPR based biosensor using antibodies immobilized to the bare gold surface via physical adsorption. Analysis was performed in a direct way. Concentration of detected cells of bacteria was on the level 10^6 CFU/ml [48]. Barlen et al. attempted to detect *Salmonella typhimurium* and *Salmonella enteritidis* simultaneously in phosphate buffer firs and then in milk using

SPR biosensor "Plasmonic". Detection was performed using "sandwich-method" and polyclonal antibodies were immobilized to the hydrophobic surface. Detection level for *S. typhimurium* in phosphate buffer was 1.25×10^5–2.5×10^6 cells/ml and for *S. enteritidis* it was 2.5×10^8–2.5×10^9 cells/ml. After milk analysis contaminated by given microorganisms, detection level was 2.5×10^5–5×10^5 and 2.5×10^8–3×10^9 cells/ml respectively [11].

Oh with group carried out a series of experiments for *S. typhimurium* and *S. paratyphi* detection using SPR biosensor. For *S. typhimurium* detection the working surface of biosensor was covered by SAM of 11-mercapto-undecanoic acid, then the layer of protein G was formed and after that, the monoclonal antibodies were immobilized. The range of device sensitivity was 10^2–10^9 CFU/ml. For *S. paratyphi* detection protein G was immobilized to the sensing surface using 2-iminohiolan and then the monoclonal antibodies to bacteria were added. Using described method the sensor sensitivity was on the range of 10^2–10^7 CFU/ml [71, 72].

Bokken et al. detected *Salmonella* of group A, B, D and E using commercial SPR based biosensor Biacore. Antibodies were immobilized to the layer of carboxymethylated dextran and for *Salmonella* serotype identification the "sandwich-method" was applied. As a result the detection level of the devise was 1.7×10^3–10^7 CFU/ml even in the presence of the other bacteria cells in quantity of 10^8 CFU/ml [15].

There are numerous articles dedicated to *Salmonella* detection using SPR biosensors based on commercial Spreeta modules. Lan et al. have determined *S. typhimurium* in a quantity of 1×10^6 CFU/ml [50]. Son with group detected *S. enteritidis*. Antibodies specific to *S. enteritidis* were immobilized to the gold surface via neutravidin. The smallest number of detected cells was about 10^5 CFU/ml [79].

Starodub with group also studied the possibility of different substances detection using SPR based biosensors and *S. typhimurium* was among them (Table 1). For the study two SPR biosensors were used: one was based on commercial Spreeta module and the other one "Plasmonotest" was designed V.M. Glushkov Institute of Cybernetics of NAS of Ukraine. As reactive part antigen-antibody interactions were used and previous working surface preparation was occurred which included several sequential steps: (a) covering of surface by polyalylamine hydrochloride (PAA) (b) immobilization of protein A from *Staphylococcus aureus*; (c) the oriented binding of the specific antibodies; (d) bovine serum albumin immobilization (BSA) for blocking free non-specific binding centers on the gold surface. For biological material immobilization on the gold surface polyelectrolytes are widely used [86]. Thin films obtained using small charged organic molecules are common since this molecules form insoluble polymer which electrostatically sorbs molecules with opposite charge [85]. In particular polyallylamin hydrochloride (PAA) immobilization on the transducer surface provides the acquisition of positive charge. To overcome the blocking of active sites of antibodies (antigen-binding sites) because of their interaction with gold surface the protein A from *Staphylococcus aureus* was previously immobilized [85]. The reason is that this protein has affinity to Fc-fragments of immunoglobulin G (IgG) and can bind its molecules without active sites participation. High quantity of free antigen-binding

Table 1 The comparison of the sensitivity of SPR based biosensors depending on the method of working surface preparation and biosensors with different types of transducer

Item	The type of biosensor and transducer modification	Analyte	Level/range of sensitivity
1.	SPR immune biosensor, physical adsorption [48]	*S. enteritidis*	10^6 CFU/ml
2.	SPR immune biosensor "Plasmonic", hydrophobic surface [11]	*S. typhimurium*, *S. enteritidis*	In phosphate buffer: 1.25×10^5–2.5×10^6 cells/ml, 2.5×10^8–2.5×10^9 cells/ml. In milk: 2.5×10^5–5×10^5 cells/ml. 2.5×10^8–3×10^9 cells/ml
3.	SPR immune biosensor; 11-mercapto-undecanoic acid, protein G[1]; 2-iminohiolan, protein G[2] [71, 72]	*S. typhimurium*[1], *S. paratyphi*[2]	10^2–10^9 CFU/ml 10^2–10^7 CFU/ml
4.	SPR immune biosensor "Biacore", carboxymethylated dextran [15]	*Salmonella* of group A, B, D and E	1.7×10^3–10^7 CFU/ml
5.	SPR "Spreeta" module based immune biosensor [50]	*S. typhimurium*	1×10^6 CFU/ml
6.	SPR "Spreeta" module based immune biosensor, neutravidin [79]	*S. enteritidis*	10^5 CFU/ml
7.	SPR "Spreeta" module based immune biosensor; PAA, protein A [70, 81, 83–84]	*S. typhimurium*	10^3–10^7 cells/ml
8.	SPR immune biosensor "Plasmonotest"; PAA, protein A [70, 81, 83–84]	*S. typhimurium*	10^2–10^6 cells/ml.
9.	TIRE based biosensor; PAA, protein A [81]	*S. typhimurium*	Several cells (less than 5) in 10 ml
10.	PhL of ZnO nanorods [96]	*S. typhimurium*	10^2–10^6 cells/ml
11.	Au and magnetic Fe_2O_3 nanoparticles modified with polyaniline [99]	*E. coli* O157:H7	10^1–10^6 CFU/ml
12.	Graphene magnetic nanosheet decorated with chitosan [2]	*P. aeruginosa*, *S. aureus*	In suspension, using fluorescence: 4.5–5.0×10^2 CFU/ml— using MALDI: 4.5–5.0×10^2 CFU/ml In blood: using fluorescence: 1.0–4.0×10^2 CFU/ml— using MALDI: 5.0–6.0×10^2 CFU/ml.
13.	ISFETs based immune biosensor; glutaraldehyde, protein A [81]	*S. typhimurium*	2–5×10^5 cells/ml.

sites improves antigen binding and provides increasing of biosensor sensitivity. In the case of Spreeta based biosensor, it has been defined that device sensitivity was on the level 10^3–10^7 cells/ml. "Plasmonotest" detection level was within 10^2–10^6 cells/ml [70]. After analyzing the above results one can conclude that the lowest limit of sensitivity of SPR based biosensors in case of *Salmonella* detection is on the level of 10^2 cells/ml. But considering the fact that the infectious dose of a pathogen such as *E. coli* O157:H7 or *Salmonella* is as low as 10 cells and the existing coliform standard for *E. coli* in water is 4 cells/100 ml [43] demonstrated technique it is not sufficient for all practice situations. The current situation makes scientists to continue their work in this field and develop more sensitive methods for bacteria detection.

4.2 TIRE Based Biosensors

One of the promising optical methods for biosensors creating is total internal reflection ellipsometry (TIRE). Ellipsometry is an optical method for studying of surfaces and environments, which is based on analysis of amplitude and phase changes of a light wave during its interaction with the investigated object. Later it was studied the possibility of using phase change of reflected light in terms of total internal reflection. This method was called total internal reflection ellipsometry and was first described in 1976 [3]. Getting ellipsometry parameters in terms of total internal reflection significantly increased sensitivity and detection level comparing with conventional ellipsometry and SPR technology. Thus, the method of TIRE provides a level of determination within 5×10^{-7} RIU while for ellipsometric measurements this index is 10^{-5} RIU [44]. The reflected wave is formed on the edge of optically contrasting environments therefore ellipsometric measurements provide the information about optical structure of surface area and these processes that affect to its optical properties [7, 9].

In the work of Starodub et al. TIRE method was applied for *S. typhimurium* detection in model solutions. For the experiment prism scheme sensor platform in combination with specific antibodies as a reactive part were used and previous surface preparation was occurred as described in [70, 81]. Biosensor based on TIRE showed higher sensitivity than the SPR based. Maximal sensitivity was on the level of several cells (less than 5) in 10 ml [70, 81].

TIRE has several advantages that make it attractive for further study and application in field of biosensing. First of all TIRE provides high sensitivity. Thus, sensitivity of biosensors based on ellipsometry can be different for different substances but usually it reaches the level of a few nanograms. At the same time, analysis does not require large amounts of reagents and can be satisfied by using of microliters of samples. Other advantages of this method are: no need of reagents labeling; no damaging effects on the object of study; the possibility of real-time analysis; large measuring range (from nM to μM) and short time of analysis [24, 33, 76].

4.3 PhL of Nanomaterials for Biosensor Application

In recent years, there is a growing interest in studying and obtaining photoluminescent nanomaterials such as nanoscale structures and quantum dots of metal oxides. This is because nanoparticles of metal oxides acquire new qualitative changes of physical and chemical properties, catalytic ability and reactivity, which are not observed in microscopic bodies of the same chemical nature. One of these properties is photoluminescence (PhL). PhL is a powerful technology for development of optical biosensors since it does not require any procedures of bioreceptors preparation, complex electrical circuits and expensive equipment. The principle of PhL based biosensor operating is to measure changes of photoluminescence spectra of nanoparticles (intensity and peak position) caused by interaction of biological components. Many different substances were successfully detected by means of nanoparticle PhL such as ions, DNA molecules,dopamine, carbohydrate antigen, bovine leukemia virus and *S. typhimurium* etc. [38, 39, 54, 77, 96].

Viter et al. described a novel approach of *S. typhimurium* detection in model solutions using biosensor based on room temperature PhL of ZnO nanorods. The metal oxide nanopowder wos dispersed in solution and then deposited on glass substrate. Immobilization of biomolecules on metal oxide surface was performed due to coordination bond and van der-Waals interaction between biomolecules and surface. As a biorecognition element specific polyclonal antibodies to *S. typhimurium* were used. Anti-salmonella antibodies were immobilized on ZnO surface via physical adsorption. To prevent non-specific adsorption of antigen BSA solution was applied after immobilization of the anti-salmonella antibodies. After washing and drying ZnO nanorod layer wos immersed for 20 min into the 0.85 % of NaCl solution containing controlled concentration of *S. typhimurium* cells. On each step PhL of ZnO nanorods was measured. It was shown that PhL intensity of ZnO nanorods increased due to interaction between adsorbed proteins (specific antibodies and BSA) and ZnO surface. After adding of *S. typhimurium* cells and Ag-Ab interaction, decrease of PhL intensity was observed (Fig. 1). The biosensor showed

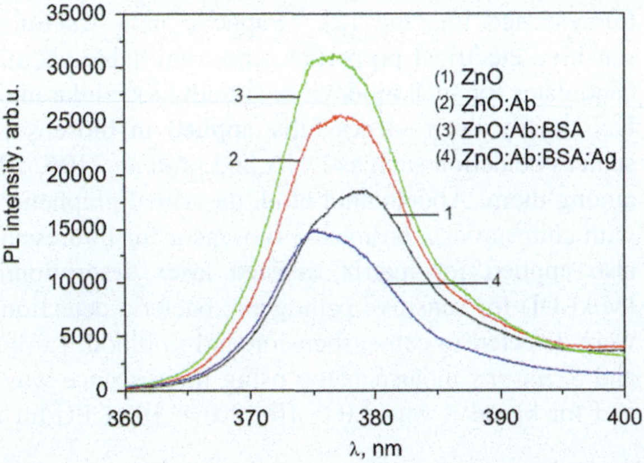

Fig. 1 PhL spectra of ZnO nanorods, after immobilization of anti-*Salmonella* Ab (ZnO:Ab), after BSA (ZnO:Ab:BSA), and after immune reaction with *Salmonella* Ag (ZnO:Ab:BSA:Ag)

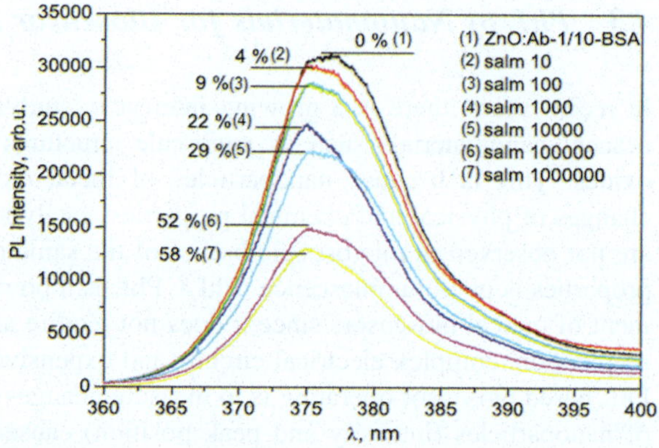

Fig. 2 Sensitivity of the ZnO biosensor to *Salmonella* Ag at different concentrations [96]

the changes of the signal in the range of 10^1–10^6 cells/ml (Fig. 2). The detection limit of biosensor was about 10^2 cells/ml [96]. Chances are to increase sensor sensitivity when surface functionalization will be provided for covalent binding of antibodies.

Gold nanoparticles in combination with magnetic nanoparticles of Fe_2O_3 were successfully combined by Wang et al. for *Escherichia coli* O157:H7. Fe_2O_3 nanoparticles were modified with polyaniline. Magnetic nanoparticles were conjugated with monoclonal antibodies to separate target *E. coli* O157:H7 cells from broth samples. Gold nanoparticles were conjugated with polyclonal Abs, and then were added to solution to form sandwich of magnetic nanoparticles- *E. coli*- gold nanoparticles. The amount of gold nanoparticles was measured through electrochemical method and the amount of the target bacteria were determined. Sensitivity of the described method was 10^1 CFU/ml with a linear range of 10^1–10^6 CFU/ml [99].

Graphene is a novel allotropic member of carbon-based nanomaterials and it has attracted worldwide interest due to excellent features such as high specific surface area, electronic conductivity, thermal conductivity, mechanical strength, cheap and easy to undergo functionalization via π–π interactions, covalent bonding, polymer blending and electrostatic interactions and it has been used for biosensing and fluorescence imaging [2]. Graphene nanostructures with their microscale area, sensitive electrical properties, and modifiable chemical functionality are excellent candidates for such biodevices at both biocellular and biomolecular scale. Graphene has already been successfully applied in biosensors creating for numerous substances detection such as DNA and proteins [106, 107] and pathogenic bacteria are among them. Abdelhamid et al. described graphene magnetic nanosheet decorated with chitosan as a promising biosensor for fluorescence spectroscopy and it can be also applied for matrix assisted laser desorption/ionization mass spectrometry (MALDI) for sensitive pathogenic bacteria detection. *P. aeruginosa* and *S. aureus* were detected in cell suspension and in blood. Limit of detection of *P. aeruginosa* and *S. aureus* in suspension using fluorescence was 5.0×10^2, 4.5×10^2 CFU/ml and for blood it was 4.0×10^2, 1.0×10^2 CFU/ml for each bacteria respectively.

MALDI provided sensitivity on the level of 5.0×10^2, 4.5×10^2 CFU/ml in suspension and 6.0×10^2, 5.0×10^2 CFU/ml in blood samples [2].

4.4 ISFETs Based Biosensors

Ion-sensitive field-effect transistors (ISFETs) represent the group of semiconductor potentiometric devices and are widely applied in biosensors. The interest to these devices only grows since 70 s of past century when P. Bergvald and T. Matsuo have proposed pH-FET as promising transducers for biosensors [20, 80, 92] and reported that these structures are suitable for highly sensitive detection of protons generated during the biochemical reaction and as elements for integrated multifunctional and multiparametric on-chip biosensors. Lately ISFET biosensors were adapted for different purposes including detection of glucose and urea, with the further use in clinical analyses. This tendency can be explained by the numerous advantages of ISFTs such as: miniature, possibility to put several electrodes on a single chip of semiconductor, low cost in mass production. The most often for biosensor creation pH-sensitive FETs are used. The numerous biological substances can be applied for modification of FETs surface to provide its selectivity depending of the task, but the largest number of works is dedicated to the use of enzymes, immunoglobulins and DNA molecules therefore biosensing FET systems are divided to enzymatic, immune and DNA FETs [14, 51]. Immune ISFET based biosensor was successfully applied by Starodub et al. for detection herbicide simazine. To the gate surface of ISFET the protein A from *Staphylococcus aureus* was immobilized to provide site-oriented binding of polyclonal antibodies specific to simazine. Detection of simazine was carried out in two ways such as "competitive" and "to saturate" way. In the first case detection level of simazine was 1,25 ng/ml and the use of "to saturate" method improved the biosensor sensitivity up to 0,65 ng/ml. The paper says that ISFET can be reused several times without signal reducing after it treatment with acid solution and water. The duration of analysis taking into account all preparatory steps was about 50 min. [80].

Starodub and Ogorodnijchuk [81] developed new type of immune biosensor based on ISFETs with CeO_x instead of Si_3N_4 gate surface, which provides high sensitivity and stability of the analysis (Figs. 3 and 4).

The biosensor was used for *S. typhimurium* detection in model solutions using immune reaction. The surface was activated twice by water solution of glutaraldehyde (GA). Next step was the protein A from *Staphylococcus aureus* immobilization and the solution of glycine was applied for blocking of GA non-bonded groups. At last, the working chip was covered by specific antiserum. Analysis was fulfilled by the "sandwich" method in the way when the immobilized specific antibodies interact with *Salmonella* cells diluted in 0.85 % solution of NaCl. Then, bounded cells on the transducer surface were treated by the specific antibodies labeled by horseradish peroxidase (HRP) which activity was registered. For HRP activity registration special buffer was used which contained 5 mM

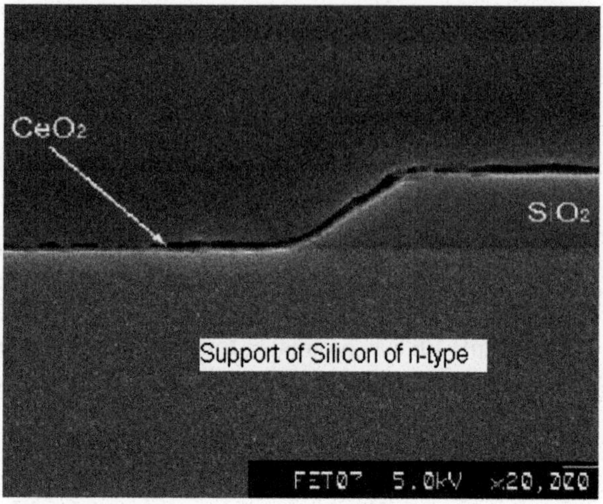

Fig. 3 Scanning electron microscopy of ISFET gate surface

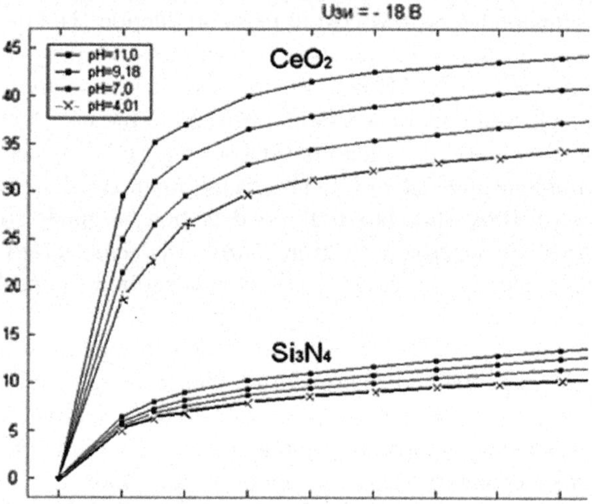

Fig. 4 Influence of pH on the output characteristics of ISFETs with the Si_3N_4 and CeO_2 dielectric structures (Starodub and Ogorodnijchuk [81])

tris-HCl (pH 7.5), 100 mM NaCl, 15 mM ascorbic acid and 5 or 10 mM H_2O_2. Substrate reaction provided local change of pH as a result forming of dehydroascorbic acid that is more alkaline than ascorbic acid. The antigen was in form of suspension of autoclaved cells of *S. typhimurium* diluted in 0.85 % NaCl with initial concentration of 10^6 cells/ml. The conjugate of specific antibodies with HRP have been prepared using previously described method [87]. It was demonstrated that the sensitivity of the analysis of salmonella was about 2–3 cells/ml with the maximal response up to 5×10^5 cells/ml. The total time of the measurements can be reduced to 15–20 min if the transducer surface was previously prepared and if one part of the specific antibodies are immobilized on it and the other one labeled by horseradish peroxidase is added to the analyzed solution. Each chip may be reused up to

5 times without signal decreasing after its treatment by 0.1 M HCl for 5 min followed by washing with 5 mM phosphate buffer, pH 7,5 [81].

The authors compare the characteristics of the immune biosensors based on the SPR, TIRE, quartz crystal acoustic wave, amperometry, chemiluminescence and on the ISFETs with CeO_x gate surface and conclude that they have similar sensitivity but the last approach may provide to achieve low cost of analysis [81]. To have the sensitivity of analysis in the respect of *Salmonella* and others bacteria on the level of infection dose [43] there is necessary to have a special system of the analyzed sample and in particular, accumulation of cells through application of bioaffinic columns as it was early demonstrated by some authors [55]. The comparison of the sensitivity of biosensors with different types of transducer is presented in Table 1.

5 Conclusions

Nowadays the need to develop rapid and highly sensitive methods for pathogens diagnostic is relevant in the world. Thousands of people die and millions illnesses occur each year in developed and developing countries caused by foodborne pathogens. High-specific and rapid methods of bacteria detection are urgently needed in all segments such as medical, environmental, food, and military. The United States Department of Agriculture (USDA) estimated that US food industry alone performs in total more than 144 million microbial tests annually. Known today methods of pathogens detection can be divided into the categories of conventional and rapid methods. Conventional methods involve enriching the food sample and performing various media-based metabolic tests (agar plates or slants). Depending on pathogen and method, tests typically require 3-7 days to obtain a result. Rapid methods, however, are based on immunochemical or nucleic acid technologies. Commercially available rapid tests can provide results in 8-48 h but results from these screening tests are presumptive and require further isolation of organism as proof of contamination. Thus, the combination of speed and sensitivity is the key for any rapid detection method. Furthermore, over the past few years, there has been a shift toward more pathogen detection on-site e.g. in food-processing plants and this demands portability and non-laboratory technics [6]. We believe that all these parameters can be successfully combined in modern instrumental devices based on the principles of biosensorics. As it shown above optical biosensors are portable and provide real-time analysis, which makes them very attractive for rapid-test systems creation. However, the lowest limit of sensitivity of SPR based biosensors for *Salmonella spp.* representatives is on the level of 10^2 cells/ml, which is insufficient since single-cell detection is required. Practical demands can satisfy TIRE based biosensors, which provides sensitivity on the level of several cells (less than 5) in 10 ml. Nanostructure-based techniques are very promising for biosensor creation and as it is shown above provide quite high sensitivity for pathogen detection both in form of deposited structures and quantum dots as well as graphene nanostructures. At the same time, these techniques are quite new in biosensorics and need further studying. ISFETs with CeO_x gate surface are the best suited for use in industrial scale such as provide high sensitivity on the level of 2–3 cells/ml, can be reused up to 5 times without signal decreasing and provide low cost of analysis comparing with other types of biosensors. In order to obtain the sensitivity on the level of infectious dose it is necessary to pay attention to the effective methods of previous food samples preparation especially to the methods of bacteria cells concentration in the solution, which will be analyzed.

Acknowledgements This work was partially supported by EU funded BIOSENSORS-AGRICULT project (Development of nanotechnology based biosensors for agriculture FP7-PEOPLE-2012-IRSES, contract N. 318520) and NATO multi-year Science for Peace Project NUKR.SFPP 984637—"Development of Optical Bio-Sensors for Detection of Bio-Toxins" as well as National University of Life and Environmental Sciences of Ukraine, projects "Development of new generation of the instrumental analytical approaches based on the principles of nano-biotechnology and biosensorics for providing systems of biosafety" and "Creation of effective methods for the construction of artificial selective sites for the instrumental analytical devices of new generation based on the principles of biosensorics", N 110/476-Appl. and №110/71-fund., respectively.

References

1. 5th World Congress Foodborne Infections and Intoxications, 7–11 June 2004, Berlin, Germany
2. Abdelhamida H, Wu H-F (2013) Multifunctional graphene magnetic nanosheet decorated with chitosan for highly sensitive detection of pathogenic bacteria. J Mater Chem B 1:3950–3961
3. Abelès F (1976) Surface electromagnetic waves ellipsometry. Surf Sci 56:237–251
4. Agroterrorism—the new threat to food security, 02 May 2013. https://www.lloyds.com/news-and-insight/news-and-features/emerging-risk/emerging-risk-2013/agroterrorism-the-new-threat-to-food-security
5. Albers WM, Vikholm-Lundin I (2011) Surface plasmon resonance on nanoscale organic films. In: Carraro S (ed) Nano-Bio-Sensing. Springer, New York. ISBN 978-1-4419-6168-6
6. Alocilja E, Radke S (2003) Market analysis of biosensors for food safety. Biosens Bioelectron 18:841–846
7. Arwin H, Poksinski M, Johansen K (2004) Total internal reflection ellipsometry: principles and applications. Appl Opt 43(15):3028–3036
8. Bachrach U, Bachrach Z (1974) Radiometric for the detection of coliform organisms in water. Appl Microbiol 28(2):169–171
9. Baleviciute I, Balevicius Z, Makaraviciute A, Ramanaviciene A, Ramanavicius A (2013) Study of antibody/antigen binding kinetics by total internal reflection ellipsometry. Biosens Bioelectron 39(1):170–176
10. Banting G, Kantor C, Kollins F et al (1990) Analysis of the genome. Methods Mir, p 246
11. Barlen B, Mazumdar S, Lezrich O, Kämpfer P, Keusgen M (2007) Detection of *Salmonella* by surface plasmon resonance. Sensors 7:1427–1446
12. Barrett DJ, Ammann AJ, Stenmark S, Wara DW (1980) Immunoglobulin G and M antibodies to pneumococcal polysaccharides detected by enzyme-linked immunosorbent assay. Infect Immun 27(2):411–417

13. Bernhard A, Field K (2000) A PCR assay to discriminate human and ruminant feces on the basis of host differences in Bacteroides-Prevotella Genes encoding 16S rRNA. Appl Env Microbiol 66(10):4571–4574
14. Turner A, Karube I, Wilson G (eds) (1987) Biosensors: Fundamentals and Application. Oxford University Press, Oxford
15. Bokken G, Corbee RJ, van Knapen F, Bergwerff AA (2003) Immunochemical detection of Salmonella group B, D and E using an optical surface plasmon resonance biosensor. FEMS Microbiol Lett 222:75
16. Borneman J, Skroch P, O'Sullivan K, Palus J, Rumjanek N, Jansen J, Nienhuis J, Triplett E (1996) Molecular microbial diversity of an agricultural soil in Wisconsin. Appl Environ Microbiol 62:1935–1943
17. Buchanan R (2004) Principles of risk analysis as applied to microbial food safety concerns. US DHHS Food and Drug Administration, Center for Food Safety and Applied Nutrition, CollegePark, MD, USA
18. Bussey D, Tsuji K (1968) Bioluminescence for USP sterility testing of phar-maceutical suspension products. Appl Environ Microbiol 5(2):349–355
19. Buxton A, Davies JM (1963) Studies on immunity and pathogenesis of salmonellosis II antibody production and accumulation of bacterial polysaccharide in the tissues of chickens infected with Salmonella gallinarium. Immunology 6:530–538
20. Caras S, Janata J (1980) Anal Chem 52:1935–1937
21. Carraway M, Tzipori S, Widmer G (1996) Identification of genetic heterogeneity in the Cryptosporidium parvum ribosomal repeat. Appl Environ Microbiol 62(2):712–716
22. Carnazza S, Foti C, Gioffrè G, Felici F, Guglielmino S (2008) Specific and selective probes for Pseudomonas aeruginosa from phage-displayed random peptide libraries. Biosens Bioelectron 23:1137–1144
23. Cheng R, Furtak T (1984) Enhanced Raman scattering. Mir, p 408
24. Cho H, Chegal W, Cho Y, Kim Y, Kim H (2005) Enhancement of biomolecular detection sensitivity by surface plasmon resonance ellipsometry. In: Islam MS, Dutta AK (eds) Proceedings of SPIE, Nanosensing: Materials and Devices II, vol 6008, p 60081F
25. Coates DA (1977) Enhancement of the sensitivity of the Limulus assay for the detection of gram negative bacteria. J Appl Bacteriol 42(3):445–449
26. Fung DYC (2004) Rapid methods for the detection and enumeration of microorganisms in water. In: Cotruvo JA et al (eds) Section VI: Analysis of Zoonotic Microorganisms, Chap. 23, Waterborne Zoonoses: Identification, Causes and Control
27. Dimmick R, Akers A (1969) An Introduction to Experimental Aerobiology, Toronto, 120 pp
28. Dittrich S, Schwill P (2003) An integrated microfluidic system for reaction, high-sensitivity detection and sorting of fluorescent all and particles. Ibid 75(21):5767–5774
29. Dzharylgasov S (1977) Device for the microbiological analysis of air. Discoveries, inventions, industrial designs, trademarks, № 20, p 72
30. Ellner P.D (1975) Detection of bacteria by bioluminescence. Microbiology, Washington, p 37–38
31. European Commission Health & Consumer Protection Directorate General. Working Document on Microbial Contaminant Limits for Microbial Pest Control Products, September 2012
32. Evdokimov L, Balashov M, Dadalova A, Krayeva V (1980) Comparative Effectiveness of samplers of air bacterial flora. Laboratory Work, № 4, pp 243–246
33. Garipcan B, Çağlayan O, Demirel G (2011) New generation biosensors based on ellipsometry. In: Serra PA (ed) New Perspectives in Biosensors Technology and Applications. InTech, Shanghi, p 458
34. Gerone P, Couch R, Keefer V et al (1966) Assessment of experimental and natural viral aerosols. Bact Rev 30:576–588
35. Glik B, Pasternak D (2002) Molecular Biotechnology: Principles and Applications. Mir, p 589
36. Grishaev M, Egorova T, Amosov A (1993) Isolation of hepatitis B virus DNA using the polymerase chain reaction. Perm (Russia), p 39

37. Grun J, Manka C, Nikitin S et al (2007) Identification of bacteria from two-dimensional resonant—Raman spectra. Anal Chem 79(14):5489–5493
38. Gu B, Xu C, Yang C, Liu S, Wang M (2011) Biosens Bioelectron 26:2720–2723
39. Guo C, Wang J, Cheng J, Dai Z (2012) Biosens Bioelectron 36:69–74
40. Gupta A, Rao K (1979) Simultaneous detection of Salmonella typhi antigen and antibody in serum by counter-immunoelectrophoresis for an early and rapid diagnosis of typhoid fever. J Immunol Methods 30:349–353
41. Homola J (2008) Surface plasmon resonance sensors for detection of chemical and biological species. Chem Rev 108:2
42. Hurek T, Egener T, Reinhold-Hurek B (1997) Divergence in nitrogenases of *Azoarcus spp.*, Proteobacteria of the b subclass. J Bacteriol 179:4172–4178
43. Ivnitski D, Abdel-Hamid I, Atanasov P, Wilkins E (1999) Biosensors for detection of pathogenic bacteria. Biosens Bioelectron 14:599–624
44. Iwata T, Maeda S (2007) Simulation of an absorption-based surface-plasmon resonance sensor by means of ellipsometry. Appl Opt 46(9):1575–1582
45. Jeffrey J, Kirk J, Atwill E, Cullor J (1998) Prevalence of selected microbial pathogens in processed poultry waste used as dairy cattle feed. Poult Sci 77:808–811
46. Kenyon CL (1975) An automated Bakteriadetektor monitor. Am Lab 7(12):35–36–38-39
47. Kihlberg B-M, Sjobring U, Kasternl W, Bjorck L (1992) Protein LG: a hybrid molecule with unique immunoglobulin binding properties. J Biol Chem 267(35):25583–25588
48. Koubová V, Brynda E, Karasová L, Škvor J, Homola J, Dostálek J, Tobiška P, Rošický J (2001) Detection of foodborne pathogens using surface plasmon resonance biosensors. Sens Actuators B: Chem 74(1–3):100–105
49. Kuts P, Tutova G, Zhavnerko N et al (1979) A method of determining the viability of the microorganisms. In: Discoveries: Inventions, № 37, p 103
50. Lan Y, Wang S, Yinc Y, Hoffmann W, Zhengd X (2008) Using a surface plasmon resonance biosensor for rapid detection of Salmonella Typhimurium in chicken carcass. J Bionic Eng 5 (3):239–246
51. Lee C-S, Kim S, Kim M (2009) Ion-sensitive field-effect transistor for biological sensing. Sensors 9:7111–7131
52. Levine M, Grados O, Gilman R, Woodward W, Solis-Plaza R, Waldman W (1978) Diagnostic value of the widal test in areas endemic for typhoid fever. Am J Trop Med Hyg 27 (4):795–800
53. Li J, Schanzle JA, Tabacco M (2004) Real time detection of bacterial contamination in dynamic aqueous environments using optical sensor. Anal Chem 76(5):1411–1418
54. Liang W, Liu Z, Liu S, Yang J, He Y (2014) Sens Actuators, B 196:336–344
55. Liebana S, Lermo A, Campoy S, Barbe J, Alegret S, Pividori M (2009) Electrochemical magneto-immunosensing of Salmonella based on nano and micro-sized magnetic particles. Anal Chem 81:5812
56. Liedberg B, Nylander C, Lundstrom I (1995) Biosensing with surface plasmon resonance—how it all started. Biosens Bioelectron 10:i–ix
57. Liu HY et al (1997) DBF2, a cell cycle-regulated protein kinase, is physically and functionally associated with the CCR4 transcriptional regulatory complex. EMBO J 16 (17):5289–5298
58. Lomke G, Meltzer R, Akzo N (1986) Electrooptical apparatus for microbial identification and enumeration. Pat. 4576916 USA, MKU C 12 M 1/36, C 12 Q 1/18. Applic. 11.07.84; Publ. 18.03.86
59. Maalouf R, Fournier-Wirth C, Coste J, Chebib H, Saikali Y, Vittori O, Errachid A, Cloarec J, Martelet C, Jaffrezic-Renault N (2007) Anal Chem 79:4879
60. Maciorowski K, Herrera P, Jones F, Pillai S, Ricke S (2007) Effects on poultry and livestock of feed contamination with bacteria and fungi. Anim Feed Sci Technol 133(1):109–136
61. Maciorowski K, Pillai S, Ricke S (2001) Presence of bacteriophages in animal feed as indicators of fecal contamination. J Environ Sci Health B 36:699–708

62. Maciorowski K (2005) Polymerase chain reaction detection of foodborne Salmonella spp. in animal feeds. Crit Rev Microbiol 31:45–53
63. Mihailovich V, Parinov S, Mazurova Y (1993) Polymerase chain reaction can serve as a reliable alternative method for the diagnosis of diphtheria infection. Perm (Russia) 83–84
64. Mihaylova S (1983) Pyrogenic impurities in drugs and methods and equipment for determination/S.N. Mikhailova//Pharmacy. NRB, vol 33, № 4, pp 3–37
65. Morello J, Digenio T, Baker E (1964) Evaluation of serological and cultural methods of chronic salmonellosis in mice. J. Bacteriol 88(5):1277–1282. American Society for Microbiology, Printed in USA
66. Nabok A, Tsargorodskaya A, Mustafa M, Székács A, Székács I, Starodub N (2009) Detection of low molecular weight toxins using optical phase detection techniques. Procedia Chem 1:1491–1494
67. Nakamizo Y, Takahashi R (1965) Latex-agglutination test in shigellosis and salmonellosis. Am J Trop Med Hyg 14(5):783–786
68. Nardiello S, Pizzella T, Rum M, Galanti B (1984) Serodiagnosis of typhoid fever by enzyme-linked immunosorbent assay determination of anti-Salmonella typhi lipopolysaccharide antibodies. J Clin Microbiol 20:718–721
69. NATO STANAG 2873 AMedP-7, Concept of Operations of Medical Support in Chemical, Biological, Radiological, and Nuclear Environments, provides guidance for planning CBRN medical operations, 6 Dec 2007
70. Ogorodniichuk I, Starodub N, Lebedeva T, Shpylovyy P (2013) Optical immune biosensors for Salmonella Typhimurium detection. Adv Biosens Bioelectron (ABB) 2(3):39–46
71. Oh BK, Lee W, Kim YK, Lee WH, Choi JW (2004) Surface plasmon resonance immunosensor using self-assembled protein G for the detection of Salmonella paratyphi. J Biotechnol 111(1):1–8
72. Oh BK, Kim YK, Park KW, Lee WH, Choi JW (2004) Surface plasmon resonance immunosensor for the detection of Salmonella typhimurium. Biosens Bioelectron 19:1497–1504
73. Onuigbo MA (1990) Diagnosis of typhoid fever in Nigeria: misuse of the Widal test. Trans R Soc Trop Med Hyg 84(1):129–131
74. Orlova N, Brusnichkin A, Fokin A et al (2005) Determination of polysaccharides using photoacoustic spectroscopy and thermal lens spectrometer. In: Analysis and Control, vol 9, № 3. pp 257–263
75. Pribil P, Fensela C (2005) Characterization of enterobacteria using MAL-DI-TOF-mass spectrometry. Ibid 77(18):6092–62095
76. Qi C, Gao G, Jin G (2011) Label-free biosensors for health applications. In: Serra PA (ed) Biosensors for Health, Environment and Biosecurity. InTech, Rijeka, p 550
77. Qian ZS, Yue-Shan X, Chai LJ, Ma JJ, Chen JR, Feng H (2014) Biosens Bioelectron 60:64–70
78. Shah J, Wilkinsa E (2003) Electrochemical biosensors for detection of biological warfare agents. Electroanalysis 15(3):157–167
79. Son JR, Kim G, Kothapalli A, Morgan MT, Ess D (2007) Detection of Salmonella enteritidis using a miniature optical surface Plasmon resonance biosensor. J Phys 61:1086–1090
80. Starodub N, Dzantiev B, Starodub V, Zherdev A (2000) Immunosensor for the determination of the herbicide simazine based on an ion-selective field-effect transistor. Anal Chim Acta 424:37–43
81. Starodub N, Ogorodnijchuk J (2012) Immune Biosensor Based on the ISFETs for express determination of Salmonella typhimurium. Electroanalysis 24:1–7
82. Starodub N, Shavanova K, Karpiuk A (2014) Complex of the new generation of the instrumental analytical approaches to prevent dangerous bioterrorism. In: Conference Paper in Proceedings of SPIE—the International Society for Optical Engineering, vol 9253, October 2014

83. Starodub NF, Ogorodnijchuk JA, Romanov VO (2011) Optical immune biosensor based on SPR for the detection of Salmonella typhimurium. In: Abstract Book: the SENSOR + TEST 2011 Conference, Nurenberg, p 7
84. Starodub N, Ogorodnijchuk J Efficiency of immune biosensor based on total internal reflection ellipsometry at the determination of Salmonella. In: Proceedings of the 14th International Meeting on Chemical Sensors in Nurnberg/Germany, 20–23 May 2012, pp 170–179
85. Starodub NF, Nabok AV, Starodub VM et al (2001) Immobilization of biocomponens for immune optical sensor. Ukr Biochem J 73(4):55–64
86. Starodub NF, Pirogova LV, Demchenko A, Nabok AV (2004) Antibody immobilisation on the metal and silicon surfaces. The use of self-assembled layers and specific receptors. Bioelectrochem, vol 66, № 1 (2):111–116
87. Starodub NF, Rachkov OE, Petik AV, Turkovskaja GV, Shul'ga NI, Balkov DI (1986) Isolation of individual mRNA and immunochemical testing of products of translation. In: Proceedings—Methods of Molecular Biology II–K.: Naukova dumka, pp 90–99
88. Starodub NF, Starodub VM (2000) Ukr Biochem Zhurn 4–5:147–163
89. Stokert J, Welaj P, Percells J (1979) Detection of microbial growth using stainless steel—electrodes. In: Abstracts, 79th Annual Meeting of the American Society for Microbiology, Los Angeles, CA, Washington, p 301
90. Svensson HG, Hoogenboom HR, Sjobring U (1998) Protein LA, a novel hybrid protein with unique single-chain Fv antibody-and Fab-binding properties. Eur J Biochem 258:8902896 (1998)
91. Valadez A, Lana C, Tu S-I, Morgan M, Bhunia A (2009) Evanescent wave fiber optic biosensor for Salmonella detection in food. Sensors 9:5810–5824
92. Van der Schoot BH, Bergveld P (1987/88) Biosensors 3:161–186
93. Van T, Moutafis G, Istivan T, Tran L, Coloe P (2007) Detection of Salmonella spp. in retail raw food samples from vietnam and characterization of their antibiotic resistance. Appl Environ Microbiol 73(21):6885–6890
94. VanBaar B, Hulst A, De Jong A, Wills E (2004) Characterization of botulinum toxins type CDE and F by matrix assisted laser desorption ionization and electrospray mass spectrometry. J Chromatogr A 1036(1):97–114
95. Verkin B, Medvedev E, Blohin V (1976) Apparatus method for detection of microorganisms in samples of dry soil. In: Questions of calculated mat. equipment. K., pp 130–135
96. Viter R, Khranovskyy V, Starodub N, Ogorodniichuk Y, Gevelyuk S, Gertnere Z, Poletaev N, Yakimova R, Erts D, Smyntyna V, Ubelis A (2014) Application of room temperature photoluminescence from ZnO nanorods for Salmonella detection. IEEE Sens J 14(6):2028–2034
97. Vladimirskiy M, Shipina L, Kalinin M (1993) Rapid and highly sensitive detection of Mycobacterium tuberculosis in diagnostic materials by polymerase chain reaction. Perm (Russia), 33
98. Walt D, Franz D (2000) Peer Reviewed: Biological Warfare Detection 72(23):738A–746. doi:10.1021/ac003002a
99. Wang Y, Alocilja E (2015) Gold nanoparticle-labeled biosensor for rapid and sensitive detection of bacterial pathogens. J Biol Eng 9:16
100. Wang Z, Dunlop K, Long S, Li L (2002) Mass spectrometric methods for generation of protein mass database used for bacterial identification. Anal Chem 74(13):3174–3182
101. Waterborne Zoonoses: Identification, Causes and Control. In: Cotruvo JA, Dufour A, Rees G, Bartram J, Carr R, Cliver DO, Craun GF, Fayer R, Gannon VPJ (eds) IWA Publishing, London 1996
102. Weary M (1986) Pyrogen testing with the Limulus amebocyte lysate test. Parm Int 7 (4):92–102
103. Weidner K, Di Cesare S, Sachs M, Brinkmann V, Behrens J, Birchmeier W (1996) Interaction between gab and the c-Met receptor tyrosine kinase is responsible for epithelial morphogenesis. Nature 384(6605):173–176

104. WHO Consultation on Pre-Harvest Food Safety 26–28 Mar 2001, Berlin, Germany
105. Wilkins I, Stoner G Detecting the presence of microorganisms. Pat. 4009078 USA, MKU C 12 K 1/04. Applic. 16.12.75, Publ. 22.02.77
106. Xue T, Cui X, Guan W, Wang Q, Liu Ch, Wang H, Qi K, Singh DJ, Zheng W (2014) Surface plasmon resonance technique for directly probing the interaction of DNA and graphene oxide and ultra-sensitive biosensing. Biosens Bioelectron 58:374–379
107. Zhang J, Sun Y, Xu B, Zhang H, Gao Y, Zhang H, Song D (2013) A novel surface plasmon resonance biosensor based on graphene oxide decorated with gold nanorod–antibody conjugates for determination of transferrin. Biosens Bioelectron 45:230–236

Biosensors for the Detection of Emerging Marine Toxins

Sandra Leonardo, Laia Reverté, Jorge Diogène and Mònica Campàs

Abstract Emerging marine toxins present in the environment are relevant for food safety issues. Researchers are currently putting special emphasis on the development of biosensors for their detection. Due to their structural complexity and the difficulty to produce the corresponding biorecognition molecules, the development of assays and biosensors for their detection has become a challenge. Compared to traditional detection techniques, biosensors can provide advantages in terms of sensitivity, specificity, design versatility, portability and multiplexed configurations. This chapter provides a critical overview of the immunosensors, receptor-based biosensors, cell-based biosensors and aptasensors that have been developed for the detection of palytoxins (PlTXs), brevetoxins (PbTXs) and tetrodotoxins (TTXs). Although only few biosensors for these emerging marine toxins have been described to date, the chapter reflects the promising advances made in this field.

Keywords Emerging marine toxins · Palytoxins · Brevetoxins · Tetrodotoxins · Immunosensors · Receptor-based biosensors · Cell-based biosensors · Aptasensors

1 Introduction

Marine toxins are secondary metabolites some of which can be produced by microalgae of the groups of dinoflagellates and diatoms. The specific role these toxins may play in the microalgae that produce them is not clear. These toxins often enter the food webs and may ultimately reach humans through food consumption or direct exposure to marine water, causing different illnesses.

As some virus, bacteria, fungi and protozoa, some marine toxins, such as saxitoxins (STXs) and tetrodotoxins (TTXs), can be considered as potential chemical warfare agents [1, 2]. Nevertheless, reported poisoning incidents related with

---

S. Leonardo · L. Reverté · J. Diogène · M. Campàs (✉)
IRTA, Carretera Poble Nou Km. 5.5, 43540 Sant Carles de La Ràpita, Spain
e-mail: monica.campas@irta.cat

D.P. Nikolelis and G.-P. Nikoleli (eds.), *Biosensors for Security and Bioterrorism Applications*, Advanced Sciences and Technologies for Security Applications,
DOI 10.1007/978-3-319-28926-7_11

marine toxins have only been due, to the best of our knowledge, to accidental ingestion of contaminated seafood or direct exposure to marine water. In a "fictive world" (movies, books) everything is plausible, including the use of marine toxins as potential weapons. In the "real world", STX, which is a potent neurotoxin produced by several species of marine dinoflagellates (e.g. *Alexandrium minutum, A. catenella, Gymnodynium catenatum*), is included in the list of chemical weapons described by the Organisation for the Prohibition of Chemical Weapons [3]. However, although being extremely potent, marine toxins seem not to be presently considered as practical weapons, probably due to the difficulty or cost to produce them in comparison with other chemical weapons. Still, as one may understand, this analysis may only be speculative since probably very little information would be available in case these toxins were presently being used as chemical weapons.

Some marine toxins are quite well described according to their structure, mechanism of action, potency and geographical distribution. This is the case for Amnesic Shellfish Poisoning (ASP, e.g. domoic acid, DA), Diarrheic Shellfish Poisoning (DSP, e.g. okadaic acid, OA) and Paralytic Shellfish Poisoning (PSP, e.g. STX) toxins for which international regulations exist that set up maximum permitted levels in food and define the official methodologies to detect them.

Nonetheless, many marine toxins are presently considered, with a certain degree of subjectivity, as "emerging" toxins including those recently discovered (e.g. pinnatoxins), those that may have recently appeared in certain areas (e.g. ciguatoxins—CTXs recently identified in fish from the Canary Islands), or those that are not yet regulated because not enough information is available regarding their toxicity or distribution (e.g. palytoxins—PlTXs, brevetoxins—PbTXs and cyclic imines—CIs), and those for which regulation exists but additional toxicological information is required (e.g. azaspiracids, AZAs) [4].

The little information and data available for some emerging marine toxins, their structural complexity and the scarcity of standards have compromised the development of methodologies for their detection. The development of biosensors requires stable biorecognition molecules such as enzymes, receptors or antibodies that will unequivocally detect the analyte. In some cases, biorecognition molecules for marine toxins have not been produced and this limits the number of biosensors developed. Nevertheless, marine toxins are nowadays awaking interest in the biosensors world, possibly due to their importance in food safety and environmental monitoring and this opens new applicability fields for biosensors.

In this chapter, we present a detailed overview of the biosensors that have been developed for the detection of PlTXs, PbTXs and TTXs. AZAs, CTXs and CIs (spirolides, gymnodimines, pinnatoxins and pteriatoxins), although also of concern as reflected in the Scientific Opinions on marine biotoxins in shellfish of the European Food Safety Authority (EFSA) Panel on Contaminants in the Food Chain [5–7], have not been included since no biosensors have been reported to date.

2 Biosensors for Palytoxins

Palytoxin (PlTX)-group toxins are complex polyhydroxylated compounds with lipophilic and hydrophilic areas (Fig. 1). They have been isolated from marine zoanthids (soft corals) of the genus *Palythoa* and are also found in benthic dinoflagellates of the genus *Ostreopsis*. PlTX-group toxins have been identified as the toxins responsible for clupeotoxism, a form of ichthyosarchotoxism caused by sardines, and are also found in seafood (fish, shellfish, gastropods or echinoderms) from Japan, Vietnam, Philippines, Malaysia, Singapore, Indonesia, Micronesia, Australia, New Zealand, Hawaii, Cook Islands, French Polynesia, Brazil, Mexico, the Caribbean sea, Madagascar, Reunion Island and the Mediterranean sea [4, 8]. Apart from the contamination with PlTX-group toxins of seafood intended for human consumption, outbreaks of *Ostreopsis* spp. have been implicated in respiratory, dermatological and ophthalmologic symptoms in coastal human populations in the Mediterranean [9, 10]. However, the available toxicological information is limited and the signs and symptoms are not well-defined. Currently there are no regulations on PlTX-group toxins in shellfish, either in the European Union (EU) [11] or in other regions of the world. Nevertheless, the National Reference Laboratories for Marine Biotoxins (NRLMB) have proposed a provisional limit of 250 μg/kg in shellfish [12].

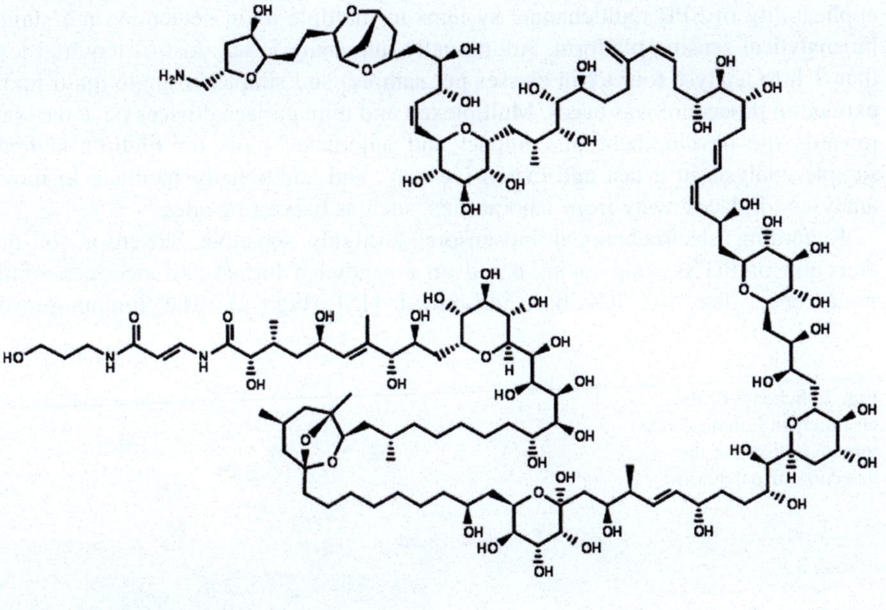

Fig. 1 Palytoxin structure

2.1 Immunosensors for Palytoxins

The relatively large size of PlTX makes easy the antibody (Ab) production via animal immunization, thereby simplifying the development of immunoassays and immunosensors. Optical surface plasmon resonance (SPR) biosensors offer the capability of Ab characterization and then incorporation of these bioreceptors into rapid, sensitive assays on the same platform. An SPR biosensor based on a direct format has been developed for the detection of PlTX [13]. After characterization of monoclonal antibody (MAb) kinetics and optimisation of the experimental parameters, the appropriate performance of the optical immunosensor was demonstrated by performing standard curves for PlTX in buffer as well as in seafood matrices. The limits of detection (LODs) obtained were 0.52, 2.8 and 1.4 ng/mL for PlTX in buffer, grouper and clam samples, respectively, which highlight the need of matrix matching for accurate determination of toxin concentration.

A substantial matrix effect was also observed in the development of a multiplexed SPR immunosensor for the detection of PlTX, STX, OA and DA [14]. Regarding PlTX, the toxin was immobilised onto the chip surface of a flow cell using the carbodiimide reaction to covalently link the amine group of the PlTX molecule to activated carboxylic groups of the chip. The low amount of antibody and toxin available—note that there are no commercial antibodies and that PlTX-group toxins standards are not always available—did not allow to perform exhaustive optimisation studies and validate the assay. Nevertheless, the work clearly demonstrated the applicability of SPR multichannel systems to multiple toxin detection on a single bioanalytical sensing platform. Additionally, the analysis was fast (it required less than 1 h to analyse four toxin classes per sample) and simple (a single multi-toxin extraction procedure was used). Multiplexed and miniaturised devices pave the way towards the development of compact and automated tools for high-throughput sample analysis in a fast and expensive way, and additionally facilitate to move analyses to places away from laboratories, such as harvesting sites.

Regarding electrochemical biosensors, a highly sensitive biosensor for the detection of PlTX-group toxins based on a sandwich format and electrochemiluminescence detection has been described [15] (Fig. 2). The immunosensor

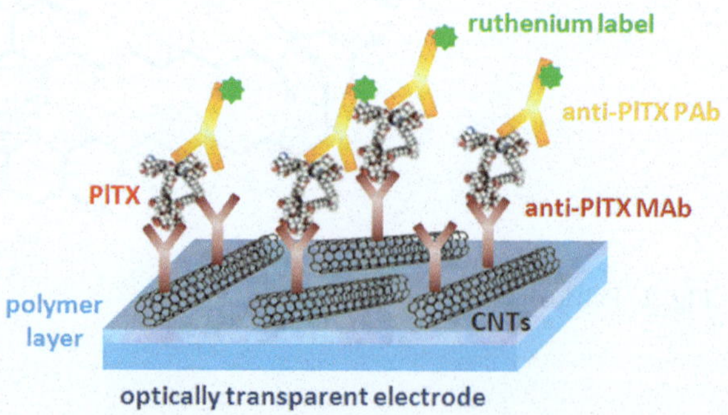

Fig. 2 Scheme of the electrochemiluminescence immunosensor for the detection of palytoxin

incorporated doubly carboxyl-functionalised multi-walled carbon nanotubes (MWCNTs), the groups on the sidewalls being used to conjugate capture anti-PlTX MAb and those on the tips being used to immobilize it on the surface of an optically transparent electrode coated by an electrochemical polymer layer. The electrochemiluminescent detection was performed by labelling the detection anti-PlTX polyclonal antibody (PAb) with a ruthenium complex. The use of CNTs increased the amount of immobilised MAb and favoured the electron transfer, providing specific and sensitive electrochemiluminescent immunosensors with an LOD as low as 0.07 ng/mL of PlTX. The applicability of the immunosensor was demonstrated by the analysis of PlTX-spiked mussels and microalgae samples. Low matrix effects were observed due to the modification of the electrode with CNTs, which minimised the non-specific adsorption, and the electrochemiluminescent transduction strategy.

2.2 *Receptor-Based Biosensor for Palytoxins*

Other biosensors for PlTX-group toxins are based on their mechanism of action. PlTX binds to the Na^+/K^+-ATPase in the cell membrane, inhibiting its activity and converting it into a permanently open ion channel [16]. As a consequence, a rapid sodium influx and potassium efflux from cell is produced, triggering with this a plethora of secondary effects. An optical SPR biosensor for the detection of PlTX based on its affinity towards Na^+/K^+-ATPase has been reported [17]. In this work, the Na^+/K^+-ATPase pump was immobilised on the SPR chip via thiol coupling, and the PlTX binding was afterwards recorded in real time. The developed biosensor showed a very low LOD (3.73 pg) and it was applied to the analysis of PlTX-group toxins in *Ostreopsis siamensis* cultures, which demonstrated the viability of the approach. It is interesting to mention that a previous work had tried to immobilise the Na^+/K^+-ATPase pump on the SPR sensor by amine coupling, but this kind of immobilisation prevented PlTX-group toxin from binding. Amine coupling is usually easy and effective because most macromolecules contain many groups that can participate in the reaction. However, in some cases, these reactive groups can be located near or on the active site of the macromolecule and the immobilisation process can result in conformational changes and loss of biological activity, as it seemed to occur when immobilising the Na^+/K^+-ATPase pump.

2.3 *Cell-Based Electrochemical Assay for Palytoxins*

Haemolysis assays for the detection of PlTX-group toxins are also based on the capacity of the toxins to interact with the Na^+/K^+-ATPase of erythrocytes and its conversion into a non-specific cation channel. Ion imbalance in red blood cells results in their haemolysis. Samples pre-treated with ouabain, a glycoside which inhibits the Na^+/K^+-ATPase pump, are used as a control to ensure the specificity of the assay towards PlTX; the presence of ouabain will reduce the haemolytic activity of PlTXs, while it will have no effect on the haemolytic effect of other haemolytic compounds. Haemolysis is usually detected by spectrophotometry, but it is interesting to describe the work performed by Volpe and co-workers [18]. These authors combined the haemolysis of sheep erythrocytes by PlTX-group toxins with the electrochemical measurement of the lactate dehydrogenase (LDH) released in the culture supernatant after cell disruption. The LDH activity was measured by adding NADH/pyruvate as enzyme substrates and PMS^+. The later reacted with the remaining NADH to produce PMSH, which reacted with the redox mediator hexacyanoferrate(III) to produce hexacyanoferrate(II). Oxidation of hexacyanoferrate(II) was then performed on the electrode surface of a strip of eight screen-printed electrodes (SPEs). The LOD depended on the haemolysis time, being 0.007 and 0.16 ng/mL for 24 and 4 h, respectively. The necessity to use matrix-standard calibration curves for accurate analysis of PlTX was observed when applying the electrochemical assay to the analysis of PlTX-spiked samples.

3 Biosensors for Brevetoxins

Brevetoxin (PbTX)-group toxins are lipid-soluble cyclic polyether compounds (Fig. 3). They are primarily produced by the dinoflagellate *Karenia brevis* (formerly *Ptychodiscus brevis*, Pb giving name to the acronym PbTX) and can accumulate in shellfish and fish. PbTX-group toxins seem to be limited to the Gulf of Mexico, the east coast of USA, and the New Zealand Hauraki Gulf region. These toxins cause neurotoxic shellfish poisoning (NSP), with symptoms such as nausea, vomiting, diarrhoea, paraesthesia, cramps, bronchoconstriction, paralysis, seizures and coma, and dermal or inhalation exposure can result in irritant effects. Moreover, they are potentially carcinogenic. However, the toxicological database for PbTX-group toxins

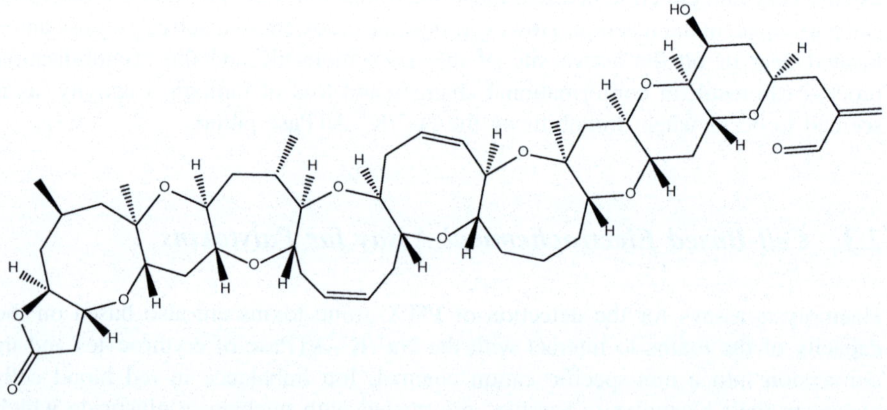

Fig. 3 Brevetoxin-1 structure

is limited. As a consequence, currently there are no regulatory limits for PbTX-group toxins in shellfish or fish in Europe [19]. Nevertheless, maximum permitted levels have been established in USA (20 mouse units (MUs)/100 g or 0.8 mg PbTX-2 equivalents/kg fish) [20], New Zealand and Australia (20 MUs/100 g, analogue not specified) [21, 22]. The discovery of new PbTX-group toxin producing algae and the apparent trend towards expansion of algal bloom distribution, suggest that PbTX-group toxins could emerge in Europe and their analysis in shellfish and fish should be considered. Further information is needed to better characterize the oral toxicity of PbTX-group toxins and their relative potencies.

3.1 Immunosensors for Brevetoxins

PbTX is a low-molecular weight toxin with very few functional groups available for cross-linking, and this usually hinders its immobilisation to develop immunosensors. The first biosensor for PbTX was reported in 1993 [23]. PbTX-3-bovine serum albumin (PbTX-3-BSA) conjugates were prepared to immobilise the toxin on a membrane through the lysine moiety of BSA. Free and immobilised PbTX-3 competed for glucose oxidase (GOx)-labelled anti-PbTX-3 Ab. Finally, the H_20_2 produced by the addition of β-D-glucose was detected amperometrically. Some years after, Kreuzer and co-workers [24] developed electrochemical immunosensors to detect different marine toxins (PbTX, OA, DA and TTX). PbTX-3-BSA conjugates were prepared via carbodiimide reaction to immobilise the toxin on the SPE surface. Nevertheless, the hydroxyl group of PbTX-3 (the only functional group available) had to be modified to make the molecule more active for coupling to BSA via carbodiimide chemistry. Once the conjugates were immobilised on the electrode surface, a competition step between immobilised and free toxin in solution for the labelled primary Ab was performed. Amperometric detection of *p*-aminophenyl phosphate produced by the enzyme label, alkaline phosphatase (ALP), was used to measure the recognition event. Sensitivity problems were observed, which were attributed to the difficulties in the conjugate synthesis and the low amount of toxin available.

The immobilisation of biomolecules on the electrode is a crucial step in the development of electrochemical immunosensors. Usually, the sensitivity of immunosensors can be improved by the control of the amount and orientation of antigens/Abs immobilised on the electrode, which have to be able to retain their biological activity. Dendrimers are hyperbranched polymers precisely engineered to carry molecules encapsulated in their interior void spaces or attached to their surface. A sensitive electrochemical immunosensor for PbTX was developed exploiting dendrimers as toxin immobilisation nanostructured supports [25]. Au NPs were simultaneously synthesised and encapsulated into amine-terminated poly (amidoamine) dendrimers, providing AuNP-PAADS conjugates that were afterwards absorbed on ester-modified gold electrodes. PbTX-2-BSA conjugates were then immobilised on the nanostructured surface. Free and immobilised PbTX-2 competed for the horseradish peroxidase (HRP)-labelled anti-PbTX-2 MAb and the immunosensor response was measured using *o*-phenylendiamine and H_2O_2. An LOD of 0.01 ng/mL was obtained, which is much lower than the LOD for the immunosensor without Au NPs or dendrimers (0.5 and 0.1 ng/mL, respectively). As demonstrated, the use of the 3D-newtwork of AuNP-PDAAs greatly increased the amount of immobilised PbTX-2-BSA and improved the conductivity of the immunosensing interface, therefore enhancing the sensitivity of the electrochemical immunosensor.

Enzyme labels, usually HRP or ALP, are widely used and well-explored in the development of immunosensors. Nevertheless, these strategies usually require an enzyme substrate and its bioactivity can decrease when the biomolecules are exposed to reactive groups and harsh reaction conditions. To tackle this issue, other electroactive species such as thionine or guanine/adenine nucleobases can be used as indicators, as well as various nanomaterials, including quantum dots (QDs), metal nanoparticles and metal ions. Chen's research group explored the use of magnetic beads to covalently immobilise anti-PbTX-2 MAb and use them as immunosensing probes for the capture of PbTX-2 [26] (Fig. 4). The recognition element was prepared by chemical modification of PbTX-2-BSA conjugates with guanine-assembled graphene nanoribbons (GGNRs). Guanine was used as label, since it can be oxidised in the presence of $Ru(bpy)_2^{3+}$. The catalytic oxidation was electrochemically detected after the competition step and entrapment of the magnetic immunocomplex on a carbon paste electrode by a magnet. The LOD of the magneto-controlled electrochemical immunoassay was 1 pg/mL of PbTX-2 and its applicability was demonstrated by the analysis of spiked mussel, clam and cockle samples, providing similar results than those obtained with a commercial ELISA kit for PbTX-2 determination. The same research group also explored the use of metal nanoclusters as labels [27]. In this work, magnetic beads were used to co-immobilise anti-PbTX-2 and anti-dinophysistoxin-1 (anti-DTX-1) MAbs by

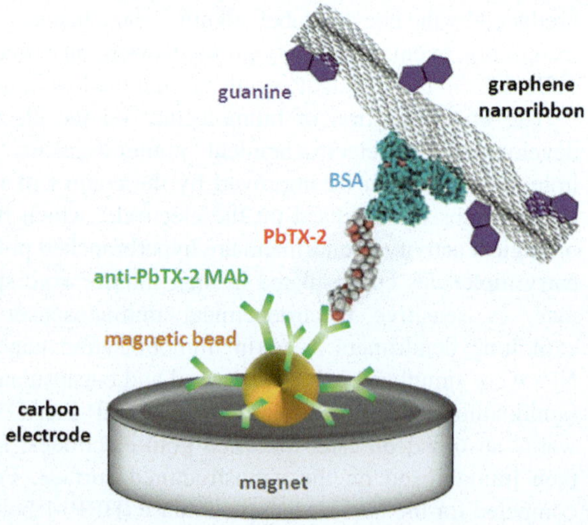

Fig. 4 Scheme of the magneto-controlled electrochemical immunosensor for the detection of brevetoxin-2

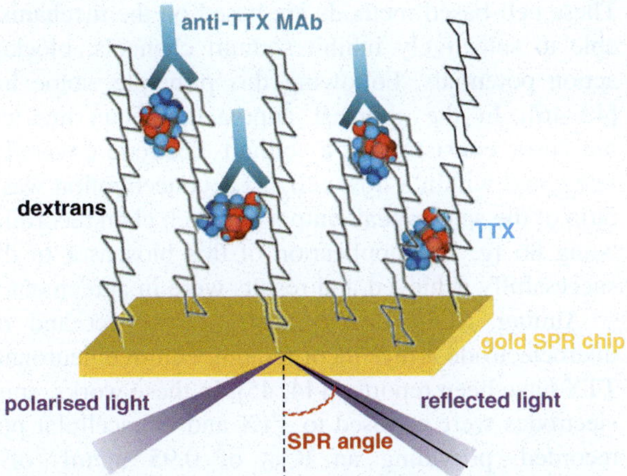

Fig. 6 Scheme of the optical SPR immunosensor for the detection of tetrodotoxin

Although, as previously mentioned, the detection of small molecules by SPR usually requires the immobilisation of the antigen, Yakes and co-workers [41] have proposed the immobilisation of the Ab on the chip and the subsequent non-competitive detection of TTX. This new configuration provided an LOD of 0.09 ng/mL, even lower than in the previous SPR immunosensors. This has been possible thanks to the advances in SPR instrumentation, including higher signal-to-noise ratio, improved fluidics with stronger vacuum pumps and higher number of antibody sites on the chips, advantages that could benefit the detection of other small analytes.

4.2 Aptasensor for Tetrodotoxins

An electrochemical aptasensor for the detection of TTX has been recently proposed [42]. In this work, glassy carbon electrodes were modified with poly (4-styrenesolfonic acid)-doped polyaniline films (PSSA/PANI) by electropolymerisation. Afterwards, an amino-terminated aptamer against TTX was covalently immobilised on the PANI/PSSA films via glutaraldehyde cross-linking. The detection of TTX was assessed by electrochemical impedance according to the charge transfer resistance of the PSSA/PANI film, providing an LOD of 0.199 ng/mL. This low LOD was attained thanks to the ordered monolayer of conductive PANI/PSSA films which improved the electron transfer.

4.3 Cell-based Sensors for Tetrodotoxins

Apart from the antibody-based methods, cell-based biosensors also deserve to be mentioned since they provide realistic models mimicking the original tissue, information which is usually complementary to that obtained from immunosensors.

These cell-based methods are based on the mechanism of action of TTX, which is able to selectively inhibit sodium channels, blocking both nerve and muscular action potentials. Following this principle, some biosensors have been reported [43–46]. In the simplest approach, Cheun and co-workers [43] developed a biosensor consisting of a sodium electrode covered with frog bladder membrane integrated within a flow cell. TTX concentration was measured from the inhibition ratio of the sensor peak output by patch clam recording, the lowest amount detected being 86 fg. The application of this biosensor to the analysis of puffer fish was successfully achieved and results were in agreement with those obtained by MBA.

Aiming at the development of compact and multiplexed devices, portable microelectrode arrays incorporating cultured neuronal networks for the detection of TTX have been reported [44, 45]. In these works, spinal cord cells cultured on gold electrodes were exposed to TTX and extracellular potentials from these cells were recorded, providing an IC_{50} of 0.95 ng/mL of TTX. Following the same biorecognition principle but a different measurement technique, a cardiomyocyte-based impedance biosensor has been recently developed for both TTX and STX, because of their common mechanism of action [46]. In this case, cardiomyocyte cells growth and beating status after TTX treatment were measured on 96-well gold electrodes plates by impedance recording. Although the LOD is higher than in other methods described for the detection of TTX (89 ng/mL vs 0.087 ng/mL for TTX and STX, respectively), this novel real-time and label-free impedimetric biosensor platform is versatile and could be applied to the detection of other neurotoxic molecules.

5 Conclusions and Perspectives

In the past years, a few biosensors for the detection of PlTXs, PbTXs and TTXs have been described. Most of them are immunosensors, but some cell-based sensors, a receptor-based biosensor for PlTX, and even aptasensors for PbTXs and TTXs and an inhibition-based sensor for PbTXs have also been recently reported, showing a breakthrough in the development of biosensors for the detection of emerging marine toxins. This is not the case for other emerging toxins such as AZAs, CTXs and CIs, for which no biosensors have been described yet. Nevertheless, given the numerous possibilities in the development of biosensors (e.g. different transduction strategies, biorecognition molecules and configurations) and the multiple advantages they offer (e.g. simplicity, rapidity, low-cost, sensitivity, multiple toxin detection and portability), it is safe to say that the future of biosensors for the detection of most marine toxins is assured.

In the field of immunosensors for the detection of emerging marine toxins, it is important to keep in mind the difficulty in some cases to obtain the recognition antibodies, since they are not always commercially available and they can be difficult to produce due to the small size of some marine toxins. It is important to mention that the interaction of toxins with antibodies is based on a structural recognition, thus not necessarily related to their toxicity. However, antibodies are robust biorecognition molecules, with high affinity and sensitivity towards their analytes, sometimes being able to detect different analogues of the same group of toxins. Additionally, their easy handling and manipulation allow their integration into different assay formats, offering a wide range of possibilities and configurations using several detection methods.

Aptasensors have been proposed to avoid some of the drawbacks associated with the development of immunosensors, such as the complicated in vivo production of antibodies. Aptasensors for PbTX and TTX have been described, demonstrating that the aptamer-based detection methods are promising. Nonetheless, it is necessary to note that the development of aptamers still require the use of toxin standards, which are not always available.

Receptor-based and cell-based biosensors provide a signal related with the toxicity of the toxins. In that sense, these biosensors have the advantage to better reflect the toxic effects that these toxins may cause to in vivo models, contrarily to structural-based sensors. However, these approaches do not allow identifying or discriminating compounds different from the target toxins but that share the same mechanism of action. Moreover, cell-based biosensors imply the use of "live" material, a factor that may increase variability in the response.

Biosensors can be considered as effective screening tools to be used in combination with confirmatory instrumental analysis methods to achieve highly specific, sensitive and fast routine monitoring of emerging marine toxins. Research on biosensor development for marine toxins should be focused on elucidating or better describing the mechanisms of action of emerging marine toxins, producing biorecognition molecules, and validating the biosensing systems with naturally-contaminated samples in order to promote their implementation. To really foster the use of biosensors in food safety and environmental monitoring, compact analysis devices, sensitive, robust, reliable and easy to handle even by non-trained personal are desired. The use of biosensors to detect and quantify emerging marine toxins is being achieved, but the implementation of such devices in daily life still requires a lot of effort. Nevertheless, progress in this area is very fast and, provided the scientific community will focus not only on the development of the bioanalytical systems but also on their validation, biosensors for emerging marine toxins could soon be implemented in routine analysis.

Acknowledgments The authors acknowledge financial support from the European Union Seventh Framework Programme (FP7/2007-2013) through the ECsafeSEAFOOD project (grant agreement n° 311820) and from the Ministerio de Economía y Competitividad (MINECO) through the DIANA (BIO2011-26311) and the SEASENSING (BIO2014-56024-C2-2-R) projects. Sandra Leonardo and Laia Reverté acknowledge IRTA—Universitat Rovira i Virgili—Banco Santander and the ECsafeSEAFOOD project, respectively, for their PhD grants.

References

1. Anderson PD (2012) Bioterrorism: toxins as weapons. J Pharm Pract 25:121–129
2. Kruger CL, Reddy CS, Conze DB et al (2014) Food safety and foodborne toxicants. In: Hayes W, Kruger CL (eds) Hayes' principles and methods of toxicology, 6th edn. CRC Press, Boca Raton, FL, pp 621–675
3. OPCW (2015) Toxins. Potential chemical weapons from living organisms.http://www.opcw.org/about-chemical-weapons/types-of-chemical-agent/toxins/ . Accessed 28 October 2015
4. Reverté L, Soliño L, Carnicer O et al (2014) Alternative methods for the detection of emerging marine toxins: biosensors, biochemical assays and cell-based assays. Mar Drugs 12:5719–5763
5. EFSA (2008) Opinion of the Scientific Panel on Contaminants in the Food chain on a request from the European Commission on marine biotoxins in shellfish – azaspiracids. EFSA J (723): 1-52.
6. EFSA (2010) Scientific opinion on marine biotoxins in shellfish—emerging toxins: ciguatoxin group. EFSA J 8(6):1627
7. EFSA (2010) Scientific opinion on marine biotoxins in shellfish: cyclic imines (spirolides, gymnodimines, pinnatoxins and pteriatoxins). EFSA J 8(6):1628
8. Aligizaki K, Katikou P, Milandri A et al (2011) Occurrence of palytoxin-group toxins in seafood and future strategies to complement the present state of art. Toxicon 57:390–399
9. Barroso GP, Rueda P, Parron CT et al (2008) An epidemic outbreak with respiratory symptoms in the province of Almeria (Spain) due to toxic microalgae exposure. Gac Sanit 22 (6): 578-584
10. Pfannkuchen M, Godrijan J, Pfannkuchen DM (2012) Toxin-producing Ostreopsis cf. ovata are likely to bloom undetected along coastal areas. Environ Sci Technol 46:5574–5582
11. EFSA (2009) Scientific opinion on marine biotoxins in shellfish—palytoxin group. EFSA J 7 (12):1393
12. CRLMB (2005) Report on toxicology working group meeting. CRLMB, Cesenatico, Italy, p 24–25
13. Yakes BJ, Deeds J, White K et al (2011) Evaluation of surface plasmon resonance biosensors for detection of tetrodotoxin in food matrices and comparison to analytical methods. J Agric Food Chem 59:839–846
14. Campbell K, McNamee SE, Huet AC et al (2014) Evolving to the optoelectronic mouse for phycotoxin analysis in shellfish. Anal Bioanal Chem 406:6867–688
15. Zamolo VA, Valenti G, Venturelli E et al (2012) Highly sensitive electrochemiluminescent nanobiosensor for the detection of palytoxin. ACS Nano 6:7989–7997
16. Hilgemann DW (2003) From a pump to a pore: how palytoxin opens the gates. Proc Natl Acad Sci USA 100:386–388
17. Alfonso A, Pazos MJ, Fernández-Araujo A et al (2014) Surface plasmon resonance biosensor method for palytoxin detection based on Na+, K+-ATPase affinity. Toxins 6:96–107
18. Volpe G, Cozzi L, Migliorelli D et al (2014) Development of a haemolytic-enzymatic assay with mediated amperometric detection for palytoxin analysis: application to mussels. Anal Bioanal Chem 406:2399–2410
19. EFSA (2010) Scientific opinion on marine biotoxins in shellfish-emerging toxins: brevetoxin group. EFSA J 8(7):1677
20. US FDA (United States Food and Drug Administration) (2001) Fish and fisheries products hazards and controls guidance, 3rd edn. Appendix 5—FDA & EPA Safety levels in regulations and guidance. http://www.fda.gov/Food/GuidanceComplianceRegulatoryInformation/GuidanceDocuments/Seafood/ucm091782.htm. Accessed 29 October 2015
21. NZFSA (New Zealand Food Safety Authority) (2006) Animal products (specification for bivalve molluscan shellfish). http://www.nzfsa.govt.nz/animalproducts/legislation/notices/animal-materialproduct/shellfish/bmsrcsspecv-16_2_signed.pdf . Accessed 29 October 2015

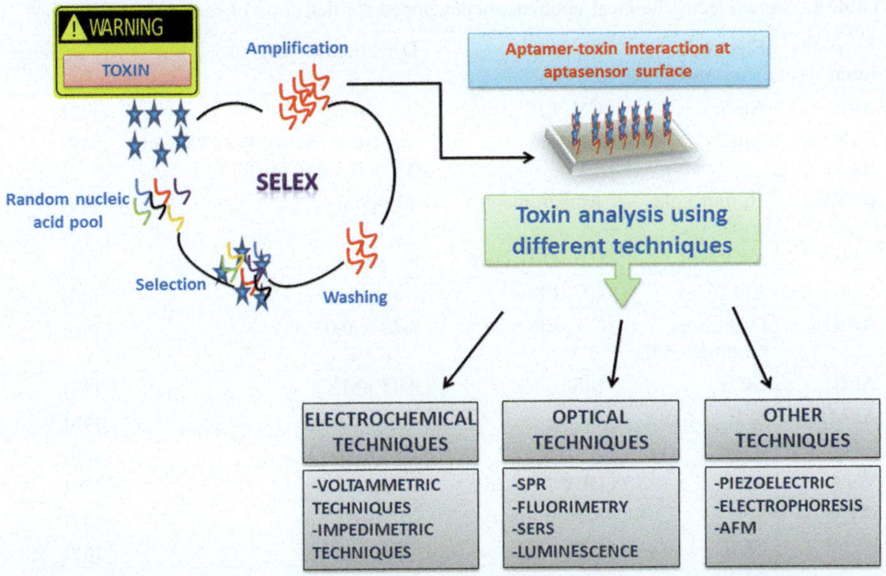

Scheme 1 A representative aptasensor technologies developed for detection of numerous toxins

[42, 63, 64]. Their importance is about medical diagnosis, environmental monitoring, and food safety surveillance [42, 44, 63, 64]. Thus, monitoring of toxins via fast, reliable, sensitive and selective detection platforms has been gained attraction by researchers. In the meantime, aptamers were introduced in the field of development of biosensing platforms. One of them is electrochemical aptasensor technologies. Some approaches in the field of electrochemical aptasensors have been progressed for detection of toxins and given in Table 1 [35, 42–50, 64–70].

Ochratoxins are well-known by-products of numerous fungal species, which can contaminate not only foods, but also beverages including, coffee, beer, and wine. They are mainly produced in the *Aspergillus* and *Penicillium* genera [71]. Due to the fact that ochratoxin A (OTA) is known as the most toxic and has hepatotoxic, nephrotoxic, teratogenic and mutagenic effects onto a wide range of mammalian species [71–73], there are many electrochemical aptasensor applications in the literature to detect OTA [35, 65–67]. Zhang et al. [66] developed an electrochemical aptasensor by using gold electrode. They immobilized single stranded thiolated DNA aptamer labelled with biotin group onto the surface of gold electrode. The interaction of OTA and its DNA aptamer was then performed at the electrode surface and the interaction was determined in the presence of the resistance against TaqaI enzyme occurred after interaction process. Then, the enzymatic reaction between streptavidin-HRP and 3,3′,5,5′-tetramethylbenzidine sulfate (TMB) was monitored by using chronoamperometry technique.

Table 1 Some electrochemical aptasensors developed for detection of toxins

Type of toxin	Electrode	Method	Detection limit (DL)	Reference
ATX	AuE	CV, EIS	0.5 nM	[42]
BoNTA, RTA	AuE	SWV	0.4 ± 0.2 nM for BoNTA and 0.7 ± 0.5 nM for RTA	[43]
BoNTA	16-unit gold array	Amperometry	40 pg/mL	[44]
	rGO/AuE	CV, DPV, EIS	8.6 pg/mL	[64]
	AuE	CV, EIS	Not reported	[45]
AFB1	Dendrimer modified AuE	CV, EIS	0.40 ± 0.03 nM	[46]
AFB1	GCE	EIS	0.05 nM	[47]
AFM1	SPE	EIS	1.15 ng/L	[48]
	IDA	CV, SWV	1.98 ng/L	[49]
VerA	AuE	DPV	10 pg/mL	[50]
OTA	SPE	Amperometry	0.05 μg/L	[35]
	GCE	CV	0.03 ng/mL	[65]
	GCE	Chronoculometry	0.4 pg/mL	[66]
	AuE	CV, DPV, EIS	0.75 pM	[67]
TOA	SPE	CV	1 nM	[68]
BTX-2	Au electrode	CV, EIS	106 pg/mL	[69]
FB-1	GCE	EIS	2 pM	[70]

Abbreviations ***Toxins:*** *OTA* Ochratoxin A, *AFB1* Aflatoxin B1, *AFM1* Aflatoxin M1, *BoNTA* Botulinum neurotoxin, *BTX-2* Brevetoxin-2, *FB-1* Fumonisin B1, *SEB* staphylococcal enterotoxin B, *TOA* toxin A, *RTA* Ricin chain A, *VerA* Versicolorin A, ZEN: ***Type of electrodes:*** *AuE* Gold electrode, *rGO/AuE* reduced graphene oxide modified gold electrode, *GCE* Glassy carbon electrode, *SPE* Screen printed carbon electrode, *IDA* interdigitated electrode array. ***Detection methods:*** *CV* Cyclic voltammetry, *SWV* square wave voltammetry, *DPV* differential pulse voltammetry, *EIS* electrochemical impedance spectroscopy

In the study reported by Rhouati et al. [35], a fully automated flow electrochemical aptasensor based on the magnetic beads (MBs) was introduced and accordingly, direct and indirect competitive electrochemical assays were developed to monitor OTA. For fabrication of this direct assay, carboxylated aptamer modified MBs were immobilized onto the surface of screen printed carbon electrode (SPCE) placed in a flow cell. After the immobilization of avidin-ALP onto the surface of the electrode, the enzymatic reaction in the presence of 1-naphthyl phosphate was occurred and the oxidation of the electro-active product 1-naphtol phosphate to 1-iminoquinone was detected by using amperometry. For fabrication of indirect assay, OTA modified MBs were immobilized onto the surface of SPCE. The free OTA molecules and the immobilized OTA molecules were competed in the solution for binding of biotinylated DNA aptamer. The avidin-ALP was then conjugated and the enzymatic reaction was utilized. A lower limit of detection

(0.05 μg/L) was obtained with the indirect flow-based aptasensor both of the electrochemical assays were tested in the presence of buffer, or beer samples.

Aflatoxins are known to be carcinogen and highly toxic secondary metabolites produced by *Aspergillus flavus* and *Aspergillus parasiticus* [46–48]. FDA limited the level of aflatoxins in nuts, seed and legumes. The monitoring and the detection of aflatoxin at low levels has become attractive in the food safety area. Therefore, some studies were reported for development of electrochemical aptasensors for detection of aflatoxins [46–49]. Nguyen et al. [49] fabricated an electrochemical aptasensor platform for monitoring of AFM1. They used Fe_3O_4 incorporated polyaniline (Fe_3O_4/PANI) film modified interdigitated electrode (IDE) as electrochemical aptasensor platform. They found the detection limit as 1.98 ng/L. In another study, an impedimetric aptasensor onto SPE surface was developed and used for detection of AFM1 [48]. The detection of AFM1 was achieved based on the changes at the charge transfer resistance (R_{ct}) even in milk samples.

Fumonisin B1 (FB-1) is primarily produced by *Fusarium moniliforme* and the most abundant and important fumonisin [74]. It has been found in maize, maize products animal feeds [75]. FB1 threats both animal and human health [76, 77]. An impedimetric aptasensor was developed by Chen and coworkers [70] for recognition of FB-1. GCE surface was modified gold nanoparticles (AuNP) and the interaction of DNA aptamer and FB-1 was investigated based on the changes at the R_{ct} value. The selectivity of the aptasensor was then tested against other toxins.

1.2 Optical Aptasensors for Detection of Toxins

Aptamers have been used as bio-probes in optical sensors based primarily on the incorporation of a fluorophore or a nanoparticle. In the case of fluorescence detection, the simplest format is to label the aptamers with both a quencher and a fluorophore. Additionally, many nano-materials, including QDs, AuNPs, CNTs, graphene oxide (GO), polymer nanobelts, and coordination polymers, have been investigated for their fluorescence-quenching effect instead of using a more traditionally quencher [78–84]. Some optical aptasensors developed for detection of toxins were summarized in Table 2.

AuNPs or several polymers that cause color changes, can be applied as novel reagents for the optical detection technique called colorimetry. The highly negatively-charged ssDNA (complementary strand of the aptamer), which is separated from the aptamer by interaction between the aptamer and the target, is stabilized against aggregation, and a color change occurs in conjunction with this phenomenon [85].

The light chain of BoNT/A (LCA) was utilized as target molecules in SELEX process. Overall, Chang et al. [86] identified three RNA aptamer species which have high binding affinity, specificity and strong inhibition activity. They showed that the endopeptidase activity was effectively inhibited by docking of aptamer to

Table 2 Some optical aptasensors developed for detection of toxins

Type of toxin	Type of substrate	Method	Detection limit (DL)	Reference
OTA	Au chip	SPR	0.005 ng/mL	[54]
LPS	AuE	SPR	–	[55]
OTA	–	Fluorescence	21.8 nM	[56]
OTA FB-1	–	Fluorescence	0.02 ng/mL 0.01 ng/mL	[57]
Abrin	–	Luminescence	1 mM	[58]
BoNT	–	Spectroflorimetry	1 ng/mL	[59]
AFB1	–	Chemilumiescence	0.11 ng/mL	[7]
Ricin	SERS substrate	SERS	10 ng/mL	[60]
Ricin	SERS substrate	SERS	25 ng/mL	[61]
Ricin B	Silicon substrate	SERS	0.32 fM	[62]

Abbreviations ***Toxins:*** *OTA* Ochratoxin A, *AFB1* Aflatoxin B1, *BoNT* Botulinum neurotoxin, *FB-1* Fumonisin B1, *LPS* Lipopolysaccharide. ***Electrodes:*** *AuE* Gold electrode. **Method:** *SPR* Surface plasmon resonance, *SERS* surface-enhanced Raman scattering

BoNT/A (LCA). Their study was the first to confirm that the aptamers for the light chain BoNT/A (LCA) could be used as therapeutic reagents against the deadly botulism [86].

1.3 Other Techniques Developed for Detection of Toxins Using Aptamer Technologies

There are some reports in the literature which can be classified as aptasensors. Nanogold modified piezoresistive microcantilevers (PZR) were used for monitoring of S*taphylococcus enterotoxin* B (SEB) which is small monomeric protein and a pathogen with high thermal and proteolytic stability [87]. PZR sensor surface was modified with DNA aptamer, then the interaction of SEB and its DNA aptamer was investigated even in milk samples.

Ricin is a plant lectin from the castor bean plant *Ricin communis* [51]. It consists of two chains, an A chain and B chain linked by a single disulfide bond and the A chain is toxic to cells [52]. Its production is relatively easy and it is a potential threat as a terrorist weapon. Capillary electrophoresis based aptasensor was reported by Haes et al. [52] for monitoring of ricin A chain. The interaction of ricin and DNA aptamer was performed in capillary surface. Detection of ricin could be achieved in nuclease-contaminated sample matrixes. In another study, atomic force microscopy (AFM) based aptasensor was developed for monitoring of ricin [53]. DNA aptamer and ricin interaction was performed at the surface of Au(111) and ricin binding sites to aptamer was predicted.

2 Conclusion

Aptamers have been utilized in biosensor area since their discovery by Tuerk and Gold [5] due to their stability against physical conditions such as ionic strength, temperature and pH and production cost. They have been alternative biorecognition elements for antibodies even their discovery is relatively new [88, 89]. Aptamers synthesized and isolated by SELEX procedure can spesifically recognize their targets even in complex matrix due to characteristic structure generated during SELEX procedure. They have been used for recognition of proteins [21, 29, 30, 32, 33, 63, 90–92], drugs [93–95] and also toxins [7, 42–62, 64, 96, 97] in combination with different detection techniques such as optic, colorimetric, electrochemical, or piezoelectric techniques. Aptasensors developed for toxin analysis have offered the advanced assays for sensitive, selective, fast, reliable and cost-effective monitoring of numerous toxins as well as their application into the real samples such as food matrices, or biological fluids.

In another aspect, aptasensors can be miniaturized and adaptable for chip technologies for development of aptasensors based on point of care systems which are portable, compatible and having an easy-to-use design. Thus, their application to the environmental or food samples such as water, milk, nuts etc. could be performed and toxins could be sensitively and selectively analyzed with on-line measurements via aptamer based chip technologies in a short time.

Acknowledgments A.E. would like to express her gratitude to the Turkish Academy of Sciences (TUBA) as an Associate member for its partial support.

References

1. Federation of American scientists special weapons primer (2015) http://www.fas.org/nuke/intro/bw/agent.htm. Accessed 22 Oct 2015
2. Sweeney MJ, White S, Dobson ADW (2000) Mycotoxins in agriculture and food safety. Irish J Agric Food Res 29:235–244
3. Richard JL, Fleetwood K (Dec 2001–Jan 2002) Current Trends in Mycotoxin Analysis Food Safety Magazine pp 18–21
4. Balaban N, Rasooly A (2000) Staphylococcal enterotoxins. Int J Food Microbiol 61:1–10
5. Tuerk C, Gold L (1990) Systematic evolution of ligands by exponential enrichment: RNA ligands to bacteriophage T4 DNA polymerase. Science 249:505–510
6. Ellington AD, Szostak JW (1990) In vitro selection of RNA molecules that bind specific ligands. Nature 346:818–822
7. Shim WB, Mun H, Joung HA et al (2014) Chemiluminescence competitive aptamer assay for the detection of aflatoxin B1 in corn samples. Food Control 36:30–35
8. Liu J, Yu J, Chen J et al (2014) Signal-amplification and real-time fluorescence anisotropy detection of apyrase by carbon nanoparticle. Mater Sci Eng C 38:206–211
9. Zhou N, Wang J, Zhang J et al (2013) Selection and identification of streptomycin-specific single-stranded DNA aptamers and the application in the detection of streptomycin in honey. Talanta 108:109–116

10. Wang J (2007) Nanoparticle-based electrochemical bioassays of proteins. Electroanalysis 19:769–776
11. Xiang Y, Zhang Y, Qian X et al (2010) Ultrasensitive aptamer-based protein detection via a dual amplified biocatalytic strategy. Biosens Bioelectron 25:2539–2542
12. Hianik T, Wang J (2009) Electrochemical aptasensors—recent achievements and perspectives. Electroanalysis 21:1223–1235
13. Numnuam A, Chumbimuni-Torres KY, Xiang Y et al (2008) Aptamer-based potentiometric measurements of proteins using ion-selective microelectrodes. Anal Chem 80:707–712
14. Xiang Y, Xie M, Bash R et al (2007) Ultrasensitive label-free aptamer-based electronic detection. Angew Chem Int Edit 46:9054–9056
15. Palchetti I, Mascini M (2012) Electrochemical nanomaterial-based nucleic acid aptasensors. Anal Bianal Chem 402:3103–3114
16. Mascini M, Ilaria P, Sara T (2011) Aptamers smart molecules for biosensing clinical samples. Chim Oggi 29:16–18
17. Polonschii C, David S, Tombelli S et al (2010) A novel low-cost and easy to develop functionalization platform. Case study: aptamer-based detection of thrombin by surface plasmon resonance. Talanta 80:2157–2164
18. Centi S, Sanmartin LB, Tombelli S et al (2009) Detection of C reactive protein (CRP) in serum by an electrochemical aptamer-based sandwich assay. Electroanalysis 21:1309–1315
19. Tombelli S, Mascini M (2009) Aptamers as molecular tools for bioanalytical methods. Curr Opin Mol Ther 11:179–188
20. Centi S, Messina G, Tombelli S et al (2008) Different approaches for the detection of thrombin by an electrochemical aptamer-based assay coupled to magnetic beads. Biosens Bioelectron 23:1602–1609
21. Bini A, Minunni M, Tombelli S et al (2007) Analytical performances of aptamer-based sensing for thrombin detection. Anal Chem 79:3016–3019
22. Centi S, Tombelli S, Minunni M et al (2007) Aptamer-based detection of plasma proteins by an electrochemical assay coupled to magnetic beads. Anal Chem 79:1466–1473
23. Tombelli S, Minunni M, Mascini M (2005) Analytical applications of aptamers. Biosens Bioelectron 20:2424–2434
24. Miodek A, Castillo G, Hianik T et al (2013) Electrochemical aptasensor of human cellular prion based on multiwalled carbon nanotubes modified with dendrimers: a platform for connecting redox markers and aptamers. Anal Chem 85:7704–7712
25. Evtugyn G, Porfireva A, Sitdikov R et al (2013) Electrochemical aptasensor for the determination of ochratoxin A at the Au electrode modified with Ag nanoparticles decorated with macrocyclic ligand. Electroanalysis 25:1847–1854
26. Castillo G, Trnkova L, Hrdy R et al (2012) Impedimetric Aptasensor for thrombin recognition based on CD support. Electroanalysis 24:1079–1087
27. Evtugyn G, Kostyleva V, Sitdikov R et al (2012) Electrochemical aptasensor based on a macrocyclic ligand bearing neutral red. Electroanalysis 24:91–100
28. Porfireva SV, Evtugyn GA, Ivanov AN et al (2010) Impedimetric aptasensors based on carbon nanotubes—poly(methylene blue) composite. Electroanalysis 22:2187–2195
29. Erdem A, Karadeniz H, Mayer G et al (2009) Electrochemical sensing of aptamer-protein interactions using a magnetic particle assay and single-use sensor technology. Electroanalysis 21:1278–1284
30. Rohrbach F, Karadeniz H, Erdem A et al (2012) Label-free impedimetric aptasensor for lysozyme detection based on carbon nanotube-modified screen-printed electrodes. Anal Biochem 421:454–459
31. Erdem A, Congur G (2014) Voltammetric aptasensor combined with magnetic beads assay developed for detection of human activated protein C. Talanta 128:428–433
32. Erdem A, Congur G (2014a) Dendrimer enriched single-use aptasensor for impedimetric detection of activated protein C. Coll Surf B 117:338–345

33. Erdem A, Congur G (2014b) Dendrimer modified 8-channel screen-printed electrochemical array system for impedimetric detection of activated protein C. Sens Actuat B-Chem 196: 168–174
34. Zhou N, Zhang J, Tian Y (2014) Aptamer-based spectrophotometric detection of kanamycin in milk. Anal Method 6:1569–1574
35. Rhouati A, Hayat A, Hernandez DB et al (2013) Development of an automated flow-based electrochemical aptasensor for on-line detection of Ochratoxin A. Sensor Actuat B-Chem 176:1160–1166
36. Kim YS, Chung J, Song MY et al (2014) Aptamer cocktails: Enhancement of sensing signals compared to single use ofaptamers for detection of bacteria. Biosens Bioelectron 54:195–198
37. Ma X, Jiang Y, Ji F et al (2014) An aptamer-based electrochemical biosensor for the detection of salmonella. J Microbiol Meth 98:94–98
38. Dong Y, Xu Y, Yong W et al (2014) Aptamer and its potential applications for food safety. Crit Rev Food Sci 54:1548–1561
39. Xing H, Hwang K, Li J et al (2014) DNA aptamer technology for personalized medicine. Curr Opin Chem Eng 4:79–87
40. Alsager OA, Kumar S, Willmott GR et al (2014) Small molecule detection in solution via the size contraction response of aptamer functionalized nanoparticles. Biosens Bioelectron 57:262–268
41. Shi H, Zhao G, Liu M et al (2013) Aptamer-based colorimetric sensing of acetamiprid in soil samples: sensitivity, selectivity and mechanism. J Hazard Mater 260:754–761
42. Elshafey R, Siaj M, Zouro M (2015) DNA aptamers selection and characterization for development of label free impedimetric aptasensor for neurotoxin anatoxin-a. Biosens Bioelectron 68:295–302
43. Fetter L, Richards J, Daniel J et al (2015) Electrochemical aptamers caffold biosensors for detection of botulism and ricintoxins. Chem Commun 51:15137–15140
44. Wei F, Bai B, Ho C-M (2011) Rapid lyoptimizing an aptamer based BoNTsensor by feedback system control (FSC) scheme. Biosens Bioelectron 30:174–179
45. Halliwell J, Savage AC, Buckley N et al (2014) Electrochemical impedance spectroscopy biosensor for detection of active botulinum neurotoxin. Sens Bio-Sens Res 2:12–15
46. Castillo G, Spinella K, Poturnayova A et al (2015) Detection of aflatoxin B1 by aptamer-based biosensor using PAMAM dendrimers as immobilization platform. Food Control 52:9–18
47. Evtugyn G, Porfireva A, Stepanova V et al (2014) Electrochemical aptasensor based on polycarboxylic macrocycle modified with neutral red for aflatoxin B1 detection. Electroanalysis 26:2100–2109
48. Istamboulié G, Paniel N, Zara L, Granados LR, Barthelmebs L, Noguer T (2016) Development of an impedimetric aptasensor for the determination of AflatoxinM1 in milk. Talanta 146:464–469
49. Nguyen B, DaiTran L, Do QP, Nguyen HL, Tran NH, Nguyen PX (2013) Label-free detection of aflatoxin M1 with electrochemical Fe3O4/polyaniline-based aptasensor. Mater Sci Eng C 33:2229–2234
50. Jiang H-L, Liu X-Y, Qiu Y, Yao D-S, Xie C-F, Liu D-L (2015) Development of an aptasensor for the fastdetection of versicolorin A. Food Control 56:202–210
51. Eitzen E (2001) Medical Management of Biological Casualties Handbook, 4th edn. U.S. Army Medical Research Institute of Infectious Diseases, Frederick, MD. ter Sci Eng C 33: 2229–2234.
52. Haes AJ, Giordano BC, Collins GE (2006) Aptamer-based detection and quantitative analysis of ricin using affinity probe capillary electrophoresis. Anal Chem 78:3758–3764
53. Wang B, Guo C, Zhang M et al (2012) High-resolution single-molecule recognition imaging of the molecular details of ricin—aptamer interaction. J Phys Chem B 116:5316–5322
54. Zhu Z, Feng M, Zuo L et al (2015) An aptamer based surface plasmon resonance biosensor for the detection of ochratoxin A in wine and peanut oil. Biosens Bioelectron 65:320–326
55. Kim SE, Su W, Cho M et al (2012) Harnessing aptamers for electrochemical detection of endotoxin. Anal Biochem 424:12–20

56. Sheng LF, Ren JT, Miao YQ et al (2011) PVP-coated graphene oxide for selective determination of ochratoxin A via quenching fluorescence of free aptamer. Biosens Bioelectron 26:3494–3499
57. Wu SJ, Duan N, Ma XY et al (2012) Multiplexed fluorescence resonance energy transfer aptasensor between upconversion nanoparticles and graphene oxide for the simultaneous determination of mycotoxins. Anal Chem 84:6263–6270
58. Tang J, Yu T, Guo L et al (2007) In vitro selection of DNA aptamer against abrin toxin and aptamer-based abrin direct detection. Biosens Bioelectron 22:2456–2463
59. Bruno JG, Richarte AM, Carrillo MA, Edge A (2012) An aptamer beacon responsive to botulinum toxins. Biosens Bioelectron 31:240–243
60. He L, Lamont E, Veeregowda B et al (2011) Aptamer-based surface-enhanced Raman scattering detection of ricin in liquid foods. Chem Sci 2:1579–1582
61. Lamont E, He L, Warriner K et al (2011) A single DNA aptamer functions as a biosensor for ricin. Analyst 136:3884–3895
62. Zengin A, Tamer U, Caykara T (2015) Fabrication of a SERS based aptasensor for detection of ricin B toxin. J Mater Chem B 3:306–315
63. Zhang Z, Yu L, Xu L et al (2014) Biotoxin sensing in food and environment via microchip. Electrophoresis 35:1547–1559
64. Chan C, Guo J, Sun C et al (2015) A reduced graphene oxide-Au based electrochemical biosensor for ultrasensitive detection of enzymatic activity of botulinum neurotoxin A, Sens Act B 220:131–137
65. Kuang H, Chen W, Xu D et al (2010) Fabricated aptamer-based electrochemical "signal-off" sensor of ochratoxin A. Biosens Bioelectron 26:710–716
66. Zhang J, Chen J, Zhang X et al (2012) An electrochemical biosensor based on hairpin-DNA aptamer probe and restriction endonuclease for ochratoxin A detection. Electrochem Commun 25:5–7
67. Yang X, Qian J, Jiang L et al (2014) Ultrasensitive electrochemical aptasensor for ochratoxin A based on two-level cascaded signal amplification strategy. Bioelectrochemistry 96:7–13
68. Luo P, Liu Y, Xia Y et al (2014) Aptamer biosensor for sensitive detection of toxin A of Clostridium difficile using gold nanoparticles synthesized by Bacillus stearothermophilus. Biosens Bioelectron 54:217–221
69. Eissa S, Siaj M, Zourob M (2015) Aptamer based competitive electrochemical biosensor for brevetoxin-2, Biosens Bioelectron 69:148–154
70. Chen X, Huang Y, Ma X et al (2015) Impedimetric aptamer-based determination of the moldtoxin fumonisin B1. Microchim Acta 182:1709–1714
71. Cruz-Aguado JA, Penner G (2008) Fluorescencepolarization based displacementassay for the determination of smallmolecules with aptamers. Anal Chem 80:8853–8855
72. Barna-Vetro I, Solti L, Teren J et al (1996) Sensitive ELISA test for determination of ochratoxin A. J Agric Food Chem 44:4071–4074
73. O'Brien E, Dietrich DR (2005) Ochratoxin A: the continuing enigma. Cr Rev Toxicol 35:33–60
74. Visconti A, Bruno Doko M, Solfrizzo M et al (1996) European inter comparison study for the determination of fumonisins in maize. Microchim Acta 123:55–61
75. Nelson PE, Desjardins AE, Plattner RD (1993) Fumonisins, mycotoxins produced by Fusarium species: biology, chemistry and significance. Annu Rev Phytopathol 31:233–252
76. Ross PF, Nelson PE, Richard JL et al (1990) Production of fumonisins by Fusarium moniliforme and Fusarium proliferatum isolates associated with Impedimetric aptamer-based determination of FB-1 1713 equine leukoencephalomalacia and a pulmonary edema syndrome in swine. Appl Eniron Microbiol 56:3225–3226
77. Yoshizawa T, Yamashita A, Luo Y (1994) Fumonisin occurrence in corn from high- and low-risk areas for human esophageal cancer in China. Appl Environ Microbiol 60:1626–1629
78. Chhabra R, Sharma J, Wang H et al (2009) Distance-dependent interactions between gold nanoparticles and fluorescent molecules with DNA as tunable spacers. Nanotechnology 20 (48):485201–485211

Electrochemical and Acoustic Biosensors Based on DNA Aptamers for Detection Mycotoxins

Tibor Hianik

Abstract Mycotoxins such as ochratoxin A and aflatoxins are dangerous food contaminants that usually occur in trace amounts from nanograms to micrograms per gram of food. Therefore high sensitive methods are necessary for their detection. The conventional methods such are high-performance liquid chromatography (HPLC), mass spectroscopy are rather expensive and time consuming, therefore biosensor technology is rather promising for rapid detection of toxicants in field conditions, far from specialized laboratories. Among biosensors those based on monoclonal antibodies and DNA/RNA aptamers are of special interest, because provide sensitivity of detection that is better than allowable quantities of toxicants in food. While antibodies are traditional receptors in biosensors, aptamers are novel biopolymers with the affinity comparable to that of antibodies. However in contrast with antibodies, aptamers are more stable and the biosensors based on aptamers can be regenerated which allowing their multiple use. This contribution reviews recent achievements in development aptamer based biosensors for detection selected mycotoxins by electrochemical and acoustic methods.

Keywords Mycotoxins · Ochratoxin A · Aflatoxin · DNA aptamers · Biosensors

1 Introduction

The natural toxins, such are mycotoxins, abrin, ricin, saxitoxin, palytoxin, batrachotoxin, botulinum neurotoxin type A and mycrocystin–RC represent considerable hazard for health and could be considered as potential warfare agents [72]. Together with mycotoxins, secondary metabolites of some fungi as well as foodborne pathogenscause an important public health problem in the world [26, 49, 59, 61, 87, 91, 119].

T. Hianik (✉)
Faculty of Mathematics, Physics and Informatics, Comenius University,
Mlynska Dolina F1, 842 48 Bratislava, Slovakia
e-mail: tibor.hianik@fmph.uniba.sk

D.P. Nikolelis and G.-P. Nikoleli (eds.), *Biosensors for Security and Bioterrorism Applications*, Advanced Sciences and Technologies for Security Applications,
DOI 10.1007/978-3-319-28926-7_13

The urgency of overall control of these factors is caused by large scale of the agriculture production and complications related to a big number of agricultural manufacturers, many of whose cannot afford adequate measures preventing biological and chemical contamination due to low production culture and lack of financial resources. The variety of conditions of cultivation and harvesting as well as long storage and transportation of raw materials and foodstuffs do not promote sufficient food safety. The globalization of world economics resulted in huge traffic flows, which create favorable conditions for fast dissemination of the contaminated products around the world so that the potential source of mycotoxins can be very far from incidents with consumers poisoning [90]. This increases the necessity of portable measurement techniques devoted to the detection of potential hazards out of well-equipped laboratories. The conventional methods such are high-performance liquid chromatography (HPLC), mass spectroscopy are rather expensive and time consuming, therefore biosensor technology is rather promising for rapid detection of toxicants in field conditions. Among biosensors those based on monoclonal antibodies and DNA/RNA aptamers are of special interest, because provide sensitivity of detection that is better than allowable quantities of toxicants in food. While antibodies are traditional receptors in biosensors, aptamers are novel biopolymers with the affinity comparable to that of antibodies. However in contrast with antibodies, aptamers are more stable and the biosensors based on DNA aptamers can be regenerated which allowing their multiple use.

Depending on the assembly, price and function, biosensors can solve many analytical tasks starting from signal on/off systems for early detection of toxins and to quantitative analysis of individual contaminants and their main metabolites in complex matrices. Their development and further progress of analytical equipment are mainly directed to increasing sensitivity of detection and improved selectivity of the response. Besides, usability and extended field of application should be taken into account. This contribution reviews recent achievements in development aptamer based electrochemical and acoustic biosensors for detection mycotoxins and contains also results obtained in author's laboratory.

2 Mycotoxins

Mycotoxins are a group of secondary metabolites of fungi that exert toxic effect on vertebrates and other living beings. Among many other toxic compounds, most important mycotoxins belong to *Aspergillus*, *Fusarium*, and *Penicillium* spp. Chemical structures of selected mycotoxins are depicted in Fig. 1 [24].

From toxicological and legislative point of view, aflatoxins, ochratoxins, some trichothecene toxins (fumonisins, deoxynivalenol, T-2, HT-2), zearalenone, patulin and citrinin are most important. Mycotoxicoses arise if environmental, social and economic conditions combine with humidity and increased temperature, which promote the growth of molds.

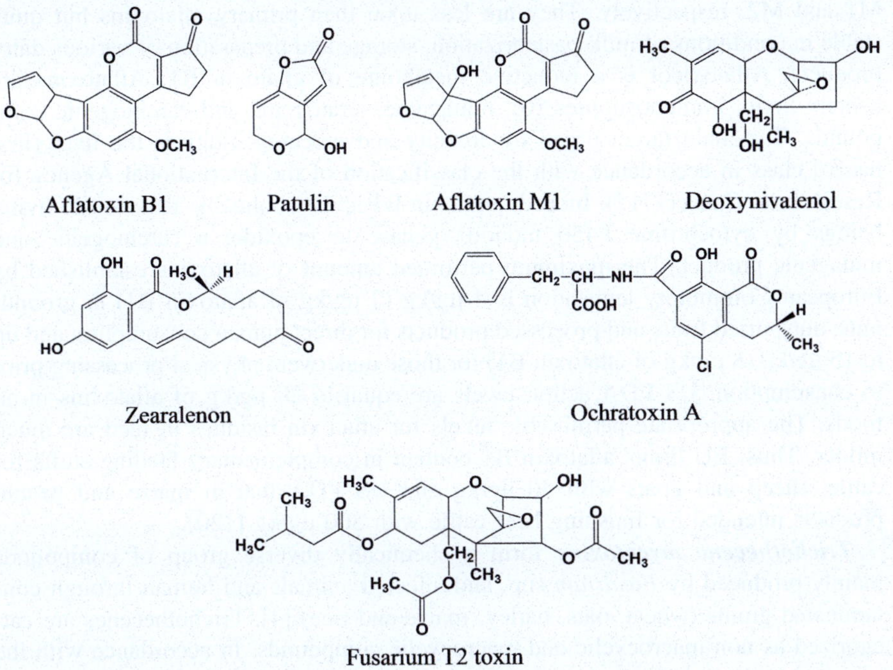

Fig. 1 Chemical structures of most important mycotoxins

Mycotoxins affect a broad range of agricultural products including cereals, cereal-based foods, dried fruits, wine, milk, coffee beans, cocoa bakery and meat products [93]. High performance liquid chromatography (HPLC) with fluorescence or mass detectors are mainly recommended for use in accredited laboratories. Analytical methods of mycotoxins determination were the subject of several recent reviews [12, 45, 120].

Ochratoxin A (OTA) is secondary metabolite produced mainly by *Aspergillus ochraceus* and *Penicillium verrucosum* sp. OTA formation occurs mainly after harvesting on insufficiently dried cereal and cereal products. It can be found in maize, barley, wheat, oats, rye, hay and mixed feed. Exposure to OTA contaminated feed reduces animal growth rates and affects animal production, especially that of pigs and poultry. OTA occurrence in human food commodities of vegetal and animal origin can result in nephropathy, renal diseases and increased frequency of tumor incidents [16]. World Health Organization proposed maximal permissible OTA content of 5 μg/kg in cereals and 3 μg/kg in cereal products. Maximal admissible levels established in EU for feed commodities vary from 50 to 250 μg/kg for cereals and feedstuffs for pigs and poultry[24].

Aflatoxins occur in nuts, cereals and rice. Two major *Aspergillus* sp. producing aflatoxins are *A. flavus* (only B aflatoxins) and *A. parasiticus* (B and G aflatoxins). Aflatoxins B1 and B2 are converted in metabolic oxidative activation to aflatoxins

M1 and M2, respectively. They are less toxic than primary aflatoxins but quite stable in conditions of milk pasteurization, storage and preparation of various dairy products. Aflatoxicol is a reductive metabolite of aflatoxin B1. Aflatoxins are acutely toxic, immunosuppressive, mutagenic, teratogenic and carcinogenic compounds. The main target organ for toxicity and carcinogenicity is the liver (first hazard class in accordance with the classification of the International Agency for Research on Cancer) [17]. Indeed, aflatoxin B1 is metabolically activated in living beings by cytochrome P450 monooxygenase to epoxide, a carcinogenic and mutagenic product. The maximum permitted amount of aflatoxins established by European Community legislation is 4 μg/kg (2 μg/kg of aflatoxin B1) in groundnuts, nuts, dried fruits and processed products for direct human consumption and up to 15 μg/kg (8 μg/kg of aflatoxin B1) for those underwent physical processing prior to consumption. US FDA action levels are equal to 20 μg/kg of aflatoxins in all foods. The appropriate permissible levels for aflatoxin residues in feed are much milder. Thus, EU limits aflatoxin B1 content in complementary feeding stuffs for cattle, sheep and goats with 50 μg/kg and US FDA that in maize and peanut products intended for finishing beef cattle with 300 μg/kg [120].

Trichothecene mycotoxins form a chemically diverse group of compounds mainly produced by *Fusarium* spp. and affected animals and human through contaminated grains (wheat, oats, barley, maize and rice) [4]. Trichothecenes are categorized as non-macrocyclic and macrocyclic compounds. In accordance with the structure and producing organisms, four groups of trichothecene mycotoxins are selected. The type A includes T-2 toxin, HT-2 toxin, neosolaniol and diacetoxyscirpenol. Deoxynivalenol, nivalenol, 3-acetyldeoxynivalenol and fusarenone belong to type B. Type C representatives contain second epoxide group (crotocin and baccharin) and those of type D macrocyclic ring system with two ester linkages (satratoxin G, H, roridin A and verrucarin A) [63].

Trichothecenes are commonly found on cereals grown in the temperate regions of Europe, America and Asia (wheat, rye, barley, oats and corn) [113]. Clinical signs of trichothecene infection in animals include feed refusal and vomiting, growth retardation, reproductive disorders, blood disorders, dermatitis and depression of the immune response. As in previous case, pigs and poultry are most sensitive toward trichothecene mycotoxins while ruminants are able to metabolize trichothecenes into less toxic metabolites [85]. Contamination with trichothecenes causes a number of illnesses in human such as anorexia, depression, inhibition on immune system function and haematoxicity.

Zearalenone is a non-steroidal estrogenic mycotoxin produced by *Fusarium* spp., which occurs in the form of four hydroxyl derivatives. Contrary to non-estrogenic trichothecenes, zaeralenone and some of its metabolites competitively bind to estrogen receptors [107]. Zearalenone is a stable compound during storage, milling and cooking of food. It is associated with reproductive problems in specific animals and possibly in humans. In 1993, it was evaluated by the International Agency or Research on Cancer and allocated together with other *Fusarium* toxins in group 3 (not classifiable as to their carcinogenicity to humans). The scientific committee on food of the European Commission on Health and Consumer Protection arrived at a temporary tolerable daily intake for zearalenone of 0.2 μg/kg of body weight, while the joint FAO/WHO Expert Committee on Food Additives recommended a provisional maximum tolerable daily intake of 0.5 ug/kg [66].

3 DNA/RNA Aptamers

Aptamers are single stranded DNA or RNA molecules with high specificity to various ligands that is comparable and in certain cases even higher than those of antibodies. In contrast to antibodies, aptamers are prepared by in vitro selection procedure developed simultaneously in early 1990s by L. Gold and A. Ellington laboratories [43]. Advantage of aptasensors in comparison with those based on antibodies consists in the possibility to develop aptamers with high specificity to large variety of compounds by means of chemical synthesis without using experimental animals. Aptamers can be developed even against toxins or low molecular weight compounds. Aptamers can be selective to different parts of target molecule. Once the aptamers sequence is identified, this can be synthesized with high purity, reproducibility and relatively low cost, Aptamers can be easily chemically modified by various chemical tags including fluorescence probes, quenchers, electrochemical indicators, nanoparticles or enzymes. This modification allows immobilization of aptamers to various solid supports, which provide their high stability, including stability of RNA against endonucleases, and to utilize various methods of detection. Therefore, aptamers can be advantageously used in the development of affinity biosensors. In contrast with antibodies, which at certain conditions can be irreversibly denaturated, aptamers-based biosensors can be regenerated without loss of integrity and selectivity [58]. Aptamers are also highly perspective for therapy. On December 2004 the anti-vascular endothelial growth factor aptamers—Macugen—was approved by US authorities for the treatment of age related macular degeneration [27].

The method of selection of aptamers was discovered independently by three groups of investigators. Robertson and Joyce [84] described the method of selection of RNA with improved enzymatic activity to cleave DNA. Tuerk and Gold [101] patented the process of selection of DNA ligands as a target for T4 RNA polymerase. This method is known as SELEX (systematic evolution of ligands by exponential enrichment). Ellington and Szostak [19] reported method of in vitro selection of RNA that specifically binds organic dyes. In this paper also the term "*aptamer*" was introduced (from the Latin *aptus*, meaning "to fit" and Greek *meros*, meaning "the part"). The identification of aptamers is based on a combinatorial approach. Specific oligonucleotides are isolated from complex libraries of synthetic nucleic acids. For this purpose, random sequence DNA libraries are obtained by automated DNA synthesis. The size of a randomized region can vary from 30 to 60 nucleotides, flanked on both sides with a specific unique DNA sequence for polymerase chain reaction (PCR) amplification. The theoretical diversity of individual oligonucleotides in these random DNA libraries is rather large. For example, in the case of oligonucleotides composed of 40 bases, it is $4^{40} = 1.2\times10^{24}$.

In practice, however, a considerably smaller library of approximately 10^{13}–10^{15} molecules is used [40]. The selection consists of DNA binding with immobilized ligands, such as proteins or low molecular compounds. The stability of complexes is characterized by apparent dissociation constant, K_D. For aptamer–protein complexes, K_D vary within the 1–100 nM range, which is similar to antibody–antigen complexes. Unbound DNA/RNA molecules are eluted from the column, while bound aptamers are isolated from the complex and then amplified by PCR. This cycle is repeated several times (6–10) and as a result the DNA or RNA sequence of a high affinity to target ligand is obtained. The SELEX technique is discussed in detail in a number of papers and reviews [96].

Although the first SELEX-related patent was filed in 1989, the potential of aptamer-based biosensors has not been fully realized due to the problems with aptamer stability during immobilization and signal transduction. Several problems related to the practical application of aptamers are still under study, for example how the immobilization of aptamers to the supported films and their microenvironment will affect the aptamer structure and aptamer–ligand interactions. Problems are connected with the application of aptamers in the complex biological systems, where interferences with other molecules could take place. In the past, mostly radio labeled aptamers were used for quantification of protein kinase [14] or in vivo detection of clots [18]. However, to be widely employed in clinical practice or in food industry, aptamers must be detected via a non-radioisotope method with a comparable sensitivity. For example aptamers can be covalently linked to an enzyme [69], or fluorescently labeled aptamers can be exploited [62]. Moreover, the most reliable and cost-effective way would be exploitation of the direct physical methods that do not require labeling of aptamers by additional chemical ligands. This highly promising direction is currently of considerable focus. It is also highly advantageous to explore the possibility of immobilization of aptamers onto novel nanomaterials such as carbon nanotubes [39], dendrimers [65], graphene [47], polymer films with nano-sized pores [44] or using nanoparticles [75] and quantum dots [32] in detection of the ligand–aptamer interactions.

Aptamers are rather flexible. Using simple molecular engineering methods it is possible to substantially improve their binding properties and to design aptamers that transduce the recognition event into a readily detectable electronic signal through binding-induced conformational changes. Biosensors based on DNA or RNA aptamers (aptasensors) represent new type of the sensor that utilize unique properties of artificial receptor—aptamers. Aptasensors are of considerable interest due to their application in detection practically unlimited kind of compounds [35]. Aptamers poses also high potential in medical therapy and food control.

3.1 *Structure of the Aptamers*

Currently, the SELEX is a highly automatized procedure and only few days are necessary for development of aptamers for certain ligands. This is much shorter in comparison with the selection of antibodies, where usually several months are required. Due to effectivity of the SELEX, the library of aptamers against various ligands has become wider. On the other hand, the primary procedure does not result in all cases in aptamers with desired affinity. Therefore, optimization of aptamer structure is required. This optimization is performed through biased library generation [40] as well as using post SELEX modifications [96]. As a result, it is possible to select aptamers with sensitivity to small ligands. Aptamers can even distinguish the chirality of molecules and their secondary structure. In principle, there is no restriction in the type of target for which the aptamer can be selected. To date, aptamers with affinity to various ligands were synthesized, including metal ions, organic dyes, drugs, amino acids, co-factors, antibiotics and nucleic acid structures [43]. Special interest was focused on the development of aptamers for the detection of proteins including enzymes, antibodies, prions, growth factors, gene regulatory factors, cell adhesion molecules and lectins. Also, aptamers for food toxins, viral particles and pathogenic bacteria were developed [43, 92]. The aptamers were initially used as therapeutic agents. For example, aptamer that selectively binds thrombin was developed with the purpose of application as an anticoagulant [6]. Only since 2001 the aptamers have been used as recognition elements in biosensors [88]. To date, the best investigated aptamers are those for thrombin. Thrombin is a multifunctional serine protease that plays an important role in procoagulant and anticoagulant functions. Thrombin converts soluble fibrinogen to insoluble fibrin that forms the fibrin gel, which is responsible either for a physiological plug or for pathological thrombus [38]. This process is catalyzed by positively charged fibrinogen binding site at the thrombin molecule. The more positive heparin-binding site is responsible for anticoagulant function of thrombin. These binding sites are spatially separated and localized at opposite poles of thrombin molecule [99]. Initially, Bock et al. [6] developed a 15-mer DNA aptamer, which selectively binds the fibrinogen site of the thrombin (Fig. 2a). This aptamer inhibits the thrombin-catalyzed fibrin clot formation. The binding constant for this structure determined by Bock et al. [6] was K_D = 200 nM, while Tasset et al. [99] reported K_D = 100 nM, which agreed with that reported by Macaya et al. [56] (K_D = 75–100 nM). However, Tsiang et al. [100] reported much lower K_D value, 1.6–6.2 nM, for this aptamer. It should be, however, noted that the dissociation constant depends both on thrombin activity as well as on experimental conditions, such as temperature, ionic strength and method of detection. In the aptamer developed by Bock et al. [6], the intramolecular G-quadruplex plays the crucial role. According to NMR studies [89] and X-ray crystallography [73], the eight guanine residues form G tetrads that are connected at one end by TT loop (T3–T4, see Fig. 2a) and at the other end by the TGT loop (T7G8T9, Fig. 2a). The non-canonical Watson–Crick pair T4–T13 is important for stabilization of the aptamer in conformation that provides binding of T3–T4 side of quadruplex to fibrinogen binding site of the thrombin. Substitution of T4 to adenine (A) resulted in loss of binding affinity. The structure of this aptamer is stabilized by K^+ ions. Later, Tasset et al. [99] developed 38-mer aptamer that selectively recognizes heparin binding site with considerably higher affinity, K_D = 0.5 nM. The G-quadruplex also represents important part of this aptamer. This aptamer had certain similarities with that developed by Bock et al. [6]. The substantial difference

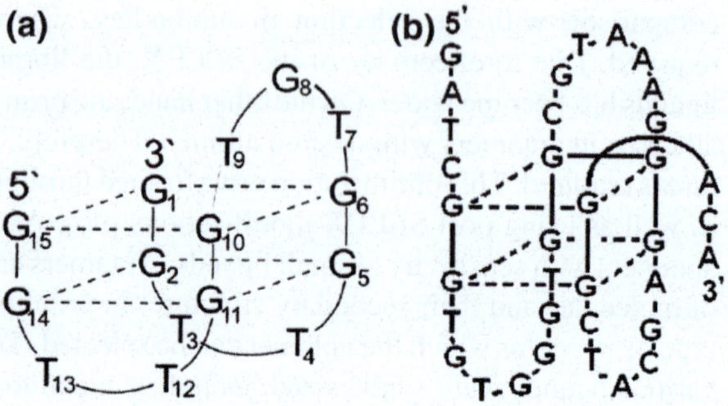

Fig. 2 **a** The structures of DNA aptamers sensitive to fibrinogen binding site at thrombin and **b** ochratoxin A

consisted in replacement of T4 with A4 in a quadruplex loop. However, the quadruplex structure remained stable due to extension of aptamer, in which four Watson–Crick pairs stabilized the aptamer conformation. It is interesting that for this aptamer K^+ ions had no influence on the binding properties.

Two-stacked guanine tetramers in the active binding site of the DNA aptamers against thrombin were typical also for other aptamers subsequently developed, such as those for insulin [114] and DNA aptamer that selectively bind Ochratoxin A (OTA) developed by Cruz-Aguado and Penner [15]. The quadruplex structure of this aptamer has been confirmed in our work using QGRS Mapper program [8] (Fig. 2b). In contrast with aptamer specific to thrombin the OTA binding aptamer is stabilized by calcium ions. Typically 10 mM Ca^{2+} is required for maintaining affinity to OTA with typical $K_D = 1.2 \pm 0.2$ nM [8]. Apart from aptamers that maintain the G-quadruplex structure of binding site, also so-called aptamer beacons were developed. These aptamers contain special complementary spacers at both 3′ and 5′ terminals. In absence of target (for example thrombin) these terminals form double helix that stabilizes the molecular beacon structure [98]. However, in presence of target, the aptamer is transformed into a typical three-dimensional structure contained G-quadruplex, allowing specific interaction with the ligand. The advantage of aptamer beacon is that it allows sensitive detection of analyte, using, for example fluorescence method. One terminal of aptamers is modified by fluorescence label, while the other terminal contains quencher. Thus, without the target the quencher and probe are in close proximity and the intensity of fluorescence is low. The addition of target resulted in conformational changes that moves fluorescence probe away from the quencher. As a result the fluorescence intensity increases [31]. This method was further used also for electrochemical detection. One end of aptamer was modified by an electrochemical indicator, such as methylene blue [110] or ferrocene [83]. Thus, conformational changes of aptamers resulted in changes of the distance of the indicator from electrode surface and as a result changes of the redox current took place. However, not all aptamers containing G-quadruplex. For example DNA aptamers against cellular prions [97] or human vascular endothelial growth factor (VEGF) [33] contains the loop parts that are important for binding the target protein.

Except the molecular beacons, another design of aptamers has been proposed recently. Hagesawa et al. [34] designed an aptamer dimer for thrombin and $VEGF_{165}$. The authors connected a 15-mer thrombin-binding aptamer sensitive to heparin binding site with a 29-mer thrombin-binding aptamer sensitive to heparin binding site of thrombin. This dimer aptamer has K_D, which was 0.1 of that for monomers. They also connected together two identical aptamers against vascular endothelial growth factor ($VEGF_{165}$), which is a homodimeric protein. Similarly like for anti-thrombin aptamer, the dimeric anti-VEGF aptamer had a much lower K_D value than that of the monomer. This study showed, that the dimerization of aptamers improves the affinities of aptamers for target proteins. In our paper we proposed another simple design based on DNA hybridization. Two thrombin binding aptamers were connected by hybridization of their complementary supporting parts. Thus, so-called aptabody was formed that contained two binding sites for thrombin. The aptasensor prepared by immobilisation of these aptabodies onto the surface of multiwalled carbon nanotubes revealed approx. 3 times higher sensitivity in comparison with traditional aptamers [36]. Even higher sensitivity was obtained for heterodimeric thrombin aptamers [37]. This approach has been further explored also in development aptasensors for detection ochratoxin A [8] and cellular prions [64]. Similar approach has been used for design of aptamers against platelet-derived growth factor BB (PDGF-BB) [117].

3.2 *Aptamers in Biosensors*

Aptamer based biosensors can be considered as an analogue of affinity biosensors utilizing antibodies, i.e. immunosensors. The aptamers as a sensing elements are immobilized onto a various support that can simultaneously serve as a transducer, that transform the affinity interactions into the electrical, acoustics or optics signal which is then analyzed by corresponding measuring device (Fig. 3). Moreover, aptamers can be modified by quantum dots, nanoparticles, electrochemical labels or fluorescence probes that can be directly used for signal generation due to conformational changes of aptamers upon analyte binding [35].

The crucial step in biosensor development is the immobilization of aptamers to a solid support. Control of surface chemistry and coverage is essential for assuring high reactivity, orientation/accessibility, and stability of the surface-bound aptamer, as well as for minimizing non-specific binding/adsorption events. In principle these methods are similar to those applied previously for the immobilization of single- or double- stranded DNA in genosensors or DNA biosensors for detection of DNA damage (see [77] for review). The methods of immobilization based on physical adsorption of DNA by means of electrostatic interactions are in general not suitable due to low stability caused by aptamers desorption from the surface. However, adsorption of aptamers onto multiwalled carbon nanotubes and their composites with electrochemical indicators, such as methylene blue (MB) or methylene green (MG) results in stable structures, especially when the immobilization process is

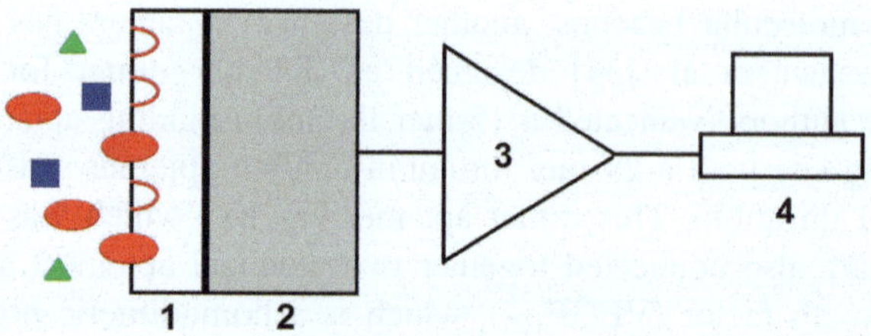

Fig. 3 Scheme of the biosensor. *1* Sensing layer with receptors, *2* transducer, *3* analyzer, *4* computer

accompanied by cycling the voltage [20]. Simple and effective method is based on chemisorption of thiol-labeled aptamers onto a gold surface [75] or on strong affinity of biotin to avidin, streptavidin or neutravidin. In latter case, one end of DNA or RNA aptamer is modified by biotin. The solid support is covered by avidin, streptavidin or neutravidin [70]. Avidin is usually chemically linked to the organic layer formed by 3,3′-dithiopropionic acid di(N-succinimidylester) (DSP) [52]. Neutravidin is very convenient because it can be directly chemisorbed on gold and does not require additional chemical modification of the surface [11, 70]. Mixed two-component alkanethiol self-assembly monolayers of recognition and shielding components (similar to those used in DNA hybridization sensors) are extremely attractive for achieving the desired balance between high loading, minimal non-specific interactions and preferred/accessible orientation. Such mixed co-assembly monolayers have been widely used in DNA hybridization sensors and are being employed for the design of electrochemical aptasensors [32, 83].

Poly(amidoamine) dendrimers (PAMAM) have been also used for aptamer immobilization. PAMAM dendrimers are globular macromolecules with amino terminal groups. Using glutaraldehyde, crosslinking of avidin to a dendrimer surface is possible, so it can be then used for immobilization of biotinylated aptamers. The advantage of dendrimers is their high stability and relatively large surface in comparison with flat electrode [65, 70]. So et al. [95] reported immobilization of thrombin aptamer onto the surface of carbon nanotubes. The nanotubes were pretreated with carbodiimidazole-activated Tween 20 (CDI-Tween). 3′- end of thrombin aptamer was modified by $-NH_2$ groups, which allowed covalent binding of aptamer to CDI-Tween. Application of conducting single walled carbon nanotubes (SWCNTs) allowed fabricating field-effect transistor—based biosensor for detection of thrombin. The unique properties of carbon nanotubes initiated their massive application in biosensing including aptasensor. SWCNTs allowed to covalent attachment of the amino-aptamers at the surface of field effect transistor [57] for effective detection of IgE. Except SWCNTs also multiwalled carbon nanotubes (MWCNTs) were used. The advantage of MWCNTs layers is their large surface in comparison with flat electrode. This allows increasing the amount of the aptamers at the surface. The aptasensor based on MWCNTs showed excellent properties, especially in combination with electrochemical quartz crystal microbalance detection [20, 36] as well as in potentiometric and impedimetric

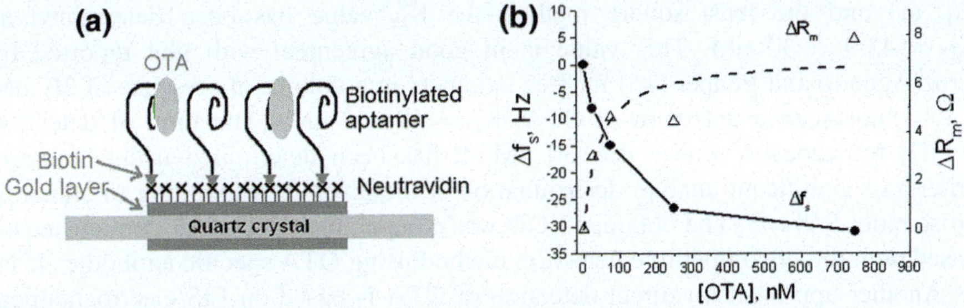

Fig. 6 **a** The scheme of immobilization biotinylated aptamers on a thin *gold* layer of the quartz crystal transducer covered by neutravidin. **b** The representative plot of the dependence of the changes of resonant frequency, Δf_s, and motional resistance, ΔR_m, as a function of OTA for TSM sensor. *Dashed line* is the fit according to Langmuir isotherm (Eq. (1)) [48]

sensor consisted in necessity of using enzyme conjugates as well as in possible non-specific interactions of conjugates with SPCE.

The direct, one step detection of OTA would be, however, more advantageous for practical applications. Recently the simple colorimetric method of OTA detection has been reported [111]. In this work the gold nanoparticles were modified by OTA sensitive aptamers. Addition of OTA resulted in removal of the aptamers from the surface of nanoparticles and after addition of salts the changes in color has been observed due to nanoparticle aggregation. This method allowed detection of OTA with LOD of 20 nM.

In our work we reported acoustic aptasensor for detection OTA using biotinylated aptamers that have been adsorbed on a surface of a thin gold layer of quartz crystal transducer covered by neutravidin (neutravidin, similarly to streptavidin has very high affinity to biotin) (Fig. 6a). This sensor allowed direct detection of OTA with LOD 30 nM [48]. In this method we used network analyzer for measurement changes of series resonant frequency, Δf_S, and the so-called motional resistance, ΔR_m. The later value is sensitive to the viscosity contribution due to the friction between the sensing layer and the buffer. This approach is important especially for detection molecules with relatively low molecular mass like OTA. For such molecules changes of thickness in a sensing layer are negligible. However due to negative charge, OTA can alter the surface properties. This has been certainly confirmed. Addition of OTA to the sensor surface resulted in increase of the R_m and decrease in resonant frequency (Fig. 6b). The shape of the frequency and resistance changes is typical for Langmuir adsorption isotherm.

This means that the OTA binds to the aptamers independently. The binding of OTA to the aptamers can be quantitatively characterized by Langmuir equation [102]. For example for changes of R_m one can write:

$$-\Delta R_m = -(\Delta R_m)_{max}[c/(K_D + c)] \quad (1)$$

where $(\Delta R_m)_{max}$ are the maximal changes of the motional resistance. The K_D value is a measure of the affinity of OTA to the aptamers at the sensor surface. Using

Eq. (1) and the least square method the K_D value has been determined as $K_D = 43.9 \pm 30$ nM. This value is in good agreement with that reported by Cruz-Aguado and Penner [15] for free aptamers in a volume at presence of 20 mM Ca^{2+} (fluorescence detection of OTA, $K_D = 49 \pm 3$ nM). The limit of detection (LOD) for acoustic sensor was 30 nM. It has been determined using common criteria of significant analyte determination at the level corresponding to signal to noise ratio, S/N = 3. The obtained LOD was comparable with QCM immunosensor based on indirect competitive detection method using OTA specific antibodies [51].

Another approach for direct detection of OTA is based on EIS electrochemical aptasensors [8]. The approach is similar to those for EIS immunosensor at presence of redox probe $[Fe(CN)_6]^{-3/-4}$. As we mentioned above, the EIS method can sensitively monitor the changes of charge transfer resistance, R_{ct}, due to the alterations at the sensor surface. The binding of negatively charged OTA to the aptamers resulted in increase of negative surface charge and in repealing of the redox probe from the surface. This causes increase in R_{ct} value. At the same time other parameters of the circuit such are capacitance and Warburg impedance changed only slightly. The changes in R_{ct} can serve as an analytical signal. This value sharply increased at relatively low OTA concentration range 0.1–3 nM with saturation at concentrations approx. 100 nM. This dependence had the shape of Langmuir isotherm and can be described by the equation analogical to that presented above (Eq. (1)). The analysis of this dependence allowed to obtain $K_D = 8.3 \pm 0.8$ nM. This value is lower in comparison with that determined by fluorescence method for free aptamers in a volume [15] as well as that determined by acoustic method [48]. But in EIS sensor the aptamers have been immobilized by different method, using chemisorption, which may affect the aptamers affinity properties. This biosensor exhibited comparable sensitivity with above-mentioned indirect assay (LOD 0.4 nM) and selectively detected OTA. The sensor was regenerable in 1 mM HCl and successfully validated in coffee and flour with recovery of 88 and 104 %, respectively for spiked samples containing 10 nM OTA. Most recently the EIS biosensor based on DNA aptamer specific to OTA covalently immobilized onto mixed Langmuir–Blodgett monolayer composed of polyanilyne-stearic acid and deposited on ITO coated glass plates has been reported [81]. This sensor revealed similar detection limit (0.24 nM), however the fabrication procedure has been more complicated in comparison with simple chemisorption used in our work. The sensor discriminated between OTA and aflatoxin, however it has not been validated in real food samples. In addition the R_{ct} changes were of opposite direction, i.e. with increased OTA concentration the R_{ct} value decreased at presence of redox probe. This is in contradiction with already published papers.

Rather high sensitivity of electrochemical detection of OTA (1 pM) has been achieved using hairpin aptamer configuration immobilized on a gold surface [118]. OTA has been, however detected indirectly using horse radish peroxidase-streptavidin conjugate, that bind to biotin modified DNA that was the part of the sensor. Wu et al. [109] reported electrochemical aptasensor using the thiol and methylene blue (MB) dual-labeled aptamer modified gold electrode. MB is electrochemical indicator that provide well resolved redox. peak. Addition of OTA resulted in conformational changes of aptamer and in increasing the distance of MB from the surface. This caused decrease of the amplitude of the current that served as sensor response. The sensor demonstrated linear range from 0.24 pM to 2.5 nM with LOD of 0.23 pM. Sensor has been validated in red wine. Detection strategy based on MB has been used in paper by Yang et al. [112]. They developed two-level cascaded signal amplification strategy where the capture DNA, aptamers, and reporter DNA functionalized-gold nanoparticles (GNPs) were immobilized on the electrode. The GNPs were used as the first-level signal enhancer. A larger number of guanine-rich DNA was bound to the GNPs' surface to provide anchoring sites for MB to achieve the second-level signal amplification. The effective sensing range from 2.5 pM to 2.5 nM was obtained with detection limit of 0.75 ± 0.12 pM.

Novel method of detection of OTA by using the integration of loop-mediated isothermal amplification (LAMP) technique and subsequently direct readout of LAMP amplicons with a signal-on electrochemiluminescent (ECL) system has been reported by Yuan et al. [115]. In this work the dsDNA composed by OTA aptamer and its capture DNA were immobilized on the electrode. After addition of OTA the OTA-aptamer complex left off the electrode, which effectively decreased the immobilization amount of OTA aptamer on electrode. Then, the remaining OTA aptamers on the electrode served as inner primer to initiate the LAMP reaction. The LOD 10 fM was achieved for this sensor. Most recently Loo et al. [54] reported electrochemical aptasensor for detection OTA utilising graphene-oxide nanoplatelets (GONPs) as electroactive labels. GONPs produced a well-defined reduction peak that was used as the analytical signal. The aptasensor was capable of detecting OTA in the concentration range of 310 fM to 310 pM, with good selectivity against typical interferences. In addition to electrochemical and acoustic method of detections the current trends are focused on application of fluorescence spectroscopy [55] and integrated evanescent wave all-fiber (EWA) biosensing platform for detection OTA [53] with LOD of 17.2 nM and 096 pM, respectively.

Aflatoxin. Aflatoxins are another dangerous and cancerogenic group of mycotoxins. They belonging to the flavus, parasiticus and nomius species of the genus Aspergillus [3]. There have been identified four types of aflatoxins: AFB1, AFB2, AFG1, AFG2, plus two additional metabolites: AFM1 and AFM2, being AFB1 classified as the most abundant and hazardous. Aflatoxins are occasionally detected in corn, peanuts, cotton seeds, nuts, almonds, figs, pistachios, spices, milk, cheese and in a variety of other food and beverages [60]; they are stable at high temperatures and consequently may resist cooking processes [5, 60]. In 2004, the Food and Agricultural Organization (FAO) report onmycotoxins revealed that as of December 2003, at least 99 countries worldwide had regulations for permitted levels of mycotoxins in food or feed [25]. According to the annual report of the Rapid Alert System for Food and Feed (RASFF) in 2012, aflatoxins were the principal hazard in border rejections in the European Union [60]. The maximum permissible level for AFB1 in food established by [13] was set up to 2 parts per billion (ppb) equivalent to 6.4 nM for peanuts, tree nuts, dried fruits and all cereals; 0.10 ppb (0.32 nM) for processed-cereal based baby food and dietary food for children; 5 ppb (16 nM) for hazelnuts, Brazil nuts, and black pepper. All of them intended for direct human consumption.

Up to date, analytical methods for detection of AFB1 in food generally consist on a first stage of extraction of the mycotoxin from the matrix, a second purification step and lastly, the revelation by a suitable technique such as high-performance liquid chromatography (HPLC) with fluorescence or mass detectors [10] Enzyme-linked immunosorbent assay (ELISA) has been also widely applied to AFB1 detection [78]. Nevertheless, there is still an increasing demand for small devices capable of rapid, accurate and reliable determination of low concentrations of AFB1 preferably in situ conditions. Responding to this requirement, various electrochemical biosensors have been reported in literature for AFB1 detection with satisfactory testing in corn [76] and barley [1]. High sensitive impedimetric immunosensors modified by silica gelionic liquid biocompatible film for detection of AFB1 in bee pollen [116], and AFB1-bovine serum albumin conjugates on polythionine-gold nanoparticles and polyaniline (PANi)-polystyrene sulphonic acid films have been also reported [71]. A high sensitive amperometric AFB1 biosensor by aflatoxin-oxidase (AFO) embedded in sol-gel and linked to multiwalled carbon nanotubes (MWCNTs) on Pt electrodes has been also purposed [50]. Immunogenic multi-arrays for simultaneous detection of various mycotoxins including AFB1 together with the detection in corn and peanuts with a limit of detection (LOD) within picogramlevels have been also achieved [105]. Additionally, the detection of AFB1 was obtained by a bifunctional gold binding protein by means of surface plasmon resonance. LOD was as low as 1 mg/mL (3.2 mM) in both, buffer and corn extracts the sensor exhibited negligible responses for other two control toxins [74]. Despite of the broad availability of antibodies and enzymatic based biosensors, the aptamers have being explored as highly promising alternatives. Recently, Neoventures Biotechnology Inc. (Canada) has patented a specific aptamer to AFB1 (Patent: PCT/CA2010/001292) [67] that has been used as molecular recognition probe for detection of AFB1. The method of detection has been based on a chemiluminescence competitive assay using hemin G quadruplex horseradish peroxidase-like DNAzyme (HRP-DNAzyme) linked to AFB1 aptamer. The assay allowed detection of AFB1 with LOD 0.35 nM and it has been validated in spiked corn samples [94]. Later on, an aptamer-based dipstick assay for the rapid and simple detection of AFB1 based on a competitive reaction of the biotinylated form of this aptamer and cy5-modified DNA probes it has been recently reported. This sensor achieved a LOD 0.32 nM and the results of AFB1 analysis in spiked corn samples and ELISA were concordant [94]. Even more sensitive method, based on configurational changes of this aptamer specific to AFB1 immobilized in PCR tubes, has been reported by Guo et al. [30]. This biosensor achieved high sensitivity (LOD 0.1 pM) and it was also tested for quantifying AFB1 levels in wild rye hay and infant rice cereal samples, demonstrating satisfactory recoveries. In our work [23] we proposed electrochemical aptasensors for detection AFB1 based on glassy carbon electrodes modified with electropolymerized Neutral red and polycarboxylated macrocyclic ligands onto which the DNA aptamers were covalently attached. The interaction with an analyte resulted in the decrease of the cathodic peak current

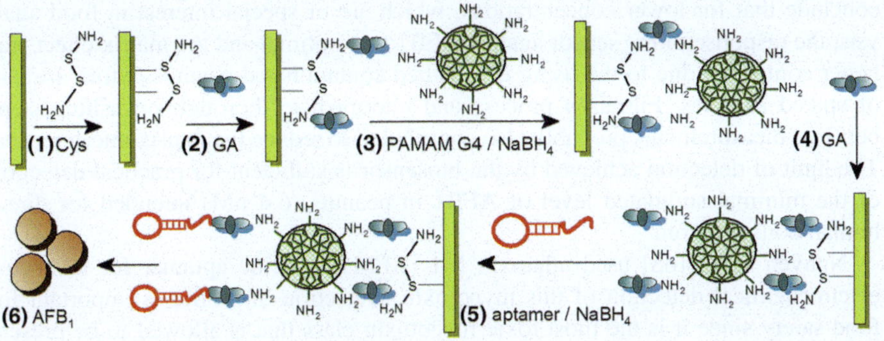

Fig. 7 Schematic representation of aptamer-based biosensor for detection of AFB1. Cystamine (Cys) was immobilized on a *gold* electrode by chemisorption of thiols *1*. Glutaraldehyde (GA) was added to the surface for linking covalently amino-groups of cystamine to amino-terminated dendrimers *2*. PAMAM G4 dendrimers layer was reacted with NaBH4 for reducing the excess of GA *3*. Next step includes addition of GA *4* that follows the anchoring of amino-modified aptamers and using $NaBH_4$ for reducing the excess of GA *5*. The sensor was ready for detection of AFB1 *6*

of the probe measured by CV and in the increase of the electron transfer resistance determined by EIS. The limit of detection was found to be 0.1 nM for CV and 0.05 nM for EIS methods, respectively. The aptasensor makes it possible to detect AFB1 in peanuts, cashew nuts, white wine and soy sauce with a recovery of 85–100 %. Recently we reported an electrochemical aptasensor for detection of AFB1 developed through layer coating of cystamine, PAMAM G4 dendrimers and DNA aptamers specific to AFB1 (Fig. 7) [9]. Electrochemistry based on EIS and CV in the presence of the redox couple $K[Fe(CN)_6]^{-3/-4}$ was employed for steeply monitoring of sensor formation and binding events. Other supporting systems consisting of one or multi sensing layers were tested in order to probe the efficacy of PAMAM dendrimers-aptamers as enhancing elements for biosensing of low molecular compounds. The LOD achieved by this sensor was 0.40 ± 0.03 nM, it was regenerable in 0.2 M glycine-HCl and it did not lose its stability up to 60 h storing at 4 °C. The specificity of aptamers to AFB1 was also studied since the potential contribution to signal detection that other mycotoxins (OTA, AFB2) could confer, especially when analyzing food samples. The aptasensor has been tested in samples of peanuts demonstrating satisfactory recovery. The effects of the matrix in spiked samples compared to previously treated contaminated samples of peanuts was also explored. For lower concentrations of AFB1: 0.1, 0.3 and 1 nM the recovery of the biosensor was 95.6, 96.5 and 90.7 % respectively. For 3 and 10 nM AFB1 the matrix effect increases leading to recoveries higher than 100 % (109.6 and 119.7 % respectively). This indicates that the sensor was able to detect not only AFB1, but other components of the samples that could also cause delaying in the electronic communication between the sensing layer and the analyte. It seems to be that the specificity of the aptamer plays a crucial role in complex samples and it should be properly studied in buffer before testing the aptasensor in food. We can conclude that for lower concentrations, which are of special interest in food analysis, the response of the sensor against AFB1 was optimal and the matrix effect was better controlled due to the using of certified contaminated peanuts extract instead of spiked samples. Filtration process and intermediate thoroughly washing steps between measurements probably also contributed to reduce the matrix interferences. The limit of detection achieved by the biosensor is sufficient for practical detection of the minimal stipulated level of AFB1 in peanuts (6.4 nM) intended for direct human consumption.

Nguyen et al. [68] used aflatoxin M1 (AFM1) specific aptamer for label-free electrochemical detection of this mycotoxin. Detection of AFM1 is important for food safety since it is the most toxic mycotoxin class that is allowed to be present on cow milk. In this paper Fe_3O_4 incorporated polyaniline (Fe_3O_4/PANi) film has been polymerized on interdigitated electrode (IDE) onto which the aptamers were immobilized. This aptasensor showed a good sensitivity to AFM1 in the range of 20–200 pM, with the detection limit of 6 pM.

Thus, the aptasensors are of high perspective even for detection small molecules such are OTA and aflatoxins. The sensitivity of detection in most cases is similar to those of antibodies [24]. Substantial advantage of aptasensors is possibility of surface regeneration which allowing their multiple use. Moreover, recent papers on application SPR, thickness shear mode acoustic method and EIS are evidence of possibility of direct detection OTA and aflatoxins without using OTA-protein conjugates or other labels.

4 Conclusion

The biosensor technology is certainly powerful tool for detection food mycotoxins such as ochratoxin A and aflatoxins. The achievements reported in this review revealed that most of the approaches allowing detect OTA and aflatoxins with high sensitivity, which is better than allowable contamination of food by these toxins. Rather perspective direction in biosensor development consisting in application of DNA aptamers. Using these novel biopolymers even most sensitive biosensor assay was demonstrated, allowing detection of OTA and aflatoxin better than in pM level. Direct detection methods such are acoustic and electrochemical impedance spectroscopy are rather challenging, due to fast response and high sensitivity which is substantial advantage over traditional methods such are HPLC or mass spectroscopy. We believe that further effort should result in appearance of low cost, portable and easy to use biosensor for detection of mycotoxins applicable in food factories and agricultural farms.

Acknowledgments This work was supported by the Slovak Research and Development Agency under the contract No. APVV-14-0267 and by Science Grant Agency VEGA (project No. 1/0152/15).

References

1. Ammida NHS, Micheli L, Piermarini S et al (2006) Detection of aflatoxin B1 in barley: comparative study of immunosensor and HPLC. Anal Lett 39:1559–1572
2. Barthelmebs L, Hayat A, Limiadi AW et al (2011) Electrochemical DNA aptamer-based biosensor for OTA detection, using superparamagnetic nanoparticles. Sens Actuators, B 156:932–937
3. Bennett JW, Keller NP (1997) Mycotoxins and their prevention. In: Anke T (ed) Fungal biotechnology. Chapman and Hall, London, pp 265–273
4. Bennett JW, Klich M (2003) Mycotoxins. Clin Microbiol Rev 16:497–516
5. Binder EM, Tan LM, Chin LJ et al (2007) Worldwide occurrence of mycotoxins in commodities, feeds and feed ingredients. Animal Feed Sci Technol 137:265–282
6. Bock LC, Griffin LC, Latham JA et al (1992) Selection of single-stranded DNA molecules that bind and inhibit human thrombin. Nature (London) 355:564–566
7. Bonel L, Vidal JC, Duato P et al (2011) An electrochemical competitive biosensor for ochratoxin A based on a DNA biotinylated aptamer. Biosens Bioelectr 26:3254–3259
8. Castillo G, Lamberti I, Mosiello L et al (2012) Impedimetric DNA aptasensor for sensitive detection of ochratoxin A in food. Electroanalysis 24:512–520
9. Castillo G, Spinella K, Poturnayová A et al (2015) Detection of aflatoxin B1 by aptamer-based biosensor using PAMAM dendrimers as immobilization platform. Food Control 52:9–18
10. Cavaliere C, Foglia P, Guarino C et al (2007) Determination of aflatoxins in olive oil by liquid chromatography-tandem mass spectrometry. Anal Chim Acta 596:141–148
11. Cavic BA, Thompson M (2002) Interfacial nucleic acid chemistry studied by acoustic shear wave propagation. Anal Chim Acta 469:101–113
12. Cigić IK, Prosen H (2009) An overview of conventional and emerging analytical methods for the determination of mycotoxins. Int J Mol Sci 10:62–115
13. Commission Regulation (EU) No 1058/2012 of 12 November 2012 amending regulation (EC) No 1881/2006 as regards maximum levels for aflatoxins in dried figs Text with EEA relevance. Official Journal L 313 of 13.11.2012, 2012, 14–15
14. Conrad R, Ellington AD (1996) Detecting immobilized protein kinase c isozymes with RNA aptamers. Anal Biochem 242:261–265
15. Cruz-Aguado JA, Penner G (2008) Determination of ochratoxin a with a DNA aptamer. J Agric Food Chem 56:10456–10461
16. Denli M, Perez JF (2010) Ochratoxins in feed, a risk for animal and human health: control strategies. Toxins 2:1065–1077
17. Do JH, Choi D-K (2007) Aflatoxins: detection, toxicity and biosynthesis. Biotechnol Bioprocess Eng 12:585–593
18. Dougan H, Hobbs JB, Weitz JI et al (1997) Synthesis and radioiodination of a stannyl oligodeoxyribonucleotide. Nucleic Acids Res 25:2897–2901
19. Ellington AD, Szostak JW (1990) In vitro selection of RNA molecules that bind specific ligands. Nature 346:818–822
20. Evtugyn G, Porfireva A, Ryabova M et al (2008) Aptasensor for thrombin based on carbon nanotubes-methylene blue composites. Electroanalysis 20:2310–2316
21. Evtugyn G, Porfireva A, Ivanov A et al (2009) Molecularly imprinted polymerized methylene green as a platform for electrochemical sensing of aptamer–thrombin interactions. Electroanalysis 21:1272–1277
22. Evtugyn G, Porfireva A, Stepanova V et al (2013) Impedimetric aptasensor for ochratoxin A determination based on Au nanoparticles stabilized with hyper-branched polymer. Sensors 13:16129–16145
23. Evtugyn G, Porfireva A, Stepanova V et al (2014) Electrochemical aptasensor based on polycarboxylic macrocycle modified with neutral red for aflatoxin B1 detection. Electroanalysis 26:2100–2109

24. Evtugyn G, Shamagsumova RV, Hianik T (2016) Biosensors for detection mycotoxins and pathogenic bacteria in food. In: Grumezesku A (ed) Nanotechnology in food industry, Vol III NanoBioSensors. Elsevier, in press
25. FAO (2004) Worldwide regulations for mycotoxins in food and feed in 2003. FAO, Rome. ISBN 92-5-105162-3, ISSN 0254-4725
26. Gehring AG, Tu S-I (2011) High-throughput biosensors for multiplexed foodborne pathogen detection. Annu Rev Anal Chem 4:151–172
27. Gold L (2006) Aptamers after fifteen years. In: Klussmann S (ed) The aptamer handbook. Wiley-VCH, Weinheim, pp 461–469
28. Green NS, Norton ML (2015) Interactions of DNA with graphene and sensing applications of graphene field-effect transistor devices: A review. Anal Chim Acta 853:127–142
29. Gronewold TMA, Glass S, Quantdt E et al (2005) Monitoring complex formation in the blood-coagulation cascade using aptamer-coated SAW sensors. Biosens Bioelectron 20:2044–2052
30. Guo X, Fang W, Nan Z et al (2014) Development of an ultrasensitive aptasensor for the detection of aflatoxin B1. Biosens Bioelectr 56:340–344
31. Hamaguchi N, Ellington A, Stanton M (2001) Aptamer beacons for the direct detection of proteins. Anal Biochem 294:126–131
32. Hansen JA, Wang J, Kawde A-N et al (2006) Quantum-dot/aptamer-based ultrasensitive multi-analyte electrochemical biosensor. J Am Chem Soc 128:2228–2229
33. Hasegawa H, Sode K, Ikebukuro K (2008) Selection of DNA aptamers against $VEGF_{165}$ using a protein competitor and the aptamer blotting method. Biotechnol Lett 30:829–834
34. Hasegawa H, Taira KI, Sode K et al (2008) Improvement of aptamer affinity by dimerization. Sensors 8:1090–1098
35. Hianik T, Wang J (2009) Electrochemical aptasensors—recent achievements and perspectives. Electroanalysis 21:1223–1235
36. Hianik T, Porfireva A, Grman I et al (2008) Aptabodies—new type of artificial receptors for detection proteins. Prot Pept Lett 15:799–805
37. Hianik T, Grman I, Karpisova I (2009) The effect of DNA aptamer configuration on the sensitivity of detection thrombin at surface by acoustic method. Chem Commun 41:6303–6306
38. Holland CA, Henry AT, Whinna HC et al (2000) Effect of oligodeoxynucleotide thrombin aptamer on thrombin inhibition by heparin cofactor II and antithrombin. FEBS Lett 484:87–91
39. Jacobs CB, Peairs MJ, Venton BJ (2010) Review: Carbon nanotube based electrochemical sensors for biomolecules. Anal Chim Acta 662:105–127
40. Jayasena SD (1999) Aptamers: an emerging class of molecules that rival antibodies in diagnostics. Clin Chem 45:1628–1659
41. Kang Y, Feng K-J, Chen J-W et al (2008) Electrochemical detection of thrombin by sandwich approach using antibody and aptamer. Bioelectrochemistry 73:76–81
42. Kikin O, D'Antonio L, Bagga PS (2006) QGRS Mapper: a web-based server for predicting G-quadruplexes in nucleotide sequences. Nucleic Acids Res 34:W676–W682
43. Klussmann S (ed) (2006) The aptamer handbook. Wiley-VCH, Weinheim
44. Kocer A, Tauk L, Déjardin P (2012) Nanopore sensors: From hybrid to abiotic systems. Biosens Bioelectron 38:1–10
45. Köppen R, Koch M, Siegel D et al (2010) Determination of mycotoxins in foods: current state of analytical methods and limitations. Appl Microbiol Biotechnol 86:1595–1612
46. Kuang H, Chen W, Xu D et al (2010) Fabricated aptamer-based electrochemical "signal-off" sensor of ochratoxin A. Biosens Bioelectr 26:710–716
47. Kuila T, Bose S, Khanra P et al (2011) Recent advances in graphene-based biosensors. Biosens Bioelectron 26:4637–4648
48. Lamberti I, Mosiello L, Hianik T (2011) Development of thickness shear biosensor based on DNA aptamers for detection of ochratoxin A. Chem Sensors 1:11–15
49. Law JW-F, Ab Mutalib N-S, Chan K-G et al (2015) Rapid methods for the detection of foodborne bacterial pathogens: principles, applications, advantages and limitations. Frontiers in Microbiol 5:770

Electrochemical Biosensors for Food Security: Allergens and Adulterants Detection

Susana Campuzano, Víctor Ruiz-Valdepeñas Montiel, Rebeca Magnolia Torrente-Rodríguez, Ángel Julio Reviejo and José Manuel Pingarrón

Abstract Food safety plays an important role in public health and thus to society as a whole. Food may become unsafe due to the presence of allergens and adulterants amongst others and hence the determination of these analytes has gained great relevance in quality control and safety of food. Electrochemical biosensor devices are emerging as one of the foremost relevant techniques for monitoring food allergens and adulterants due to their quick, specific, sensitive and reliable performances, ease of mass fabrication, economics and field applicability. These electrochemical biosensors, based on the use of existing or recently developed bioreceptors in combination with nanomaterials and surface fabrication techniques are able to offer attractive and efficient platforms which may become a viable option for the development of simpler, faster, cheaper and more robust and reliable analytical methods for the detection of allergens and adulterants in food. This Chapter describes recent analytical strategies developed so far using electrochemical biosensors for the determination of potentially hazardous adulterants and allergenic food residues to ensure food safety. The main progresses achieved to date are presented, highlighting general considerations and potential prospects for the future. The variety of electrochemical biosensors that have appeared in recent years shows that it is a booming research area with still many challenges but also great opportunities to develop sensitive, reliable, robust, and cost effective food allergens and adulterants biosensing methodologies.

Keywords Electrochemical biosensors · Food security · Allergens · Adulterants

---

S. Campuzano (✉) · V.R.-V. Montiel · R.M. Torrente-Rodríguez · Á.J. Reviejo
J.M. Pingarrón (✉)
Departamento de Química Analítica, Facultad de Ciencias Químicas, Universidad Complutense de Madrid, Avda. Complutense S/N, 28040 Madrid, Spain
e-mail: pingarro@quim.ucm.es

D.P. Nikolelis and G.-P. Nikoleli (eds.), *Biosensors for Security and Bioterrorism Applications*, Advanced Sciences and Technologies for Security Applications,
DOI 10.1007/978-3-319-28926-7_14

1 Food Security

Ensuring food safety is nowadays a top priority of authorities and professional players in the food supply chain. Between the different targets of importance in food safety control: pathogens, allergens, adulterants, toxins and other forbidden contaminants, adulterants and allergens are growing food concerns.

Food allergy is nowadays regarded as a problem of public health importance due to its widespread among most countries and the concern raised by allergic consumers inadvertently exposed to the offending ingredient through allergen containing foods. Even little intake of offending allergen can trigger allergic reactions in allergic consumers; therefore the only effective way of prevention is the avoidance of the allergen containing food from the diet [1]. The risks associated with the presence of hidden allergens in the food chain have generated a high demand for fast, sensitive and reliable methods to trace food allergens in different commodities to be implemented by food industries [1, 2].

During the past decade, a considerable interest has been paid also to dairy products quality and methods of production due to the recent crises and scandals in food industry, which have seriously undermined consumer confidence. The need for rapid analytical techniques to determine the authenticity and to detect adulteration is greater than ever. The determination of authenticity of milk and dairy products requires appropriate analytical tools for analysis both during production and storage [3].

Then, one of the key challenges to determine the safety of food and guarantee a high level of consumer protection is the availability of fast, sensitive and reliable analytical methods to identify these specific risks associated to food before they become a health problem. The limitations of existing methods have encouraged the development of new technologies, such as biosensors, which represent an interesting alternative that may address a number of analytical measurement problems with a great potential for food safety. As compared to conventional analytical methods, biosensors are easy to handle, portable, quick and the user does not require special skills [4, 5].

Particularly, electrochemical biosensors have been emerging as powerful alternatives to standard methods for the detection and quantification of allergens and adulterants in foods because there are able to perform rapid and simple determinations with high sensitivity and selectivity and low-cost instrumentation. Moreover, they are easy miniaturized and automated, ideal to perform routine and decentralized analysis.

Because of these interesting advantages, the purpose of this chapter is to provide an overview of the variety of electrochemical biosensors-based analytical strategies developed so far for the detection of allergens and adulterants present in food, either accidentally or deliberately, which may be not acceptable to certain consumers. This chapter does not attempt to provide an exhaustive list of all the strategies developed given the numerous scientific papers already published in this area.

Instead, it will provide different trends in electrochemical biosensing strategies based on distinct approaches and what are the future prospects and potential barriers exist with respect to commercialization of such devices within the food sector.

2 Electrochemical Biosensors for Food Allergens

Nowadays, allergic diseases are a public health worldwide problem, getting diseases such as asthma, atopic eczema and respiratory or food allergies are no longer classified as rare diseases and to be considered common diseases in industrialized countries, affecting largely at children.

Food allergy is a type of adverse reaction to food, produced by an immunological mechanism resulting from ingestion, contact or inhalation of certain substances presented in food, known as allergens. Food allergy is nowadays regarded as a problem of public-health relevance, the main concern being the unintentional exposure of allergic consumers to the offending ingredient through allergen containing food. Food allergy affects about 1–10 % of the world population, being a growing public health problem in countries like US where more than 15 million affected people [6] and are suffered by about 4 % of adults and 8 % of children and responsible of about 30,000 cases of medical emergencies and between 150 and 200 deaths per year. Since today there is no cure for allergies, the strict elimination of these allergens from the diet and the availability of reliable methods for detection and quantification of food allergens are the only options to ensure compliance with food labeling and improve consumer protection [2]. A pivotal issue in food allergen quantification is the impossibility to define a useful threshold (a limit below which a stimulus causes no reaction) and valuable determination limits (LOQs). Since the sensitivity of a patient to a given allergen varies from one patient to another and over the time, it is difficult to define threshold doses for allergenic foods [7] and therefore, without well-established thresholds, these quantification methods must be as sensitive, accurate and reliable as possible.

In addition, currently there is a variety of allergenic foods in which different allergenic proteins of different properties were found; which further complicates the assessment of the allergenic potential of certain foods. While more than 160 foods can cause allergic reactions in people, there are eight major foods or food groups believed to account for 90 % of food allergies: cow's milk, eggs, peanuts, soybeans, wheat, tree nuts, fish and crustacean shellfish although a higher number of allergenic compounds have been identified so far [1]. On the other hand, food allergens can become part of the food through unintended exposure such as through contamination which can occur by raw materials or during the production process [8]. Therefore, analytical testing systems are needed by businesses in the food industry to enable them to test whether allergens are present in their raw materials and in the finished products and whether production lines have been correctly sanitized, by food inspection authorities for market surveillance, and by academia to enable and stimulate research into food allergy and allergen detection [9].

At present, the main analytical techniques used to detect food allergens can be classified into protein-based or DNA-based assays. Protein-based assays detect either a specific target food allergen, using enzyme-linked immunosorbent assays (ELISAs), or total soluble allergenic proteins. DNA-based techniques detect the presence of allergens by amplifying a specific DNA fragment of the encoding gene for the target allergenic protein through polymerase chain reaction (PCR) [1]. However, false positive results (due to cross-reactivity with other co-extracted allergenic proteins) as well as large difference in quantitative results between available ELISA kits and the high numbers of replicates for samples or an external standard required by all the PCR methods described have hindered their application to processed foods or complex food matrices. Additional limitations of ELISA for food allergens detection include the possibility to evade detection in some ELISA formats of protein hydrolysates, heat-processed proteins or protein residues with lipophilic nature [10]. Furthermore, even though some level of automation has been achieved in the recent years, ELISAs remain laborious, time consuming and expensive, particularly when multiple targets need to be screened [2]. An additional limitation of DNA-based methods resides in the likelihood to find residual DNA encoding for the allergenic protein in the final processed food, PCR sensitivity strongly depending on the amount and the quality of the DNA isolated [1].

Therefore there is an urgent need to improve the robustness of the available analytical methods and to develop new standardized methods which must be fast, more sensitive, more accurate and more specific for better reliability to allow unambiguous identification of the allergens [7].

An emerging technology in food allergen analysis is the use of biosensors consisting of an integrated receptor-transducer device, which is capable of providing selective quantitative or semi-quantitative information using a biological recognition element [1]. Among these, the electrochemical biosensors have been well established as valid alternatives to classical analysis methods for detection of allergens, by offering the advantages of being easy to use, rapid, robust and often cheap multi-analyte testing [2, 11]. They hold great promise for reliable *point-of-care* sensor (POC) for household use, in particular, in applications where minimizing size and cost is crucial (e.g. *on-line* allergen-contamination monitoring). Electrochemical biosensors can be classified into potentiometric, amperometric, voltammetric and impedance types. Bioreceptors most commonly used in electrochemical biosensors for food allergen management include specific antibodies raised against an allergen [12–14], single-stranded DNA molecules capable of hybridizing with allergen-specific DNA fragments [15, 16], aptamer selected to recognize the target allergen directly [17, 18] and living cells which specifically recognize the target allergen and convert the recognition into a signal that can be electrochemically recorded and quantified [19]. Although the first applications of electrochemical biosensors to food-allergen management date back to 2008 [20, 21], in the following sub-sections, the most relevant electrochemical biosensors described for food allergens detection in the last 10 years, have been selected and classified according to the type of target and bioreceptor type employed.

2.1 Immunosensors

Already in 2008, Huang et al. [21] developed an impedimetric immunosensor for determination of peanut protein Ara h 1 by immobilizing a monoclonal antibody to Ara h 1 onto an Au electrode through amide bond formation to a carboxylate-terminated 11-mercaptoundecanoic acid (MUA) self-assembled monolayer (SAM). The detection limit (LOD) of this reagentless biosensor was estimated to be less than 0.3 nM.

Singh et al. reported the development of a nanopore immunosensor by covalent immobilization of peanut antibody within gold thin-film coated pores of commercial polycarbonate membranes [22]. In this approach protein Ara h 1 was detected by measuring the decrease in the pore conductivity as the peanut protein concentration increases due to the partial pore occlusion by antigen binding.

Direct casein detection was described by Cao et al. [23] by using an electrochemical immunosensor based on a glassy carbon electrode (GCE) functionalized with a gold nanoparticles (AuNPs) and poly(l-Arginine)/multi-walled carbon nanotubes (P-L-Arg/MWCNT) composite film through electropolymerization of L-arginine. Subsequently, AuNPs were adsorbed on the modified electrode in order to immobilize the capture antibody. Under optimized conditions and due to the formation of antibody–antigen complex on the modified electrode, peak currents of the redox couple (ferricyanide) obtained by differential pulse voltammetry (DPV) decreased linearly with increasing casein concentrations (in the range from 1×10^{-7} to 1×10^{-5} g mL^{-1}). This electrochemical immunosensor has a low LOD of 5×10^{-8} g mL^{-1} and was applied to the determination of casein in cheese samples with satisfactory results.

An electrochemical magnetoimmunosensor for the quantification of gliadin or small gliadin fragments in natural or pretreated food samples was described by Laube et al. [24]. The immunological reaction was performed on tosyl-activated magnetic beads (MBs) as solid support by the oriented covalent immobilization of the protein gliadin and using a specific HRP-labeled antibody for the detection. The modified MBs are then captured onto the surface of a magnetoelectrode based on graphite-epoxy composite (m-GEC) to perform the amperometric detection using H_2O_2 and HQ. Excellent LODs (in the order of μg L^{-1}) were achieved, according to the legislation for gluten-free products. The performance of this magnetoimmunosensor was successfully evaluated using spiked gluten-free foodstuffs (skimmed milk and beer), obtaining excellent recovery values.

A novel label-free voltammetric immunosensor for sensitive detection of β-lactoglobulin (β-LG) based on electrografting of aryl diazonium salt organic film on graphene modified screen printed electrodes has also been developed [25]. The derivatization of the graphene electrode surface was achieved by electrochemical reduction of in situ generated 4-nitrophenyl diazonium cations in aqueous acidic solution, followed by electrochemical reduction of the terminal nitro groups to amines, which were activated using glutaraldehyde and used for the covalent immobilization of the capture antibody. Results demonstrated that the DPV reduction peak current of $[Fe(CN)_6]^{3-/4-}$ decreased linearly with increasing

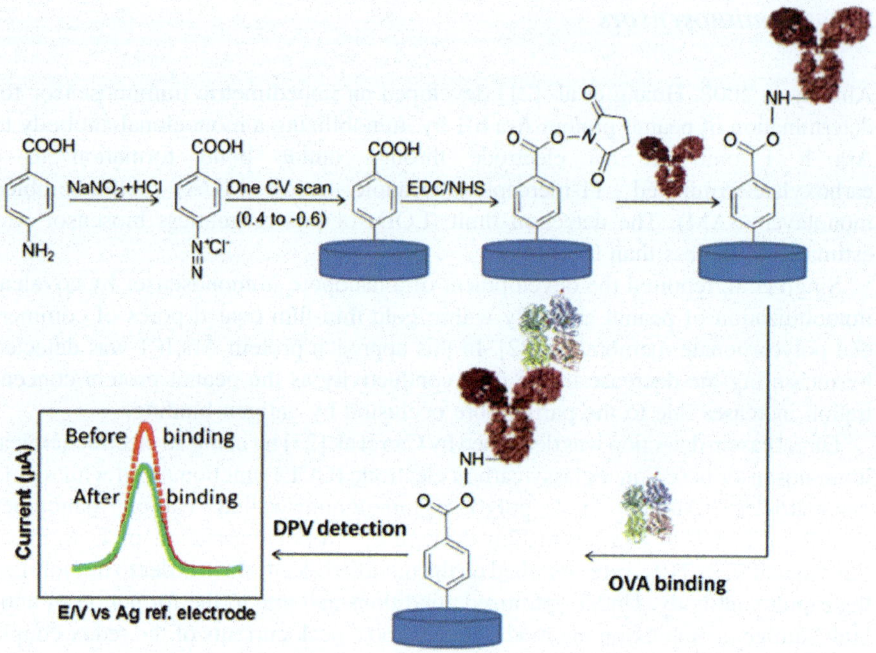

Fig. 1 Scheme of the ovalbumin immunosensor based on the use of carboxyphenyl modified graphene (Reprinted from [26] with permission. Copyright 2013 RSC)

concentrations of the target protein due to the formation of antibody–antigen complex on the modified electrode surface. This label-free voltammetric immunosensor enabled a LOD of 0.85 pg mL^{-1} and a dynamic range from 1 pg mL^{-1} to 100 ng mL^{-1}. Noteworthily, the immunosensor was also evaluated for the determination of β-LG in several milk-containing food products (cake, cheese, snacks and biscuits) and results obtained showed excellent correlation with those provided by a commercial ELISA kit. These authors reported also a label-free voltammetric immunosensor for ovalbumin by using carboxyphenyl modified graphene. In this approach, graphene-modified screen printed carbon electrodes were covalently functionalized using electrochemical reduction of in situ generated aryl diazonium salt forming a carboxyphenyl film on the graphene surface (Fig. 1) [26]. The terminal carboxylic groups on the graphene surface were activated using EDC/NHS and used to immobilize the capture antibody using the decrease in the $[Fe(CN)_6]^{3-/4-}$ DPV reduction peak current after the immunochemical reaction for the ovalbumin detection. The developed immunosensor has been used for ovalbumin detection in the concentration range of 1 pg mL^{-1} to 0.5 mg mL^{-1} with a LOD of 0.83 pg mL^{-1} and applied to the determination of spiked cake extracts.

Ruiz-Valdepeñas Montiel et al. proposed a disposable amperometric magnetoimmunosensor for the determination of β-LG using a sandwich configuration, SPCEs and HOOC-modified MBs (HOOC-MBs) [12]. This methodology involved the immobilization of the capture antibody onto the activated carboxylic-modified MBs (HOOC-MBs) and successive incubation steps of the modified MBs with the analyte and a specific HRP-labeled detector antibody. The resulting modified MBs are captured by a magnet placed under the surface of a disposable carbon screen-printed electrode (SPCE) and the amperometric responses are measured at −0.20 V (vs. Ag pseudo-reference electrode), upon addition of hydroquinone (HQ) as electron transfer mediator and H_2O_2 as the enzyme substrate. The β-LG magnetoimmunosensor exhibited a wide range of linearity (2.8–100 ng mL^{-1}) and a low LOD of 0.8 ng mL^{-1}. The magnetoimmunosensing platform was successfully applied for the detection of β-LG in different milk samples without any matrix effect after just a sample dilution. The results correlated properly with those provided by a commercial ELISA spectrophotometric kit offering a truthful analytical screening tool.

Same authors reported also a similar configuration for the sensitive determination of Ara h 1 in only 2 h [27]. This Ara h 1 magnetoimmunosensor exhibits a wide range of linearity (20.8–1000.0 ng mL^{-1}) Ara h 1, a LOD of 6.3 ng mL^{-1}, a great selectivity and a useful lifetime of 25 days. The usefulness of the immunosensor was demonstrated by determining the endogenous Ara h 1 in different matrices (diluted food extracts and undiluted saliva samples). The results correlated properly with those provided by a commercial ELISA method offering a reliable and promising analytical screening tool in the development of user-friendly devices for on-site determination of Ara h 1.

Other MBs-based electrochemical immunosensor has been developed for ovalbumin determination [13]. A sandwich-type immunocomplex formed by specific anti-ovalbumin immunoglobulin G, ovalbumin and secondary anti-ovalbumin antibodies conjugated with HRP was performed onto the surface of activated carboxylate-modified MBs. The electrochemical signal proportional to the enzymatic reaction of HRP during the reduction of H_2O_2 with thionine as electron mediator was measured by linear sweep voltammetry at screen-printed platinum electrodes. This method was suitable for quantification of ovalbumin in the range of 11 to 222 nM and a LOD of 5 nM.

Alves et al. have developed recently integrated voltammetric immunosensors for the determination of Ara h 6 [14] and Ara h 1 [28]. AuNPs-modified screen-printed carbon electrodes were used to develop sandwich-type immunoassays using specific capture antibodies and ALP-labeled detector antibodies. The electrochemical detection of the antibody-antigen interaction was performed by applying an anodic (stripping) voltammetric potential scan of the enzymatically deposited silver in the presence of 3-indoxyl-phosphate (3-IP) and silver ions. Both methodologies were reproducible, presented a good repeatability, provided accurate results, very low LOD (0.27 and 3.8 ng mL^{-1} of Ara h 6 and Ara h 1, respectively) and were successfully applied to determination in complex food matrices, such as cookies and chocolate.

A portable indirect competitive amperometric immunosensor using rabbit polyclonal anti-bovine β-casein, goat anti-rabbit IgG-HRP conjugate antibodies and a set of 8 carbon working electrodes screen printed on alumina able to perform β-casein detection in eight different samples has been recently published by

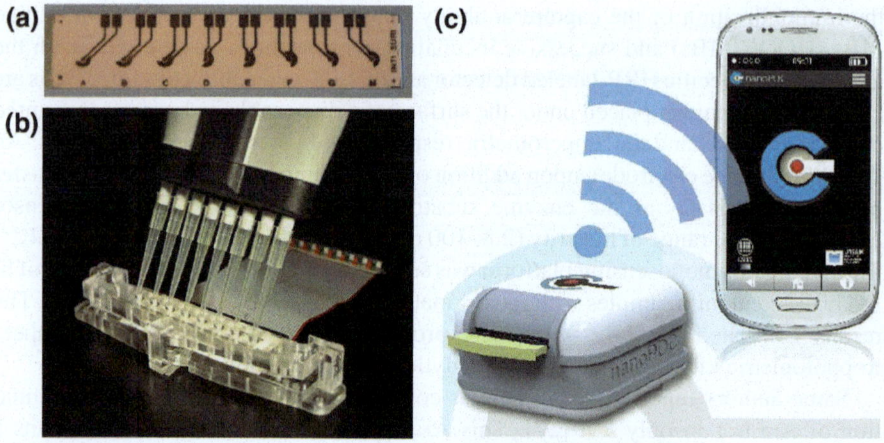

Fig. 2 Set of working and auxiliary electrodes screen printed on alumina, electrochemical cells and portable potentiostat connected to a smartphone and controlled by bluetooth. (Reprinted from [29] with permission. Copyright 2015 Creative Commons Attribution Licence)

Molinari et al. [29]. Commercial β-casein is covalently immobilized onto the electrode surface using the EDC/NHS method after the functionalization of the sensor surfaces with carboxylic groups by plasma treatment and an indirect competitive reaction between soluble and electrode surface immobilized antigen to the binding sites of the anti-bovine β-casein antibodies which are then recognized by HRP-conjugated secondary antibodies, took place. The amperometric signal detected, due to the enzymatic reduction of the redox mediator after adding H_2O_2, is inversely proportional to the β-casein concentration present in the sample solutions. Although this approach was not particularly sensitive (0–10 ppm range), no complex electronic instrumentation such as computer was required in this methodology since the eight electrochemical cells are integrated into a small-size portable potentiostat controlled by a smartphone via Bluetooth communication (Fig. 2).

2.2 *DNA Sensors*

Electrochemical DNA-based biosensors have been widely used for detecting specific genes associated with target allergens. Betazzi et al. [20] developed an electrochemical genosensor platform for the simultaneous, sensitive and specific detection of PCR amplicons obtained from cDNA of the major hazelnut protein allergens (Cor a 1.04 and Cor a 1.03) in foodstuffs. The used platform was a low density array of eight individually addressable gold working electrodes, enabling the simultaneous analysis of different samples. Unmodified PCR products were captured at the sensor interface via sandwich hybridization with thiol-tethered DNA capture probes assembled on the gold electrodes and biotinylated signaling probes.

The resulting biotinylated hybrids were coupled with a streptavidin–alkaline phosphatase conjugate and DPV was finally used to detect the α-naphthol signal produced by enzymatic reaction from α-naphthyl phosphate. LODs of 0.3 and 0.1 nmol L^{-1} were obtained for Cor a 1.03 and Cor a 1.04, respectively. Storage stability studies demonstrated that the capture probe-modified chip provided a stable response for up to 4 months. The genosensor was applied to detect the presence of these allergens in several commercial foodstuffs, obtaining a good agreement with the results obtained with a standard ELISA kit.

Sun et al. [30] developed a very sensitive DNA sensor, based on the use of a dually-labeled stem-loop probe, for detecting peanut-allergen Ara h 1 using a probe. The probe (modified with a thiol group at its 5′ end and a biotin tag at its 3′ end) assembled on a gold electrode was "closed" when the target oligonucleotide was absent, and "open" after hybridization with the target, thus moving away from the electrode surface the biotin group at its 3′ end. The electron transfer efficiency changes resulting from the detachment of biotin tags from the electrode surface were measured by faradaic EIS. This method with a linear response in the range of 10^{-15}–10^{-10} M, a LOD of 0.35 fM and the ability to discriminate a single-base mismatch, was successfully applied to detect the target allergen in a peanut-milk beverage.

A sandwich selective electrochemical genosensor has been developed also for the detection of an 86-mer DNA peanut sequence encoding part of the allergen Ara h 2 (conglutin-homolog protein) [31]. The method is based on the use of thiolated capture and biotinylated detector probes and screen-printed gold electrodes. The electrochemical signal monitored corresponded to the oxidation of 1-naphthol generated through the hydrolysis of 1-naphthyl phosphate in the presence of Strep-ALP (use to label the biotinylated detector probe attached to the electrode surface). The proposed sensor showed a sensitivity as high as 3 mA nM^{-1}, a linear range from 5×10^{-11} to 5×10^{-8} M and a LOD of 10 pM.

Sun et al. developed a stem-loop DNA probe biosensor for peanut allergen Ara h 1 using a multilayer graphene–gold nanocomposite as a signal amplification material [15]. In this case a thiolated hairpin DNA-biotin probe was immobilized onto a multilayer graphene–gold nanocomposite prepared by a cycle-alternate electrodeposition method to deposit monolayers of graphene and AuNPs on a GCE. The prepared biosensor, based on an "off" state in the presence of the target DNA demonstrated a linear response ranging from 10^{-16} to 10^{-13} M, an ultrasensitive LOD of 0.041 fM, a one-base mismatch selectivity and successful recoveries (86.8–110.4 %) of the Ara h 1 gene from a peanut milk beverage.

Very recently, the first electrochemical genosensor able to quantify gluten in food products was developed [16]. This approach used a sandwich-based format, involving the use of a capture probe immobilized onto the screen-printed gold electrode, and a signaling probe functionalized with fluorescein isothiocyanate (FITC), both able to hybridize with the target analyte, a fragment encoding the immunodominant peptide of α2-gliadin amplified by a tailored PCR. The sensor was able to reliable detect as low as 0.001 % (w/w) of wheat flour in an inert matrix and to analyze processed food samples, without complicated pretreatment protocols, giving results in agreement with the Codex recommended method.

Among the DNA-based biosensors, aptameric ones have attained also great attention in the determination of food allergens and will be discussed in the following subsection [5].

2.2.1 Aptasensors

Aptamers are small single-stranded DNA or RNA sequences which inherently adopt stable three dimensional sequence-dependent structures and bind a target ligand with high affinity [32].

Amaya-Gonzalez et al. [33] developed a competitive electrochemical aptaassay based on MBs for determination of gliadin in real samples. Two different aptamer sequences (Gli1 and Gli4) against the immunodominant peptide from wheat gliadin, the 33-mer, that also recognize celiac disease related proteins from barley, rye and oat were selected [34]. The competition is established between the biotinylated peptide immobilized on the surface of streptavidin-modified MBs (Strep-MBs) and gliadin in solution in the presence of a fixed amount of the specific biotinylated aptamer. The electrochemical signal was achieved by chronoamperometry using the system H_2O_2/TMB after enzymatic labeling with streptavidin-HRP of the biotinylated aptamer fraction bound to the 33-mer modified MBs. In the presence of increasing concentrations of gliadin, the amount of free aptamer available to bind the 33-mer immobilized peptide on the Strep-MBs diminishes resulting in a decreasing analytical signal with analyte concentration. This competitive electrochemical magnetoassay did not show any cross-reactivity with non-triggering celiac disease proteins from soy, rice or maize. Moreover, the results achieved demonstrated that the assays based on the two aptamer sequences, Gli1 and Gli4 are complementary: while the assay using the Gli1 aptamer is suitable for quantifying gluten in hydrolysed samples (LOD of 4.9 ppm), the method based on the Gli4 aptamer is more sensitive (LOD of 0.5 ppm) and allows the quantification of gluten in heated foods thus providing answers to one of the main challenges associated with analytical determination of gluten, namely detection of gliadin in heated and hydrolysed foods. Aptamer-based sensors for the label-free determination of Ara h 1 have been developed by immobilizing on gold surfaces a thiolated aptamer [17] or an amino (NH_2)-terminated 77-base DNA aptamer covalently to a carboxylated SAM employing carbodiimide chemistry [35]. Both platforms are based on the use of an 80-base DNA aptamer and impedimetric detection. The fast and low-cost aptameric platform based on the amino-modified aptamer can detect Ara h 1 with a LOD of $\sim$1 nM and up to concentrations of $\sim$15 nM.

A novel aptasensor for lysozyme detection was reported based on the immobilization of the aptamer onto the surface of a screen-printed carbon electrode by covalent binding via diazonium salt chemistry [18]. The developed aptasensor exhibited a very good linearity (0.025–0.8 μM) and a LOD of 25 nM far below the maximum amount allowed by the International Organization of Vine and Wine

(OIV). This disposable platform demonstrated promising results in the analysis of spiked wine samples. Same authors developed also other impedimetric aptasensor for lysozyme detection based on a sandwich format using a capture aptamer immobilized covalently via diazonium salt chemistry on a screen-printed carbon electrode and a biotinylated detector antibody [36]. Detection was performed by monitoring by DPV the electrochemical oxidation signals of 1-naphthol after labeling the biotinylated antibody with avidin marked with alkaline phosphatase and using 1-naphthyl phosphate as the enzymatic substrate. This aptasensor, characterized by a wide detection range, from 5 fM to 5 nM and a LOD of 4.3 fM was also applied to the determination in spiked wine samples.

2.3 Whole Cell-Based Sensors

Jiang et al. developed a novel cell-based impedimetric biosensor to quantify major shrimp allergen Pen a 1 (tropomyosin) [37]. This aptasensor was based on the immobilization of anti-shrimp tropomyosin IgE pre-sensitized rat basophilic leukemia (RBL-2H3) living mast cells, encapsulated in type I collagen, on a self-assembled L-cysteine/AuNPs (AuNPsCys)-modified gold electrode (see Fig. 3). In the presence of antigen, the sensitized immobilized cells lead to initiation of signaling cascade resulting in the degranulation of secretory vesicles and releasing of inflammatory molecules, which can be monitorized by the increase in the impedance value (R_{et}). Results presented demonstrated that R_{et} values increased with the concentration of purified shrimp allergen Pen a 1 in the range of 0.5–0.25 μg mL^{-1} with an LOD of 0.15 μg mL^{-1}.

Same authors developed also a very attractive fluorescent magnetic nanobeads-based mast cell biosensor for electrochemical detection of activated cells that were employed to quantify both shrimp allergen tropomyosin (Pen a 1) and fish allergen parvalbumin (PV) [19]. In this approach cationic magnetic fluorescent nanoparticles (CMFNPs) were used for transfecting RBL-2H3 cells and the resulting CMFNP-transfected RBL-2H3 cells (pre-sensitized by anti-Pen a 1 IgE or Anti-PV IgE) were magnetically captured on a magnetic glassy carbon electrode (MGCE). The response induced by the activated cells in the presence of the target allergen was monitorized by EIS. Results show high detection accuracy for these targets, with linear ranges from 0.5 to 10 ng mL^{-1} and 0.1 to 6 μg mL^{-1} and LODs of 0.16 μg mL^{-1} and 0.03 ng mL^{-1} for the PV and Pen a 1, respectively.

2.4 Other Biosensors

Apart from the most commonly bioreceptors mentioned in Sect. 2, lectins have been also applied to develop electrochemical biosensors for food allergens determination. Xu and co-workers [38] developed an electrochemical biosensor for the direct

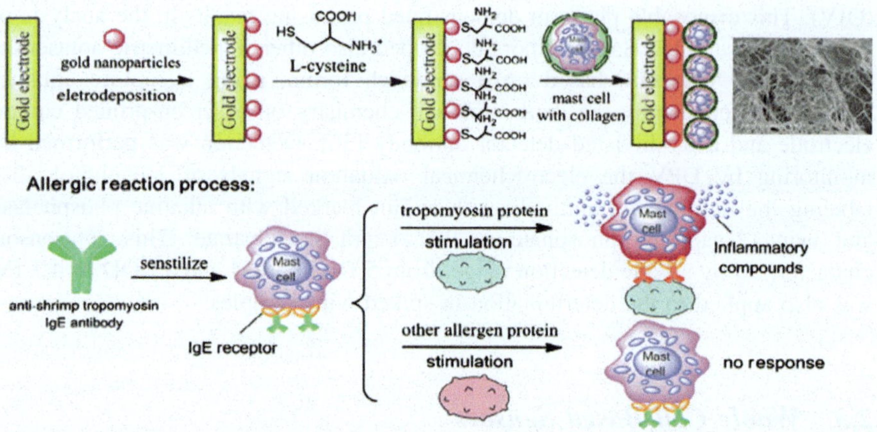

Fig. 3 Schematic illustration of the preparation of modified electrode (above) and principle of allergen reaction process (below). (Reprinted from [37] with permission. Copyright 2013 Elsevier)

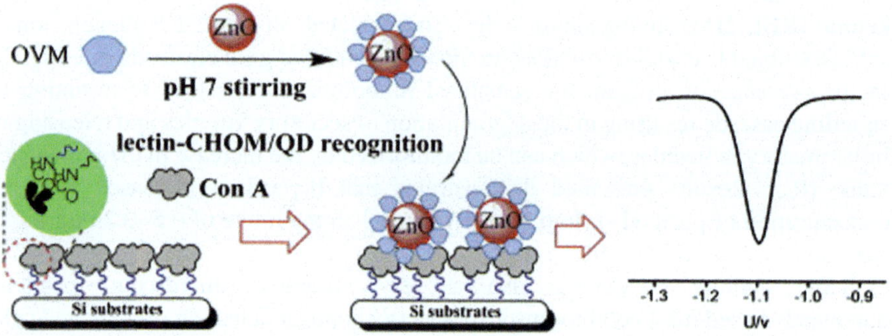

Fig. 4 Scheme of assembly and electrochemical strategy for detection of CHOM. (Reprinted from [38] with permission. Copyright 2011 Elsevier)

detection of chicken ovomucoid (CHOM) using ZnO quantum dots (QDs) bioconjugates and Concanavalin A as a recognition element. The novel bioconjugates were synthetized by self-assembling of CHOM onto ZnO QDs through electrostatic interaction between positively charged QDs surface and negatively charged protein. This ZnO-QD/CHOM bio-conjugate was captured on a Concanavalin A-modified electrode and the extent of the biorecognition was accomplished by square wave voltammetric (SWV) of captured QDs after acidic dissolution (see Fig. 4). The zinc SWV peaks monitored were dependent on the CHOM concentration in the 1–140 ng mL^{-1} range. The proposed method, with a LOD of 0.1 ng mL^{-1}, offers a simpler platform for the detection of other glycoprotein allergens.

Recently Sugawara et al. [39] developed an electrochemical biosensor to detect ovalbumin (OVA) by using a specific peptide probe (peptide-1) conjugated with

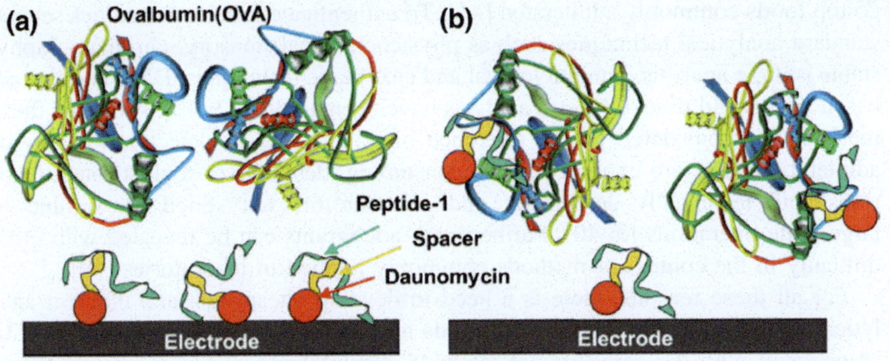

Fig. 5 Principle of the OVA detection by using a peptide-based biosensor. (Reprinted from [39] with permission. Copyright 2015 Elsevier)

daunomycin adsorbed on a GCE. The DPV peak current of the daunomycin moiety decreased as the concentration of OVA increased due to the binding between the OVA and the peptide probe which was released from the electrode surface (Fig. 5) The calibration curve of the OVA using the peptide probe was linear and ranged from 1.5×10^{-11} to 3.0×10^{-10} M. Furthermore, this method could be applied to the electrochemical sensing of the OVA in egg whites and in fetal bovine serum.

3 Electrochemical Biosensors for Food Adulterants

Food adulteration, which induced lost large of money as well as the confidence of consumers, has been practiced since a long time ago and becomes increasingly in the last years more sophisticated. Nowadays European consumers are increasingly demanding information and reassurance not only on the origin but also on the content of their food. Protecting consumer rights and preventing fraudulent or deceptive practices such as food adulteration are important and challenging issues facing the European food industry, as manufacturers are required to provide and confirm the authenticity and point of origin of food products and their components [40].

Non-authentic food products arise from the adulteration and fraud. The replacement of original substance partially or completely with more easily available and cheap substance is the most common procedure performed by defrauders such as the addition of: (i) flavors/aromas to improve the value of cheap products; and/or (ii) cheap substances to the food products. Foods and ingredients presenting high-value are the most vulnerable for adulteration. Therefore, nowadays, the determination of the authenticity and the detection of adulteration of dairy products is a major concern in order to: (i) assure the traceability system from milk to fork; and (ii) ensure that dairy products are correctly labeled in terms of which animals are actually processed for consumption [3]. Indeed, milk and meat are between the

25 top foods commonly adulterated [41]. To authenticate food commodities several standard analytical techniques such as physicochemical, sensory, chromatography, stable isotope analysis, immunological and enzymatic techniques, DNA and protein based assays and triacylglycerol analysis have been applied [3, 40]. Although these methods are considered as the reference ones and able to detect low levels of adulteration, they are expensive, time consuming, destructive, sophisticated, laborious, and technically demanding and are therefore not suited for on-line or large-scale operations [3, 40]. Furthermore, adulterants can be revealed with great difficulty in the context of methods commonly applied in laboratories [40].

For all these reasons, there is a need to develop cheap, fast and efficient analytical methods for the detection of frauds and authentication of food products. In comparison with the conventional methods, electrochemical biosensors constitute very promising tools to determine and/or to detect the authenticity and adulteration of milk and dairy products; they are rapid, non-destructive, effective and reliable, require only basic training in a user-friendly software, and low cost per assay.

This book chapter will provide also a comprehensive overview of the applications of different electrochemical biosensors to determine the authenticity and to detect the adulteration of dairy products during the last 10 years. Actual examples illustrating the utilization of these techniques mainly in laboratory environments will be discussed as well as their advantages and disadvantages.

Electronic aptamer based (E-AB) sensors were developed for the determination of cocaine [42, 43]. E-AB sensors comprise an aptamer that is attached at one end to an electrode surface. The distal end of the aptamer probe is modified with an electroactive redox marker for signal transduction [43]. These sensors are based on the use of a thiolated DNA strand assembled on gold electrodes and modified at the other end with methylene blue (MB). In the absence of target, the aptamer is thought to remain partially unfolded, in the presence of target the rearrangement of the aptamer structure brings the redox tag in close proximity of the electrode thus resulting in an increase of the MB reduction peak. While Baker et al. fabricated the sensor on a ~ 1 mm^2 gold electrode, White et al. compared the performance of three different E-AB architectures (aptamer, concaptamer and pseudosandwich) on array-based chips containing 36 working gold electrodes. Baker et al. demonstrated that the "signal-on" sensor responded rapidly (seconds) and specifically to micromolar cocaine in adulterated samples and was regenerated via a brief, room temperature wash.

Cao et al. developed a novel and simple electrochemical method for determination of melamine based on oligonucleotides ($d(T)_{20}$) film modified gold electrodes [44]. The proved interactions between oligonucleotides and melamine, via electrostatic and hydrogen-bonding interactions, lead to the increase in the peak currents of ferricyanide, measured by differential pulse stripping voltammetry (DPSV), which could be used for electrochemical sensing of melamine. This approach provided a linear range from 3.9×10^{-8} to 3.3×10^{-6} M and a LOD of 9.6×10^{-9} M. The proposed electrochemical biosensor is rapid, convenient and low-cost for effective sensing of melamine and was applied successfully to the determination of melamine in milk products, with a recovery of 95 %.

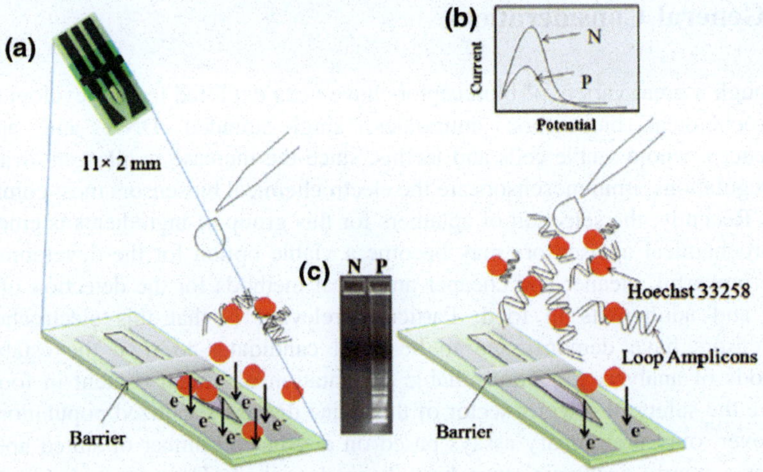

Fig. 6 Species-specific identification using a LAMP amplification-based electrochemical genosensor. (Reprinted from [46] with permission. Copyright 2010 Elsevier). **a** Without amplification product (N). **b** With amplification product (P). **c** Gel electrophoresis analysis of loop amplicons

Singh et al. described the use of electrochemically prepared PANi/ClO_4 doped films onto ITO electrodes to immobilize DNA as biosensing platform for the detection of sanguinarine in adulterated multard oil [45]. The ds-DNA was covalently immobilized on the PANi film using EDC and NHS and the basis of this methodology was the decrease of sanguinarine electrochemical signal after its binding to ds-DNA by intercalation with insertion between adjacent base pairs of DNA duplex strand. This genosensor showed acceptable linearity from 1 to 40 µg and the recovery experiment results found recoveries between 89 and 121 % in spiked edible mustard oil sources.

An easy, rapid and sensitive method of detection of the presence of meat species in raw or processed foods was developed by Ahmed et al. [46]. This strategy, based on the use of loop-mediated isothermal amplification (LAMP) and disposable electrochemical genosensors, detected by Linear Sweep Voltammetry (LSV) the change in the anodic peak current of the DNA binder H33258 after its interaction in solution with the generated loop amplicons (Fig. 6). The comparison with the multiplex-PCR (M-PCR) detection method demonstrated that this approach was more specific, reduced the cross-reactivity and avoided the formation of non-specific amplicons in the detection of meat species in meat containing raw and processed foods. Moreover, this method, which gave LODs of ~20.33 ng μL^{-1}, 78.68 pg μL^{-1} and 23.63 pg μL^{-1} for pork, chicken and bovine species, respectively, took only an hour and being isothermal may be a good candidate to become a portable biosensor for on-site monitoring of meat species identification in raw and processed foods.

4 General Considerations

Although a great variety of bioreceptors have been explored to the development of electrochemical biosensors: antibodies, single-stranded DNA and aptamer sequences, whole viable cells and lectins, since the increase in allergen awareness and regulations, immunosensors are the electrochemical biosensors most commonly used. Recently, the selection of aptamers for this group of ingredients is emerging. Electrochemical aptasensors may become a viable option for the development of more powerful, cleaner and cheaper analytical methods for the detection of allergens and adulterants in food. Particular relevant is that this electrochemical aptasensors have demonstrated to be good candidates to rival the established methods of analysis for more reliable quantitation of gluten content in food and ensure the safety of a wider sector of the celiac disease sensitized population [33]. However, only preliminary assays based on a reduced number of suited aptamers and at a purely academic stage have been described. There is a clear need for selection of new aptamers against the great variety of allergenic and adulterant residues that can be found in food, which is continuously increasing.

It is important to mention also that although quantification of food allergens using genosensors is a challenging task that only rarely devices have addressed and results in terms of DNA concentration are meaningless for consumers and often used as yes/no tools, a genosensor that allows for the first time to quantify the gluten content in food products has been described recently [16].

In addition, given the sensitivity of persons with allergen sensitivity it can be predicted that biosensor devices with improved sensitivity based on the use of nanomaterials, such as QDs [39], AuNPs [14, 19, 23, 28, 37] carbon nanotubes [23] and graphene [15, 25, 26] will find widespread utility. Moreover, although very scarcely explored until now, the use of lectins as recognition elements is very promising to develop simple and cheap biosensing platforms for glycoprotein allergens determination.

Regarding the electrochemical biosensors for determining adulterants it is important to mention that the type of bioreceptors applied for these target analytes is limited to DNA sequences and aptamers. In order to cope with the crucial step of DNA-probe immobilization, surface-confined stem-loop DNA structures have been designed and proposed as capture probes with performance superior to linear probes in terms of ability to discriminate mismatches [15, 30]. Although the applicability of food adulterants electrochemical biosensors have been less explored, they have demonstrated great potential to detect melamine in milk and to identify meat species of particular interest in the food industry.

Worth to mention also the increasing tendency to fabricate these electrochemical biosensors onto screen-printed electrodes providing the advantages of being disposable and mass produced, with a low manufacturing cost and require low volume of reagents and samples. Magnetobiosensors have also been described for food allergens determination [12, 19, 24, 27, 33] making use of the advantages of MBs in the final performance of electrochemical biosensors in terms of sensitivity, reduced assay time and minimization of matrix effects which is essential for the analysis in complex matrices such as food extracts. Regarding the formats of assay while direct, sandwich and competitive formats have been employed in electrochemical immunosensors and aptasensors developed for allergens determination; all the approaches reported for adulterants are based on direct assays.

5 Conclusions and Future Prospects

Food quality and safety is of paramount importance from health as well as economic point of view. Particularly, the detection of food allergens and adulterants is an upcoming field, since standard techniques are not able to fulfill current standards in their determination. The present chapter has reviewed the most recent literature (over the past 10 years) on electrochemical biosensors development for food-allergens and adulterants management using a bioreceptor-based classification and highlighting the most important achievements and the new research trends.

From the discussed literature it is clear that electrochemical biosensors can play a very helpful role in the detection of food allergens and adulterants and a vital role in assuring food safety and help to take quicker preventive actions when required. Featuring high speed of execution, good sensitivity and selectivity, robustness, ease of use and high degree of automation and portability for in-field applications, they have demonstrated all the potential for direct, real-time, on-line, monitoring of food allergens and adulterants along the production chain. However, despite the great efforts performed in the last years, there are still some important challenges that should be faced in this innovative application field for electrochemical biosensors. Off-line measurements, long incubation times and sensitivity to non-specific binding from matrix components are common open issues, which often constrain the use of electrochemical biosensors to preliminary proof-of-concept investigations on standard solutions. Most of the described assays have proved the quantification of targets in aqueous solutions and only a few faces the analysis of spiked samples. Moreover, in some cases, the contaminant was added in an intermediate step or even at the end of sample preparation. There are two relevant issues associated with real sample analysis: possible electrochemical interferences (1) and efficient extraction of target analyte from the complex food matrix (2). To avoid electrochemical interferences, surface chemistry needs to be carefully optimized, in conjunction with sample pre-treatment and cleanup. Sample preparation and efficient extraction of the target analytes remain being the limiting factors for the total analysis time and final performance of the electrochemical biosensor. While for liquid samples often times a simple filtration and dilution is sufficient, for solid samples several steps of extraction and cleanup are necessary. The development of online sample preparation systems or easy to use kits for rapid sample extraction and cleanup will also be required in order to facilitate the development of commercial devices and alleviate current problems related to complex and time demanding sample pre-treatment protocols and the presence of false negative results (attributed to non-effective extractions from the food matrix). Moreover, in order to validate the claim that electrochemical biosensors are reliable alternatives to existing methods, more efforts should be directed towards participation in proficiency testings, comparative studies using current approaches, analysis of certified reference materials and more closely consideration of the reproducibility issue.

A great deal of effort should be devoted in these directions in order to exploit all the attractive features and great potential of electrochemical biosensors in screening for food allergens and adulterants. Analogously, nanomaterial-based biosensors were shown to be promising tools for improving sensing performance but there are still in their infancy in their application to food matrices. Detailed investigations on the interferences in real sample analysis and evaluation of technological issues related to the final application to food must be addressed before biosensors can fully benefit from integrating nanotechnology.

Apart from facing all these challenges, recent trends in the development of electrochemical biosensors for food field quality and safety include the development of portable biosensors, with multiplexing capabilities and amplified analytical signals.

Finally, giving the on-going demand for devices that can be used outside the laboratory environment to assess the safety and quality of foods on-site, next years will witness a steep increase in the number of research studies devoted to electrochemical biosensors for this purpose, where the start of the race to bring to market aptamer-based assays for different targets of importance in food safety control will occur. However, it is worth to mention that the transfer of this attractive technology to market will require also overcoming the resistance of the agri-food sector, a conservative sector mainly relying on well-known processes and not focused on emerging technologies. But as food analysts are facing increasingly complex challenges, they will need the best available technology and this is where electrochemical biosensors will find their niche.

Acknowledgements The financial support of the Spanish Ministerio de Economía y Competitividad Research Projects, CTQ2015-64402-C2-1-R, and the NANOAVANSENS Program from the Comunidad de Madrid (S2013/MT-3029) are gratefully acknowledged. R. M. Torrente-Rodríguez acknowledges a FPI fellowship from the Spanish Ministry of Education.

References

1. Monaci L, Visconti A (2010) Immunochemical and DNA-based methods in food allergen analysis and quality assurance perspectives. Trends Food Sci Technol 21:272–283
2. Pilolli R, Monaci L, Visconti A (2013) Advances in biosensor development based on integrating nanotechnology and applied to food-allergen management. TrAC Trends Anal Chem 47:12–26
3. Kamal M, Karoui R (2015) Analytical methods coupled with chemometric tools for determining the authenticity and detecting the adulteration of dairy products: a review. Trends Food Sci Technol 46:27–48

4. Narsaiah K, Jha SN, Bhardwaj R, Sharma R, Kumar R (2012) Optical biosensors for food quality and safety assurance—a review. J Food Sci Technol 49:383–406
5. Amaya-González S, de-los-Santos-Álvarez N, Miranda-Ordieres AJ, Lobo-Castañón MJ (2013) Aptamer-based analysis: a promising alternative for food safety control. Sensors 13:16292–16311
6. Wammes AEM, van Delft FL, van Hest JCM, Löwik DWPM (2015) Surface plasmon resonance (SPR) sensors for detecting allergens in food, A volume in Woodhead Publishing Series in Food Science, Technology and Nutrition, 229–244
7. Kirsch S, Fourdrilis S, Dobson R, Scippo M-L, Maghuin-Rogister G, De Pauw E (2009) Quantitative methods for food allergens: a review. Anal Bioanal Chem 395:57–67
8. Immer U, Lacorn M (2015) Enzyme-linked immunosorbent assays (ELISAs) for detecting allergens in food. Handbook of Food Allergen Detection and Control, p 199–217
9. Paschke A, Ulberth F (2009) Allergens in foods. Anal Bioanal Chem 395:15–16
10. Taylor SL, Nordlee JA, Niemann LM, Lambrecht DM (2009) Allergen immunoassays—considerations for use of naturally incurred standards. Anal Bioanal Chem 395:83–92
11. Arkan E, Saber R, Karimi Z, Shamsipur M (2015) A novel antibody–antigen based impedimetric immunosensor for low level detection of HER2 in serum samples of breast cancer patients via modification of a gold nanoparticles decorated multiwall carbon nanotube-ionic liquid electrode. Anal Chim Acta 874:66–74
12. Ruiz-Valdepeñas Montiel V, Campuzano S, Conzuelo F, Torrente-Rodríguez RM, Gamella M, Reviejo AJ, Pingarrón JM (2015) Electrochemical magnetoimmunosensing platform for determination of the milk allergen β-lactoglobulin. Talanta 131:156–162
13. Cadková M, Metelka R, Holubová L, Horák D, Dvoráková V, Bílková Z, Korecká L (2015) Magnetic beads-based electrochemical immunosensor for monitoring allergenic food proteins. Anal Biochem 484:4–8
14. Alves RC, Pimentel FB, Nouws HPA, Correr W, González-García MB, Oliveira MBPP, Delerue-Matos C (2015) Detection of the peanut allergen Ara h 6 in foodstuffs using a voltammetric biosensing approach. Anal Bioanal Chem 497:7157–7163
15. Sun X, Jia M, Guan L, Ji J, Zhang Y, Tang L, Li Z (2015) Multilayer graphene–gold nanocomposite modified stem-loop DNA biosensor for peanut allergen-Ara h 1 detection. Food Chem 172:335–342
16. Martín-Fernández B, de-los-Santos-Álvarez N, Martín-Clemente JP, Lobo-Castañón MJ, López-Ruiz B (2016) Challenging genosensors in food samples: The case of gluten determination in highly processed samples. Talanta 146:490–495
17. Trashin S, de Jong M, Breugelmans T, Pilehvar S, DeWael K (2015) Label-free impedance aptasensor for major peanut allergen Ara h 1. Electroanal 27:32–37
18. Ocaña C, Hayat A, Mishra RK, Vasilescu A, del Valle M, Marty J-L (2015) Label free aptasensor for Lysozyme detection: a comparison of the analytical performance of two aptamers. Bioelectrochem 105:72–77
19. Jiang D, Zhu P, Jiang H, Ji J, Sun X, Gu W, Zhang G (2015) Fluorescent magnetic bead-based mast cell biosensor for electrochemical detection of allergens in foodstuffs. Biosens Bioelectron 70:482–490
20. Bettazzi F, Lucarelli F, Palchetti I, Berti F, Marrazza G, Mascini M (2008) Disposable electrochemical DNA-array for PCR amplified detection of hazelnut allergens in foodstuffs. Anal Chim Acta 614:93–102
21. Huang Y, Bell MC, Suni II (2008) Impedance biosensor for peanut protein Ara h 1. Anal Chem 80:9157–9161
22. Singh R, Sharma PP, Baltus RE, Suni II (2010) Nanopore immunosensor for peanut protein Ara h1. Sens Actuat B-Chem 145:98–103
23. Cao Q, Zhao H, Yang Y, He Y, Ding N, Wang J, Wu Z, Xiang K, Wang G (2011) Electrochemical immunosensor for casein based on gold nanoparticles and poly(l-Arginine)/multi-walled carbon nanotubes composite film functionalized interface. Biosens Bioelectron 26:3469–3474

24. Laube T, Kergaravat SV, Fabiano SN, Hernandez SR, Alegret S, Pividori MI (2011) Magneto immunosensor for gliadin detection in gluten-free foodstuff: towards food safety for celiac patients. Biosens Bioelectron 27:46–52
25. Eissa S, Tlili C, L'Hocine L, Zourob M (2012) Electrochemical immunosensor for the milk allergen b-lactoglobulin based on electrografting of organic film on graphene modified screen-printed carbon electrodes. Biosens Bioelectron 38:308–313
26. Eissa S, L'Hocine L, Siaj M, Zourob M (2013) A graphene-based label-free voltammetric immunosensor for sensitive detection of the egg allergen ovalbumin. Analyst 138:4378–4384
27. Ruiz-Valdepeñas Montiel V, Campuzano S, Pellicanò A, Torrente-Rodríguez RM, Reviejo AJ, Cosio MS, Pingarrón JM (2015) Sensitive and selective magnetoimmunosensing platform for determination of the food allergen Ara h 1. Anal Chim Acta 880:52–59
28. Alves RC, Pimentel FB, Nouws HPA, Marques RCB, González-García MB, Oliveira MBPP, Delerue-Matos C (2015) Detection of Ara h 1 (a major peanut allergen) in food using an electrochemical gold nanoparticle-coated screen-printed inmunosensor. Biosens Bioelectron 64:19–24
29. Molinari J, Moina C, Ybarra G (2015) Electrochemical immunosensor for the determination of β-casein. J. Electrochem. Sci. Eng. 5:9–16
30. Sun X, Guan L, Shan X, Zhang Y, Li Z (2012) Electrochemical detection of peanut allergen Ara h 1 using a sensitive DNA biosensor based on stem—loop probe. J Agric Food Chem 60:10979–10984
31. Sánchez-Paniagua M, López G, Frutos Cabanillas MJ, Lobo Castañón MJ, López-Ruiz B (2014) Development of a genosensor for peanut allergen ARA h 2 detection and its optimization by surface response methodology. Biosens Bioelectron 62:350–356
32. Tombelli S, Minunni M, Mascini M (2005) Analytical application of aptamers. Biosens Bioelectron 20:2424–2434
33. Amaya-Gonzalez S, de-los-Santos-Alvarez N, Miranda-Ordieres AJ, Lobo-Castañón MJ (2014) Aptamer binding to celiac disease-triggering hydrophobic proteins: a sensitive gluten detection approach. Anal. Chem. 86:2733–2739
34. Amaya-Gonzalez S, de-los-Santos-Alvarez N, Miranda-Ordieres AJ, Lobo-Castañón MJ (2013) Aptámeros Específicos Contra el Gluten y Método de Detección del Gluten Asociado (in Spanish). International Patent Application PCT/ES2013/000133
35. Peeters M, Jiménez-Monroy KL, Libert C, Eurlings Y, Cuypers W, Wackers G, Duchateau S, Robaeys P, Nesládek M, van Grinsven B, Pérez-Ruiz E, Lammertyn J, Losada-Pérez P, Wagner P (2014) Real-Time monitoring of aptamer functionalization and detection of Ara h 1 by electrochemical impedance spectroscopy and dissipation-mode quartz crystal microbalance. J Biosens Bioelectron 5:3. doi:10.4172/2155-6210.1000155
36. Ocaña C, Hayat A, Mishra RK, Vasilescu A, del Valle M, Marty J-L (2015) A novel electrochemical aptamer–antibody sandwich assay for lysozyme detection. Analyst 140: 4148–4153
37. Jiang D, Ji J, An L, Sun X, Zhang Y, Zhang G, Tang L (2013) Mast cell-based electrochemical biosensor for quantification of the major shrimp allergen Pen a 1 (tropomyosin). Biosens Bioelectron 50:150–156
38. Yang C, Gu B, Xu C, Xu X (2011) Self-assembled ZnO quantum dot bioconjugates for direct electrochemical determination of allergen. J Electroanal Chem 660:97–100
39. Sugawara K, Kadoya T, Kuramitz H (2015) Construction of a peptide with an electroactive daunomycin like a pendant arm to detect ovalbumin. Anal Chim Acta 857:71–78
40. Ropodi AI, Pavlidis DE, Mohareb F, Panagou EZ, Nychas GJE (2015) Multispectral image analysis approach to detect adulteration of beef and pork in raw meats. Food Res Int 67:12–18
41. Moore JC, Spink J, Lipp M (2012) Development and application of a database of food ingredient fraud and economically motivated adulteration from 1980 to 2010. J Food Sci 77: R118–R126
42. Baker BR, Lai RY, Wood MS, Doctor EH, Heeger AJ, Plaxco KW (2006) An electronic, aptamer-based small-molecule sensor for the rapid, label-free detection of cocaine in adulterated samples and biological fluids. J Am Chem Soc 128:3138–3139

43. White RJ, Plaxco KW (2009) Engineering new aptamer geometries for electrochemical aptamer-based sensors. Proc SPIE Int Soc Opt Eng 7321:732105. doi:10.1117/12.820419
44. Cao Q, Zhao H, Zeng LX, Wang J, Wang R, Qiu XH, He YJ (2009) Electrochemical determination of melamine using oligonucleotides modified gold electrodes. Talanta 80:484–488
45. Singh RP, Kang DY, Choi JW (2010) Electrochemical DNA biosensor for the detection of sanguinarine in adulterated mustard oil. Adv Mat Lett 1:48–54
46. Ahmed MU, Hasan Q, Hossain MM, Saito M, Tamiya E (2010) Meat species identification based on the loop mediated isothermal amplification and electrochemical DNA sensor. Food Control 21:599–605

Redox Labeling of Nucleic Acids for Electrochemical Analysis of Nucleotide Sequences and DNA Damage

Miroslav Fojta

Abstract Electrochemical methods have been established as potent tools to analyze nucleic acids (NA) and their interactions, including DNA or RNA hybridization, DNA-protein interactions, DNA damage, interactions of NA with diverse small molecules including drugs, pollutants and toxins, and so forth. Electrochemical NA sensors have been designed and applied for various purposes related to molecular diagnostics, clinical and environmental analysis, as well as prompt detection of chemical or biological weapons. These devices and techniques may work as label-free ones, utilizing intrinsic electrochemical properties of the NA, or employ electrochemically active and/or enzyme labels to improve their analytical parameters (namely, sensitivity and selectivity). A palette of electrochemically active moieties has been introduced into NA either via chemical modification of natural NA components (such as thymine bases forming redox active adducts upon treatment with osmium tetroxide reagents, or terminal ribose in RNA reacting with six-valent osmium complexes), or via polymerase incorporation of modified nucleotides. The latter technique, utilizing modified deoxynucleotide triphosphates as substrates for the polymerases, represents particularly versatile approach to sequence-specific construction of various modified NA and for multipotential redox coding of nucleobases by organometallics (ferrocene), transition metal chelates or electrochemically reducible or oxidizable organic moieties. A combination of DNA modification with affinity tags (biotin, digoxigenin) with subsequent attachment of enzymes (peroxidase, alkaline phosphatase) has successfully been utilized for signal amplification via biocatalysis.

Keywords DNA labeling · DNA hybridization · DNA damage · Electrochemistry · Biosensors · Multipotential redox coding · Signal amplification

---

M. Fojta (✉)
Institute of Biophysics Academy of Sciences of the Czech Republic,
Královopolská 135, 612 65 Brno, Czech Republic
e-mail: fojta@ibp.cz

D.P. Nikolelis and G.-P. Nikoleli (eds.), *Biosensors for Security and Bioterrorism Applications*, Advanced Sciences and Technologies for Security Applications,
DOI 10.1007/978-3-319-28926-7_15

1 Introduction

Due to their central role in preservation, transfer and expression of genetic information, nucleic acids (NA) belong to most prominent analytes on one hand, or analytical tools on the other, in a broad spectrum of purpose-designed biosensors and bioassays. Based on the detection of specific nucleotide sequences, various organisms, including pathogenic bacteria or viruses, can be identified and detected in biological samples. Application areas of techniques developed to detect nucleic acids sequences (see Sect. 1.1) include those utilized in medicinal diagnostics, in food control (here, detection of transgenes to identify genetically modified crops represents another typical application besides pathogens) as well as detection of biological (bacteriological) weapons (various aspects of these techniquers have recently been reviewed [1–6]). Chemical and/or structural changes in DNA (such as strand breaks, abasic sites or nucleobase adducts) arising from exposure of the genetic material to different environmental or endogenic agents and collectively being referred to as "DNA damage" [7, 8], represent another class of entities for which specific biosensors and bioassay have been designed. A number of genotoxic species interact with (usually double-stranded, ds) DNA non-covalently and in such cases the dsDNA is applied, in various analytical modes, as an affinity substrate to enrich and/or detect the respective substances. Specific class of the NA-based affinity sensors and assays are featured by those employing DNA or RNA aptamers, synthetic in vitro-selected oligonucleotides binding various analytes (from drugs and other small molecules to proteins) with a high selectivity and affinity [9–11]. Unrepaired DNA damage may give rise to mutations i.e., hereditary changes in the genetic information featured by substitution, deletion or insertion of one or more base pairs. Detection of mutations is another application area of DNA sensors and assays with respect to their relation to severe healthy risks. Regardless of the particular application, these approaches can employ various detection platforms (such as optical including fluorescence [12, 13] or surface plasmon resonance (SPR) [4], microgravimetric [14], or electrochemical [2, 5, 6]) and may be designed as either label-free (utilizing solely intrinsic physico-chemical properties of NA or substances interacting with them) or employing various labels or indicators to improve their analytical parameters. This chapter is focused on techniques of NA labeling for the analysis of nucleotide sequences and DNA damage using electrochemical methods.

1.1 Analysis of Nucleotide Sequences

Apart from de novo sequencing of full genomes, nowadays employing fast, high-throughput so called "next generation DNA sequencing methods" developed in the late 1990s [15, 16], the majority of techniques developed for analyzing NA sequences use NA hybridization principle based on the inherent ability of two complementary NA strands to form the DNA double helix. In the analytical sense, NA hybridization refers to the formation of the duplex NA of a probe (designed to display a specific nucleotide sequence) and a target DNA or RNA strand complementary to the probe [5, 17–19]. The probe can be featured by a piece (oligonucleotide, typically 20–30 base pairs) of DNA, RNA or a synthetic NA analogue (e.g., peptide nucleic acid, PNA [20]; or locked nucleic acid, LNA [21]) developed to improve selectivity of the hybrid formation (such as sensitivity to the presence of base mismatches). Depending on detection platform used in the given assay, the probe or the target can be modified with a suitable label (e.g., radionucleotide, fluorophore, electrochemically active group or affinity tag such as biotin; for more details see Sect. 3). A typical DNA hybridization assay involves immobilization of one of the hybridizing strands (target sequence or the capture probe) on a surface (nitrocellulose membrane as in classical Southern blot assays [22, 23], glass plate as in gene arrays employing fluorescence labeling [13], metal-coated glass prism as in SPR biosensors [4], electrode surface in electrochemical DNA hybridization biosensors [2, 6], or magnetic beads in techniques combining magnetic separation and target enrichment with, in principle, any detection platform [11, 19]). After additional steps typically involving washing and blocking of the surface against non-specific adsorption of the other hybrid-forming strand (signaling probe or target NA) or any other component of the detection system, the hybridization step is performed, followed by another washing cycle and detection. Some detection platforms allow monitoring of all these steps in real time (e.g., SPR [4]).

Another approach of the detection of specific DNA sequence stretches eploys amplification of specific DNA fragments by the polymerase chain reaction (PCR) [24, 25], cyclic DNA replication in vitro. It should be emphasized that the PCR application is essentially based on the DNA hybridization principle as well; in this case, the sequence specificity is rendered by hybridization (annealing) of PCR primers with corresponding sites in the template DNA, delimiting the fragment to be amplified. Amplification products can be detected by various techniques either after running certain number of PCR cycles or in real time (more exactly, cycle by cycle). The real time PCR techniques are well-suited for quantification of the initial concentration of template i.e., specific DNA sequence to be determined in a sample [26, 27]. This technique is frequently utilized for quantifying RNA sequences to determine transcription of specific genes. In this case, RNA sequences are first converted into complementary DNA (cDNA) by RNA-dependent RNA polymerase (reverse transcriptase) which is then subjected to PCR amplification [28, 29].

Besides the detection of a DNA or RNA sequence stretches e.g., to identify and/or detect pathogenic microorganisms, transgenes, or to monitor gene expression, changes in the nucleotide sequences featuring mutations are commonly applied. One group of these techniques applied to detect point mutations (exchange, insertion or deletion of a single base pair) is based on mismatch-sensitive DNA hybridization, utilizing reduced stability of duplexes (heteroduplexes) formed of a wild-type probe and a mutated target (or vice versa) [30, 31]. Under specific conditions, particularly optimized hybridization temperature and medium composition, the heteroduplexes are not stable in contrast to perfectly matched homoduplexes. In additon, even under less stringent conditions allowing existence of the heteroduplexes, the mismatches can be detected on the basis of altered structure of the DNA double helix at these sites [32, 33] (an example will be discussed in Sect. 3.1). The other group of techniques applied for the detection of mutations is based on polymerase "minisequencing" or "resequencing" of the mutation sites and typically employ labeled nucleosides to incorporate them, on the basis of Watson-Crick base pairing, opposite of the expected site of mutation [34, 35]. Examples will be given in Sect. 3.2.

1.2 Analysis of DNA Damage

The most common products of DNA damage include abasic sites (due to hydrolysis of the N-glycosidic bonds linking the base residue to deoxyribose), strand breaks (sb; i.e., interruptions of the DNA sugar phosphate backbone), 8-oxoguanine (8-OG) arising from oxidative DNA damage or base deamination products (e.g., uracil or hypoxanthine) induced by some known mutagens (such as nitrous acid or bisulfite) [7, 8]. Among others, pyrimidine dimers arise from UV-irradiation of DNA induced by sunlight. Different bulky adducts may be formed due to DNA interactions with metabolically activated carcinogens such as aromatic amines or polycyclic aromatic hydrocarbons. Intrastrand or interstrand cross-links are formed via DNA reactions with bifunctional agents such as antitumor agent cisplatin or a chemical weapon yperite. To analyze damaged DNA and to identify specific products of DNA interactions with the genotoxic species, various analytical techniques were developed. Basically, two groups of techniques can be distinguished. In the first, the analyzed DNA is hydrolyzed into monomeric components (mononucleotides, nucleosides or even free bases, deoxyribose and phosphate) and the DNA damage products are identified and quantified by chromatographic, electrophoretic and/or mass-spectroscopic techniques [36–38]. A variety of DNA adducts have been analyzed by means of so called ^{32}P-postlabeling technique [39, 40] involving adduct-selective enzymatic DNA hydrolysis followed by introduction of ^{32}P radionucleotide, thin layer chromatographic separation and autoradiographic detection.

The other group of techniques is used to measure changes of features of whole DNA molecules upon their damage. Most of these approaches utilize, as the final step, determination of DNA sb by means of gel electrophoresis. Using agarose gel electrophoresis, single (ssb) or double strand breaks (dsb) can be determined with a high sensitivity via relaxation or linearization of supercoiled (sc) plasmid DNA molecules [41, 42]. "Comet" assay is a technique developed to assess the abundance of dsb in genomic DNA of a single cell exposed to DNA damaging conditions [43]. A modification of this technique, alkaline elution assay in which the electrophoresis is run under alkaline denaturing conditions [44, 45], reveals the presence of ssb and alkali-labile sites (together with the dsb), the latter featuring sites of nucleobase lesions. Combination of the electrophoretic methods with DNA digestion by lesion-specific DNA repair enzymes (N-glycosylases and/or endonucleases) has been employed to detect specifically e.g., abasic lesions, uracil or 8-OG [45, 46]. Abundance of 3'-OH ends of DNA i.e., strand breaks formed by the action of specific endonucleases (such as those involved in apoptosis or DNA repair) can be estimated by the "TUNEL" test [47], a technique based on labeling of free 3'-OH polynucleotide termini through a reaction catalyzed by a terminal nucleotide transferase. Many of these principles can be combined with other detection techniques, including electrochemical methods.

2 Electrochemical NA Sensing

Among other detection techniques, the electrochemical ones proved a great potential in NA analysis (reviewed e.g., in [2, 48, 49]). Natural NA as well as their synthetic oligonucleotide (ON) analogues possess intrinsic electrochemical activity owing to the presence of nitrogenous nucleobases. Irreversible electrochemical reduction of cytosine and adenine moieties at mercury and silver amalgam [50] electrodes is manifested in a single common signal (peak CA, in random NA sequence). In addition, chemically reversible reduction/oxidation of guanine on mercury and amalgam electrodes gives rise to an anodic peak specific for this base (peak G) [51, 52]. The mercury-containing eletrodes have also been utilized to measure specific tensammetric (capacitive) signals connected with adsorption/desorption (reorientation) processes undergone by the NA at negatively charged electrode surface (reviewed in [2, 49]). The latter signals, as well as the cathodic peak CA, are strongly sensitive to changes in the DNA structure, and have been widely utilized in electrochemical studies of DNA structural transitions and in sensing of DNA damage (reviewed in [7]). All nucleobases have been demonstrated to undergo irreversible electrochemical oxidation on various types of carbon electrodes, and especially the oxidation peak of guanine has found wide application in label-free NA electroanalysis, detection of DNA damage [7], as well as DNA hybridization [6]. Besides the label-free approaches, utilizing solely the electrochemical activity of natural DNA components (albeit sometimes mediated by suitable redox species, such as ruthenium complexes used for catalytical oxidation of guanine residues [53]), a number of techniques based on applications of redox labels or indicators or labels have been introduced [54–57]. Techniques of electrochemical DNA labeling and their applications are reviewed in Sect. 3.

2.1 *Adsorptive Stripping of NA and NA-Modified Electrodes*

Electrochemical analysis of nucleic acids usually utilizes stripping techniques to reach a high sensitivity of NA detection by adsorptive preconcentration at the electrode surface. Owing to a high adsorptivity of NA and their analogues on mercury, amalgam or carbon electrodes, a high surface concentration of the NA can be achieved after a short accumulation time (one or several minutes) even when its solution concentration is low [49]. Moreover, adsorption of NA at the above mentioned electrodes is firm enough to resist medium exchange [2, 58]. Thus, NA can be adsorbed at the electrode from a small (several microliters) drop of sample solution to create a NA-modified electrode. Responses of the adsorbed NA are then measured in a usual electrochemical cell in blank (lacking any NA) background electrolyte. This procedure, called adsorptive transfer stripping (AdTS), medium exchange or ex situ electrochemical analysis, has extended application possibilities of nucleic acids electrochemistry: (i), low requirements of the sample volume has made AdTS procedures suitable for analyzing samples that are usually not available in large quantities, including sample series in typical magnetic beads-based NA assays (see below); (ii) some limitations inherently connected with the conventional electrochemical measurements (such as the necessity of dissolving the analyte in media in which the given electrode process can take place) are circumvented via the medium exchange; (iii) an electrode with adsorbed NA layer can be used as a simple electrochemical biosensor. DNA modified mercury, amalgam or carbon electrodes have successfully been applied as biosensors for DNA damage [7, 42]. In the area of DNA hybridization sensors, carbon electrodes with probe or target DNA simply adsorbed onto the electrode surface have been successfully applied [28, 59, 60] in addition to other solid electrodes modified with covalently immobilized capture probes (e.g., thiolated ONs chemisorbed on gold or carbon electrodes with ON probes attached via carbodiimine chemistry [2, 61, 62]).

2.2 Techniques Combining Separation on Magnetic Beads with Electrochemical Microanalysis

Magnetic beads (MB) modified with various biomolecules (nucleic acids, streptavidine, antibody-binding proteins, antibodies etc.) are commercially available tools frequently used for convenient affinity capturing, enrichment and isolation of specific biomolecules from complex matrices (such as biological samples) [11, 19]. Among other applications, the MB have been applied in techniques developed for detecting DNA hybridization [18, 63–65], DNA-protein interactions [33, 66–68], studies of chemical modification of DNA [69] or preparative strand separation [35, 70, 71]. These procedures can conveniently be connected with AdTS electrochemical analysis. Compared to "classical" concepts of electrochemical DNA sensors, using the same (electrode) surface for probe immobilization, binding of the affinity partner and detection, the MB-based techniques have been named by Palecek et al. "double-surface" i.e., using two different surfaces for interaction and detection [19]. The latter concept allows in principle much easier optimization of both steps because, being spatially separated, they do not affect one another. The carrier surface need not be electrically conductive and possess other properties critical for analytically applicable working electrodes. On the other hand, the detection electrode and electrochemical technique can be chosen only with respect to the particular analyte and given electrode process: one electrode type may be best suited for label-free NA detection while another for determination of e.g., modified signaling probes labeled with electrochemically active tags or for detection of electroactive indicators generated in techniques employing enzyme labels (see below). In such applications, the working electrodes do not require specific laborious pretreatment and can include any of the electrodes suitable for AdTS electrochemical analysis of NA or simple organic compounds generated as indicator products in enzyme-linked techniques (e.g., mercury, silver amalgam and various types of carbon electrodes, such as pyrolytic graphite, carbon paste, pencil graphite, or screen printed electrodes).

3 Labeling of NA for Electrochemical Sensing and Its Applications

To improve sensitivity and selectivity of electrochemical analysis of NA, various extrinsic electroactive species have been applied. One group of such species include soluble redox indicators that selectively interact with dsDNA non-covalently, typically via intercalative or groove binding [54, 55] and provide discrimination between duplex (target-probe hybrid or intact dsDNA) and single stranded (probe or target alone) or damaged DNA. Another group includes covalently bound redox moieties that are used to label a nucleotide sequence (hybridization signaling probe [72] or a sequence-specific DNA target for DNA-protein interaction studies [67]), or even to encode a specific nucleotide (e.g., for the detection of point mutations or single nucleotide polymorphisms). The following sections are devoted to techniques of covalent redox labeling of NA and their selected applications.

3.1 Modification of Nucleic Acids with Oxoosmium Complexes

3.1.1 Modification of Nucleobases with Osmium Tetroxide Reagents

Osmium tetroxide reacts with alkenes to give vicinal diols in a reaction involving [3 + 2] addition of osmium tetroxide on the C = C double bond giving rise to an osmic acid diester glycolate, that is subsequently hydrolyzed to corresponding glycol moiety and osmate (reviewed in [56]). Analogous reactions are given by various compounds possessing the C = C double bonds, including pyrimidine nucleobases (at C5 = C6). It has been further shown that tertiary amines such as 2,2'-bipyridine (bipy), 1,10-phenanthroline (phen) derivatives, or N,N,N',N'-tetramethyl ethylenediamine (TEMED), stabilize the osmate glycolates due to coordination of the central osmium atom by the nitrogenous ligands. Products of modification of the pyrimidines with osmium tetroxide in the presence of the nitrogenous ligands are stable adducts retaining the osmium moiety and the given ligand (Fig. 1a). In contrast, nucleobases do not react with analogous oxo-osmium (VI) complexes (see Sect. 3.1.2) under anaerobic conditions [73, 74]; however, in the presence of ambient oxygen or other oxidants, such as hexacyanoferrate(III) or N-methylmorpholine N-oxide [75], the above mentioned products of reactions at the C = C double bonds are formed.

In nucleic acids, reactions with the osmium tetroxide complexes (further referred to as Os(VIII),L) are strongly selective for pyrimidine nucleobases (reviewed in

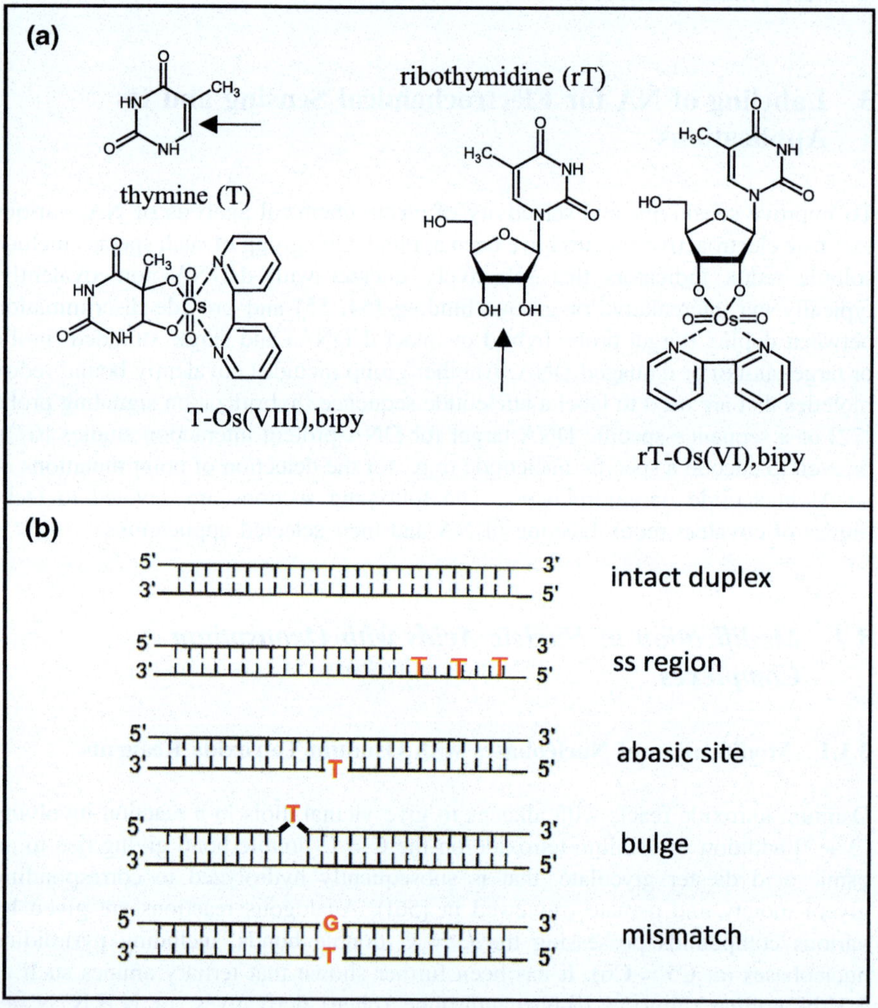

Fig. 1 Modification of nucleic acids with oxoosmium complexes. **a** Reaction products of Os (VIII),bipy with thymine nucleobase and of Os(VI),bipy with ribose. **b** Schematic representation of sites in DNA comprising thymine residues reactive to Os(VIII),bipy (in *red*)

[56]). Remarkable differences in the reactions rates with OsO4 or Os(VIII),L have been observed among natural pyrimidines differing in substituents at position 4 and 5. In general, keto group at C4 and methyl group at C5 facilitate modification at C5 = C6 of thymine (4-keto, 5-methyl) modification proceeds by an order of magnitude faster, compared to modification of cytosine (4-amino, no methyl), while reaction rates observed for uracil (4-keto, no methyl) and 5-methylcytosine (4-amino, 5-methyl) are intermediate [56, 76].

The DNA-Os,L adducts exhibit distinct electrochemistry due to the presence of the osmium central atom which gives several analytically useful faradaic signals due to consecutive redox processes [77]. Moreover, the final reduction step at mercury electrodes is coupled to catalytic hydrogen evolution [77] allowing determination of low concentrations of the osmium-labeled DNA (or low levels of DNA modification). Besides the mercury drop electrodes, the osmium catalytic processes were also observed at glassy carbon electrodes covered with mercury film [78] and at solid silver (AgSAE) or copper amalgam electrodes modified with mercury meniscus [79]. At carbon electrodes, faradaic current signals due to reversible redox processes of the osmium moiety can be measured [63, 72, 80]. Redox potentials of the DNA-Os,bipy adducts differ from those of the free Os(VIII),bipy reagent, which in principle allows to determine the DNA-Os,bipy in the presence of unbound complex. Further, the redox potentials of the osmium DNA markers depend on the nitrogenous ligands, making it possible to use different ligands for "multicolor" (multipotential) DNA coding [72]. The faradaic responses of the osmium-modified DNA have also been measured with gold electrodes [81–85].

Applications related to DNA hybridization. Os(VIII),L modification of signalling (reporter) probes or target DNAs has been introduced as a convenient approach to analyzing DNA sequences via DNA hybridization [63]. ON probes comprising a 5'- or 3'-terminal oligo(T) tail attached to a recognition sequence have been designed, in which the osmium labels were naturally accumulated in the oligo(T) tail. Using longer tails (20–30 thymines) adopting corresponding number of the osmium labels it has been possible to reach remarkable signal amplification resulting in low detection limits. To obtain an osmium-labeled RP capable of hybridization (i.e. with intact recognition sequence), modification of nucleobases within the probe recognition segment has to be precluded. When the probe contains a homopurine recognition stretch and the oligo(T) tail, it can be safely treated with Os(VIII),L in its ss form thanks to a highly preferential modification of the thymine residues and very low reactivity of purine bases towards the osmium reagent [63, 72]. Probes possessing cytosine (but not thymine) residues within the recognition sequence can be modified without losing their hybridization capacity using milder reaction conditions. However, in the case of probes designed for hybridization with target DNA sequences involving all four nucleobases, the recognition stretch of the probe must be protected during the modification reaction by "protective strands"—ON complementary to the probe recognition sequence. Then only thymine residues in the oligo(T) tail (forming a ss overhang) are exposed to the reagent, while those within the specific sequence are hidden within duplex DNA and do not react with Os(VIII),L. Separation of the labeled probe strand from a terminally biotinylated protective ON can be attained using MB covered with streptavidine [56]. In this way, signalling probes labeled with various Os(VIII),L reagents (differing in the ligand L to "tune" redox potentials of the probes) were prepared and applied in parallel analysis of multiple target DNA sequences. An alternative approach developed by Flechsig et al. [84] has been based on the application of protective strands involving several mismatches with respect to the recognition sequence in a target DNA. Due to these mismatches, strand exchange between the protective oligonucleotide and the fully complementary capture probe (attached via thiol linkage to a gold electrode) is favored towards formation of the more stable fully matched hybrid and thus the extra strand separation step can be omitted. The latter technology has been used in a variety of applications to detect synthetic as well as natural PCR-amplified target DNAs [82, 83, 85–87], including sequences related to transgenes present in genetically modified crops [87].

Applications related to DNA damage. As mentioned above, modification of thymine residues with Os(VIII),bipy in dsDNA (where they are base-paired with opposite adenines) is precluded, in contrast to facile modification of unpaired thymines in ssDNA (Fig. 1b). Hence, DNA lesions resulting in disruption of the base pairs, as well as those convertible into ssDNA, can in principle by detected via chemical probing with Os(VIII),bipy to improve structure selectivity and sensitivity of electrochemical DNA damage detection. For example, label-free detection of DNA strand breakage via measurements of guanine oxidation signal at carbon electrodes is inherently much less sensitive, compared to label-free determination of DNA strand breaks with mercury electrodes [7]. While down to one break per 10^5 nucleotides (i.e., a ssb in one per cent of plasmid DNA molecules) can easily be detected by voltammetry at mercury electrodes [42], using carbon electrodes (which are otherwise better suited for applications in biosensing due to their non-toxicity, mechanical stability, disposability and low cost) it is practically possible to detect a ssb per one plasmid DNA molecule [88]. Nevertheless, combination of DNA treatment with E. coli exonuclease III (EXOIII) to generate terminal or internal ss regions in dsDNA and their subsequent Os(VIII),bipy modification has provided a highly sensitive detection of DNA damage with carbon electrodes [89]. Covalently closed circular duplex DNA molecules lacking any ends are not substrate for the EXOIII enzyme. Once the DNA molecule contains a ssb, the enzyme degrades one of the dsDNA strands from its 3'-OH terminus, generating ssDNA stretch in the other strand which then accumulates the osmium tags. Other types of lesions can be detected in this way as well after their conversion to the strand breaks by suitable DNA repair glycosylases and/or nucleases.

A basic lesions featured in sites of missing adenine residues, leaving opposite thymines unpaired, represent another class of entities reactive towards Os(VIII), bipy [90] (Fig. 1b). It was demonstrated that, besides simple direct voltammetric determination of the DNA-bound osmium, also indirect immunochemical approaches can be applied. Binding of an anti-DNA-Os,bipy antibody to the osmium-DNA adducts can be monitored either via an enzyme-linked electrochemical assay (see Sect. 4 for more details), or via acoustic thickness shear mode detection. Similarly, thymines in single-nucleotide bulges (resulting from insertion mutations) or base mismatches (resulting from single base substitutions) can easily be detected through introduction of the osmium marker at the unpaired or mispaired thymine followed by AdTS voltammetry at the pyrolytic graphite electrode [32].

3.1.2 Modification of Terminal Sugar Residues in RNA with Six-Valent Osmium Complexes

In general, vicinal diols, including those present in carbohydrate moieties, undergo condensation reactions with six-valent osmium (osmate) complexes Os(VI),L, yielding osmate glycolates [91] (Fig. 1a). The Os(VI),L readily react with cis-1,2-diols, and usually do not react with trans-1,2-diols; nevertheless, exceptions from this rule were identified for e.g., some glucose derivatives. Application of Os (VI),L modification to introduce redox labels into oligo- and polysaccharides [92, 93], as well as glycan components of glycoproteins [94] were reported. Attention has been also paid to modification of ribose moieties in ribonucleosides and ribonucleotides [74, 95, 96]. $K_2OsO_2(OH)_4$, in the presence of suitable nitrogenous ligands, reacted with the sugar residue of all usual ribonucleotides. Os(VI),bipy showed specific modification of ribose in thymine and adenine ribosides in comparison with unreactive deoxynucleosides; no modification of thymine base with the six-valent osmium was observed under anaerobic conditions. In contrast, Os(VIII),bipy reagent modified only the thymine base; the two reagents thus offer exclusive modification of NA components on either ribose sugar, or on pyrimidine nucleobase [74]. Specific 3'-OH terminal labeling with Os(VI),bipy has been utilized in the development of MB-based electrochemical techniques for the detection of microRNAs, important biomarkers related to severe diseases including cancer [17, 97].

3.2 Polymerase Incorporation of Redox-Labeled Nucleotides

A potent and highly versatile approach to preparation of labeled (or, generally, chemically functionalized) nucleic acids consists in the incorporation of modified nucleotides into NA (ON) by polymerase enzymes, using corresponding modified deoxynucleotide triphosphates (dNTP) as precursors (reviewed in [57, 98]). The external conjugate groups may be attached to the nucleotides on sugar moieties or on nucleobases, and may comprise fluorophores [99], chemically reactive groups for further conjugation reactions [100–102], bioaffinity tags (biotin or digoxigenin) [28, 34] as well as electrochemically (redox) active moieties [55, 57]. While some of the modified dNTPs are commercially available (e.g., those labeled with fluorescein, biotin, digoxigenin, electroactive ferrocene or azide for subsequent coupling reactions via click chemistry), development of new types of labels and labeling approaches represents a dynamically developing field of NA research.

Development of electrochemically labeled dNTPs in recent years opened new possibilities in electrochemical NA sensing. Incorporation of the labeled nucleotides into DNA is potentially useful not only for preparation of labeled hybridization probes, but also for approaches based on sequence-specific introduction of the electroactive tags (e.g., detection of expected mutations via DNA minisequencing [35, 71, 103]). Moreover, application potential of these techniques can be further extended by application of electrochemically or chemically switchable labels [104, 105] or by ratiometric analytical approaches [70], as presented and discussed in the following paragraphs.

3.2.1 Base-Modified Redox-Labeled dNTPs

Recently developed facile synthesis of base-modified dNTPs using aqueous cross-coupling reactions (reviewed in [98]) has resulted in considerable extension of the portfolio of redox-labeled nucleotides. Depending on the nature of the external conjugate groups and related reaction conditions, either direct cross-coupling of these moieties to dNTPs, or modification of nucleosides followed by their triphosphorylation, have been applied. With respect to compatibility of base modifications with subsequent polymerase reactions (see Sect. 3.2.2), either pyrimidines substituted at position 5, or 7-deazapurines substituted at position 7 have been selected for these studies (Fig. 2) (C8-substituted purine dNTPs are not accepted by the polymerases as useful substrates). The dNTP modifications are typically prepared via Sonogashira cross coupling reaction of corresponding iodinated nucleosides or dNTPs and ethynyl derivatives of the label to be attached, or via Suzuki-Miyaura reactions of the iodinated precursors with arylboronic acids.

Redox active moieties coupled to the dNTPs in this way include "classical" reversible redox systems such as ferrocene, tris-bipyridine complexes of $Os^{3+/2+}$ or $Ru^{3+/2+}$ (in which the redox electrochemistry is rendered by the metal component) [35] or athraquinone (quinone/hydroquinone redox couple) [104]. Another group of newly developed redox labels include organic nitrogenous and/or oxygenous moieties which are electrochemically reducible (e.g., nitroderivatives [71], benzofurazane [70], hydrazones [100, 106, 107], aldehydes [100], azidophenyl [105]) or oxidizable (e.g., aminoderivatives [71], methoxyphenol or dihydrobenzofuran [108]). Albeit electrochemical processes and signals yielded by these substances are usually irreversible, some of them exhibit other convenient properties such as high electron yields upon their electrochemical transformation, offering a high sensitivity of detection (e.g., the nitro group undergoes four-electron reduction on mercury and carbon electrodes, while benzofurazane is reduced with six electrons on merury and apparently with four electrons on carbon; both labels give very well developed and analytically useful signals [70]).

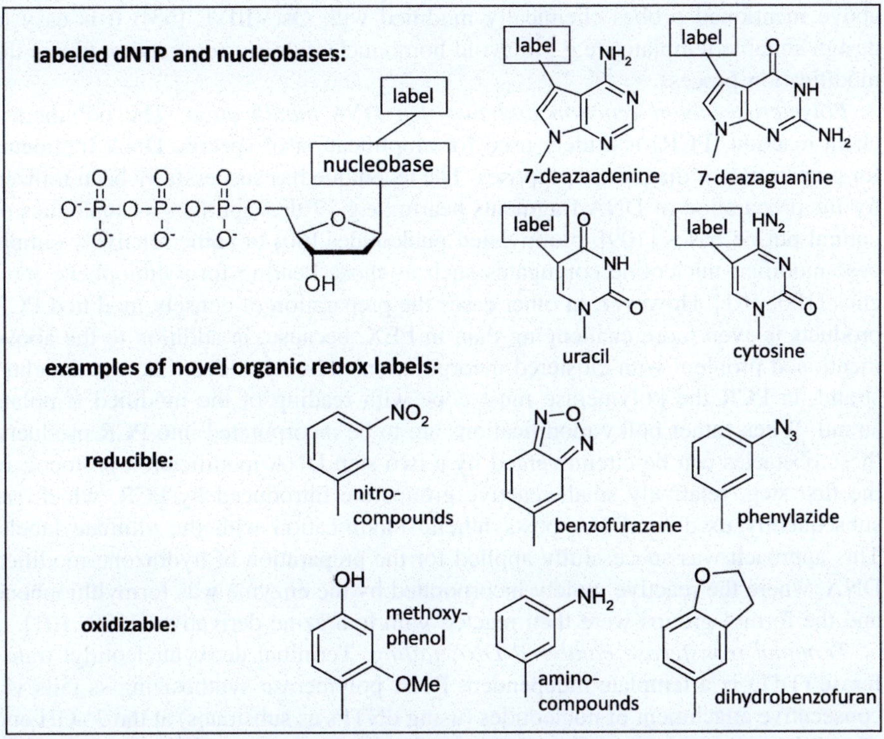

Fig. 2 Scheme of base-modified deoxynucleoside triphosphates and examples of nitrogenous and/or oxygenous organic compounds designed as electrochemically reducible or oxidizable labels

3.2.2 Construction of labeled DNA using polymerases and modified dNTPs

Primer extension. DNA-dependent DNA polymerases, key enzymes involved in DNA replication as well as repair processes, work on the principle of template sequence-driven attachment of a nucleotide (using dNTP as an activated precursor) to terminal 3‘–OH group of RNA or DNA primers (Fig. 3a). Using this primer extension (PEX) reaction and suitable modified dNTPs, labeled DNA can easily by prepared with the modified nucleotides incorporated sequence-specifically according to the template (complementary) sequence [57, 98]. It was shown that Klenow fragment of bacterial DNA polymerase I and a number of thermostable DNA polymerases belonging to the “B-family” (e.g., Vent, Pwo, KOD) tolerate the presence of a number of modifications on sugar as well as at 4-position of pyrimidines or 7-position of 7-deazapurines. Some limitations in the PEX efficiency may arise from steric effects when two or more base-modified nucleotides bearing bulky external substituents are incorporated in adjacent positions [35]. Nevertheless, for a preparation of multiply tail-modified probes (in analogy to the above mentioned probes chemically modified with Os(VIII),L [63]) it is easy to design suitable templates (e.g., to avoid homonucleotide clusters) and optimize the modification process.

Polymerase chain reaction and two-step DNA modification. The polymerase chain reaction (PCR) is widely used for amplification of specific DNA fragments for preparative or diagnostic purposes. The technique has successfully been utilized for the preparation of DNA fragments bearing e.g., 7-deazapurines as analogues of natural purine bases [109], biotinylated nucleotides [60] or some relatively simple base-modified nucleotide conjugates such as those bearing formylthiophene moieties [100, 101]. However, in other cases the preparation of densely modified PCR products is even more challenging than in PEX, because, in addition to the above mentioned problem with clustered incorporation of bulky adduct into the daughter strand, in PCR the polymerase must cope with reading of the modified template strand. When rather bulky modifications are to be incorporated into PCR products, these obstacles can be circumvented by a two-step DNA modification protocol: in the first step, relatively small reactive groups are introduced by PCR, which are subsequently used for DNA postsynthetic modification with the ultimate labels. This approach was successfully applied for the preparation of hydrazone-modified DNA, where the reactive moiety incorporated by the enzyme was formylthiophene and the formyl groups were then reacted with hydrazine derivatives [100, 101].

Terminal transferase catalyzed DNA tailing. Terminal deoxynucleotidyl transferase (TdT) is a template-independent DNA polymerase synthesizing ss tails via consecutive attachment of nucleotides (using dNTPs as substrates) at the 3'-OH end of ss or ds primers (initiators, Fig. 3b). TdT has been used in the area of recombinant DNA construction to add sticky ends to cloned DNA fragments [110]. It has been demonstrated that the enzyme can also be used for synthesis of modified DNA tails to introduce multiple redox tags into hybridization probes or substrates for protein-DNA interaction studies [111].

3.2.3 Utilization of specific features of the redox DNA labels

Multipotential redox coding of nucleobases and SNP typing. Since reduction of natural nucleobases takes place at relatively highly negative potentials (on mercury/amalgam electrodes) and oxidation at rather positive potentials (on carbon electrodes), there is a relatively wide region of potentials available for measuring electrochemical signals of various DNA labels without interference with the intrinsic DNA signals (Fig. 4). The labels can be chosen according to potentials of their reduction and/or oxidation so that their signals do not mutually overlap and the signals can be measured independently on each other. Hence, a combination of different redox labels can be utilized in electrochemical typing of SNPs, as demonstrated previously for a combination of nitrophenyl, $[Os(bipy)_3]^{3+/2+}$ and aminophenyl to encode three different nucleobases [35] or, quite recently, for a combination of ferrocene, anthraquinone, methylene blue and phenothiazine to complete the set of all four DNA bases [103]. Besides differences between the relation to two-step DNA labeling techniques, is one of typical examples: due to such conversion, the formyl-specific reduction signal decreases while signals due to electroactive moieties present in the reaction products appear (i.e., the hydrazone C = N bond and groups attached via the hydrazone linkage—nitro groups or and/or benzofurazane, as presented in [100]). From relative changes of the primary and secondary signals, the conversion degree can be deduced. Another example comprises polymerase-incorporated phenylazide group (yielding a strong reduction signal on mercury-based electrodes) click-transformable into corresponding substituted triazoles in reactions with acetylene derivatives (Fig. 5b) [105]. The triazole group has been found to be electrochemically "silent", hence after the click reaction, the modification-specific (phenylazide) signal disappears. When an acetylene derivative bearing another redox-active moiety (such as nitrophenyl) is used for the click reaction, the resulting product yields a signal specific for the latter; thus, similarly as above, the conversion degree can be assessed from relative changes of the disappearing (phenylazide) and appearing (nitrophenyl) signals. This system has been utilized to monitor shielding effects of a DNA-binding protein [105].

4 DNA Labeling with Enzymes and Biocatalytical Signal Amplification

Enzymes as biocatalytic labels were introduced in connection with various immunoassays, employing various enzyme-antibody conjugates, usually in sandwich detection systems (e.g., enzyme-linked immunosorbent assays commonly well-known as ELISA, or western immunoblot assay). Enzymes applied in these techniques include, for example, horse radish peroxidase, β-galactosidase or alkaline phosphatase, and substrates chosen according to the detection technique applied, with colorimetry or enhanced chemiluminiscence featuring typical examples. The enzyme-linked techniques can also be coupled to electrochemical detection of product of the enzyme reactions [113, 114]. Since one molecule of the enzyme can catalyze conversion of many molecules of (electrochemically silent) substrate into many molecules of (electroactive) indicator, the enzyme-linked detection systems are inherently offering a high sensitivity of the target molecule detection. Typical examples of these indicators are electrochemically oxidizable phenols (e.g., 1-naphthol) or reversible redox p-aminophenol, both released by alkaline phosphatase from corresponding inactive phosphoesters. To attach the enzymes to NA, bioaffinity (biotin-streptavidin or antibody-hapten) interactions are usually utilized.

a. DNA modification with biotin for enzyme-linked electrochemical assays
DNA hybridization assays employing biotin-labeled signaling probes with subsequently attached streptavidin-alkaline phosphatase conjugate (SALP) have been successfully used in various experimental modes. One group of techniques have been based on application of magnetic beads, on which the capture probe-biotinylated target DNA-SALP conjugate was assembled and the

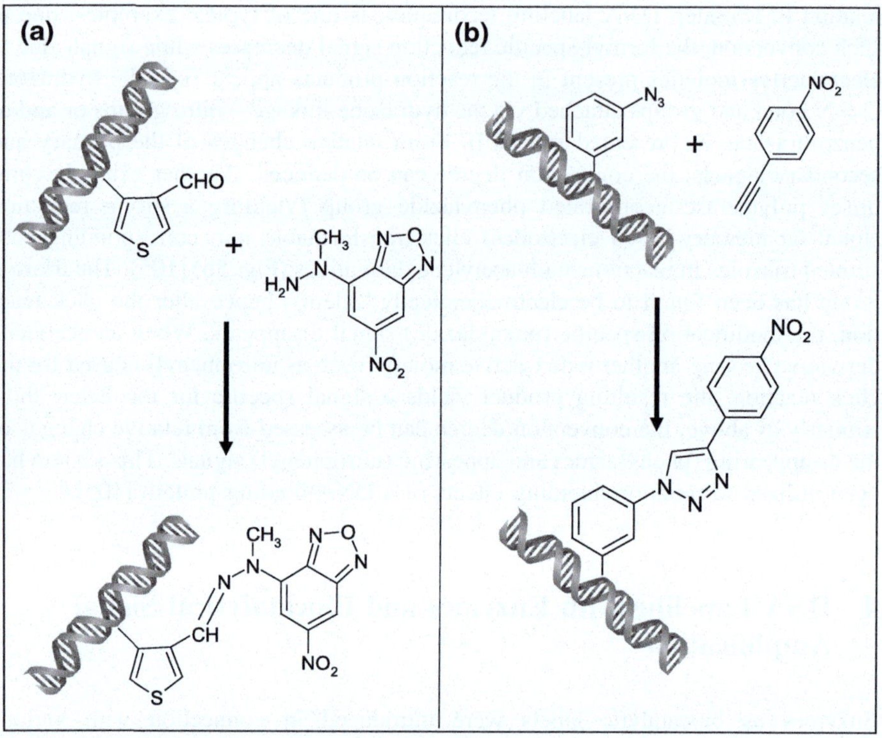

Fig. 5 Postsynthetic modification of DNA functionalized with reactive groups: **a** hydrazone formation between formylthiophene enzymatically incorporated into DNA (see Fig. 3a) and a hydrazine derivative bearing nitrobenzofurazane as a redox label; **b** click reaction to convert azidophenyl moiety on DNA into triazole bearing electrochemically active nitrophenyl group

electroactive indicator generated, being subsequently determined voltammetrically in the supernatant [18, 34]. Other techniques have involved adsorption of target DNA at a carbon (pyrolytic graphite or screen-printed) electrode surface, hybridization with biotinylated probe, attachment of SALP and enzymatic generation of the soluble indicator directly at the electrode, with subsequent voltammetric measurement directly in the substrate solution. The latter approach appeared useful for a fast, simple and relatively inexpensive analysis of real PCR-amplified DNA samples, including amplified cDNA resulting from reverse transcription of RNA to monitor gene expression [28]. Besides utilization of commercially available biotinylated ON probes, the biotin affinity tags can be incorporated into DNA using any of polymerase-based techniques mentioned in Sect. 3.2. and biotin-labeled dNTPs. Hence, it has been possible to apply analogous techniques to detect SNP in real PCR-amplified cDNA via sequence-specific PEX incorporation of biotinylated uridine or cytidine [34], or to monitor PCR reactions via incorporation of biotinylated nucleotides [60]. Biotin-based techniques have been proposed also for the detection of DNA damage e.g., in a variant of the TUNEL test [47].

27. Rynans S, de Walthoffen SW, Dzieciatkowski T, Mlynarczyk G (2015) Postepy Mikrobiol 54:75–82
28. Horakova-Brazdilova P, Fojtova M, Vytras K, Fojta M (2008) Sensors 8:193–210
29. Provenzano M, Mocellin S (2007) Adv Exp Med Biol 593:66–73
30. Hooyberghs J, Van Hummelen P, Carlon E (2009) Nucleic Acids Res 37
31. Naef F, Lim DA, Patil N, Magnasco M (2002) Phys Rev E 65
32. Kostecka P, Havran L, Bittova M, Pivonkova H, Fojta M (2011) Anal Bioanal Chem 400:197–204
33. Masarik M, Cahova K, Kizek R, Palecek E, Fojta M (2007) Anal Bioanal Chem 388:259–270
34. Horakova P, Simkova E, Vychodilova Z, Brazdova M, Fojta M (2009) Electroanalysis 21:1723–1729
35. Vrabel M, Horakova P, Pivonkova H, Kalachova L, Cernocka H, Cahova H, Pohl R, Sebest P, Havran L, Hocek M, Fojta M (2009) Chem-Eur J 15:1144–1154
36. Collins A, Gedik C, Vaughan N, Wood S, White A, Dubois J, Rees JF, Loft S, Moller P, Cadet J, et al (2003) Free Radical Biol Med 34:1089–1099
37. Guetens G, De Boeck G, Highley M, van Oosterom AT, de Bruijn EA (2002) Crit Rev Clin Lab Sci 39:331–457
38. Helbock HJ, Beckman KB, Shigenaga MK, Walter PB, Woodall AA, Yeo HC, Ames BN (1998) Proc Natl Acad Sci USA 95:288–293
39. Koskinen M, Vodicka P, Vodickova L, Hemminki K (2001) Biomarkers 6:175–189
40. Stiborova M, Rupertova M, Hodek P, Frei E, Schmeiser HH (2004) Collect Czech Chem Commun 69:476–498
41. Fojta M, Kubicarova T, Palecek E (1999) Electroanalysis 11:1005–1012
42. Fojta M, Palecek E (1997) Anal Chim Acta 342:1–12
43. Collins AR (2004) Mol Biotechnol 26:249–261
44. Faust F, Kassie F, Knasmuller S, Boedecker RH, Mann M, Mersch-Sundermann V (2004) Mutat Res Rev Mutat 566:209–229
45. Collins AR (2005) Am J Clin Nutr 81:261S–267S
46. Gedik CM, Boyle SP, Wood SG, Vaughan NJ, Collins AR (2002) Carcinogenesis 23:1441–1446
47. Loo DT (2002) Methods Mol Biol 203:21–30
48. Bowater RP, Cobb AM, Pivonkova H, Havran L, Fojta M (2015) Monatsh Chem 146:723–739
49. Fojta M, Jelen F, Havran L, Palecek E (2008) Curr Anal Chem 4:250–262
50. Fadrna R, Kucharikova-Cahova K, Havran L, Yosypchuk B, Fojta M (2005) Electroanalysis 17:452–459
51. Studnickova M, Trnkova L, Zetek J, Glatz Z (1989) Bioelectrochem Bioenerg 21:83–86
52. Vidlakova P, Pivonkova H, Kejnovska I, Trnkova L, Vorlickova M, Fojta M, Havran L (2015) Anal Bioanal Chem 407:5817–5826
53. Thorp HH (1998) Tibtech 16:117–121
54. Ferapontova EE (2011) Curr Anal Chem 7:51–62
55. Fojta M, Havran L, Pivonkova H, Horakova P, Hocek M (2011) Curr Org Chem 15: 2936–2949
56. Fojta M, Kostecka P, Pivonkova H, Horakova P, Havran L (2011) Curr Anal Chem 7:35–50
57. Hocek M, Fojta M (2011) Chem Soc Rev 40:5802–5814
58. Palecek E, Postbieglova I (1986) J Electroanal Chem 214:359–371
59. Fojta M, Brazdilova P, Cahova K, Pecinka P (2006) Electroanalysis 18:141–151
60. Haronikova L, Spacek J, Plucnara M, Horakova P, Pivonkova H, Havran L, Erdem A, Fojta M (2015) Monatsh Chem 146:849–855
61. Chang H, Wang Y, Li J (2011) Curr Org Chem 15:506–517
62. Zauner G, Wang Y, Lavesa-Curto M, Macdonald A, Mayes AG, Bowater RP, Butt JN (2005) Analyst 130:345–349
63. Fojta M, Havran L, Kizek R, Billova S, Palecek E (2004) Biosens Bioelectron 20:985–994

64. Palecek E, Billova S, Havran L, Kizek R, Miculkova A, Jelen F (2002) Talanta 56:919–930
65. Palecek E, Kizek R, Havran L, Billova S, Fojta M (2002) Anal Chim Acta 469:73–83
66. Nemcova K, Havran L, Sebest P, Brazdova M, Pivonkova H, Fojta M (2010) Anal Chim Acta 668:166–170
67. Nemcova K, Sebest P, Havran L, Orsag P, Fojta M, Pivonkova H (2014) Anal Bioanal Chem 406:5843–5852
68. Palecek E, Masarik M, Kizek R, Kuhlmeier D, Hassmann J, Schulein J (2004) Anal Chem 76:5930–5936
69. Horakova P, Tesnohlidkova L, Havran L, Vidlakova P, Pivonkova H, Fojta M (2010) Anal Chem 82:2969–2976
70. Balintova J, Plucnara M, Vidlakova P, Pohl R, Havran L, Fojta M, Hocek M (2013) Chem-Eur J 19:12720–12731
71. Cahova H, Havran L, Brazdilova P, Pivonkova H, Pohl R, Fojta M, Hocek M (2008) Angew Chem Int Edit 47:2059–2062
72. Fojta M, Kostecka P, Trefulka MR, Havran L, Palecek E (2007) Anal Chem 79:1022–1029
73. Trefulka M, Ferreyra N, Ostatna V, Fojta M, Rivas G, Palecek E (2007) Electroanalysis 19:1334–1338
74. Trefulka M, Ostatna V, Havran L, Fojta M, Palecek E (2007) Electroanalysis 19:1281–1287
75. Choudary BM, Jyothi K, Kantam ML, Sreedhar B (2004) Adv Synth Catal 346:45–48
76. Reske T, Surkus AE, Duwensee H, Flechsig GU (2009) Microchim Acta 166:197–201
77. Havran L, Fojta M, Palecek E (2004) Bioelectrochemistry 63:239–243
78. Kostecka P, Havran L, Pivonkova H, Fojta M (2004) Bioelectrochemistry 63:245–248
79. Yosypchuk B, Fojta M, Havran L, Heyrovsky M, Palecek E (2006) Electroanalysis 18:186–194
80. Fojta M, Havran L, Kizek R, Billova S (2002) Talanta 56:867–874
81. Flechsig GU, Reske T (2007) Anal Chem 79:2125–2130
82. Duwensee H, Jacobsen M, Flechsig GU (2009) Analyst 134:899–903
83. Mix M, Reske T, Duwensee H, Flechsig GU (2009) Electroanalysis 21:826–830
84. Reske T, Mix M, Bahl H, Flechsig GU (2007) Talanta 74:393–397
85. Surkus AE, Flechsig GU (2009) Electroanalysis 21:1119–1123
86. Jacobsen M, Flechsig G-U (2013) Electroanalysis 25:373–379
87. Mix M, Rueger J, Krueger S, Broer I, Flechsig G-U (2012) Electrochem Commun 22:137–140
88. Cai X, Rivas G, Farias PAM, Shirashi H, Wang J, Fojta M, Palecek E (1996) Bioelectrochem Bioenerg 401:41–47
89. Havran L, Vacek J, Cahova K, Fojta M (2008) Anal Bioanal Chem 391:1751–1758
90. Bartosik M, Gajdos V, Kostecka P, Fojta M, Palecek E, Volkov E, Oretskaya T, Hianik T (2009) Electroanalysis 21:295–302
91. Resch JF, Meinwald J (1981) Tetrahedron Lett 22:3159–3162
92. Trefulka M, Palecek E (2010) Electroanalysis 22:1837–1845
93. Trefulka M, Palecek E (2012) Bioelectrochemistry 88:8–14
94. Trefulka M, Palecek E (2014) Electrochem Commun 48:52–55
95. Bartosik M, Trefulka M, Hrstka R, Vojtesek B, Palecek E (2013) Electrochem Commun 33:55–58
96. Trefulka M, Bartosik M, Palecek E (2010) Electrochem Commun 12:1760–1763
97. Bartosik M, Hrstka R, Palecek E, Vojtesek B (2014) Electroanalysis 26:2558–2562
98. Hocek M (2014) J Org Chem 79:9914–9921
99. Riedl J, Pohl R, Ernsting NP, Orsag P, Fojta M, Hocek M (2012) Chem Sci 3:2797–2806
100. Raindlova V, Pohl R, Klepetarova B, Havran L, Simkova E, Horakova P, Pivonkova H, Fojta M, Hocek M (2012) Chempluschem 77:652–662
101. Raindlova V, Pohl R, Sanda M, Hocek M (2010) Angew Chem Int Edit 49:1064–1066
102. Dadova J, Orsag P, Pohl R, Brazdova M, Fojta M, Hocek M (2013) Angew Chem Int Edit 52:10515–10518

103. Debela AM, Thorimbert S, Hasenknopf B, Ortiz M, O'Sullivan CK Chem Comm (in press) DOI 10.1039/x0xx00000x
104. Balintova J, Pohl R, Horakova P, Vidlakova P, Havran L, Fojta M, Hocek M (2011) Chem-Eur J 17:14063–14073
105. Balintova J, Spacek J, Pohl R, Brazdova M, Havran L, Fojta M, Hocek M (2015) Chem Sci 6:575–587
106. Danhel A, Raindlova V, Havran L, Barek J, Hocek M, Fojta M (2014) Electrochim Acta 129:348–357
107. Danhel A, Raindlova V, Havran L, Pivonkova H, Hocek M, Fojta M (2014) Electrochim Acta 126:122–131
108. Simonova A, Balintova J, Pohl R, Havran L, Fojta M, Hocek M (2014) ChemPlusChem 79:1703–1712
109. Pivonkova H, Horakova P, Fojtova M, Fojta M (2010) Anal Chem 82:6807–6813
110. Adachi Y, Fukuhara C (2012) Anal Biochem 431:66–68
111. Horakova P, Macickova-Cahova H, Pivonkova H, Spacek J, Havran L, Hocek M, Fojta M (2011) Org Biomol Chem 9:1366–1371
112. Vidlakova P, Pivonkova H, Fojta M, Havran L (2015) Monatsh Chem 146:839–847
113. Wang J (2002) Electrophoresis 23:713–718
114. Wang J (2005) Electroanalysis 17:7–14
115. Stejskalova E, Horakova P, Vacek J, Bowater RP, Fojta M (2014) Anal Bioanal Chem 406:4129–4136
116. Buzek J, Kuderova A, Pexa T, Stankova V, Lauerova L, Palecek E (1999) J Biomol Struct Dyn 17:41–50

Biosensing of Neurotoxicity to Prevent Bioterrorist Threats and Harmful Algal Blooms

Arkadiy Eremenko, Taisiya Prokopkina, Vadim Kasatkin, Vladislav Zigel, Anna Pilip, Iana Russkikh, Zoya Zhakovskaya and Ilya Kurochkin

Abstract The chapter discusses the different types of amperometric sensors based on the inhibition of butyrylcholinesterase for the analysis of neurotoxins. Analytical characteristics of sensors based on the manganese dioxide allow determining the neurotoxins in the nanomolar concentration range. The authors of this publication examined in detail the application of such sensors for monitoring of water samples. The data obtained indicates the presence of anatoxin A(s) in the analyzed samples. The study revealed the potential of the developed sensors to monitor the neurotoxic cyanobacteria at water "blooms".

Keywords Amperometric biosensors · Manganise dioxide · Nanoparticles · Thiol compounds · Neurotoxines · Harmful algal blooms

1 Introduction

Enzymes of the cholinesterase family (ChE) are widely used in the development of various biosensor devices for highly sensitive and selective monitoring of toxicity in environmental studies, the agriculture and food industries, as well as military and

A. Eremenko · I. Kurochkin (✉)
Emanuel Institute of Biochemical Physics, Russian Academy of Sciences, Moscow, Russia
e-mail: ikur@genebee.msu.su

T. Prokopkina · I. Kurochkin
Faculty of Chemistry, Department of Chemical Enzymology, Lomonosov Moscow State University, Moscow, Russia

V. Kasatkin
Frumkin Institute of Physical Chemistry and Electrochemistry, Russian Academy of Sciences, Moscow, Russia

V. Zigel · A. Pilip · I. Russkikh · Z. Zhakovskaya
Saint-Petersburg Scientific-Research Center for Ecological, Russian Academy of Sciences, St. Petersburg, Russia

D.P. Nikolelis and G.-P. Nikoleli (eds.), *Biosensors for Security and Bioterrorism Applications*, Advanced Sciences and Technologies for Security Applications,
DOI 10.1007/978-3-319-28926-7_16

biomedical applications [1]. These devices are based on ChE inhibition by neurotoxins, namely, organophosphates (OP) and carbamates. The main driving force for biosensor development is the search of reliable methods other than instrumental analytical methods used in laboratories. It is worth mentioning that instrumental analytical methods are powerful tools for toxic organophosphate and carbamate pesticide detection, but they have a number of significant drawbacks. The equipment is usually expensive, sample preparation is time consuming and complicated, and users should be highly qualified, which make real-time measurements unachievable. Moreover, it is impossible to obtain any information about the potential toxicity of a sample using instrumental analytical methods.

Biosensor devices allow one to determine not only the presence and amount of certain chemical compounds, but the biological effects caused by them [2], for example, neurotoxicity. In some cases information other than the chemical composition of the analyzed sample is more important. As a result, currently, special attention is paid to bioanalytical systems based on enzyme inhibition for the determination of neurotoxicity of organophosphates and carbamate pesticides which inhibit cholinesterases [3]. The most promising and sensitive devices for the detection of neurotoxins are electrochemical sensors. Therefore, our attention has been focused on the development of different types of electrochemical biosensors based on cholinesterase inhibition for their application in various fields. In recent years, scientists of the Department of Chemical Enzymology, Faculty of Chemistry, Lomonosov Moscow State University and the Institute of Biochemical Physics RAS, together with colleagues from other institutions have developed various types of biosensor and amperometric devices for the detection of neurotoxins.

2 Amperometrc Sensors for Neurotoxins

The amperometric neurotoxin analyzer (*EasyChEck*, v3.05X) is a specially developed automatic flow-injection system for quantitative and qualitative analysis of OPs and carbamates [4]. The assay performed with EasyChEck is based on inhibition of butyrylcholine esterase (BChE) in solution by neurotoxins. Flow cell of analyzer consisted of combined Pt-Ag/AgCl electrode and biosensor based on choline oxidase. The enzyme choline oxidase (ChO) was incorporated into the polymer complex with poly-4-vinylpyridine modified on 24 % with lauryl bromide and sandwiched between two layers of cellulose acetate membrane in a plastic holder (**sensor type I**). BChE activity measurement was carried out after hydrolysis of butyrylcholine to choline, followed by its biocatalytic oxidation by choline oxidase generating hydrogen peroxide. Electrochemical analyzer response was proportional to the current of anodic oxidation of hydrogen peroxide with platinum working electrode at a potential of +600 mV versus Ag/AgCl reference electrode. This **type I** of biosensor provides a measure of BChE activity in a linear range up to 0.7n Menzyme with a sensitivity of 12.6 nA^{-1} nM^{-1}. Reliable analytical biosensor response, as well as good technical characteristics allow it suse for monitoring for

Table 3 The analytical characteristics of thiocholine sensitive sensors **type IV**

Range of linearity (M)	5×10^{-6}–2×10^{-4}
Limit of detection (M)	1×10^{-5}
Sensitivity (μA (M·cm^2)$^{-1}$)	1650
Change of response per one measurement (%)	1

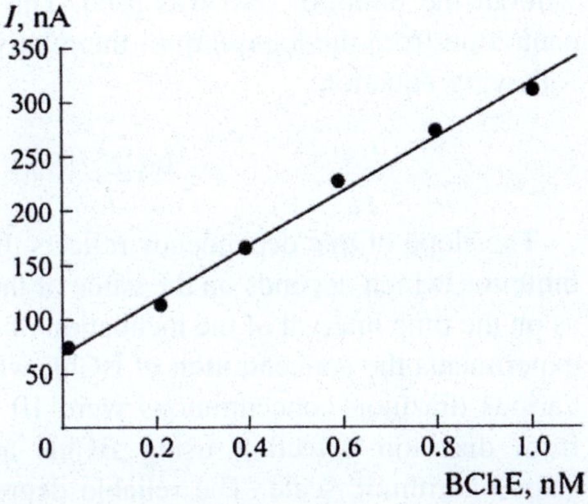

Fig. 4 The dependency of sensor response on BChE concentration

sensitivity, and the operational stability, which is the change of the response per one measurement, are shown in Table 3.

The characteristics of thiocholine sensitive **sensors type IV** that were obtained allow one to measure the thiocholine concentrations with the necessary sensitivity for the determination of BChE activity using the following procedure. In a 1 mL cell 900 μL of HEPES buffer solution containing BSA (1 mg mL^{-1}) and 100 μL of enzyme solution of the required concentration were added. After 10 min 900 μL of the overall solution was placed in the cell containing of butyrylthiocholine solution. The incubation reaction was 10 min. The response was registered after 80 s at applied of potential. Figure 4 reflects the calibration curve of the sensor response on BChE concentration. The range of concentrations studied was 1×10^{-10}–1×10^{-9} M. The dependency has linear shape with $R2 = 0.998$. The calculated enzyme detection limit was 1×10^{-10} M.

In the current work, only an irreversible inhibitor was studied. The organophosphate pesticide diazinon was selected as a model of an irreversible inhibitor. Diazinone (DZ) was preliminarily converted to the oxon form, which reacts better with the enzyme by its oxidation with bromine water. In a 1 mL test tube 700 μL of HEPES containing BSA (1 mg mL^{-1}), pH 7.5 was added, along with 200 μL of an enzyme solution with a 5×10^{-9} M concentration (final concentration of the enzyme in the tube was 1×10^{-9} M) and 100 μL of the inhibitor solution of the required concentration. After 10 min 900 μL of the overall solution was placed in another tube containing of the substrate solution. Incubation with the substrate was performed for

10 min. In this time interval enzymatic hydrolysis reaction has a linear dependence on time. The electrode was then placed in the cell and after 80 s the current was recorded.

The concentration of diazinon was determined by the degree of decrease of the initial rate of enzymatic reaction of butyrylthiocholine hydrolysis after preliminary incubation of BChE with a sample containing inhibitor during a defined period of time [17]. In this case the ratio of the initial rate of the enzymatic reaction (*E*) after preliminary incubation with the inhibitor to the initial rate of enzymatic reaction without the inhibitor (E_0) was used. The linear dependency of the inhibitor concentration from the logarithm of the relative enzymatic activity ln (E/E_0) obeys the following equation:

$$\ln([E]/[E]_0) = -k_I[I]_0\tau$$

The slope of this dependency reflects the detection sensitivity of the irreversible inhibitor, which depends on the value of the constant of the inhibition rate k_I as well as on the time interval of the incubation of enzyme with inhibitor τ. In the inhibition experiments the concentration of BChE was 5×10^{-10} M and incubation time with various diazinon concentrations were 10 min. Figure 5 shows the data obtained from diazinon detection using BChE and thiocholine sensitive sensors on a semi-logarithmic scale. The reliable degree of BChE inhibition by diazinon was calculated from the error in the detection of the enzyme without inhibitor (E_0) and was found to be 15 %. This allowed us to determine the detection limit for diazinon of 1×10^{-9} M. A similar detection limit for diazinon was obtained when the measurements were performed in a kinetic mode with stirring [17]. Nevertheless, the method described above requires more effort and additional equipment as compared to the one offered in the current work. It should be noted that developed sensors with a mediator layer of manganese dioxide could be used numerous times, with a minimum of ten times, for the analysis of BChE inhibitors without a notable sensitivity change.

Thus, in the current work we demonstrated that combined planar electrodes modified by manganese dioxide could be utilized as electrochemical mediators for the measurement of thiocholine concentration, the activity of butyrylcholinesterase,

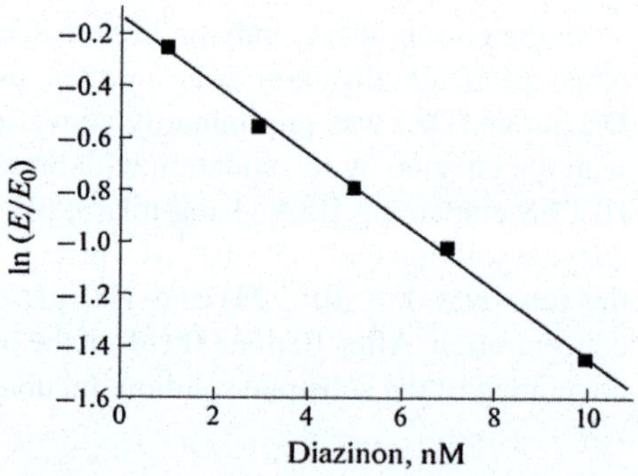

Fig. 5 The dependency of ln (E/E_0) on the diazinon concentration in the incubation mixture

and a highly sensitive analysis of its inhibitors. Sensors and measurement methods that have been developed described above greatly simplify the analysis of inhibitors, and have been applied in the areas of water quality and environmental monitoring.

2.1 *Biosensors Monitoring of Cyanobacterial Water "Bloom"*

The anthropogenic pollution of water bodies and the river flow control in common practice are the reasons of the quick eutrophication extension in water basins. It contributes to the phytoplankton biomass amplification and its species composition shifts to the increased cyanobacteria rate [18, 19]. Harmful algae "blooms" have become a major problem of the last decades, being observed in the Europe, America and Russia water bodies [20–24]. The most extending and frequently occurring water "blooms" are recorded in European Russia in the reservoirs of Volga basin (Tsimlianskoe, Izhevskoe etc.) [25–27]. The potentially hazardous cyanobacteria species presence is recorded also for most Russian North-West water bodies [24].

Cyanobacteria can synthesize a wide range of secondary metabolites, varying on structure, toxicity and effects. Some metabolites are known for neurotoxic effect [28]. Cyanobacterial neurotoxins are divided in three basic groups: (i) anatoxin-a and homoanatoxin-a (low molecular alkaloids), are potent agonists of muscular and neuronal nicotinic acetylcholine receptors; (ii) saxitoxins (carbamate alkaloids), are a group of structurally related moleculars that block voltage-gated sodium channels; (iii) anatoxin-a(s) (N-hydroxyguanidinemethylphosphate ester), is a potent irreversible inhibitor of cholinesterases [28, 29]. This group shows different degrees of toxicity (i.p. mouse LD_{50}) is ranging from 10 to 450 mkg g^{-1} [30]. Neurotoxins must be considered because of their high toxicity (the group of "fast-death factor").

The neurotoxic water "blooms" are mostly caused by species from *Anabaena*, *Aphanizomenon*, *Cylindrospermopsis*, *Lyngbya* and *Planktothrix* (*Oscillatoria*) genus [31].

The negative health effect of cyanotoxins is produced not only by consuming of the polluted water but also during recreational use of blooming water areas. As some cyanotoxins (such as saxitoxins) are stable under field conditions for a long time, even the low concentrations might pose a serious health threat during recreational use of "blooming" water zones [32, 33].

Neurotoxins synthesized by cyanobacteria have been considered as a reason of some recorded birds, dogs and domestic animals deaths in Canada [34], USA [35], Scotland [36], France [37, 38], Denmark [39], New Zealand [40], Kenya [41] and Brazil [42, 43].

Methods used for cyanobacteria neurotoxic metabolites detection in environmental water can be conventionally classified as biological, biochemical and physicochemical. Biotesting methods as well as bioassays method are of long duration and non-targeted [44]. Among other biological methods we can mark immunologic methods [45], including immunoenzymometric strip analyzers, and cell bioassay [46]. Physicochemical methods, especially mass-spectrometry, are widely used for neurotoxins detection such as anatoxin-a (homoanatoxin) and saxitoxins [47–50]. However anatoxin-a(s) detection using physicochemical methods is difficult by some reasons. Anatoxin-a(s) structure differs from other neurotoxins by phosphate group presence and chromophoric group absence, thus UV-detection cannot be used for its detection [51]. Also the development of anatoxin-a(s) detection methods is significantly complicated by the absence of its (anatoxin-a(s)) commercially available standards.

At the present day biochemical methods based on anatoxin-a(s)'s inhibition effect toward cholinesterase group enzymes activity are the most efficient way of anatoxin-a(s) detection, due to the specificity of detection of the neurotoxically active substances. Amperometric biosensor system Easy Check for detection of neurotoxins is one of the reliable equipment for running such studies [4]. The results of using amperometric biosensor systems based on manganese oxide nanoparticles allowing to measure the cholinesterase inhibition degree in water samples eventually containing corresponding inhibitors are presented in this article.

The investigations were conducted using the phytomass samples from SestroretskyRazliv Lake (Saint-Petersburg, Russia), which is a surveillance object exposed to massive recreational loading. SestroretskyRazliv Lake (N 60 5.200' E 30 0.950) is the largest artificial reservoir situated on the territory of Saint-Petersburg. The lake surface is 11.04 square km, the average depth is 1.6 m. The lake catchment area is 566 square km.

The limnologic studies were run similar to described in our previous work [27]. The biomass of phytoplankton was estimated from the total volume of algae according to the counted cell density and measured average cell density. During the research period from 2008 to 2014 major hydrochemical parameters varied in the wide ranges. The maximum gradients for ammonia nitrogen (0.6–2.0 mg N L^{-1}) and phospates (0.002–0.600 mg P L^{-1}) are typical. The pH values varied significantly (6.2–8.5) during the study period. According to the data of limnologic studies high phytoplankton biomass (from 2.34 mg L^{-1} to 62.9 mg L^{-1}), including cyanobacteria (from 0.07 mg L^{-1} to 44.9 mg L^{-1}) observed in the SestroretskyRazliv Lake during the period from 2008 to 2014. Toxic species of Cyanobacteria which can produce neurotoxins such as *Anabaena flos-aqua*, *Anabaena lemmermannii*, *Planktothrixagardhii*, *Aphanizomenonflos-aquae* were noted in the phytoplankton biomass.

For the purposes of cyanobacteria neurotoxic effect screening in Sestroretsky Razliv Lake, biomass samples were taken in the surface layer twice a month during the warm season (from June to September-October). The cyanobacteria phytomass samples were collected using a plankton net and then filtered using the Supelco filtration equipment. The filtered phytomass samples were lyophilized and analyzed for the presence of neurotoxic substances (organophosphorous and carbamate pesticides, heavy metal and cyanobacterial metabolites) by different methods.

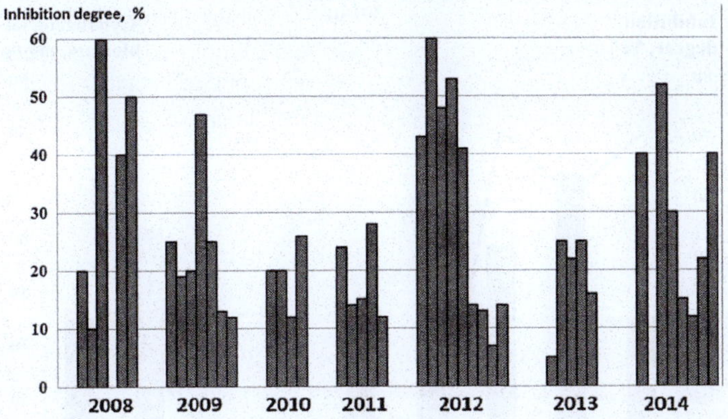

Fig. 6 The registered cholinesterase inhibition degree over the period from 2008 to 2014

For the amperometric bioassay the lyophilized samples were resuspended with processing buffer solution (2 mg g^{-1}), and then centrifuged at 4000 rpm 10 min prior the measurements. The ChE activity was assessed before and after its inhibition using IPC-micro neurotoxin amperometric analyzer (EasyCheck device) with the integrated MnO_2 modified planar electrode.

According the obtained data the effect of cholinesterase inhibition was observed in biomass samples each warm season during cyanobacterial water “bloom” (Fig. 7). The degree of inhibition varied from 15–20 % (May–June) to 50–60 % (July-August). The dynamics differed between seasons. The maximum inhibition degrees were recorded in 2008 and 2012, and the minimum were obtained in 2010 and 2011. Some samples, mostly collected in the beginning of warm season, (sampling dates were 03-Jun-2008, 22-Jun-2012, 26-Jun-2014, 30-Sep-2014) have shown the inhibition degree exceeding 50 %. However, low toxicity degrees were noted in most of studied samples (Fig. 6).

During the study period we observed the dependence of neurotoxicity (inhibition degree, %) of biomass samples versus the biomass (mg L^{-1}) of neurotoxins producing cyanobacteria toxigenic genus found in the SestroretskyRazliv lake (*Anabaena flos-aqua, Anabaena lemmermannii, Planktothrixagardhii, Aphanizomenonflos-aquae*). The Fig. 7 shows an example of such a patterns obtained in the warm season of 2014.

Also we noticed the dependence of inhibition degree (%) in studied phytomass samples versus the ratio of biogenic elements (nitrogen and phosphorus) in water (Fig. 8). It might be connected with the dependence of cyanobacteria vegetation from the ratio of these nutrients. The biomass of cyanobacteria capable of producing neurotoxins also might be increased. The Fig. 8 shows this dependence obtained in 2012.

Investigation of the effect of the reactants mixing order on the value of the inhibitory effect of the samples showed it has a significant impact on the degree of inhibition, which indicates mostly irreversible inhibition.

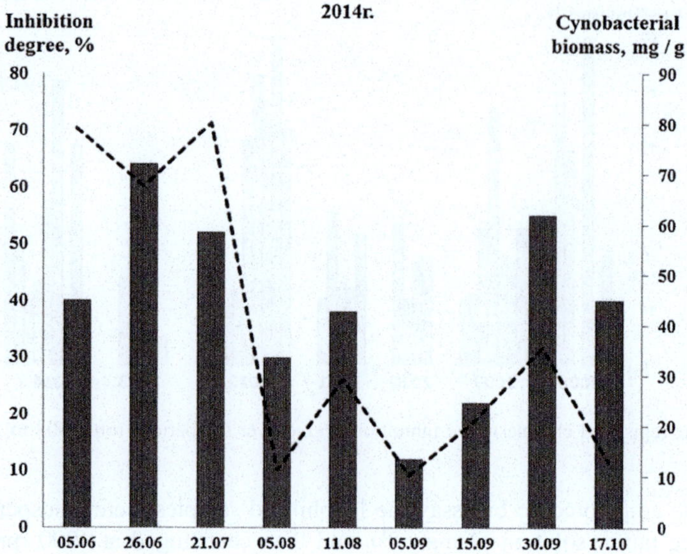

Fig. 7 The dependence of inhibition degree (%) versus biomass of the concentration of toxigenic cyanobacteria capable of producing neurotoxins (mg g^{-1}) during the growing season 2014

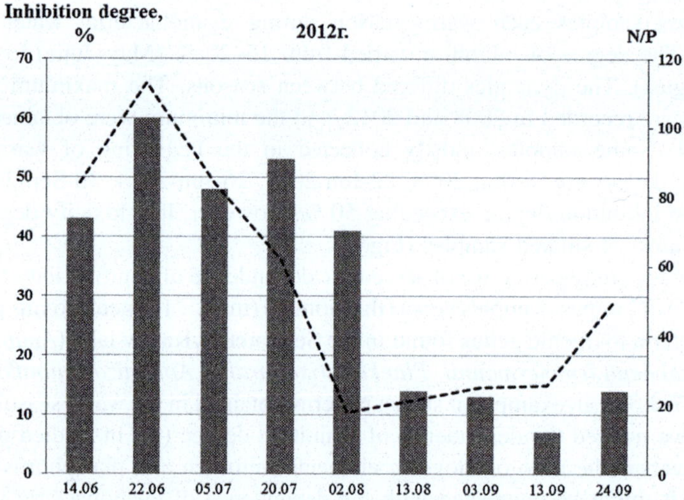

Fig. 8 The dependence of the inhibition degree (%) versus biogenic elements ratio (N/P) and of biomass of toxigenic neurotoxin producing cyanobacteria

The anatoxin-a(s) presence in biomass samples was confirmed by MALDI-MS/MS method. The protonated molecular ion with m/z 253 ($[M + H]^+$) was used as a precursor-ion for the fragmentation in MS/MS regime [51]. Diagnostic ion-products with m/z 58, m/z 59, m/z 96, m/z 98 and m/z 141 were chosen with the help of software Mass Frontier 5.1 data base (Thermo Scientific). Fragmentation pattern is confirmed by some published works [52], studying the *A. flos-aquae*strain NRC 525-17 for detection of the chemical structure of this natural organophosphates.

Mass-spectrometry analyses was carried out using Axima Performance time-of-flight mass-spectrometer with the MALDI ions source, equipped with UV-laser (337 nm) set up as follows: positive ion detection mode (phosphonylated peptides), reflectron ion source settings: potential 20 kV, lens potential 6.5 kV, reflectron potential (Ref.)—24.38 kV. Ions were detected in the range 50–5000 m/z in reflectron mode. Mass-spectra were registered with use of MALDI-MS Shimadzu Biotech program (Shimadzu, Japan). Calibration in positive ion mode was run using the peptides calibration mix: angiotensin 2 ($[M + H]^+$ m/z = 1046,542), N-acetylrenin ($[M + H]^+$ m/z = 1800,943), adenocorticotropic hormone (1–17) ($[M + H]^+$ m/z = 2093,086).

The data of MS/MS analysis for m/z 253 are presented on the Fig. 9. The fragment mass-spectra shown on this figure contains signals which are diagnostic ones for anatoxin-a(s) (m/z 159, m/z 141, m/z 98, m/z 96), confirming the anatoxin-a(s) presence in the biomass extract.

As most of organophosphorous and carbamate pesticides as well as a number of heavy metals can produce an inhibition effect toward the cholinesterase enzymes, we studied the collected samples for the presence of the above mentioned substances using different appropriate methods.

ICP-MS method was used for the analysis of filtered suspended lyophilized biomass samples in order to prevent heavy metals contribution in the total biomass neurotoxicity. The obtained concentrations of detected metals in studied samples were significantly lower than concentrations inducing cholinesterase inhibiting (data are not shown). The collected samples were analyzed for the presence of carbamate and organophosphate pesticides, capable to produce cholinesterase inhibition (carbofuran, phenmedipham, diazinon, isopropylphenilcarbamate, isopropylchlorophenilcarbamate, chlorpyriphos, tributylphosphate, chlorophos) by the method of high resolution chromato-mass-spectrometry with tandem mass-spectrometry (the LTQ

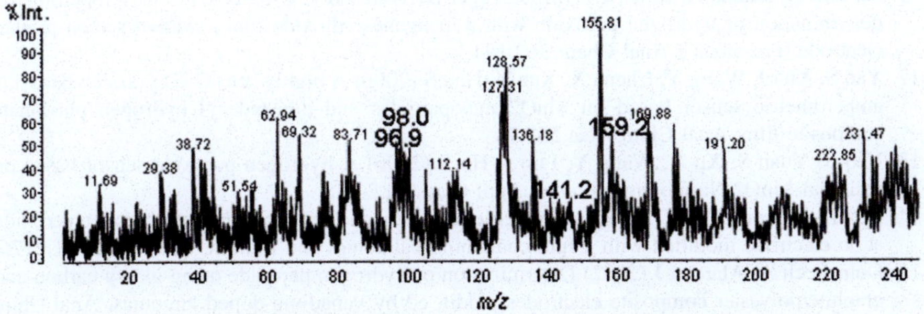

Fig. 9 Fragment of the MS/MS spectrum for protonated molecule with m/z 253 (underlined are signals with m/z 98, 96, 141, 159)

Orbitrap, ThermoFinnigan).The presence of above mentioned compounds were not detected. This fact allowed excluding the pesticides contribution to the registered neurotoxicity.

Obtained data of all the analysis indicates the presence of anatoxin-a(s) in analyzed samples. Our study showed the potential of the neurotoxins biosensor systems for monitoring neurotoxic cyanobacterial water "blooms".

References

1. Andreescu S, Marty JL (2006) Twenty years research in cholinesterase biosensors: from basic research to practical applications. Biomol Eng 23:1
2. Lee JH, Mitchell RJ, Kim BC, Cullen DC, Gu MB (2005) A cell array biosensor for environmental toxicity analysis. Biosens Bioelectron 21:500
3. Rodriguez-Mozaz S, de Alda MJL, Barcelo D (2006) Biosensors as useful tools for environmental analysis and monitoring. Anal Bioanal Chem 386:1025
4. Kurochkin IN, Eremenko AV, Makhaeva GF, Sigolaeva LV, Dubacheva GV, Richardson RJ (2009) Multi-strip assay and multimodal biosensors for environmental and medical monitoring of neurotoxicants. In: Dishovsky CH, Pivovarov A (eds) Counteraction to chemical and biological terrorism in the East Europe countries. NATO science for peace and security, series A: chemistry and biology. Springer, Dordrecht
5. Mannino S, Cosio MS, Ratti S (1993) Cobalt (II, III)-oxide chemically modified electrode as amperometric detector in flow injection systems. Electroanalysis 5:145
6. Salimi A, Hallaj R, Soltanian S, Mamkhezri H (2007) Nanomolar detection of hydrogen peroxide on glassy carbon electrode modified with electrodeposited cobalt oxide nanoparticles. Anal Chim Acta 594:24
7. Karyakin AA, Karyakina EE (1999) Prussian Blue-based 'artificial peroxidase' as a transducer for hydrogen peroxide detection. Application to biosensors. Sens Actuators B Chem 57:268
8. Ricci F, Amine A, Palleschi G, Moscone D (2003) Prussian Blue based screen printed biosensors with improved characteristics of long-term lifetime and pH stability. Biosens Bioelectron 18:165
9. Ricci F, Palleschi G (2005) Sensor and biosensor preparation, optimisation and applications of Prussian Blue modified electrodes. Biosens Bioelectron 21:389
10. Schachl K, Alemu H, Kalcher K, Jeozkova J, Svancara I, Vytoras K (1997) Amperometric determination of hydrogen peroxide with a manganese dioxide-modified carbon paste electrode using flow injection analysis. Analyst 122:985
11. Schachl K, Alemu H, Kalcher K, Moderegger H, Svancara I, Vytoras K (1998) Amperometric determination of hydrogen peroxide with a manganese dioxide film-modified screen printed electrode. Fresenius J Anal Chem 362:194
12. Yao S, Xua J, Wang Y, Chena X, Xua Y, Hua S (2006) A highly sensitive hydrogen peroxide amperometric sensor based on MnO2 nanoparticles and dihexadecyl hydrogen phosphate composite film. Anal Chim Acta 557:78
13. Yao S, Yuan S, Xu Y, Wang Y, Luo J, Hu S (2006) A hydrogen peroxide sensor based on colloidal MnO2/Na-montmorillonite. Appl Clay Sci 33:35
14. Cox JA, Jaworski RK (1989) Voltammetric reduction and determination of hydrogen peroxide at an electrode modified with a film containing palladium and iridium. Anal Chem 61:2176
15. Domenech A, Alarcon J (2002) Determination of hydrogen peroxide using glassy carbon and graphite/polyester composite electrodes modifi ed by vanadium-doped zirconias. Anal Chim Acta 452:11

Biosensors for Detection of Anticholinesterase Agents

Gennady Evtugyn

Abstract Cholinesterase biosensors based on various transducers and enzyme carriers have been considered in terms of inhibitor determination. The mechanism of inhibition and influence of immobilization on biosensor performance are briefly considered. The assembly of biosensors for inhibitor detection and measurement conditions are summarized for the period from 2011 to 2015 with particular emphasis to the influence of the sample matrix and immobilization protocol on the sensitivity of inhibitors detection. Finally, the prospects of cholinesterase biosensors are considered, especially those related to miniaturization and operation in extreme environment.

Keywords Enzyme sensor · Biosensor · Inhibitor detection · Anticholinesterase agent · Pesticide determination · Aflatoxin B1 determination

1 Introduction

There is an urgent need in the development of fast and reliable sensors for the detection of most toxic compounds that are formed in natural and industrial processes and can cause severe consequences for human health. Among many other species, inhibitors of acetylcholinesterase are in the focus of many investigations due to variety of chemical structures and large scale of industrial production. Some anticholinesterase agents, e.g., sarin, soman and VX gas, were specially developed as chemical warfare with extremely low toxic exposures and lethal consequences for the soldiers and civil population. Although the accumulated stockpiles of chemical weapons are mostly destructed following the Chemical Weapons Convention, some incidents related to the use of anticholinesterase agent shave been reported during the civil wars in Iraq and Syria. The use of home-made sarin in Tokyo subway by Aum

G. Evtugyn (✉)
A.M. Butlerov' Chemistry Institute of Kazan Federal University,
18, Kremlevskaya Street, Kazan 420008, Russian Federation
e-mail: Gennady.Evtugyn@ksu.ru

D.P. Nikolelis and G.-P. Nikoleli (eds.), *Biosensors for Security and Bioterrorism Applications*, Advanced Sciences and Technologies for Security Applications,
DOI 10.1007/978-3-319-28926-7_17

Shinrikyo in 1995 is most known incident related to nerve gases after the Second World War. Meanwhile the threats related to the production and application of anticholinesterase agents by terrorists exists up to present days. In addition to chemical warfare, organophosphorus and carbamate pesticides irreversibly inhibit cholinesterase activity and can cause poisoning of agriculture workers and contamination of some foodstuffs. These hazards call for the further efforts in the development of appropriate biosensors devoted to detection of anticholinesterase agents in the levels allowing the use of personal protection equipment and hence decrease in the number of potential victims. In this review, the progress in the detection of anticholinesterase species based on biosensing technologies is considered with particular emphasis to the results obtained within last 10 years.

2 Cholinesterase Inhibition

Cholinesterases belong to the class of hydrolases and catalyze the reaction of hydrolysis of a natural neurotransmitter, acetylcholine (Scheme 1).

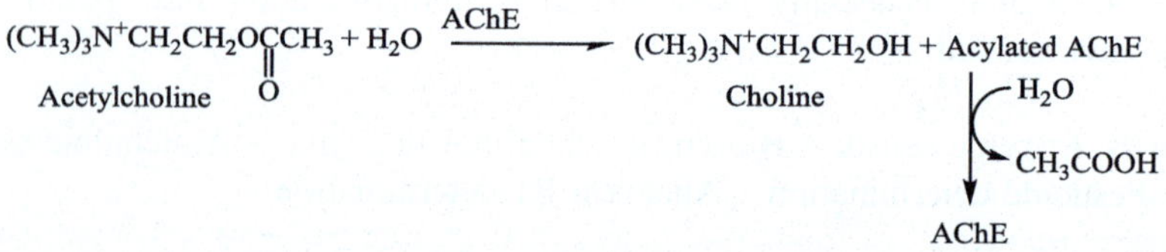

(1)

They are widely present in various biological species, mainly vertebrates, with insignificant variations of the amino acid residues responsible for the substrate binding. In accordance to the relative rate of the hydrolysis of different choline esters, cholinesterases are subdivided into several groups, from which acetylcholinesterase (AChE, EC 3.1.1.7) and butyrylcholinesterase (BChE, EC 3.1.1.8) are preferably used in the enzymatic assay of inhibitors [1].

AChE is mainly bonded to the cellular membranes of excitable tissues and erythrocytes. It hydrolyzes the acetylcholine to terminate the transmission of the neural impulse in presynaptic cell to somatic neuromuscular junction. In case of the AChE inhibition, accumulation of acetylcholine overstimulates autonomic nervous system. The poisoning symptoms include vomiting, abdominal cramps, nausea, salivation, and sweating. Also, the excitation of somatic nervous system results in involuntary muscle twitching, convulsions, respiratory failure and coma.

BChE, serum cholinesterase, was found in plasma, liver and muscle tissues. The biochemical function of BChE is yet not clear but probably is related to scavenging of organophosphate and carbamate inhibitors before they reach AChE. BChE is also involved in regulation of cholinergic transmission in the absence of AChE. Both enzymes participate in biochemical conversion of some drugs, e.g., aspirin, amitriptyline and bambuterol and in metabolic transformation of cocaine and heroin [2].

The AChE active site involves so-called anionic and esteratic centers different in their affinity toward the substrate and choline. The distance between the above centers is not changed in the enzyme acylation. Rather small conformational changes of the enzyme globule corresponded to the substrate binding result in high efficiency of the substrate conversion: the AChE molecule catalyzes breakdown of about 10000 acetylcholine molecules per second. The decoding of the first AChE structure showed that the active site is placed at the bottom of a 20Å deep gorge lined with tryptophan residues. Actually, no anionic sites are involved in acetylcholine binding. The peripheral "anionic site" responsible for acetylcholine sequester lies at the entrance to the gorge and interacts with a substrate by π-cation interactions with the tryptophan residues whereas the carbonyl of the acetyl group forms weak hydrogen bond to tyrosine residue placed further down the gorge. Three amino acid residues, namely, glutamate, histidine and serine, also known as a catalytic triad, are involved in the formation of covalent bond between serine hydroxyl and >C=O double bond of acetyl group of the substrate. This "oxyanion" is stabilized by release of choline leaving acetylserine (acetylated AChE). The serine—acetate bond is then hydrolyzed to the acetate and initial enzyme ready to the further substrate conversion [3].

First cholinesterase based devices have been developed for early detection of chemical warfare and utilized BChE from horse and human serum which showed high sensitivity toward warfare agents and rather high stability of preparations available at that moment [1]. Later on, AChE from *electric eel* became dominating due to higher specific activity and ability to detect a broader range of organophosphate and carbamate pesticides [4]. Nowadays, most cholinesterases commercially available are produced by genetically modified microorganisms, e.g., *Escherichia coli*. Modification of enzyme active site makes some difference in the sensitivity of the activity toward various groups of inhibitors. This can be used for differentiation of the response of appropriate biosensors especially in inhibitor mixtures [5, 6].

Most of the analytes determined with cholinesterase biosensors exert irreversible inhibition of immobilized enzyme. Such compounds mimic the structure of the enzyme active site and substitute acetylcholine on the first stage of the formation of serine ester. However, the product of esterification, e.g., phosphorylated cholinesterase in case of organophosphate pesticides and nerve gases, is quite stable toward hydrolysis and stops the following regeneration of the enzyme active site (Scheme 2).

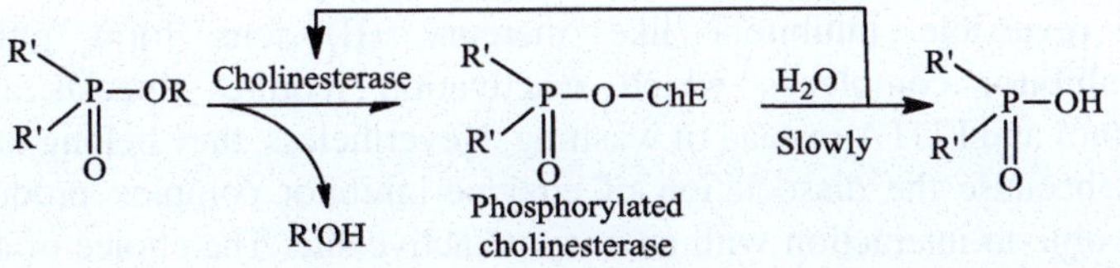

(2)

The appropriate stage called as spontaneous regeneration requires the time sufficiently longer than that monitored with biosensor so that it does not affect the decay in the enzyme activity applied for the analyte determination. Similar reaction of carbamylated cholinesterase formed in the presence of carbamate pesticide is much faster so that maximal decay of enzyme activity observed in the determination of carbamates can be below 100 %. For this reason, the influence of carbamates on cholinesterases is named in some sources as reversible inhibition. This seems incorrect because the product of reactivation of carbamylated cholinesterase does not react with free enzyme and has the structure different from initial carbamate. From the point of view of physical chemistry, such interactions are considered as irreversible.

Apart spontaneous reactivation, the regeneration of initial enzyme activity is accelerated by some reagents added to the sample after inhibition. Such reagents, e.g., aldoxime derivatives, are also used as antidotes in case of poisoning with anticholinesterases. 2-PAM and TMB-4 are mostly mentioned in the operation of cholinesterase biosensors [7, 8]. It should be mentioned that such reactivators exert their own weak reversible inhibition on cholinesterases. For this reason, they should be carefully washed out from the immobilized enzyme preparation prior to measurement of residual enzyme activity. Besides, regenerated enzyme shows the sensitivity toward inhibitors different from that of enzyme prior to first contact with the sample tested. This is mainly attributed to different accessibility of enzymes immobilized in inert matrix on the transducer. As a result, a reactivator reacts first with the inhibited enzyme molecules that are positioned near the solution and to a less extent with those placed deeply inside the biolayer. Additional washing steps and protective effect of excessive reactivator quantities (see below) also complicate repeated use of biosensor for multiple inhibitor detection.

Heavy metals, quaternary ammonia compounds and biogenic amines exert reversible inhibition on cholinesterases [9–12]. In this case, the enzyme-inhibitor complex can dissociate to initial enzyme and inhibitor molecule/ion so that the detection of inhibition can be performed only in the presence of an inhibitor. Maximal decay of enzyme activity is sufficiently lower than 100 % and washing the biosensor with buffer solution can restore the signal recorded prior to the contact of the enzyme with an inhibitor. Most of the reversible inhibitors mimic positively charged ammonia group of acetylcholine. The concentrations detected are significantly higher than those of irreversible inhibitors. For this reason, the cholinesterase biosensors cannot be used for direct detection of reversible inhibitors on their permissible levels. The determination of some pharmaceuticals applied as myorelaxants or drugs against Alzheimer's disease [12] is an exception.

Some reversible inhibitors like mercury (II) ions form rather stable enzyme-inhibitor complexes which reactivation requires special agents like dithiotreithol and EDTA instead of washing. Nevertheless, they belong to reversible inhibitors because the dissociation of enzyme-inhibitor complex produces initial metal ion able to interaction with an enzyme active site. The choice of appropriate reactivator and monitoring of reactivation efficiency provides information which might be useful for preliminary identification of an inhibitor nature especially for complex samples including potential hazards of different nature.

Chemicals exerting denaturation effect and disturbing the protein steric structure also alter the activity of cholinesterases. Urea, surfactants and polar organic solvents suppress the activity of free enzyme to an extent depending on their nature, concentration and contact time [13]. It was shown that the influence of polar solvents is additive to that of dissolved irreversible inhibitor if the total inhibition remains rather low (about 20–50 %) [14]. However, the possible interference of the solvents on the detection of extracted pesticides and other anticholinesterases needs further investigation. Acetonitrile exert activating effect on immobilized cholinesterase and increases the sensitivity of pesticide detection [15]. Immobilization of cholinesterases decreases the influence of denaturation agents due to stabilizing effect of the matrix. This allows adapting the sample treatment schemes developed for HPLC and applying direct detection of an inhibitor in diluted extract with no evaporation of organic solvent.

2.1 *Irreversible Inhibition*

Irreversible inhibition is measured by consecutive incubation of the enzyme, either free or in the assembly of biosensor, in the inhibitor solution followed by the addition of the substrate which stops interaction of the enzyme with inhibitor. After that, the rate of enzymatic reaction is measured. For immobilized enzyme, additional washing is often used to remove the access of an inhibitor. This makes it possible to vary the conditions of incubation and signal detection, which are not always the same especially in pH region and electrolyte content. Both AChE and BChE are irreversibly inhibited by organophosphates and carbamates that form esteric bonds with serine -OH group in enzyme active site. Chemical structures of common inhibitors applied as pesticides are presented in Fig. 1.

Relative decay of the rate of enzymatic reaction caused by irreversible inhibition depends on the incubation period and the inhibitor concentration in accordance with the Ellman' Eq. (3) [16].

$$\ln \frac{v_0}{v_t} = k_{II} C_I \tau \tag{3}$$

Here, v_0 and v_t are the reaction rates prior to and after the contact of the enzyme with an inhibitor, C_I is the inhibitor concentration and τ the incubation period. The bimolecular inhibition constant k_{II} depends on the nature of an inhibitor, the enzyme source and incubation conditions but not on the inhibitor concentration. It is used for comparison of the cholinesterases from various sources or in various measurement conditions against the same inhibitor. The Ellman equation assumes the inhibitor concentration to be sufficiently higher than that of an enzyme active site. In the opposite case, the decay of the reaction rate linearly depends on the inhibitor concentration. The use of Eq. (3) gives estimate of the limit of detection (LOD) for irreversible inhibitor. For 15 % inhibition often considered as minimal

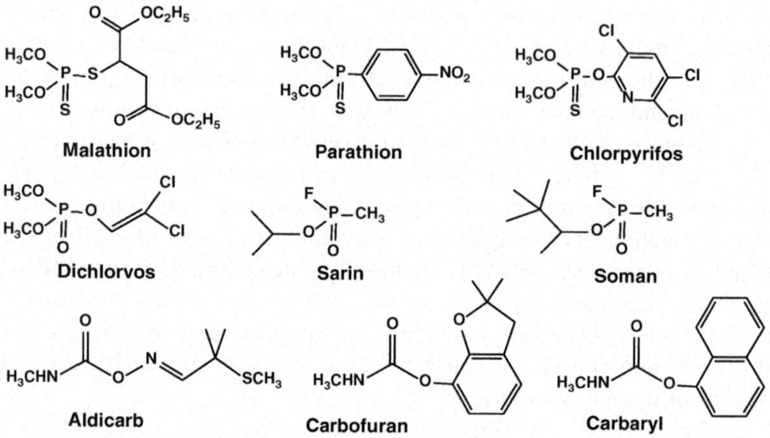

Fig. 1 Irreversible inhibitors commonly detected with cholinesterase biosensors

level reliably detected, the LOD is equal to $0.13/(k_{II}{\cdot}\tau)$. The k_{II} values vary from $n \times (10^6 \div 10^8)$ M^{-1} s^{-1} for chemical warfare to $n \times (10^3 \div 10^5)$ $M^{-1}s^{-1}$ for organophosphate pesticides [17, 18].

The following criteria of irreversible inhibition are mostly used: (1) the proportional increase of the inhibition with incubation period; (2) full inhibition of an enzyme with increased inhibitor concentration or incubation period; (3) no recovery of enzyme activity after addition of increasing substrate concentration after inhibition step. The reactivation with oximes is also a sign of irreversible inhibition caused by organophosphates and carbamates because reversible inhibition does not assume the formation of esteric bonds in the active site. It should be mentioned that about all of these criteria are not absolute. Slow establishment of equilibrium in the reaction media and fast spontaneous reactivation can mask irreversible character of enzyme—inhibitor interaction.

The biosensor format of the inhibition measurement with immobilized cholinesterase changes the quantitative relations from those corresponded to Ellman' equation. Commonly, the sensitivity of inhibitor determination obtained with a biosensor is lower than that of free cholinesterase. Meanwhile, some examples reported recently demonstrate significant decrease of the limits of detection (LODs) against homogeneous reaction conditions. The reasons of such a behavior of cholinesterase biosensors can be summarize as follows [19]:

- Immobilization matrix effect. The charge and buffering properties of the immobilization matrix affect the transfer of cationic acetylcholine in the surface layer and hence shift the signal of a biosensor. To some extent, the permeability of the surface layer for inhibitor molecules can be also taken into account.
- Kinetic limitations. Decrease in the rate of the substrate transfer into the surface layer can prevent saturation of the enzyme active site with substrate. As a result,

inhibition is partially compensated for by involvement of free active sites in the substrate hydrolysis [20].

- pH shift. Both AChE and BChE exert maximal activity in weakly basic media. Any reactions resulted in release of hydrogen ions or their neutralization will shift the local pH in the surface layer and hence the parameters of both substrate hydrolysis and enzyme inhibition. Slow transfer of H^+ ions from the bulk solution aggravates the situation. This is especially important for potentiometric biosensors based on pH measurement [21].
- Unspecific transfer of reactants in the surface layer. Adsorption or aggregation of a substrate/inhibitor in the surface layer or on the transducer surface change their distribution in the reaction layer and hence affect experimental values of inhibition.
- Side reactions of reactants. Non-enzymatic hydrolysis of acetylcholine can be accelerated by matrix material. The same can happen with inhibitors able to hydrolyze (organophosphates and carbamates) or aggregate in inactive complexes (heavy metal ions).

As could be seen, most of these factors decrease inhibitory effect of irreversible inhibitors. However, sorptional accumulation of an inhibitor and changes in the substrate transfer can improve the performance of cholinesterase biosensors with no respect of biochemical backgrounds of inhibition quantification.

2.2 *Reversible Inhibition*

Reversible inhibitors exert a variety of mechanisms referred to the true equilibria of the stages with the enzyme participation [22]. Most of the reaction schemes assume Michaelis-Menten kinetics and possibility to form triple enzyme—inhibitor—substrate complexes. Contrary to irreversible inhibitors, the binding of reversible inhibitor allows producing the final product of the substrate conversion. The reversible inhibitors of cholinesterases can mimic cationic part of a natural substrate, acetylcholine, nut they cannot form covalent bond with the amino acid residues of the active site. Even involved in the reaction, they do not fully suppress the access of the substrate. For AChE, reversible inhibitors can attack allosteric center of regulation of enzyme activity so that the enzyme-inhibitor complex formed retains the possibility to interact with the substrate. From the point of view of detection of chemical hazards, the determination of reversible inhibitors is not very interesting because the appropriate analytes are sufficiently less dangerous than irreversible inhibitors presented above or their detectable concentrations are much higher than limited threshold values. Nevertheless, the investigation of reversible inhibition is important if the cholinesterase biosensors are intended for the assessment of total contamination of environmental subjects. Besides, reversible inhibitors decrease the influence of irreversible inhibitors due to so called protecting effect. The reversible enzyme–inhibitor complex cannot react with irreversible

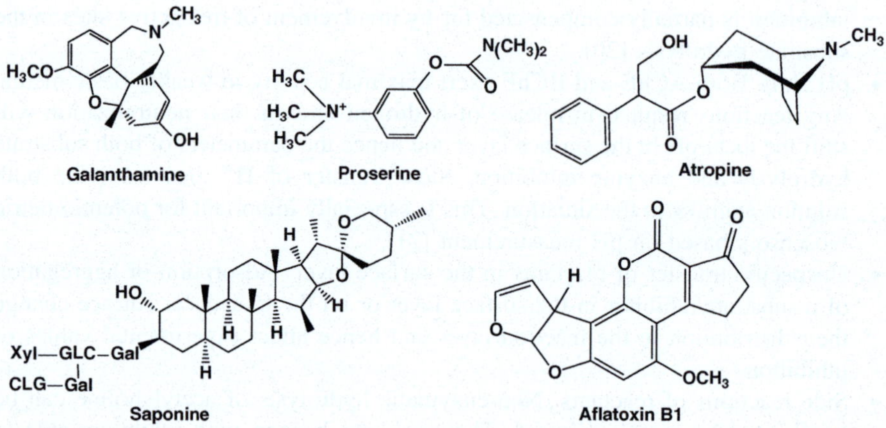

Fig. 2 Reversible inhibitors of cholinesterase determined with biosensors

inhibitor but after incubation of the biosensor it can be destroyed by addition of a substrate followed by restoration of the enzyme activity. Protecting effect is widely used for antidote therapy and prevention of acute poisoning with anticholinesterase species. Some of the organic reversible inhibitors studied with cholinesterase biosensors are presented in Fig. 2.

They include some drugs applied as myorelaxants and for Alzheimer disease treatment and toxins produced by potato (saponine) and molds (aflatoxin B1). The latter one is of special interest because of carcinogenicity of the pollutant and low residual concentrations allowed in foodstuffs. The use of AChE is an alternative of immuno- and aptasensors that show comparable sensitivity but higher specificity of the response. On the other hand, the use of immobilized enzyme makes possible multi-use of the biosensor and decreased measurement time.

Kinetic consideration of reversible inhibition is mainly based on some simplifications found more convenient for application.

In *competitive inhibition* (Scheme 4), the inhibitor and substrate both compete for the same active site of an enzyme and the enzyme-inhibitor complex cannot catalyze the substrate conversion.

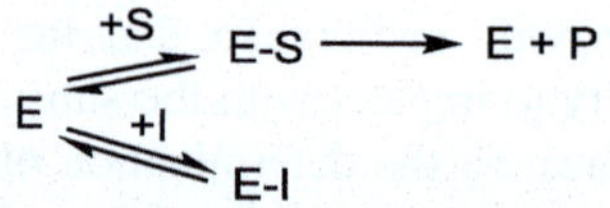

(4)

Irreversible inhibition of cholinesterase performed in the presence of the substrate corresponds to this type of inhibition. For such a reaction, the maximal rate of the reaction does not depend on the concentration of the competitive inhibitor whereas Michaelis constant increases with inhibitor content.

Non-competitive inhibition corresponds to the case when the substrate and inhibitor do not interfere with each other in enzyme binding. The interaction with an inhibitor affects the affinity of an enzyme towards a substrate but not the reactivity of the enzyme-substrate complex (Scheme 5).

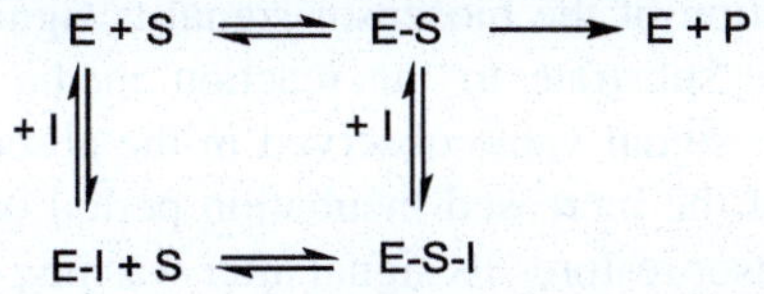

(5)

Non-competitive inhibition can be observed for heavy metal ions which either bind allosteric center of AChE cholinesterase or affect the protein conformation due to participation in the charge distribution and pH shifts within the protein globule.

Uncompetitive inhibition assumes the interaction of an inhibitor only with the enzyme-substrate complex. It is rarely observed for cholinesterases. The same can be said about *mixed inhibition*, but both types of inhibition can be indicated by the formal kinetics analysis for immobilized enzyme with complicated reactant transfer stages.

In all the mechanisms of reversible inhibition, the equilibrium of enzyme–inhibitor interaction is quantified with equilibrium constant called as inhibition constant (K_i). The increase in the inhibition constant decreases the sensitivity of an enzyme towards an inhibitor. The value of inhibition constant corresponds to the inhibitor concentration resulting in a 50 % decrease of the rate of the enzymatic reaction, or a biosensor signal. However, the Ki values can be compared only for inhibitors interacting with cholinesterase in the same way.

For kinetically 'pure' reversible inhibition, the conclusion on the mechanism of enzyme–inhibitor interaction is made from the experiments with varied concentrations of the substrate/inhibitor. The shape of the kinetic curves and their relative position against each other provide the criteria of particular inhibition mechanism described above. Besides, changes in the Michaelis constant and maximal reaction rate can be calculated in a common manner for different content of an inhibitor in the reaction media (see kinetic analysis in more detail in [19]).

The knowledge on mechanism of reversible inhibition makes it possible to optimize the reaction conditions. Thus, for reversible inhibitors, the substrate concentration is of critical importance for the sensitivity of the assay. The lower is the substrate concentration the higher is the slope of calibration curve. Contrary to that, non-competitive inhibition is insensitive to the substrate concentration chosen. Uncompetitive inhibition depends on the substrate concentration only if that is below the Michaelis constant. At upper substrate content, the inhibition does not change with variations of this parameter [14, 19].

As in the case of irreversible inhibitors, immobilization of the enzyme affects the changes in the biosensor response. The sensitivity of inhibitor detection becomes lower due to additional electrostatic interaction and steric limitations of an inhibitor transfer to the enzyme active site. Besides, the use of appropriate substrate concentrations far from optimal values is dictated by the accuracy reasons and properties of appropriate transducer. The reversible character if inhibition can be established from the following experimental evidences:

- The maximal inhibition of the biosensor signal is significantly below 100 %;
- The addition of the substrate to the reaction media containing an inhibitor partially restores the signal value observed in the absence of an inhibitor;
- There is no effect of the increased incubation period on the inhibition;
- The inhibited biosensor restores its signal after washing in the substrate solution.

A short overview of the inhibition mechanisms and their use for interpretation of a biosensor response shows a variety of the properties that complicate the comparison of various inhibitors in accordance of their ability to suppress the inhibition activity. The calculation of inhibition kinetics requires data on the mechanism and stoichiometry of enzyme–inhibitor interaction. To simplify such comparison, empirical parameters are mostly used for characterization of a biosensor performance. Thus, relative decay of biosensor response (inhibition degree) against that measured prior to the contact with the sample is frequently applied for calibration graphs. The I_{50}, an inhibitor concentration corresponded to 50 % inhibition of the signal, characterizes the sensitivity of the response similarly to k_{II} of irreversible inhibitor and K_i of reversible inhibition.

3 Cholinesterase Biosensor Assembling

The development of cholinesterase biosensor is based on the development of two protocols, i.e. (i) that of the enzyme immobilization, and (ii) measurement of the signal related to the cholinesterase activity. Although both problems mentioned are interconnected, some universal solutions exist. Below, the approaches to the development of cholinesterase biosensors are briefly considered with particular emphasis on the specificity of inhibition measurement.

3.1 *Cholinesterase Immobilization*

The immobilization assume the procedure that fixes the enzyme on appropriate carrier or directly to the transducer surface. First, the immobilization was directed to the multiple use of a biosensor to decrease the cost of expensive enzyme preparations. However, inhibition decreases the signal and the following application of biosensor once contacted with an inhibitor is often not possible. For this reason, the immobilization of cholinesterase solves other problems like ease operation of biosensor and stability of the signal. To some extent, the requirements to immobilization of cholinesterase are contradictory. The protocol chosen should provide long operation of biosensor and meanwhile high its sensitivity toward an inhibitor.

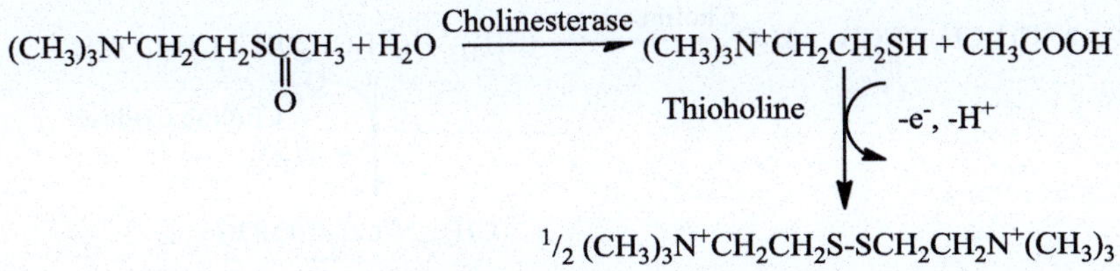

(9)

First, the reaction was monitored in galvanostatic regime with Pt net electrodes by appropriate shifts of the potential related to two processes, i.e. thiocholine oxidation and iodide oxidation [23]. The latter one was used as a counter ion of thiocholine ester due to much lower hygroscopicity against appropriate chloride and bromide. The formation of mixed potential increased the sensitivity of the response referred to the mediation function of I_2/I^- pair [53]. Later on, voltammetric and amperometric transducers have been introduced for the same reaction detection. However, in these works, chloride counter ion was used to avoid complication of parallel oxidation of thiocholine and iodide. As was shown on bare metal and carbon electrodes, thiocholine is oxidized with rather high overvoltage at about 0.6–0.8 V. The reaction can result in partial passivation of metal electrodes due to formation of insoluble sulfides. For this purpose, now the reaction is performed in the presence of mediators decreasing working potential of appropriate biosensor to 0.15–0.40 V. Prussian blue [54–56], Co phtalocyanine [57] 7,7,8,8-tetracyanoquinodimethane (TCNQ) [58, 59], silver nanoparticles [38, 60–63] can be used for amplification of the signal related to thiocholine oxidation. In many cases, the mediators are assembled on CNTs or carbon black that exert their own catalytic activity. The application of mediators excludes chemisorption of oxidation products and simplifies recovery of the biosensor after signal measurement. Interpretation of the results obtained with thiocholine esters should take into account that the rate of its enzymatic hydrolysis is lower than that of acetylcholine by 1.5–1.8 times. In addition, thiocholine can release H^+ ions with the dissociation constant comparable with that of acetic acid. This increases the pH shift during the reaction run.

Potentiometric biosensors utilize pH-metric glass electrodes [63] or polyaniline layers which potential is pH dependent [64, 65]. Few publications are devoted to the application of membrane ion-selective electrodes sensitive to choline or acetylcholine [66–68]. More recently, pH sensitive field-effect transistors (ISFETs) [69, 70] and conductometric devices [71, 72] have been described as transducers for the detection of cholinesterase inhibitors. Although it could be stated that potentiometric biosensors are less sensitive in inhibitor detection than voltammetric sensors, in comparable conditions and with similar enzymatic membranes the performance of such biosensors is comparable [19].

Second approach to the measurement of AChE/BChE biosensor signal uses enzymatic conversion of choline resulted in formation of betaine and hydrogen peroxide (10). The rate of the second step is estimated by the concentration of hydrogen peroxide which is either oxidized or reduced on the electrode.

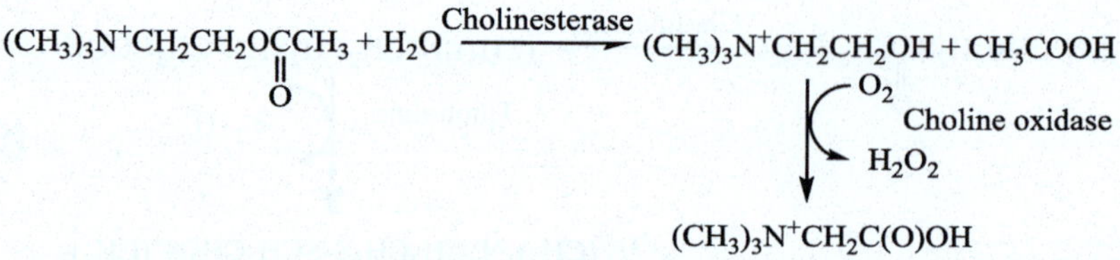

(10)

The oxidation of the hydrogen peroxide is promoted by mediators of electron transfer, e.g., Pt nanoparticles [73], MnO_2 [74] and phtalocyanine complex [75]. Horseradish peroxidase (HRP) [76, 77], Prussian blue [78] and CdTe quantum dots [79] catalyze reduction of hydrogen peroxide.

Choline oxidase differs from cholinesterase by specific activity and pH maximum of activity. This complicates their co-immobilization on the same support. To overcome the difficulties, AChE is used in soluble form whereas the residual activity after inhibitor addition is measured with choline oxidase biosensor.

3.3 Analytical Characteristics of Inhibitor Determination

Performance of cholinesterase biosensors developed for inhibitor determination is summarized in Table 1 for the period from 2011 to 2015. Previous publications as well as alternative approaches to toxicant detection were reviewed elsewhere [4, 14, 80]. Main attention is paid to the signal transduction and sensitivity of inhibitor determination.

As could be seen, many of the biosensors developed were tested on standard inhibitors, i.e., paraoxon as a model of organophosphates and carbaryl as a representative of carbamate pesticides. Relative sensitivity of response toward other inhibitors commonly coincides with the values of appropriate k_{II} values determined for free enzyme. This rule does not work if inhibitors are accumulated on the carbonaceous supports, e.g., CNTs or carbon black. Thus, relative sensitivity of malaoxon and paraoxon determination with appropriate biosensors is opposite to that expected from inhibition kinetics [38, 62]. The same reason explains significant decrease in the detection limits of irreversible inhibitors observed in recent publications in this area.

AChE from *E. eel* is mainly applied for inhibitor detection due to its high specific activity and broad selectivity toward various substrates. The application of BChE from horse serum increases relative response toward organophosphates against that of carbamates. For this reason, this enzyme source has some advantages in detection of nerve gases or in attempts to increase the selectivity of the response in the mixtures of irreversible inhibitions.

As was mentioned before, all the commercial enzymes are produced by genetically modified microorganisms so that their prize tends to decrease. For the same reason, the aspects of enzyme purity and special purification are not considered in appropriate publication devoted to biosensor design.

Table 1 Analytical characteristics of cholinesterase inhibitors detection (2011–2015)

Inhibitor	Detection mode	Sample	Detection characteristics	References
Nerve gases				
Sarin, soman, tabun, VX	Screen-printed electrode, AChE in gelatin cross-linked with glutaraldehyde, chronoamperometry	Standard solutions	LOD 7.41×10^{-12} (sarin), 6.31×10^{-12} (soman), 6.17×10^{-11} M (tabun), 2.19×10^{-11} M (VX); 10 min measurement	[81]
Sarin, diisopropyl-fluorophosphate	Ferule graphite electrode modified with single-walled CNTs and nafion with AChE cross-linked with glutaraldehyde, chronoamperometry	Standard solutions	25–35 nM of sarin and 15–20 nM of diisopropylfluorophosphate (30–40 % inhibition)	[33]
Nerve gas simulant	Screen-printed carbon electrodes modified with Prussian blue and nafion with BChE cross-linked with BSA by glutaraldehyde, chronoamperometry	Gas samples	Alarm at 1 ppm	[82]
Sarin, soman, cyclosarin, tabun, VX	Cloth with inmmobilized AChE and paper with acetylthiocholine and Ellman reagent, colorimetry	Gas samples	Distinguishing pairs of chemical warfare agents by artificial neural net based on inhibition/reactivation efficiency	[83]
Sarin, soman, VX, tacrine	AChE immobilized on cellulose, indoxylacetate, visual detection (semi-quantitative estimation of color intensity)	Standard samples in 2 % isopropanol	LODs from 10 to 100 nM with 40 µL sample volume	[84]
Sarin	Screen-printed carbon electrode modified with Co phtalocyanine in microfluidic chip, AChE cross-linked with BSA by glutaraldehyde, chronoamperometry	Standard samples	LOD below 1 nM within 5 min measurement	[85]

(continued)

Table 1 (continued)

Inhibitor	Detection mode	Sample	Detection characteristics	References
Pesticides				
Paraoxon	Gold screen-printed electrode modified with cysteamine self-assembled monolayer and AChE, chronoamperometry in the presence of ferricyanide ion	Drinking water	LOD 2 ppb, upper quantification limit 40 ppb (15 min incubation period)	[86]
	Paper dipstick, indoxylacetate, visual semi-quantitative detection of color	Standard solutions	LOD 0.1 μM	[46]
	AChE-chitosan deposition followed by glutaraldehyde cross-linking, thiocholine chronoamperometric signal	Standard solutions	LOD 7.5 (25 % methanol), 100 nM (15 % acetonitrile) and 2.5 μM (100 % cyclohexane)	[31]
	Screen-printed carbon electrode modified with Ni nanoparticles and genetically modified AChE with histidine tags for chelate immobilization, thiocholine signal in chronoamperometric and cyclic voltammetry mode	Standard solutions	LOD 10^{-13}, conc range 10^{-8}–10^{-12} M, 20 min incubation	[87]
	Electroacoustic resonator with AChE immobilized to self-assembled monolayer via carbodiimide binding, frequency measurement in the presence of an inhibitor	Standard solutions	LOD 1.8×10^{-11} M	[88]
	Glassy carbon electrode modified with aminated CNTs and adsorbed AChE, thiocholine signal measured by differential pulse voltammetry	Standard solutions	LOD 0.08 nM, conc. range 0.2–1 and 1–30 nM, 10 min incubation	[89]
	Screen-printed carbon electrode modified with carbon black and BChE cross-linked with glutaraldehyde, thiocholine chronoamperometric signal	Waste waters	LOD 5 μg/L, upper quantification level 30 μg/L, 20 min incubation	[90]

(continued)

Table 1 (continued)

Inhibitor	Detection mode	Sample	Detection characteristics	References
	Bulk acoustic resonator covered with AChE by self-assembling on Au thin film, resonant frequency changes	Vapors and standard solutions	LOD 5.3×10^{-11} M, conc. range 10^{-5}–10^{-10} M, 40 min incubation	[91]
	Glassy carbon electrode covered with multi-walled CNTs—ZnO—graphene sheets and physically adsorbed AChE, thiocholine signal measured by cyclic voltammetry	Standard solutions	LOD 1 pM, conc. range 1–26 nM, incubation period not reported	[92]
	Carbon paste electrode covered with multi-walled CNTs and ACHE implemented in siloxane gel, thiocholine signal accumulated by multiple scanning the potential in flow-through conditions	Standard solutions	LOD 1.7×10^{-7} mg/mL, conc. range $2.5 \times 10^{-7} - 0.05$ mg/mL, 12 min incubation	[93]
Methyl-parathion, monocrotophos	Carbon paste electrode covered with AChE in siloxane polymer, thiocholine chronoamperometric signal	Orange, tomato, banana	LOD 0.04 (methyl parathion), 47 ppb (monocrotophos), conc. range 0.1–1.0 (methyl parathion), 90–590 ppb (monocrotophos), spiked fruits (orange, tomato, banana): recovery 73–94 %	[94]
Paraoxon, monocrotophos, dichlorvos	Pt disk electrode covered with polyacrylonitrile membrane modified with AChE immobilized in BSA by cross-linking with glutaraldehyde, flow-injection analysis of thiocholine chronoamperometric signal	Standard solutions	LOD 0.87×10^{-11} (paraoxon), 1.22×10^{-10} (dichlorvos), 1.08×10^{-11} M (monocrotophos), conc. range 1.0×10^{-7}–2.5×10^{-6} (paraoxon), 1.0×10^{-8}–1.0×10^{-5} (dichlorvos), 1.0×10^{-9}–3.0×10^{-6}, 1.0×10^{-5}–1.0×10^{-4} M (monocrotophos), 10 min incubation	[95]
Paraoxon, dichlorvos	ITO glass covered with CdSe@ZnS quantum dots, graphene and AChE, immobilized by inclusion in polyelectrolyte complex with polyethylene imine	Apple	LOD 1×10^{-14} (paraoxon), 1×10^{-12} (dichlorvos), conc. range 1×10^{-12}–1×10^{-6} (paraoxon, dichlorvos)	[96]

(continued)

Table 1 (continued)

Inhibitor	Detection mode	Sample	Detection characteristics	References
Paraoxon, parathion, chlorfenvinphos	Glassy carbon electrode covered with carboxylated multi-walled CNTs and electropolymerized 4-(2,5-di(thiophen-2-yl)-1H-pyrrol-1-yl)benzenamine, AChE immobilized by carbodiimide binding, thiocholine chronoamperometric signal	Tap water	LOD 2.46 (paraoxon), 0.542 (parathion), 4.90 ng/mL (chlorfenvinphos), conc. range 0.005–0.1 (paraoxon), 0.001 – 0,01, 0.01–7 (parathion), 0.005–0,1, 0.1–12.5 μg/mL (chlorfenvinphos), 5 min incubation	[97]
Paraoxon-methyl, chlorpyrifos	Screen-printed carbon electrode modified with CNTs and TCNQ, AChE immobilized by sol-gel method in polysiloxane matrix, chronoamperometry	Tap water	LOD 30 pM (paraoxon-methyl), 0.4 nM (chlorpyrifos), conc. range 0.1–500 nM (paraoxon-methyl), 1–100 nM (chlorpyrifos), 30 min incubation	[59]
Paraoxon, malaoxon	Glassy carbon electrode modified with single-walled CNTs and Co phtalocyanine, AChE immobilized by carbodiimide binding, thiocholine signal by chronoamperometry	Sparkling and tape water	LOD 3 (paraoxon) and 2 ppb, conc. range 5–50 (paraoxon), 2–50 ppb (malaoxon), 15 min incubation	[57]
Paraoxon, malaoxon, carbofuran, aldicarb	Glassy carbon electrode modified with carbon black, silver nanoparticles and AChE covalently attached by carbodiimide binding	Standard solution	LOD 0.1 (malaoxon), 0.05 (paraoxon), 0.1 (carbofuran), 10 nM (aldicarb); conc. range 0.4 nM–0.2 μM (malaoxon), 0.2 nM–0.2 μM (paraoxon), 0.2 nM–2.0 μM (carbofuran), 10 nM–0.20 μM (aldicarb), 10 min incubation	[38]
Malaoxon, methyl-paraoxon, carbofuran, aldicarb	Glassy carbon electrode modified with carbon black and pillar[5]arene, AChE immobilized by carbodiimide binding, thiocholine signal by chronoamperometry	Peanut, beetroot	LOD 4×10^{-12} (malaoxon), 5×10^{-9} (methyl-paraoxon), 2×10^{-11} (carbofuran), 6×10^{-10} M (aldicarb), conc. range 1×10^{-11}–1×10^{-6} (malaoxon), 1×10^{-8}–7×10^{-6} (methyl-paraoxon), 1×10^{-10}–2×10^{-6} (carbofuran), 7×10^{-9}–1×10^{-5} M (aldicarb), 10 min incubation	[62]

(continued)

Table 1 (continued)

Inhibitor	Detection mode	Sample	Detection characteristics	References
Paraoxon, chlorpyrifos-oxon, malaoxon	Screen-printed carbon electrode modified with Co phtalocyanine, FIA mode, chronoamperometry	Milk	LOD 5 pM (chlorpyrifos-oxon), 5 nM (paraoxon), 0.5 nM (malaoxon) in milk. Conc. range 5 pM–5 μM (10 min incubation)	[98]
Chlorfenvinfos	Screen-printed carbon electrode modified with Co phtalocyanine, thiocholine signal	Wheat, cabbage, apple, orange and cherry	Spiked methanol extracts containing 10^{-5} and 10^{-7} M chlorfenvinfos, recovery 78 and 93 %, respectively	[99]
Malathion	Au electrode modified with carbon nanotubes and Fe_3O_4, chronoamperometric thiocholine signal	Non-fat dry milk, whole milk	Recovery 106–109 % against malathion calibration curve in water solution, spiked samples contained 20, 40, and 80 nM of pesticide	[100]
Chlorpyrifos	Screen-printed carbon electrode covered with MnO_2, poly(dimethyldiallylammonium chloride) and choline oxidase electrostatically accumulated in polyelectrolyte layer, BChE was placed in inhibitor solution	Standard solutions	LOD 50 pM	[74]
Chlorpyrifos	Glassy carbon modified with exfoliated graphite nanoplatelets	Broccoli	LOD $1.58 \cdot 10^{-10}$ M (water solution), conc. range 10^{-8}–10^{-7} M (spiked ethyl acetate extract from broccoli), recovery against HPLC 95 %	[101]
Monocrotophos	Glassy carbon electrode covered with Au nanoparticles stabilized with poly (dimethyldiallylammonium chloride) and Prussian blue, chronoamperometry	Standard solution, garlic	LOD 0.8 pg/mL, conc. range 1–1000 pg/mL and 1–10 ng/mL (10 min incubation)	[55]

(continued)

Table 1 (continued)

Inhibitor	Detection mode	Sample	Detection characteristics	References
Paraoxon, carbofuran	Pt electrode potentiostatically covered with polypyrrole with entrapped AChE; thiocholine signal in chronoamperometric and cyclic voltammetry mode	Standard solutions	LOD 1.1 (paraoxon), 0.12 ppb (carbofuran), conc. range 0.1–12.5, 12.5–150 (paraoxon), 0.025–2, 2–60 ppb (carbofuran), incubation 60 min	[36]
Malathion, carbofuran	Glassy carbon electrode with electrodeposited Au nanoparticles, reduced graphene, β-cyclodextrin followed by Prussian blue—chitosan deposition and AChE immobilization; thiocholine signal measured by cyclic and differential pulse voltammetry	Standard solution	LOD 4.14 (malathion), 1.15 pg/mL (carbaryl), conc range 7.98–0.002 (malathion), 4.3–0.001 pg/mL	[54]
Methylparathion, carbofuran	Glassy carbon electrode modified with SnO_2 nanoparticles, carboxylated graphene and nafion. AChE immobilized by carbodiimide binding; thiocholine signal by chronoamperometry	Standard solution	LOD 5×10^{-14} (methylparathion) 5×10^{-14} M (carbofuran), conc. range 10^{-13}–10^{-10}, 10^{-10}–10^{-8} (methylparathion) and 10^{-12}–10^{-10}, 10^{-10}–10^{-8} M (carbofuran), 10 min incubation	[102]
Carbaryl	Gold electrode covered with aminated PAMAM G4 dendrimer with AChE covalently bonded by carbodiimide, cyclic voltammetry	Standard solution	LOD 0.032 μM, conc. range 1.0-9.0 μM (incubation 5 min).	[103]
	Polycrystalline Au electrode modified with mercaptopropionic acid and AChE and choline oxidase, chronoamperometry in the presence of hydroquinone	Standard solution	LOD 5.96 nM, conc. range 10–1000 nM (incubation 8 min)	[104]
	Glassy carbon electrode covered with Prussian blue—chitosan film and AChE physically adsorbed, cyclic voltammetry	Standard solutions	LOD 3 nM, conc. range 0.01–0.4 and 1.0–5.0 μM, 10 min incubation	[105]

(continued)

Table 1 (continued)

Inhibitor	Detection mode	Sample	Detection characteristics	References
	Glassy carbon electrode modified with reduced graphene oxide and multi-walled CNTs and AChE covalently attached by carbodiimide binding, thiocholine chronoamperometric signal	Cabbage	LOD 1.7 nM, conc. range 5–5000 nM, 3 min incubation	[106]
	Interdigitated Au electrode covered with chitosan and physically adsorbed AChE, impedance measurement in the presence of ferricyanide ions	Standard solutions	LOD 0.78 ng/mL, conc. range 1–100 ng/mL	[107]
Carbaryl, methomyl	Glassy carbon electrode modified with electropolymerized polyaniline—multi-walled CNTs composite and physically adsorbed AChE, thiocholine chronoamperometric signal	Standard solutions, apple, broccoli, cabbage	LOD 1.4 (carbaryl), 0.95 μM (methomyl), conc. range 10–50 (carbaryl), 5–30 μM (methomyl), recovery in spiked apple juice 94-103 %	[108]
Methomyl	Carbon paste including Co phtalocyanine covered with AChE immobilized by cross-linking with glutaraldehyde, thiocholine chronoamperomeric signal	Standard solutions, waste waters	LOD 30.4 μg/L, conc. range up to 1×10^{-5} M, 12 min incubation	[109]
Carbofuran, oxamyl, methomyl, carbaryl	Fenugreek hydrogel—agarose composite with entrapped Au nanoparticles and self-assembled AChE, optic density detection	Orange, tomato, ber, cabbage	LOD 2 (carbofuran), 21 (oxamyl), 113 (methomyl), 236 nM (carbaryl), recovery in fruit and vegetable samples 53–105 %	[110]
Other inhibitors				
Aflatoxin B1	AChE immobilized in gelatin film on the bottom of 96-well microplate, spectrophotometric detection with Ellman' reagent	Standard solutions	LOD 2.75 ppb	[111]
	AChE immobilized by physical adsorption on self-assembled monolayer of Au film of SPR chip,	Standard solutions	LOD 0.94 ng/mL	[112]

(continued)

Table 1 (continued)

Inhibitor	Detection mode	Sample	Detection characteristics	References
	Conductometric microcell with interdigitated Au film electrodes, AChE immobilized in BSA matrix by cross-linking with glutaraldehyde	Standard solutions	LOD 0.05 μg/mL, conc. range up to 5 μg/mL	[113]
	pH-Sensitive field effect transistor AChE immobilized in BSA matrix by cross-linking with glutaraldehyde	Sesame, walnut, pea	Conc. range 0.2–40 μg/mL	[114]
α-Saponine and α-chaconine	Carbon paste electrode with Co phtalocyanine and AChE and BChE immobilized in polyvinylalcohol based photocross-linkable polymer, chronoamperometric thiocholine signal	Standard solutions	LOD 30 (α-saponine), 15 nM (α-chaconine) conc. range up to 30 (α-saponine), 20 nM (α-chaconine)	[115]
Galantamine	Potentiometric measurements with choline selective sensor, acetylcholine as substrate	Drug preparations	LOD 5.4×10^{-9} M, conc. range 2×10^{-8}–1×10^{-6} M	[116]
Anti-dementia drugs	Graphite electrode coated with copolymer of 5,6-bis(octyloxy)-4,7-di(thiophen-2-yl)benzo[c][1,2,5] oxadiazole with (2-(((9H-fluoren-9-yl) methoxy) carbonylamino) acetic acid, AChE and choline oxidase co-immobilized by carbodiimide binding, chronoamperometry	Drug preparations	LOD 0.027 (donepezil), 0.559 μg/L (neostigmine)	[117]
Codeine	Screen-printed carbon electrode modified with tetrathiafulvalene, AChE immobilized in BSA matrix by cross-linking with glutaraldehyde, chronoamperometric thiocholine signal	Drug preparations	LOD 9 μM, conc. range 20–200 μM	[118]
Al(III)	Screen-printed carbon electrode modified with Au nanoparticles, AChE immobilized by carbodiimide binding. Thiocholine chronoamperometric signal	Standard solutions	LOD 2.1 μM, conc. range 3.6–30 μM	[119]

The variety of inhibitors detected does not fully correspond to the capabilities of appropriate biosensors. Thus, the use of cholinesterases from various sources for heavy metal detection was rather popular in the past decades due to attempts to create the systems for the total assessment of the sample toxicity [119–122]. Among others, Hg(II), As (III, V), Ag(I), Cu(II), Ni(II) are mainly mentioned as cholinesterase inhibitors [123]. Although the above mentioned metals exert inhibitory effect on cholinesterase on ppm-ppb level, their influence is non-selective and the detectable concentrations are higher than their limited threshold values and characteristics achieved by atomic spectroscopy. Nevertheless, the interest exists in the part related to improvement of biosensor based detection of more dangerous contaminants. Metal ions as reversible inhibitors are able to partially suppress irreversible inhibition and hence to underestimate toxic species content. In multisensory systems, the inhibition of cholinesterases can be compared with that of urease, the enzyme more sensitive toward metals and less sensitive than cholinesterases toward organophosphates [122, 124]. Alternative approach assumes the use of cholinesterase from several sources and chemometric treatment of the results [125, 126].

In addition to heavy metals, some other non-specific inhibitors have been determined with cholinesterase biosensors. Thus, surfactants affect the enzyme activity measured by conductometric sensor [122, 127] or potentiometrically [128]. Hypochlorite ions irreversibly damage AChE globule estimated by the potentiometric biosensor response [129].

The detection systems described during the past 5 years utilize the well-known principles proposed earlier. Most of the articles describe amperometric transducers with mediated thiocholine oxidation monitored either continuously or in batch regime. High reliability of the detection system as well as suppression of interfering chemisorption of the products on the electrode made it possible to develop microfluidic systems and lab-on-chip devices with electrochemical [85, 130, 131] and fluorescent [132] detection of the signal. Although the enzyme activity was found to be stable in the flow-through conditions, the sensitivity of inhibitor detection was sufficiently lower than that in batch conditions. Thus, the LOD of 420 nM of malathion [131] and 10 nM of carbaryl [132] were achieved in flow regime. This is significantly higher than the characteristics of conventional devices (see the results presented in Table 1 for comparison).

Although most cholinesterase biosensors utilize common transducers, a number of new approaches can be mentioned. Thus, the phenomenon of biometallization was proposed to amplify the AChE signal toward the substrate and inhibitors [133]. In this optical biosensor, liquid crystals of AChE are used. The thiocholine released from the substrate hydrolysis reacts with the $AuCl_4^-$ ion to form Au nanoparticles that disrupt the orientational arrangement of liquid crystals and hence changes the optical properties of the surface layer. The biosensor was tested on sub-nanomolar concentrations of organophosphate inhibitor.

In the second approach, the association and dissociation of polyelectrolyte complex between polycarboxylated polythiophene and myristoylcholine was recorded by absorbance spectroscopy in the presence of AChE [134]. The addition of myristoylcholine in the system resulted in disaggregation of the thiophene derivative whereas the hydrolysis of the additive in the presence of the enzyme returned the reaction to aggregation. The detection system allows visual detection of very small AChE activity and can be potentially applied for enzyme inhibition quantification.

The sensitivity of thiophosphate detection significantly depends on the sample pre-treatment. Thiophosphorus pesticides, e.g., malathion, parathion, etc., exert only weak reversible effect on human AChE. In the insects, such compounds are metabolically converted in highly toxic oxon derivatives which are real anticholinesterase agents. The application of thiophosphates referred to the second generation of insecticides is explained by lower acute toxicity of such chemicals against first generation pesticides (paraoxon, malaoxon, etc.) for warm-blooded organisms. For this reason, the appropriate pesticides should be first oxidized prior to their contact with the cholinesterase biosensor. This can be made by mixing the standard solution or extract with bromine [38, 62, 74] or bromosuccinimide [135]. The oxidant can be generated by electrolysis of bromides or chlorides performed in the same working cell as the inhibition of a biosensor [136]. The excess of the reagent can be easily removed by addition of formic acid or thiosulfate. In some publication devoted to cholinesterase biosensors such a stage of a pesticide 'activation' is assumed but not described in detail. This might be a reason of difference in the sensitivity of the biosensors utilizing the same enzyme and inhibitor. The electrolysis with soluble Al anode was also suggested for removal of phenolic compounds precipitated in the salt forms. This precipitation decreased their interference with cholinesterase detection of organophosphates residuals in grape must and wine [137].

As was mentioned in Sect. 2, inhibition degree is mostly used for the inhibition quantification and inhibitor determination. In semi-logarithmic plots, the appropriate graph is linear within 2–3 orders of magnitude. In many cases, the whole curve is approximated by two linear pieces which dramatically differ in their slope. The idea to increase the range of concentration determined by such mathematical treatment requires consideration in each particular case. An original approach to data processing is suggested in [138] where quantification of inhibitory effect of captan, a new fungicide, is described. The results obtained are processed using principal component analysis followed by establishment of non-linear regression models. The results show rather high accuracy of the prediction of inhibitor concentration for fully non-linear calibration plots.

Low selectivity of irreversible inhibition of cholinesterases remains a main weak point of appropriate biosensors in the detection of potential hazards. Regarding chemical warfare, extremely high toxicity of nerve gases makes it possible to avoid possible influence of other anticholinesterases by sample dilution. As to insecticides, the assessment of their mixtures is limited by some reasons related both to the measurement protocol and inhibition mechanism. Ellman' Eq. (3) shows non-linear contribution of individual inhibitors in resulting decay of enzymatic reaction rate. The difference in k_{II} values of inhibitors does not allow quantifying their content by a single biosensor. To some extent, distinguishing the signal can be reached by

39. Marinov I, Gabrovska K, Velichkova J, Godjevargova T (2009) Immobilization of acetylcholinesterase on nanostructure polyacrylonitrile membranes. Int J Biol Macromol 44:338–345
40. Pedrosa VA, Caetano J, Machado SAS, Freire RS, Bertotti M (2007) Acetylcholinesterase immobilization on 3-mercaptopropionic acid self assembled monolayer for determination of pesticides. Electroanalysis 19:1415–1419
41. Ivanov Y, Marinov I, Gabrovska K, Dimcheva N, Godjevargova T (2010) Amperometric biosensor based on a site-specific immobilization of acetylcholinesterase via affinity bonds on a nanostructured polymer membrane with integrated multiwall carbon nanotubes. J Mol Catal B 63:141–148
42. Bucur B, Danet AF, Marty J-L (2005) Cholinesterase immobilisation on the surface of screen-printed electrodes based on concanavalin a affinity. Anal Chim Acta 530:1–6
43. Istamboulie G, Andreescu S, Marty J-L, Noguer T (2007) Highly sensitive detection of organophosphorus insecticides using magnetic microbeads and genetically engineered acetylcholinesterase. Biosens Bioelectron 23:506–512
44. Pohanka M (2015) Determination of acetylcholinesterase and butyrylcholinesterase activity without dilution of biological samples. Chem Pap 69:1044–1049
45. Pohanka M (2015) Photography by cameras integrated in smartphones as a tool for analytical chemistry represented by an butyrylcholinesterase activity assay. Sensors 15:13752–13762
46. Pohanka M (2012) Acetylcholinesterase based dipsticks with indoxylacetate as a substrate for assay of organophosphates and carbamates. Anal Lett 45:367–374
47. Pohanka M, Drtinova L (2013) Spectrophotometric methods based on 2,6-dichloro-indophenol acetate and indoxylacetate for butyrylcholinesterase activity assay in plasma. Talanta 106:281–285
48. Hossain SMZ, Luckham RE, McFadden MJ, Brennan JD (2009) Reagentless bidirectional lateral flow bioactive paper sensors for detection of pesticides in beverage and food samples. Anal Chem 81:9055–9064
49. Ellman GL, Courtney KD, Andres V, Featherstone RM (1961) A new and rapid colorimetric determination of acetylcholinesterase activity. Biochem Pharmac 7:88–95
50. Sabelle S, Renard P-Y, Pecorella K, de Suzzoni-Dézard S, Créminon C, Grassi J, Mioskowski C (2002) Design and synthesis of chemiluminescent probes for the detection of cholinesterase activity. J Am Chem Soc 124:4874–4880
51. Rajan CS, Gupta BD (2007) Surface plasmon resonance based fiber-optic sensor for the detection of pesticide. Sens Actuators B 123:661–666
52. Liron Z, Zifman A, Heleg-Shabtai V (2011) Surface-enhanced Raman scattering detection of cholinesterase inhibitors. Anal Chim Acta 703:234–238
53. Guilbault GG, Kramer DN, Goldberg P (1963) The application of modified Nernstian equations to the electrochemical determination of enzyme kinetics. J Phys Chem 67:1747–1749
54. Zhao H, Ji X, Wang B, Wang N, Li X, Ni R, Ren J (2015) An ultra-sensitive acetylcholinesterase biosensor based on reduced graphene oxide-Au nanoparticles-β-cyclodextrin/Prussian blue-chitosan nanocomposites for organophosphorus pesticides detection. Biosens Bioelectron 65:23–30
55. Wu S, Lan X, Zhao W, Li Y, Zhang L, Wang H, Han M, Tao S (2011) Controlled immobilization of acetylcholinesterase on improved hydrophobic gold nanoparticle/Prussian blue modified surface for ultra-trace organophosphate pesticide detection. Biosens Bioelectron 27:82–87
56. Song Y, Zhang M, Wang L, Wan L, Xiao X, Ye S, Wang J (2011) A novel biosensor based on acetylecholinesterase/prussian blue–chitosan modified electrode for detection of carbaryl pesticides. Electrochim Acta 56:7267–7271
57. Ivanov AN, Younusov RR, Evtugyn GA, Arduini F, Moscone D, Palleschi G (2011) Acetylcholinesterase biosensor based on single-walled carbon nanotubes—Co phtalocyanine for organophosphorus pesticides detection. Talanta 85:216–221

58. Zamfir L-G, Rotariu L, Bala C (2013) Acetylcholinesterase biosensor for carbamate drugs based on tetrathiafulvalene–tetracyanoquinodimethane/ionic liquid conductive gels. Biosens Bioelectrons 46:61–67
59. Rotariu R, Zamfir L-G, Bala C (2012) A rational design of the multiwalled carbon nanotube–7,7,8,8-tetracyanoquinodimethan sensor for sensitive detection of acetylcholinesterase inhibitors. Anal Chim Acta 748:81–88
60. Zhang Y, Liu H, Yang Z, Ji S, Wang J, Pang P, Feng L, Wang H, Wu Z, Yang W (2015) An acetylcholinesterase inhibition biosensor based on a reduced graphene oxide/silver nanocluster/chitosan nanocomposite for detection of organophosphorus pesticides. Anal Methods 7:6213–6219
61. Huang X, Du D, Gong X, Cai J, Tu H, Xu X, Zhang A (2008) Composite assembly of silver nanoparticles with avidin and biotinylated AChE on gold for the pesticidal electrochemical sensing. Electroanalysis 20:402–409
62. Shamagsumova RV, Shurpik DN, Padnya PL, Stoikov II, Evtugyn GA (2015) Acetylcholinesterase biosensor for inhibitor measurements based on glassy carbon electrode modified with carbon black and pillar[5]arene. Talanta 144:559–568
63. Crochet KL, Montalvo JG (1973) Enzyme electrode system for assay of serum cholinesterase. Anal Chim Acta 66:259–269
64. Ivanov AN, Evtugyn GA, Lukachova LV, Karyakina EE, Budnikov HC, Kiseleva SG, Orlov AV, Karpacheva GP, Karyakin AA (2003) New polyaniline-based potentiometric biosensor for pesticides detection. IEEE J 3:333–340
65. Snejdarkova M, Svobodova L, Evtugyn G, Budnikov H, Karyakin A, Nikolelis DP, Hianik T (2004) Acetylcholinesterase sensors based on gold electrodes modified with dendrimer and polyaniline: a comparative research. Anal Chim Acta 514:79–88
66. Cuartero M, Ortuño JÁ, García MS, García-Cánovas F (2012) Assay of acetylcholinesterase activity by potentiometric monitoring of acetylcholine. Anal Biochem 421:208–212
67. Cuartero M, García MS, García-Cánovas F, Ortuño JÁ (2012) New approach for the potentiometric-enzymatic assay of reversible-competitive enzyme inhibitors. Application to acetylcholinesterase inhibitor galantamine and its determination in pharmaceuticals and human urine. Talanta 110:8–14
68. Khaled E, Hassan HNA, Mohamed GG, Ragab FA, Seleim AEA (2010) Disposable potentiometric sensors for monitoring cholinesterase activity. Talanta 83:357–363
69. Hai A, Ben-Haim D, Korbakov N, Cohen A, Shappir J, Oren R, Spira ME, Yitzchaik S (2006) Acetylcholinesterase–ISFET based system for the detection of acetylcholine and acetylcholinesterase inhibitors. Biosens Bioelectron 22:605–612
70. Soldatkin AP, Arkhypova VN, Dzyadevych SV, El'skaya AV, Gravoueille J-M, Jaffrezic-Renault N, Martelet C (2005) Analysis of the potato glycoalkaloids by using of enzyme biosensor based on pH-ISFETs. Talanta 65:28–33
71. Tekaya N, Saiapina O, Ouada HB, Lagarde F, Ouada HB, Jaffrezic-Renault N (2013) Ultra-sensitive conductometric detection of pesticides based on inhibition of esterase activity in *Arthrospira platensis*. Environ Pollut 178:182–188
72. Dzyadevych SV, Soldatkin AP, Arkhypova VN, El'skaya AV, Chovelon J-M, Georgiou CA, Martelet C, Jaffrezic-Renault N (2005) Early-warning electrochemical biosensor system for environmental monitoring based on enzyme inhibition. Sens Actuators B 105:81–87
73. Upadhyay S, Rao GR, Sharma MK, Bhattacharya BK, Rao VK, Vijayaraghavan R (2009) Immobilization of acetylcholinesterase-choline oxidase on a gold-platinum bimetallic nanoparticles modified glassy carbon electrode for the sensitive detection of organophosphate pesticides, carbamates and nerve agents. Biosens Bioelectron 25:832–838
74. Dontsova EA, Zeifman YS, Budashov IA, Eremenko AV, Kalnov SV, Kurochkin IN (2011) Screen-printed carbon electrode for choline based on MnO_2 nanoparticles and choline oxidase/polyelectrolyte layers. Sens Actuators B 159:261–270
75. Ciucu AA, Negulescu C, Baldwin RP (2003) Detection of pesticides using an amperometric biosensor based on ferophthalocyanine chemically modified carbon paste electrode and immobilized bienzymatic system. Biosens Bioelectron 18:303–310

76. Espinosa M, Atanasov P, Wilkins E (1999) Development of a disposable organophosphate biosensor. Electroanalysis 11:1055–1062
77. Ghindilis AL, Morzunova TG, Barmin AV, Kurochkin IN (1996) Potentiometric biosensors for cholinesterase inhibitor analysis based on mediatorless bioelectrocatalysis. Biosens Bioelectron 11:873–880
78. Ricci F, Amine A, Palleschi G, Moscone D (2003) Prussian Blue based screen printed biosensors with improved characteristics of long-term lifetime and pH stability. Biosens Bioelectron 18:165–174
79. Zheng Z, Li X, Dai Z, Liu S, Tang Z (2011) Detection of mixed organophosphorus pesticides in real samples using quantum dots/bi-enzyme assembly multilayers. J Mater Chem 21:16955–16962
80. Kumar P, Kim K-H, Deep A (2015) Recent advancements in sensing techniques based on functional materials for organophosphate pesticides. Biosens Bioelectron 70:469–481
81. Pohanka M, Adam V, Kizek R (2013) An acetylcholinesterase-based chronoamperometric biosensor for fast and reliable assay of nerve agents. Sensors 13:11498–11506
82. Arduini F, Neagu D, Dall'Oglio S, Moscone D, Palleschi G (2011) Towards a portable prototype based on electrochemical cholinesterase biosensor to be assembled to soldier overall for nerve agent detection. Electroanalysis 24:581–590
83. Hoskovcová M, Dubina P, Halámek E, Kobliha Z (2011) Identification of pairs of organophosphorus warfare agents through cholinesterase reaction. Anal Lett 44:2521–2529
84. Pohanka M, Vlcek V (2014) Preparation and performance of a colorimetric biosensor using acetylcholinesterase and indoxylacetate for assay of nerve agents and drugs. Interdisc Toxicol 7:215–218
85. Tan HY, Loke WK, Nguyen N-T, Tan SN, Tay NB, Wang W, Ng SH (2014) Lab-on-a-chip for rapid electrochemical detection of nerve agent Sarin. Biomed Microdevices 16:269–275
86. Arduini F, Guidone S, Amine A, Palleschi G, Moscone D (2013) Acetylcholinesterase biosensor based on self-assembled monolayer-modified gold-screen printed electrodes for organophosphorus insecticide detection. Sens Actuators B 179:201–208
87. Ganesana M, Istarnboulie G, Marty J-L, Noguer T, Andreescu S (2011) Site-specific immobilization of a $(His)_6$-tagged acetylcholinesterase on nickel nanoparticles for highly sensitive toxicity biosensors. Biosens Bioelectron 30:43–48
88. Chen D, Wang J, Xu Y, Zhang L (2012) A thin film electro-acoustic enzyme biosensor allowing the detection of trace organophosphorus pesticides. Anal Biochem 429:42–44
89. Yu G, Wu W, Zhao Q, WeiX L (2015) Efficient immobilization of acetylcholinesterase onto aminofunctionalized carbon nanotubes for the fabrication of high sensitive organophosphorus pesticides biosensors. Biosens Bioelectron 68:288–294
90. Arduini F, Forchielli M, Amine A, Neagu D, Cacciotti I, Nanni F, Moscone D, Palleschi G (2015) Screen-printed biosensor modified with carbon black nanoparticles for the determination of paraoxon based on the inhibition of butyrylcholinesterase. Microchim Acta 182:643–651
91. Chen D, Wang J, Xu Y, Li D, Zhang L, Li Z (2013) Highly sensitive detection of organophosphorus pesticides by acetylcholinesterase-coated thin film bulk acoustic resonator mass-loading sensor. Biosens Bioelectron 41:163–167
92. Nayak P, Anbarasan B, Ramaprabhu S (2013) Fabrication of organophosphorus biosensor using ZnO nanoparticle-decorated carbon nanotube-graphene hybrid composite prepared by a novel green technique. J Phys Chem C 117:13202–13209
93. Ebrahimi B, Shojaosadati SA, Daneshgar P, Norouzi P, Mousavi SM (2011) Performance evaluation of fast Fourier-transform continuous cyclic-voltammetry pesticide biosensor. Anal Chim Acta 687:168–176
94. Raghu P, Reddy TM, Swamy BEK, Chandrashekar BN, Reddaiah K, Sreedhar M (2012) Development of AChE biosensor for the determination of methyl parathion and monocrotophos in water and fruit samples: A cyclic voltammetric study. J Electroanal Chem 665:76–82

95. Marinov I, Ivanov Y, Vassileva N, Godjevargova T (2011) Amperometric inhibition-based detection of organophosphorus pesticides in unary and binary mixtures employing flow-injection analysis. Sens Actuators B 160:1098–1105
96. Li X, Zheng Z, Liu X, Zhao S, Liu S (2015) Nanostructured photoelectrochemical biosensor for highly sensitive detection of organophosphorous pesticides. Biosens Bioelectron 64:1–5
97. Kesik M, Kanik FE, Turan J, Kolb M, Timur S, Bahadir M, Toppare L (2014) An acetylcholinesterase biosensor based on a conducting polymer using multiwalled carbon nanotubes for amperometric detection of organophosphorous pesticides. Sens Actuators B 205:39–49
98. Mishra RK, Dominguez RB, Bhand S, Muñoz R, Marty J-L (2012) A novel automated flow-based biosensor for the determination of organophosphate pesticides in milk. Biosens Bioelectron 32:56–61
99. Crew A, Lonsdale D, Byrd N, Pittson R, Hart JP (2011) A screen-printed, amperometric biosensor array incorporated into a novel automated system for the simultaneous determination of organophosphate pesticides. Biosens Bioelectron 26:2847–2851
100. Chauhan N, Pundir CS (2011) An amperometric biosensor based on acetylcholinesterase immobilized onto iron oxide nanoparticles/multi-walled carbon nanotubes modified gold electrode for measurement of organophosphorus insecticides. Anal Chim Acta 701:66–74
101. Ion I, Ion AC (2012) Determination of chlorpyriphos in broccoli using a voltammetric acetylcholinesterase sensor based on carbon nanostructure-chitosan composite material. Mater Sci Eng C 32:1001–1004
102. Zhou Q, Yang L, Wang G, Yang Y (2013) Acetylcholinesterase biosensor based on SnO_2 nanoparticles-carboxylic graphene-nafion modified electrode for detection of pesticides. Biosens Bioelectron 49:25–31
103. Santos CS, Mossanha R, Pessôa SA (2015) Biosensor for carbaryl based on gold modified with PAMAM-G4 dendrimer. J Appl Electrochem 45:325–334
104. Hatefi-Mehrjardi A (2013) Bienzyme self-assembled monolayer on gold electrode: an amperometric biosensor for carbaryl determination. Electrochim Acta 114:394–402
105. Song Y, Zhang M, Wang L, Wan L, Xiao X, Ye S, Wang J (2011) A novel biosensor based on acetylecholinesterase/prussian blue-chitosan modified electrode for detection of carbaryl pesticides. Electrochim Acta 56:7267–7271
106. Liu Q, Fei A, Huan J, Mao H, Wang K (2015) Effective amperometric biosensor for carbaryl detection based on covalent immobilization acetylcholinesterase on multiwall carbon nanotubes/graphene oxide nanoribbons nanostructure. J Electroanal Chem 740:8–13
107. Gong Z, Guo Y, Sun X, Cao Y, Wang X (2014) Acetylcholinesterase biosensor for carbaryl detection based on interdigitated array microelectrodes. Bioprocess Biosyst Eng 37:1929–1934
108. Cesarino I, Moraes FC, Lanza MRV, Machado SAS (2012) Electrochemical detection of carbamate pesticides in fruit and vegetables with a biosensor based on acetylcholinesterase immobilised on a composite of polyaniline-carbon nanotubes. Food Chem 135:873–879
109. Caetano J, Dragunski DC, Pedrosa VA, Machado SAS (2013) Quantification of methomyl levels in vabbage, tomato, and soya milk using a renewable amperometric biosensor. Int J Electrochem Sci 8:7795–7805
110. Kestwal RM, Bagal-Kestwal D, Chiang B-H (2015) Fenugreek hydrogeleagarose composite entrapped gold nanoparticles for acetylcholinesterase based biosensor for carbamates detection. Anal Chim Acta 886:143–150
111. Pohanka M (2013) Spectrophotomeric assay of aflatoxin B1 using acetylcholinesterase immobilized on standard microplates. Anal Lett 46:1306–1315
112. Puiu M, Istrate O, Rotariu L, Bala C (2012) Kinetic approach of aflatoxin B1-acetylcholinesterase interaction: A tool for developing surface plasmon resonance biosensors. Anal Biochem 421:587–594
113. Soldatkin O, Burdak OS, Sergeyeva TA, Arkhypova VM, Dzyadevych SV, Soldatkin AP (2013) Acetylcholinesterase-based conductometric biosensor for determination of aflatoxin B1. Sens Actuators B 188:999–1003

114. Stepurska KV, Soldatkin OO, Arkhypova VM, Soldatkin AP, Lagarde F, Jaffrezic-Renault N, Dzyadevych SV (2015) Development of novel enzyme potentiometric biosensor based on pH-sensitive field-effect transistors for aflatoxin B1 analysis in real samples. Talanta 144:1079–1084
115. Espinoza MA, Istamboulie G, Chira A, Noguer T, Stoytcheva M, Marty J-L (2014) Detection of glycoalkaloids using disposable biosensors based on genetically modified enzymes. Anal Biochem 457:85–90
116. Cuartero M, García MS, García-Cánovas F, Ortuño JÁ (2013) New approach for the potentiometric-enzymatic assay of reversible-competitive enzyme inhibitors. Application to acetylcholinesterase inhibitor galantamine and itsd etermination in pharmaceuticals and human urine. Talanta 110:8–14
117. Turan J, Kesik M, Soylemez S, Goker S, Kolb M, Bahadir M, Toppare L (2014) Development of an amperometric biosensor based on a novel conducting copolymer for detection of anti-dementia drugs. J Electroanal Chem 735:43–50
118. Asturias-Arribas L, Alonso-Lomillo MA, Domínguez-Renedo O, Arcos-Martínez MJ (2013) Screen-printed biosensor based on the inhibition of the acetylcholinesterase activity for the determination of codeine. Talanta 111:8–12
119. Soldatkin OO, Pavluchenko OS, Kukla OL, Kucherenko IS, Peshkova VM, Arkhypova VM, Dzyadevych SV, Soldatkin AP, El'skaya AV (2009) Application of enzyme multibiosensor for toxicity analysis of real water samples of different origin. Biopolym Cell 25:204–209
120. Evtugyn GA, Rizaeva EP, Stoikova EE, Latipova VZ, Budnikov HC (1997) The application of cholinesterase potentiometric biosensor for preliminary screening of the toxicity of waste waters. Electroanalysis 9:1124–1128
121. Czolkos I, Dock E, Tønning E, Christensen J, Winther-Nielsen M, Carlsson C, Mojzíková R, Skládal P, Wollenberger U, Nørgaard L, Ruzgas T, Emnéus J (2016) Prediction of wastewater quality using amperometric bioelectronic tongues. Biosens Bioelectron 75:375–382
122. Stepurska KV, Soldatkin OO, Kucherenko IS, Arkhypova VM, Dzyadevych SV, Soldatkin AP (2015) Feasibility of application of conductometric biosensor based on acetylcholinesterase for the inhibitory analysis of toxic compounds of different nature. Anal Chim Acta 854:161–168
123. Domínguez-Renedo O, Alonso-Lomillo MA, Arcos-Martínez MJ (2013) Determination of metals based on electrochemical biosensors. Crit Rev Environ Sci Technol 43:1042–1073
124. Volotovskky V, Kim N (2003) Ion-sensitive field effect transistor-based multienzyme sensor for alternative detection of mercury ions, cyanide, and pesticide. J Microbiol Biotechnol 13:373–377
125. Danzer T, Schwedt G (1996) Multivariate evaluation of inhibition studies on an enzyme electrodes system with pesticides and mixtures of mercury and pesticides. Anal Chim Acta 325:1–10
126. Danzer T, Schwedt G (1996) Chemometric methods for the development of a biosensor system and the evaluation of inhibition studies with solutions and mixtures of pesticides and heavy metals. Part 1. Development of an enzyme electrodes system for pesticide and heavy metal screening using selected chemometric methods. Anal Chim Acta 318:275–286
127. Kucherenko IS, Soldatkin OO, Arkhypova VM, Dzyadevych SV, Soldatkin AP (2012) A novel biosensor method for surfactant determination based on acetylcholinesterase inhibition. Measur Sci Technol 23:065801
128. Evtugyn GA, Budnikov HC, Nikolskaya EB (1996) Influence of surface-active compounds on the response and sensitivity of cholinesterase biosensors for inhibitor determination. Analyst 121:1911–1915
129. Soldatkin AP, Gorchkov DV, Martelet C, Jaffrezic-Renault N (1997) New enzyme potentiometric sensor for hypochlorite species detection. Sens Actuators B 43:99–101
130. Yoon Y-J, Li KHH, Low JZ, Yoon J, Ng SH (2014) Microfluidics biosensor chip with integrated screen-printed electrodes for amperometric detection of nerve agent. Sens Actuators B 198:233–238

131. Wang J, Satake T, Suzuki H (2015) Microfluidic device for coulometric detection of organophosphate pesticides. Anal Sci 31:591–595
132. Hadd AG, Jacobson SC, Ramsey JM (1999) Microfluidic assays of acetylcholinesterase inhibitors. Anal Chem 71:5206–5212
133. Liao S, Qiao Y, Han W, Xie Z, Wu Z, Shen G, Yu R (2012) Acetylcholinesterase liquid crystal biosensor based on modulated growth of gold nanoparticles for amplified detection of acetylcholine and inhibitor. Anal Chem 84:45–49
134. Li Y, Bai H, Li C, Shi G (2011) Colorimetric assays for acetylcholinesterase activity and inhibitor screening based on the disassembly-assembly of a water-soluble polythiophene derivative. ACS Appl Mater Interfaces 3:1306–1310
135. Del Carlo M, Pepe A, Sergi M, Mascini M, Tarentini A, Compagnone D (2010) Detection of coumaphos in honey using a screening method based on an electrochemical acetylcholinesterase bioassay. Talanta 81:76–81
136. Evtyugin GA, Stoikova EE, Iskanderov RR, Nikol'skaya EB, Budnikov GK (1997) Electrochemical sample preparation for the enzymatic determination of cholinesterase inhibitors. J Anal Chem 52:2–5
137. Ivanov A, Evtugyn G, Budnikov H, Ricci F, Moscone D, Palleschi G (2003) Cholinesterase sensors based on screen-printed electrodes for detection of organophosphorus and carbamic pesticides. Anal Bioanal Chem 377:624–631
138. Nesakumar N, Sethuraman S, Krishnan UM (2015) Rayappan JBB (2015) Cyclic voltammetric acetylcholinesterase biosensor for the detection of captan in apple samples with the aid of chemometrics. Anal Bioanal Chem 407:4863–4868
139. Bachmann TT, Schmid RD (1999) A disposable multielectrode biosensor for rapid simultaneous detection of the insecticides paraoxon and carbofuran at high resolution. Anal Chim Acta 401:95–103
140. Bachmann TT, Leca B, Vilatte F, Marty J-L, Fournier D, Schmid RD (2000) Improved multianalyte detection of organophosphates and carbamates with disposable multielectrode biosensors using recombinant mutants of *Drosophila* acetylcholinesterase and artificial neural networks. Biosens Bioelectron 15:193–201
141. Rhouati A, Istamboulie G, Cortina-Puig M, Marty J-L, Noguer T (2010) Selective spectrophotometric detection of insecticides using cholinesterases, phosphotriesterase and chemometric analysis. Enzyme Microb Technol 46:212–216
142. Mwila K, Burton MH, Van Dyk JS, Pletschke BI (2013) The effect of mixtures of organophosphate and carbamate pesticides on acetylcholinesterase and application of chemometrics to identify pesticides in mixtures. Environ Monit Assess 185:2315–2327
143. Mishra RK, Alonso GA, Istamboulie G, Bhand S, Marty J-L (2015) Automated flow based biosensor for quantification of binary organophosphates mixture in milk using artificial neural network. Sens Actuators B 208:228–237

Efficiency of Non-label Optical Biosensors for the Express Control of Toxic Agents in Food

Nickolaj F. Starodub and Nelja F. Shpirka

Abstract This chapter is devoted to the analysis of the efficiency of a different types of the immune biosensors for the control of toxic agents among of environmental objects, The main attention is paid to the non-labeled immune biosensors and, in particular, optical ones. Among them the immune biosensors based on the porous silicon (PS), surface plasmon resonance (SPR) and total reflection internal ellipsometry (TIRE) are detailed considered. In additional to, the immune biosensors based on calorimeter and thermistors as well as on the piezocristals are described. As model of toxic elements the synthetic chemicals as pesticides and nonylethoxylates as well as the nature biological substances, in particularly, a number of mycotoxins: T2, aflatoxins, patulin and others are used. It is necessary, to underline that the analysis was fulfilled with the model solutions and with real samples: some corn, vegetables and fruits. At the end of the chapter the perspectives of the developed instrumental analytical devices based on the principles of biosensorics are analyzed. Especially it draws attention to the development of multi-parameter portable biosensors based on the basis of the nano-structured porous silicon (nano-PS) from one side and on the artificial selective template surface, calyx(4)arenas and aptamers.

Keywords Toxins · Environmental objects · Control · Immune biosensors · Optical · Non-labeled

1 Introduction

Scientific-technical progress gives a lot of new possibilities to sharply increase the life of people but it brings, unfortunately, simultaneous a number of difficulties for human which should be overcome with the including of the specific measures for

---

N.F. Starodub (✉) · N.F. Shpirka
Laboratory of Biosensorics, National University of Life and Environmental Sciences of Ukraine, 5, Geroyiv Oborony, Kyiv 03041, Ukraine
e-mail: nikstarodub@yahoo.com

D.P. Nikolelis and G.-P. Nikoleli (eds.), *Biosensors for Security and Bioterrorism Applications*, Advanced Sciences and Technologies for Security Applications,
DOI 10.1007/978-3-319-28926-7_18

the adaptation of the living organisms to the new conditions. At first it means the application of very wide specters of chemical substances. According to classification from the position of ecology and sources of manufacturing the chemical substances (ChS) may be divided on: pesticides, products of industry (polychlorinated biphenyl's, hexachlorbensens) and concomitant substances (dioxins, furans and polycyclic hydrocarbons) [101]. The production of similar substances in USA only has achieved in 1997 $\approx 5 \times 10^9$ tons [52]. In spite of the fact that controlling units in all countries are recommend using biodegradable surfactants but their mineralization can not be complete and they accumulate gradually in natural water sources [18]. Acute effects of ChS are well known and the cases of their mass use in big concentrations as results of disasters are described in details [11, 24]. Chronic sub-lethal effects of ChS are associated with the worsening reproductive performance, influence on skin, immune, neural and endocrine systems [8, 15, 49]. ChS may be absorbed by phytoplankton, filtrated organisms and vegetables [25, 40, 98].

Pesticides are as the essential part of the modern agro technologies. Due to their use we have possibility to keep 50 % of the crop capacity. However, in consequence of high toxicity of pesticides there is needed a complex system to eliminate their negative effect and first of all it concerns the constant control of these substances in different classes of environmental objects.

Other aspect of the scientific-technical progress is in the intensification of the production of ChS and their application as detergents in different fields of the human activity and especially in the domestic conditions. As a rule, such ChS has not direct toxic effect but after biodegradation some substances are formed from it which may be delivered to so-called "endocrine disrupting factor". Nonylphenol ethoxylates (nPhEthO) are as one group of such factors [79, 80].

The next third of ChS belongs to biological derivatives which naturally are widespread among products and, moreover, may be used in purified form as instrument for attack of bioterrorist [23, 85]. The problem with weapons including bio- and chemical substances exists a long time. One of group of biological toxins is presented by mycotoxins which include more than 300 individual substances produced by different fungi strains [47]. T2 mycotoxin, aflatoxins, searelenone, patulin and others cause a great interest since they are widespread and characterized by high level toxicity. T2 mycotoxin has more toxic effect (in 400 times) than mustard gas and lewisite [47, 68]. It is well-known that mycotoxin T2 was packed into rockets, bombs, cisterns, some explosive cylinders, hand-grenade and they were applied in Laos and Afghanistan as yellow rain [51, 68]. This mycotoxin may be prepared by very simple way. Both circumstances (simplicity obtaining and high level of toxicity) form a very serious problems since these toxic elements may be as instrument for bioterrorists. There is necessity to mention that the use of toxins in generally among others groups of biological weapon by terrorists is the most probably since viruses and bacteria present big danger not for victims only but for executors of terrorist act too.

The list of ChS which are produced and widely used may be continued and continued but at last we would like to pay attention nano-particles. In recent decades, it was created several thousands of varieties of nanomaterials, and therefore the possibility of exposure nanoparticles on animals, humans and the environment in whole is increased [63]. A number anthropogenic sources, such as metallurgical, cement industry; combustion of coal, polymeric compounds, oil, gas, diesel fuel, and other processes have significantly increased contents of nanoparticles in the environment [70]. It is generally recognized that changes in the physical properties of the material at the transition to form nanoparticles naturally accompanied by changes in its biological effects. In particular, a substantial their accumulation, in the lung, penetration to the tissue, overcoming skin barrier, ability to have the so-called "inflammatory potential" and to interact with different biological molecules, including nucleic acids as carrier of genetic information [27, 96]. The level of hydrophobic properties and the presence of electrical charge increase nanoparticles to binding with biomolecules and to their accumulation and in organisms since the immune system often is not able to recognize presence of such complexes [31].

Despite the dramatic increase in the use of nano sized materials, little information is available on their potential toxic effects on the environment. Their potential deleterious effects on ecological health should be identified to allow their safe use. Most current literature on the toxicity of nanoparticles come from mammalian studies that focus on respiratory exposure or from in vitro assays with mammalian cells [6]. In last time it was taken the attention to the estimation of nano-particles on the reproductive function of animals which is under effect of these substances [9, 33]. The ecotoxicological studies of nanoparticles are much more limited, with only a few reports focusing on the acute toxic effects of nanoparticles on the aquatic biotas [32].

Especially there is necessary to underline that for the an effective as well as safe use of nano-particles and different composites with their participation the detailed and comprehensive analysis should be done not only concerning general toxicity of these substances and in respect of their genotoxicity since the last effect may be reflected in form of cancerogenesis for living organisms and more significantly in arousing genetically mutations in next generations. That is why there is a very important to control of the level of genotoxicity of nanomaterials.

Taken into account the global dispersion of ChS and extraordinary dangers which they present for living organisms there arises necessity to resist imminent catastrophe. In really such situation is formed when a large scale of production and consumption of ChS is exist.

Now, there is a very important question: how to prevent non-desirable consequences, to preserve live and health of people. To counteract this undesirable situation it arises a urgent necessity in the organization of the permanent control since, as a rule, the ChS are toxic, stable, and have the ability for bioaccumulation and long term transferring in atmosphere [101]. These substances may influence on the living organisms far from places of their production and on remote territories. So, discovering chemicals in Arctic where they are never been produced and where their concentration are increased up to level which is threatening for wild nature and humanities is confirming the global character of their dispersion [4, 54].

In all cases there is necessary to have instrumental analytical devices for the express, sensitive and simple monitoring of the contamination of the environmental objects by the concrete toxins.

Certainly, to prevent non-desirable effect and to do effective this monitoring we developed some family of biosensors for the fulfillment of control on the third levels. At first it is foreseen the determination of the general toxicity of the analyzed objects, than the revealing what is group toxins presented in the contaminated material and last the discovering of the individual chemical substance. Unfortunately, the existing conventional methods of analysis often do not satisfactorily solve these practical needs, since they are sophisticated, time-consuming, expensive and difficult to be applied under on-line regime and on-field conditions. There is thus a specific need to use a new generation of instrumental analytical devices, in particular, based on the principles of biosensorics. Nevertheless, among them, some developed variants do not meet the requirements of practice due to their complexity, high prices and others disadvantages or to lack of adaptability in the analysis of required indicators.

To overcome this gap between biosensor development and practical demand, we have started studying a number of optical and electrochemical biosensors capable to fulfill analysis under real time and on-field conditions as well as to check simultaneously a large number of samples and analytical parameters. This article is devoted to analysis of biosensors which were created by us. Among the different types of proposed biosensors, we have paid special attention to portable optical systems suitable for working under on-field conditions and obtaining results in on-line regime without the application of any additional labels, which are used, as a rule, in these cases. Recently, it has been demonstrated that achieving such goals is possible by applying nano-structured porous silicon as a transducer surface with the registration of the specific signal through measuring photoluminescence or photoconductivity. Such biosensors are very effective for screening observation. We pay attention too that the results of analysis (especially mycotoxins) depend on the conditions of the sample preparation.

2 Creation of Non-labelled Immune Biosensors for the Determination of Low Weight Toxic Synthetic Substances in Environment [79, 80]

In this section we would like to demonstrate efficiency of the immune biosensor based on the porous silicon (PS) at the determination of such low weight toxins as nonylphenol (NP) and 2,4-dichlorophenoxyacetic acid (2,4-D). Moreover, it will compared the obtained results with that which were demonstrated by the immune sensor based on the surface plasmon resonance (SPR).

2.1 *The Overall Characteristics of PS Obtaining and Specific Signal Registration*

To prepare PS samples the monocrystalline (MK) silicon plate (*p*-type with the specific resistance (ρ) = 10 Ohm/centimeter) was treated at first by computer controlled and in-focus beam from YAG: Nd (yttrium-aluminum garnet, alloyed Nd^{3+}) laser (wave-length $\lambda = 1.06$ μm, pulse duration $\tau = 2 \times 10^{-4}$ s and energy E = 0.3 J) in the regime of surface scanning. Then this plate was immersed in the mixture of HF: HNO_3: H_2O = 1:3:5 (in volume ration) for 3–10 min. At last the obtained plates of PS was cut into sections with the dimensions in 4 × 4 mm^2. The total scheme of PS preparation is given in Fig. 1. The dimensions of pores (diameter and depth) were controlled by atom force microscopy (AFM).

To observe visible PhL the PS plate was placed in quartz cell with the volume of 2 ml and was exposed to He-Cd laser beam (wave-length λ = 440 nm, power = 0.001 Watt. The intensity of PhL was measured with the use of the set included standard monochromator, photoelectric multiplayer (FEU-83) and personal computer (PC). The overall scheme of excitation and measurement of intensity of PS PhL is shown in Fig. 2. Before immobilization of biological components the surface of PS was washed by ethanol and several times by distilled water and then PS samples were dried at the room temperature in laminar-cleaned conditions. Immobilization of biological components was accomplished by passive sorption. For this purpose PS samples were dipped in the solution of one of immune component, which was taken in the appropriate concentration and they were incubated during some time (the concentration of solutions and time of immobilization were determined by experimentally way). After immobilization of biological components PS surface was washed by the phosphate buffer (20 mM, pH 7.3) contained NaCl (140 mmol/l) and then it was dried.

The observed depth of pores was in range 10–200 nm and their diameter (d) and width of rising (silicon crystallite) was from 50 to 400 nm. In concordance with the classification of PS on the micro- (d < 2 nm), meso- (2 nm < d < 50 nm) and macro-

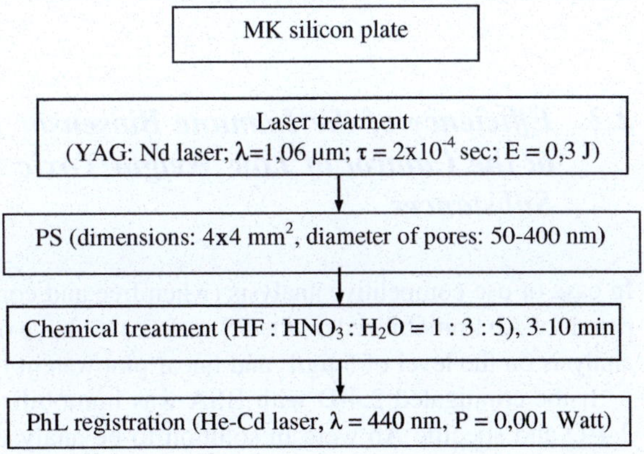

Fig. 1 The overall scheme of PS preparation

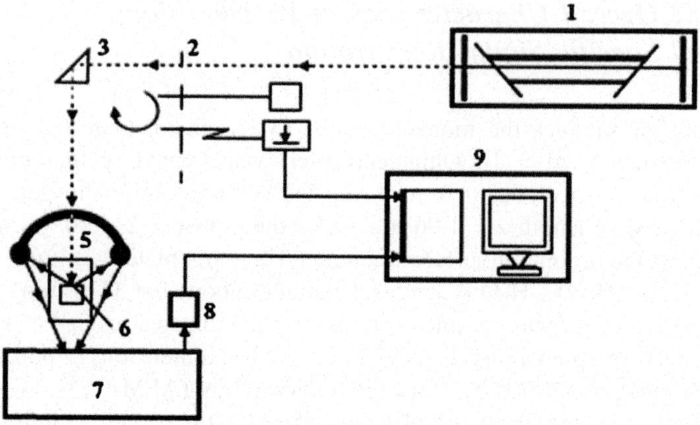

Fig. 2 Block scheme of device for PS PhL measurement. *1*—laser, *2*—modulator, *3*—prism, *4*—aluminum mirror, *5*—quartz cell, *6*—PS sample, *7*—UM-2 monochromator, *8*—FEU, 9—PC,—laser beam,—PhL quanta, —electrical connections

(d > 50 nm) porous [72] and according to our AFM analysis the obtained PS samples may be characterized as macro porous structures. PS had maximal level of fluorescence at wavelength of 650 nm with semi-band ~400 meV. The intensity of its fluorescence increased at the increasing of laser beam density from 11.5 to 33.5 J/cm^2. That is why we kept this index at the level of 33.5 J/cm^2. Exponent describes the spontaneous decreasing of PhL intensity at the temperature of 300 K and this effect is occurred during ≥200 μs. Some details of the determination of the specific immune complex formation were described early [75–78].

Three main types of analysis were compared: (a) the competitive way when the conjugate of 2,4-D or NP with bovine serum albumin (BSA) or soybean inhibitor of trypsin (STI), or ovalbumin (Ova) immobilized on the PS and free appropriate substance compete for specific antibodies (Ab) in the solution; (b) the same way as in point "a" but free 2,4-D or NP and their conjugates compete for specific Ab immobilized on the PS and (c) the direct way when appropriate specific Ab were immobilized on the PS and free 2,4-D or NP were alone in solution.

2.2 *Efficiency of the Immune Biosensor Based on the PS at the Control of Low Weight Toxic Synthetic Substances*

In case of use competitive analysis (when free and conjugated with BSA 2,4-D competed for Ab immobilized on the PS surface) we have obtained the sensitivity of 2,4-D analysis on the level of 5 μg/L and linear plot was in frame of 5–200 μg/L (Fig. 3).

If the conjugated 2,4-D with BSA was immobilized on the PS surface and free 2,4-D and specific Ab were in solution to be analysed the sensitivity analysis was

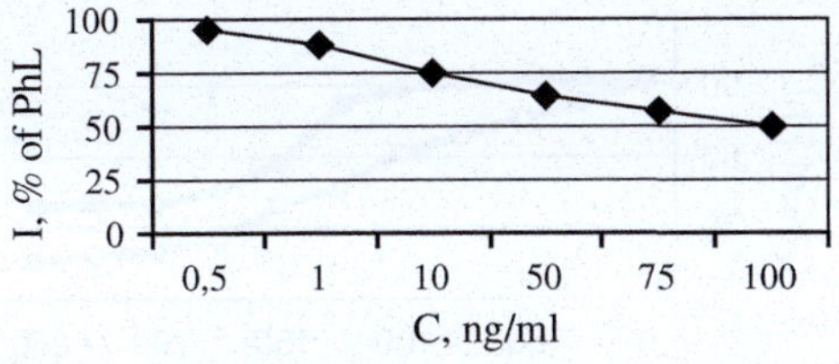

Fig. 3 Changes of intensity of PS ChL at the analysis of free 2,4-D in solution. 2,4-D-BSA was immobilized on the surface. Time of incubation 30 min

higher, approximately in 5 times (up to 1 μg/L). The optimal conditions of NP analysis were in general the same as for 2,4-D but in this case we used antiserum (not pure Ab). That is why first of all the optimal dilution of antiserum was determined. It was shown that this dilution should be about 1:7500 since in this case we can have about 50 % of maximal sensor response (Fig. 4). The sensitivity of NP analysis by competitive way and in case of the immobilisation of NP-BSA on the PS was 10 μg/L (Fig. 5). So, the developed immune sensor allows providing rapid, sensitive and selective analysis of individual pesticides.

We would like to pay attention on the possible mechanism of the work of these immune biosensors. According to our hypothesis the work of the immune sensor based on PS PhL is connected with hydrogen proton interruption from PS surface by the formed immune complex. It is accompanied by sharply decreasing intensity of ChL. The last process is directly proportional to the intensity of immune complex formation. The obtained results (Fig. 5) testify that two process, namely, interruption of hydrogen from surface and its blocking by biological molecules may effect on the PS PL.

The binding of small molecules (2,4-D and NP) with immobilised specific Ab has an influence on the PS PL at their high concentration only. At the same time conjugates of these molecules with some protein are able to change of parameters of

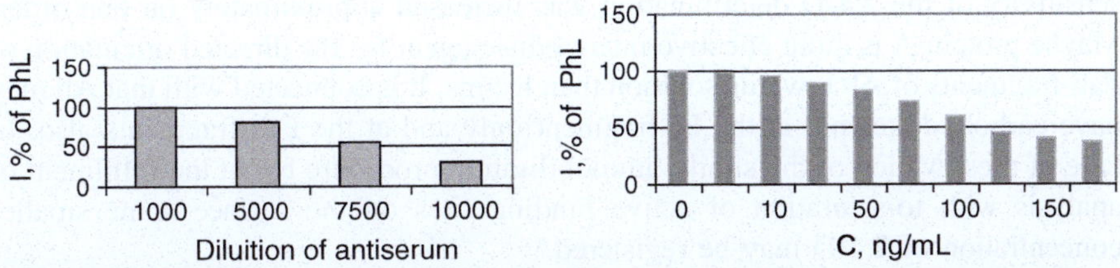

Fig. 4 *Left* The level of sensor signal at the different dilution of specific antiserum. The concentration of free NP was 100 ng/ml. The time of incubation of PS surface with mixture of antiserum and free NP was 30 min. *Right* The dependence of immune sensor signal on the concentration of NP in solution to be analyzed

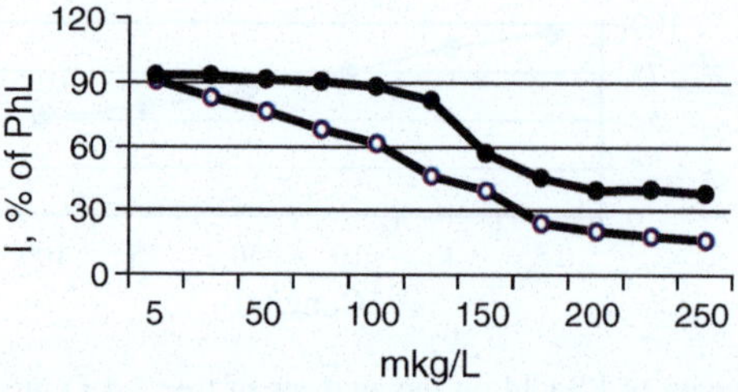

Fig. 5 The changes of intensity of PS PhL at the analysis of free 2,4-D by the competitive (o) and direct (•) analysis. Time of incubation 15 min. Concentration of 2,4-D-BSA—100 mg/L

PS PL more effective. Maybe the presence of a big mass of target molecules at the low weight substances causes the additional blocking of PS recombinant centres.

It was compare the efficiency of NP and 2,4-D analysis by the immune biosensors based on the PS and SPR. In last case the glass plate with 20 nm thickness of the evaporated gold film were as transducer. We have too analysed in detail three main variants of approaches: (a) conjugate (NP or 2,4-D with some protein—BSA, or STI, or Ova) was direct immobilised on the gold surface and free NP or 2,4-D with appropriate antiserum were in solution; (b) the specific Ab from antiserum was immobilised on the gold surface through intermediate layer from *Staphylococcal* protein A or some lectin and free NP or 2,4-D as well as NP or 2,4-D conjugated with some protein were in solution; (c) the same as in "b" but Ab immobilised and oriented toward solution reacted with free NP or 2,4-D and then with appropriate their conjugate (approach with to saturation of active binding sites on the surface). In all cases the surface was preliminary treated by polyelectrolites self assembled (PESA) as it was described early [55, 56]. The obtained results are shown in Fig. 6). It was shown that in case of the immobilisation of specific Ab from antiserum on the gold surface covered by PESA or PESA with lectins we obtained the same sensitivity of 2,4-D determination, namely: about 1 ng/ml. If we used procedure of the immobilisation of specific Ab from antiserum on the gold surface covered by PESA with protein A and fulfilled competitive analysis the sensitivity of the 2,4-D determination was increased approximately on one order. Maybe protein A is more effective intermediate agent for the directed orientation of Fab-fragments of Ab towards solution than lectins. It is connected with that Ab may have carbohydrates not at the Fc-fragment only and at the Fab-fragments also. In case of preservation of the similar immobilisation procedure but at the fulfilment of analysis with to-saturation of active binding sites on the surface some smaller concentration of 2,4-D may be registered.

Nevertheless in model experiments with the corn at the same conditions we can determine 2,4-D in concentration about 0.1 ng/ml only. The model experiments were fulfilled in two ways. (1) 2,4-D in different quantities (mass ratio) was added to corn and carefully mixed. Then corn was washed by the solution of ethanol. (2) Corn was reduced to fragments and extracted by tris-HCl buffer with pH of 7.5.

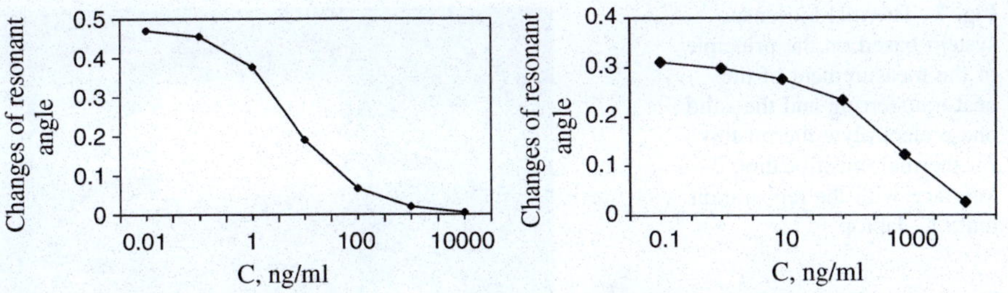

Fig. 6 SPR immune sensor response in the presence of different concentrations of 2,4-D in solution. *Left*—Conditions: Ab (antiserum) were immobilised on the surface covered by PESA and protein A. 2,4-D-BSA saturated bounds left free after interaction with 2,4-D solution. *Right*—Conditions: specific Ab (antiserum) were immobilised on the surface covered by PESA and protein A, NP-STI saturated bounds left free after interaction with NP solution

Then 2,4-D was added to extract in different concentrations. Probably in both cases some part of 2,4-D was adsorbed on the corn particles and as result of it the concentration of this substance was sharply reduced. At the immobilisation of specific Ab from antiserum on the gold surface covered by PESA with protein A and at the fulfilment of analysis with to-saturation of active binding sites on the surface the sensitivity of NP determination may be on the level of 1 ng/ml and lower.

2.3 *Efficiency of the Calorimetric Immune Biosensor at the Control of Low Weight Toxic Synthetic Substances*

This biosensor was designed in two different forms: as micro calorimeter and thermistor based devices. Efficiency of immune biosensor for the determination of low molecular substances we will demonstrate by the results obtained in the experiments with NPh [21, 22]. The portable biosensor based on the microcolorimeter and thermistors are shown in Fig. 7.

For successful development of the calorimetrical biosensor at first it was necessary to set the optimal concentration of antiserum (for example antiserum to NPh). For this purpose 150 μl antiserum in different concentrations was brought in a measuring cell and incubated during 15 min for establishment of base line (for this time a temperature in a barn was set on an optimum level). Then 50 μl solution of NPh in concentrations of 1; 5 and 10 μg/ml were brought into the cell. So, it was set that the optimum concentration of antiserum was about 5 mg of protein in 1 ml.

For the determination of NPh in solutions with the help of thermal biosensor it was necessary to build-up corresponding calibration curve. For this purpose 150 μl of antiserum (in concentration 5 μg of protein in 1 ml) was brought into a

Fig. 7 Thermal biosensor system based on the principle of the measurement of the heat transferring and the solid phase electrolyte thermistors. *1*—thermal sensitive unit; *2*—interface with the registration unit; *3*—laptop

measuring cell and then 50 μl of NPh in a range of concentration from 0.5 up to 10 μg/ml was pumped into the measuring cell. Thus, it was demonstrated an opportunity of "direct" detection of NPh by calorimetrical biosensor with the sensitivity about 1 μg/ml. The overall time of analysis is about 20–30 min.

Certainly that the sensitivity of the determination of NPh by thermal immune biosensor is much less than in case of application of SPR or TIRE biosensor but it is necessary to mention simplicity of measurement fulfilment. Maybe thermal biosensor could be used for the screening of toxic elements in environmental objects with next verification of results of analysis by optical immune biosensors.

2.4 Conclusion

At the optimisation of all parameters of the analysis by the proposed immune sensors it is possibility to reveal the concentrations of 2,4-D and NP on the level of practice demands, in particular, less than 0.1–1 μg/L, respectively. The total time of analysis was about 6–10 min only. The developed immune sensors allow providing rapid, sensitive and selective analysis of individual pesticides. These immune sensors may be created as stationary and portable devices. The last model of biosensors may be used for fulfilment of analysis in field conditions. The main principle of work of the immune sensor based on the SPR is common known. According to our hypothesis the work of the immune sensor based on PS PhL is connected with hydrogen proton interruption from PS surface by the formed immune complex and blocking of active silicon sites. It is accompanied by sharply decreasing intensity of ChL. The last process is directly proportional to the intensity of immune complex formation. There is necessary to underline that the immune biosensor based on the PS is more simple and more cheap than that based on SPR and it can be used for wide screening observation of environmental objects.

3 Creation of Non-labelled Immune Biosensors for the Determination of Low Weight Biological Origin Toxic Substances in Environment

In this section it will be given the main attention to the immune biosensors for control of level of mycotoxins in different environmental objects.

3.1 *General Characteristics of Mycotoxins and Methods of Their Analysis*

Different generations of micromycetes are able for the synthesis of a big set of mycotoxins: aflatoxins, rubratoxins, ochratoxins, fumonisines, trichothecenes and others [14]. Mycotoxins may be found in desert, in alkaline lands and in highland regions although they are more characteristic for middle latitudes [17]. These toxins are accumulated in feeds and foods at the appropriate temperature which promotes to grow of fungi's. They present a special group of low-molecular weight and non-immunogenic substances which as a rule have thermo stability. Different types of mycotoxins may cause intoxication with the destruction of organs and tissues: liver, kidney, oesophagus, bowels, brain and genitals. That is why, mycotoxins are included into list of substances which content in foods, feeds and row materials should be regulated. According to State standard of Ukraine (DSTU 3768-98) the maximal content of T2 in corn for the bread and technical grain is permitted on the level of 0.1 and 0.2 mg/kg, respectively. In Czech Republic the determination of mycotoxins is included in the system of the environment monitoring [60].

3.2 *Specificity of Biosynthesis of Mycotoxins*

The mechanisms of the formation of mycotoxins from the primary metabolites depend on producer type and biosynthesis conditions. Among them it is distinguished next: (1) polyketide specific for aflatoxin, ochratoxin, patulin and others; (2) terpene which is correspond to trichothecenes; (3) mediated trough cycle of tricarboxylic acids (in case of rubratoxins); (4) amino-acid when initial substances are presented by some structures (ergoalcaloids, cyclopiasonic acid and others) and (5) hybrid (composition of two or more mechanisms) which is typical for cyclopiasonic acid [83]. The main way for the biosynthesis of big group mycotoxins is polyketide mechanism. Depends on quantity of the incorporated C_2-units mycotoxins which are synthesized according to first mechanism are divided on tetraketide (patulin), pentaketide (ochratoxin), hexaketide, heptaketide, octaketide, monoketide (zearalenone) and decaketide (aflatoxins).

3.3 Chemical and Physical Abilities of Trichotecenes

They are belong to chemical substances named as sesquiterpenoids and trichothecene ring contended olefinic bond between C-9 and epoxy group in field of C-12, 13 is their speciality [16]. There are up to 150 derivates of such structure. Depends on structure of trichothecene center mycotoxins are divided on four groups: A—when substances have H or OH groups at C_8; B—in case of the presence of carboxylic group at C_8; C—micro cyclic trichothecene and D—when the second epoxy group is at C_7 [2].

Trichothecenes are low-molecular substances (250–55 Da), non volatile and bad soluble in water but they have a good solubility in acetone, ethyl acetate, chloroform, dimethyl sulfoxide, ethanol, methanol and propylene glycol [16]. Pure trichothecenes are characterized by the low level of the steam tension but they evaporate at the heating with organic solvents. High purified mycotoxins have wait color and fixed melting point [2].

Micotoxins may be preserved in solution or crystalline form at that they are stable at the light and oxygen effects. They do not inactive at the influence of 480 ° C during 10 min or at 260 °C during 30 min. For chemical neutralization of T2 mycotoxin 3–5 % of solution of sodium hypo chloride is used the efficiency of which may be improve by some quantity of alkali [104].

3.4 Biotransformation of Trichotecenes

As opposite to others mycotoxines trichotecenes do not require a preliminary metabolic activation for the manifestation of their toxicity [12]. At the application on the skin or in case of use with food trichothecenes cause immediately irritation of skin or mucous of bowels. In experiments with single cells or their cultures it was shown that trichthecenes cause disintegration of polyribosome and as result of it the inhibition of protein synthesis starts (Ciegler and Bennett [14, 104]. Lipophilic character of trichotecenes gives them to sorb simple and fast on the skin and mucous of tissues. At the “per os” introduction of mycotoxin to mousc the peak of its concentration appears in the range of one hour [46, 106]. A different cells transform T2 toxin trough removal of oxygen from epoxy ring at the C-12, 13 position with the formation of binary carbonic bound and with the oxygenation of isovaleric side chains [95, 109]. Pancreas and bowels tissues are as main organs which metabolize trichotecenes. Pharmacokinetics investigations are shown that T2 mycotoxin appears in blood serum aside from way of introduction in organism. Up to 95 % of T2 mycotoxin introduced by intravenously is removed form organism through urine and excrements in ratio of 3:1 [103].

3.5 Some Biological Effects of Trichotecenes

Mycotoxins effect on a many organs and the most toxicity was revealed for A type. Nevertheless, mycotoxins of D type have a small toxicity in spite of the presence of two epoxy groups. The reduction of double bond leads to minor decreasing of the toxicity but at the lost of epoxy ring at C_9–C_{10} stipulates sharply its reduction [62].

Semi lethal dose (LD_{50}) at the T2 inhalation is 200–5800 mg*min/m^3 [106] the similar results may be achieved for lewisite and mustard gas at the dose of 1500–1800 mg*min/m^3 [105]. At the application of T2 mycotoxin on the skin LD_{50} is on the level of 2–12 mg/kg and it is higher than in case of mustard gas (100 mg/kg) [31]. The process intoxication may be increased much more if T2 mycotoxin will be used as solution based on dimethyl sulfoxide [95]. At the low concentrations of T2 mycotoxin it is observed the inhibition of growth of young animals and the decreasing of their stability to diseases as the result of immune system destroying. In particular, experiments with birds have shown that this mycotoxin may effect on the immunocompetent organs [41]. Firstly, T2 mycotoxin was revealed at the cow toxicosis and it was stated that *Fusarium tricinctum* is as main producer. *Mycothecium, Trichoderma, Trichothecium, Cephalosporium and Stachybotrys* may provide synthesis of this toxin but in much less level [3]. It was registered the intoxication of horses in Japan and symptoms of the intoxication were revealed at the concentration of 1 mg/kg [100]. The similar disease accompanying by murrain was observed among goslings and turkey poultry [42]. The incident of the people disease as result of using of foods fabricated from moldy wheat was described [5]. During number years in Bashkiria and Orenburg district it was registered accidents of people death as result of using of foods fabricated from grain (millet, wheat and barley) which was passed the winter and effected by *Fusarium sporotrichioides* [71]. It is known a number of pathological states stipulated by the chronic intoxication: toxicosis, disease of red mould, hemorrhagic disease, alimentary toxic aleukia and others [29].

3.6 Traditional Methods of the Determination of Mycotoxins

The identification of mycotoxins in different objects of environment is not simple procedure but in spite of it the special system was developed today. The standard analytical methods include a several procedures: sampling, extraction, purification, identification, determination and, at last, verification of results. At first, after beginnings of problem with mycotoxins the biotests were proposed and for this purpose a wide set of biological objects were used: from cell cultures to the vertebrate animals [30]. Biotest in which the mortality of chick-embrio was as the criteria of toxicity may be considered as traditional. Biotests with the application of different strains of yeasts obtained a greatest development. The limit of the sensitivity to T2 varies for individual strains and it may be in frame from 20 to 50 ng [10]. Simplified variants of such biotests gives possibility to control the contamination of corn by trichothecenes with the analyze fulfillment over the time of 24 h [7]. Biotests were used for the investigation of the dependence of toxicity on the peculiarities of mycotoxin structure [45, 69]. In what follows the methods based on the technology of the thin layer chromatography (TLCh) were elaborated. These methods provide high specificity but they have much less sensitivity (about 0.2–5 μg/ml) and demand preliminary concentration of the analyzed extract in 10–1000 times.

Implementation of idea about "bioautographic" approach combined advantages of both methods based on the microbial analysis and TLCh allow control several mycotoxins in the same sample simultaneously. Potential possibilities of biotests are not yet realized and till today a new their variants with the improved characteristics are appeared.

The application of gas liquid chromatography (GLCh) allows to increasing the sensitivity of analysis. Unfortunately, the application of this method demands multistage procedure of the sample preparation and its chemical modification [67]. The sensitivity of the T2-mycotoxin determination in serum blood by this method is on the level 30 ng/ml with the reproducibility of results upon the average of 95.5 ± 8.6 % [108]. GLCh is used as one of reliable method for the verification of results of analysis [99]. The same role may be fulfilled by the high-performance liquid chromatography (FPLC) and high pressure liquid chromatography (HPLC).

After obtaining of the specific antibodies (Ab) to some types of mycotoxins there is appeared possibility to use immune chemical approaches including the radio immune (RIA), immune chemiluminescence (IChL) and fluorescence (IFL) analyses, as well as the ELISA-method [13]. So, RIA is able to reveal T2-mycotoxin in maize and wheat at 1.0 and 2.5 μg/kg, respectively [43]. In case of others types of biological liquids the sensitivity of T2-mycotoxin analysis was from 2 to 5 ng/ml [26].

The development of the immune chemical methods was accomplished by two ways: obtaining of the specific antibodies and development of special algorithms of analysis [28]. In other case at the application of the monoclonal antibodies conjugated with horse radish peroxidase and at the realization of competitive variant of analysis it was achieved the sensitivity no less as 10 ng/ml [66]. More complicate homogenous competitive analysis with the use liposome contended some substances labeled by the fluorochrome allows to obtaining sensitivity of T2-mycotoxin determination at the concentration of 2 ng/ml [44].

Today, for the characterization, identification and confirmation of the results of the analysis the methods based on the mass spectroscopy, nuclear magnetic resonance or in combination with the GLCh, HPLC gives possibility to analyze mycotoxins at the picogram level [48, 102]. Unfortunately, in not all cases there is possible to avoid such complicate procedure of the sample preparation as process of the obtaining of derivates.

3.7 *Instrumental Analytical Approaches Based on the Principles of Biosensorics at the Determination of Some Mycotoxins*

3.7.1 Optical Immune Biosensor [57, 58, 64, 83, 84, 86–90, 92]

For the determination of some mycotoxins in the model solution it was used the immune biosensor based on the total internal reflection ellipsometry (TIRE). A typical set of $\Delta(\lambda)$ experimental spectra are shown on Fig. 8.

It demonstrates the spectral shift caused by consecutive adsorption of layers of polyamine hydrochloride, protein A (both as intermediate layers at the immobilization of selective structures) and specific Ab as well as by binding different concentrations of T2-mycotoxin to Ab (from 0.15 up to 300 ng/ml).

Ellipsometry data fitting allows the evaluation of thickness values of the adsorbed layer. Since the refractivity increments caused by adsorption of different biomolecues represent only 0.1–0.14 % of the refractive index, the spectral changes were associated mainly with the thickness. The resulted calibration curve for T-2 mycotoxin showed a possibility of detection of T-2 mycotoxin in concentrations down to 0.15 ng/ml (or 0.15 ppb).

The calibration curves (i.e. thickness changes vs. mycotoxin concentration) obtained from the TIRE experiments for the other two mycotoxins: zearalanone and aflatoxin are shown in Fig. 9. The response to aflatoxin appeared to be about 3 times less than that for zearalanone; and both are smaller than that for T-2 (compare with Fig. 9). This could be due to the limited concentration (or activity) of antibodies. Another explanation may be related to the hydrophobicity of the above mycotoxins and thus their abilities to form aggregates in aqueous solutions [84–87].

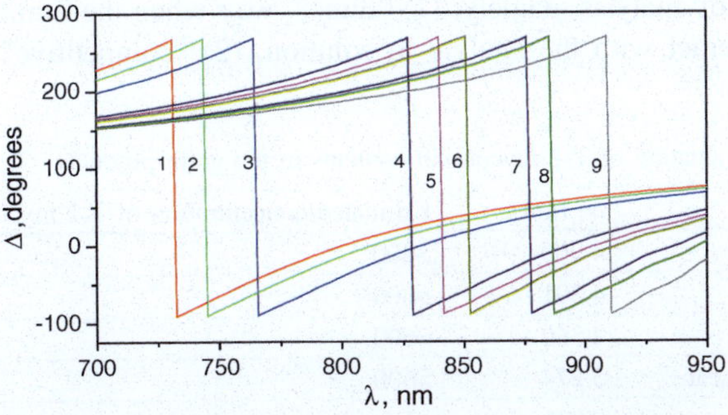

Fig. 8 Typical spectra of Δ for the bare Au surface (*1*), after consecutive adsorption of PAH (*2*), Protein A (*3*), antibodies to T-2 (*4*), and after binding T-2 mycotoxin in different concentrations of 0.15 mg/ml (*5*), 1.5 mg/ml (*6*), 7.5 mg/ml (*7*), 75 mg/ml (*8*), 300 mg/ml (*9*)

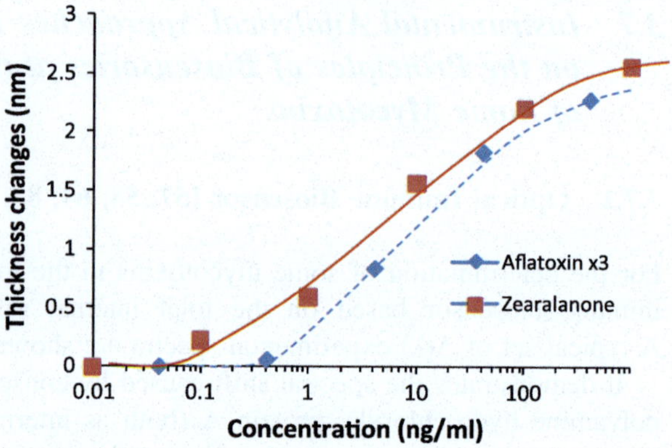

Fig. 9 TIRE calibration curves for direct immunosensors for zearalanone and aflatoxin. The thickness values for aflatoxin were three fold increased

The results showed that mould grain products contain T-2 mycotoxin in dangerously high concentrations of more than 100 ng/ml. Even stale grain products may contain up to 15 ng/ml of T-2 which is higher than the allowed amount; while fresh products do not contain any traces of T-2.

Similarly to T-2, the minimal detected concentrations for both zearalanone and aflatoxin are about 0.1–0.4 ng/ml, respectively, which is quite remarkable achievement (no other optical direct immunoassays provide such high sensitivity) [59].

The TIRE based immune biosensor was used for the control of T2 mycotoxin in real samples [84, 85]. The results of TIRE experiments (direct immunoassay) carried out on the samples of grain products are presented in Table 1. The obtained data points for fresh, stale, and mould grain product samples were added to the calibration curve for T-2 mycotoxin. The calibration was obtained earlier by TIRE direct immunoassay using commercial (from Sigma) monoclonal antibodies to T-2 mycotoxin [57, 58, 84].

The patulin and T-2 mycotoxin were determined in model solution by immune biosensor based on the SPR [57, 58, 64, 84, 86, 87]. There were realized a number of algorithms of analysis, namely: (1) "direct" way when the immobilized specific antibodies interact with the analyte in solution, (2) "competitive" way when the

Table 1 The evaluation of T-2 mycotoxin contents in the grain products

Sample	δd (nm)	Estimated concentration of T-2 mycotoxin (ng/ml)
Mouldy fodder	4.900	>600
Mouldy bread	4.391	>600
Mouldy maize	4.369	>600
Mouldy buckwheat	5.274	>600
Stale bread	0.031	<1.5
Stale muesli	0.298	<1.5
Stale buckwheat	1.199	7.5–15
Fresh muesli	0.000	None

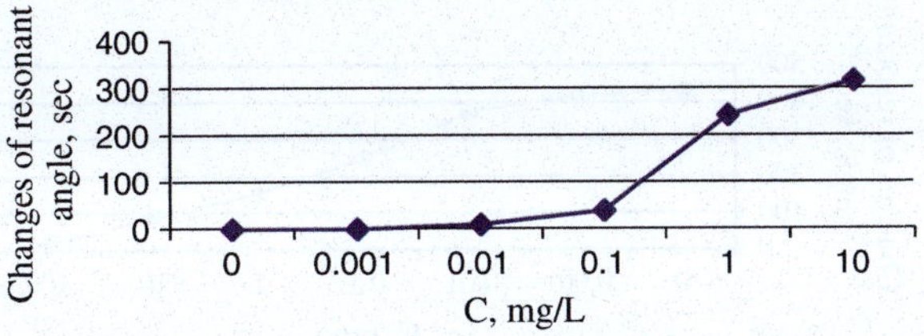

Fig. 10 Calibration curve for the patulin determination by SPR immune biosensor at "direct" way

immobilized conjugate of analyte with some protein compete with free analyte for the specific antibodies in solution. In last case it is possible to realize third variant when the specific antibodies are immobilized on the transducer surface and the appropriate conjugate and free analyte compete for the free binding centers. Next variant of analysis is "to saturate" one when the immobilized specific antibodies interact at first with the analyte and than with the appropriate conjugate. So, it was shown that the "direct" analysis by SPR immune biosensor allows to us revealing patulin at the concentration of 1 mg/L (Fig. 10). It is not practically significant level. The "competitive" way (variant 3) was much more sensitive (up to 2 orders). It can provide the determination of patulin at the concentration in the range from 0.01 to 10 mg/L (Fig. 11). The similar results were obtained at the realization of "to saturate" way at the patulin determination.

To provide optimal regime of the immune biosensor analysis we have examined a different approaches for the antibody immobilization of the transducer surface [81]. There was necessary to achieve high density of the immobilized selective structures and to expose maximally binding centers towards solution. For this purpose we used direct immobilization of the antibodies on the gold surface or through intermediate layers from dodecanthiol, or such polyelectrolyte as polyalylamine hydrochloride. I was shown that the formation of the above mentioned intermediate layer increase the analysis sensitivity and the application of polyelectrolyte and dodecanthiol demonstrates the advantages of first before second one. Mainly it is connected with the stability of results between measurements.

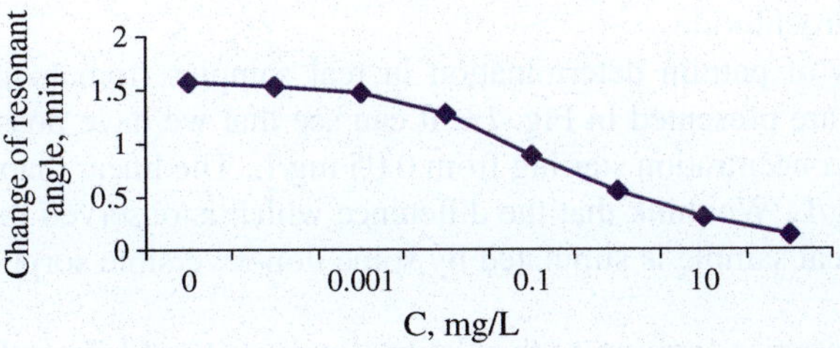

Fig. 11 Calibration curve for the patulin determination by SPR immune biosensor at "competitive" analysis

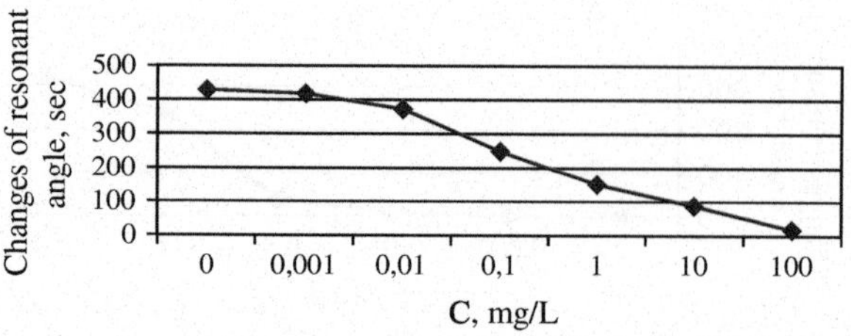

Fig. 12 Calibration curve for the T-2 mycotoxin determination by SPR immune biosensor at "to saturate" way of analysis

The same algorithms of analysis as in respect of patulin were fulfilled with T-2 mycotoxin [57, 58, 64, 84, 86, 87]. There is necessary to underline that it was revealed the similar regularities including the sensitivity. The minimal quantity which can be determined by the "direct" way of analysis is 100 ng/ml. At the "competitive" and "to saturate" ways the sensitivity are on the level of 10 ng/ml (Fig. 12. At the same time the similar results were obtained in case of the use of mono- and polyclonal antibodies.

SPR based immune biosensor was developed for the determination of aflatoxin too. It was examined in two algorithms of analysis; "competitive" way (variant 2) and "to saturate" one [19]. In both cases the polyclonal antibodies were used. The linear field of the biosensor responses were in the frame of 12–25000 ng/ml and 3–98 ng/ml, respectively for the "competitive" and "to saturate" ways. The specificity of the analysis depended on the specificity of antibodies. To fulfillment of the repeated analysis the transducer surface was treated by the solution of 1 M ethanolamine with 20 % of acetonitrile (pH 12). Authors confirmed our results that the total time of the above mentioned analysis is taken about 20–30 min. This time may be shortened at the "direct" way of analysis (up to 10 min) but its sensitivity will be 50 ng/ml only.

To determine deoxynivalenol the "competitive" analysis was developed when the mycotoxin conjugated with casein was immobilized on the SPR surface and the specific monoclonal antibodies and free analyte were in solution [53]. The optimal controlled concentrations were in the frame of 2.5–30 ng/ml. For the fulfillment of the repeated analysis (up to 500 times) the transducer surface was treated by 6 M guanidine hydrochloride.

The results of patulin determination in real samples (tomato juice) by "competitive" way are presented in Fig. 13. It can see that we have possibility to reveal patulin at the concentration starting from 0.05 mg/L. The linear diapason is in range of 0.05–10 mg/L. We think that the difference which is observed between standard solution and real sample is stipulated by some non-reversible sorption of patulin in organic phase.

At the subsistence farming of the National university of Life and Environmental Sciences of Ukraine the sample of maize and bran affected by fungi's were chosen. With the help of the ELISA-method it was stated that the concentration of T-2

Acknowledgments This work was partially supported by EU funded BIOSENSORS-AGRICULT project (Development of nanotechnology based biosensors for agriculture FP7-PEOPLE-2012-IRSES, contract N. 318520) and NATO multi-year Science for Peace Project NUKR.SFPP 984637—"Development of Optical Bio-Sensors for Detection of Bio-Toxins" as well as National University of Life and Environmental Sciences of Ukraine, projects "Development of new generation of the instrumental analytical approaches based on the principles of nano-biotechnology and biosensorics for providing systems of biosafety" and "Creation of effective methods for the construction of artificial selective sites for the instrumental analytical devices of new generation based on the principles of biosensorics", N 110/476-Appl. and № 110/71-fund., appropriate.

References

1. Banza R, Rois A, Gomes-Hens A, Valcarcel M (2004) Supercritical fluid immunoextraction: a new approach for immunoassay automation. Anal Chem Acta 518:151–156
2. Bamburg JR (1976) Mycotoxins and other fungal related food problems. Washington, D.C., pp 144–162
3. Bamdurg JR, Riggs NV, Strong FM (1968) The structure of toxins from two strains of Fusarium tricinctum. Tetrahedrom 24:3329–333
4. Barrie LA, Gregor D, Hargrave B, Lake R et al (1992) Arctic contaminants: sources, occurrence and patways. Total Environ 122:1–74
5. Bhat VR, Ramakrishna J, Sashidhar RB (1989) Outbreak of mycotoxicosis in Kashmir Valley. Nutr News India 10(1):1–3
6. Blum JL, Xiong JQ, Hoffman C, Zelikoff JT (2012) Cadmium associated with inhaled cadmium 32 oxide nanoparticles impacts fetal and neonatal development and growth. Toxicol Sci 126(2):478–486
7. Binder J (1999) A yeast bioassay for trichothecenes. Nat Toxins 7(N6):401–406
8. Birnbaum LS (1995) Workshop on potential exposure to dioxin-like compounds:V. immunologic aspects. Environ Health Persp 103:157–159
9. Boisen AM, Shipley T, Jackson P et al (2012) NanoTiO(2) (UVTitan) does not induce ESTR mutations in the germline of prenatally exposed female mice. Part Fibre Toxicol 9:9–19
10. Boltjanskaja EV, Kuvajeva IB, Krojakova EA (1988) Screening of yeast strains which are sensitive to T2-mycotoxin. Probl Nutr N 6:67–68
11. Brouwer A, Morse DC, Lans MC, Schuur AG et al (1998) Interactions of persistent environmental organohalogens with the thyroid hormone system: mechanisms and possible consequences for animal and human health. Toxicol Ind Health 14:59–84
12. Busby WF Jr, Wogan GN (1980) Trichothecenes. In: Shank RC (ed) Mycotoxins and N-Nitroso compounds: environmental risks, vol 2. Fleering Center: CRC Pressm, Boca Raton, pp 29–41
13. Chu FS, Grossman S, Wei RD, Microcha CJ (1979) Production of antibody against T-2 toxin. Appl Environ Microbiol 37:104–108
14. Ciegler A, Bennett JW (2014) Mycotoxins and mycotoxicoses. Bioscience 30(N8):512–515
15. Colborn T (1991) J. Epidemiology of Great Lakes bald eagles. Tococol Environ Health 33:395–453
16. Cole RJ, Cox RH (1981) The trichothecenes. In: Cole RJ, Cox RH Handbook of toxic fungal metabolites. Academic Press, New York, pp 152–263
17. Committee (1983) Committee on Protection Against Mycotoxins, Board on Toxicology and Environmental Health Hazards, Commission on Life Sciences, National Research Council. Protection Against Trichothecene Mycotoxins. National Academy Press, Washington, DC

18. Council Directive 82/242 EEC of March 31, 1982 on the approximation of the laws of the Member States relating to methods of testing degradability of non-ionic surfactants on amending Directive 73/404/EEC
19. Daly SJ, Keating GJ, Dillon PP et al (2000) Development of surface plasmon resonance-based immunoassay for aflatoxin B_1. J Agric Food Chem 48(N11):5097–5104
20. De Koe WJ (1995) Natural toxins. Chairman, Panel discussion., No 3, p 280
21. Demchenko AV, Mel'nik VG, Starodub NF (2007) Thermal biosensor for detecting nonylphenol in the environment. Ukr Biochem J 79(5):173–176
22. Demchenko AV, Starodub NF (2007) Modification of immune SPR biosensor surface at nonylphenol analysis. Biopolym Cell 23(2):143–147
23. Ember LR (1984) Yellow rain. Chem Eng News 62:8–34
24. Eriksson P (1997) Developmental neurotoxicity of environmental agents in the neonate. Neurotoxicology 18:719–726
25. Eriksson G, Jensen S, Kylin H, Strachan W (1989) The pine needle as a monitor of atmospheric pollution. Nature 341:42–44
26. Fan TSL, Zhang GS, Chu FS (1984) An indirect enzyme-linked immunosorbent assay for T-2 toxin in biological fluids. J Food Protect 47(N12):964–967
27. Gomes SI, Soares AM, Scott-Fordsmand JJ, Amorim MJ (2013) Mechanisms of response to silver nanoparticles on Enchytraeus albidus (Oligochaeta): survival, reproduction and gene expression profile. Hazard Mater 15:254–255
28. Gendloff EH, Pestka JJ, Swanson SP, Hart L. P. (1984) Detection of T-2 toxin in *Fusarium sporotrichioides*—infected corn by enzyme-linked immunosorbent assay. Appl Environ Microbiol 1161–1163
29. Hassanin N, Gabal MA (1990) Biological and chemical characterization of metabolites of Fusarium moniliforme isolates. Vet Hum Toxicol 32(N6):536–540
30. Hayes MA, Schiefer HB (1990) A method for accurate measurement of cutaneous irritancy of trichothecenesJ. Environ Pathol Toxicol Oncol 10(3):103–105
31. Hooper HL, Jurkschat K, Morgan AJ et al (2011) Comparative chronic toxicity of nanoparticulate and ionic zinc to the earthworm Eisenia 6 veneta in a soil matrix. Environ Int 37(6):1111–1117
32. Houk VS, Waters MD (1996) Genetic toxicology and risk assessment of complex environmental mixtures. Drug Chem Toxicol 19:187–219
33. Hsu PC, O' Callaghan M, Al-Salim N, Hurst MR (2012) Quantum dot nanoparticles affect the reproductive system of Caenorhabditis elegans. Environ Toxicol Chem 31(10): 2366–2374
34. Jodldauer J, Maier NM, Linder W (2002) To wards ochratoxin A selective molecularly imprinted polymers for solidphase extraction. J Chromatogr 945A:45–63
35. Kachinskaya TS, Melnichenko MM, Starodub MF, Shmyryeva OM (2008a) Nanobiosensors for the determination of the concentration of mycotoxins. In Proceedings of the Electronic and connection. Issue: "Problems of Electronics", part 1. "medical devices and systems", pp 153–159
36. Kachinskaya TS, Melnichenko MM, Starodub MF et al (2008b) Production of biosensor controls on the basis of nanostructured silicon. In: Abstract of international symposium on surface science and nanotechnology, 9–13 Nov 2008, International Conference Center, Waseda University, Tokyo, Japan, p 550
37. Kalchenko OI, Solovyov AV, Cherenok SA et al (2003) Complexation of Calix are nephosphonous acids with 2,4-dichlorophenoxyacetic acid and atrazine in water. Incl Phenom Microcyclic Chem 46:19–25
38. Kataoka H (2002) Automated sample preparation using in#tube solidphase microextraction and its application. Rev Anal Bioanal Chem 373:31–45
39. Kononenko GP, Burkin AA, Soboleva NA, Zotova EV (1999) Enzyme immunoassy for determination of T-2 toxin in contaminated grain. Appl Biochem Microbiol 35:457–462
40. Kipopoulou AM, Manoli E, Samara C (1999) Bioconcentration of polycyclic aromatic hydrocarbons in vegetables grown in an industrial area. Environ Pollut 106:369–380

41. Kotik AN (1999) Micotoxicosis of avian. Publ. "Donets", Borki, 267 pp
42. Kotik AN, Truchanova VA (1997) Events of micotoxicosis among farming birds in Ukraine in 1974–96 years. In: Breeding, interdepartmental proceedings (Borki, Kharkiv destrict), B. 47, pp 92–100
43. Lee S, Chu FS (1981) Radioimmunoassay of T-2 toxin in corn and wheat. J Assoc Anal Chem 64(1):156–161
44. Ligler FS, Bredehorst R, Talebian A et al (1987) A homogeneous immunoassay for the mycotoxin T-2 utilizing liposomes, monoclonal antibodies, and complement. Anal Biochem 163(2):369–375
45. Madhyastha MS, Marquardt RR, Abramson D (1994) Structure-activity relationships and interactions among trichothecene mycotoxins as assessed by yeast bioassay. Toxicon 32 (9):1147–1152
46. Matsumoto H, Ito T, Ueno Y (1978) Toxicological approaches to the metabolites of fusaria, XII: Fate and distribution of T-2 toxin in mice. Jpn J Exp Med 5(N48):393–399
47. Mirocha CJ, Pawlosky RA, Chatterjee K, Watson S, Hayes W (1983) Analysis for Fusarium toxins in various samples implicated in biological warfare in Southeast Asia. J Assoc Anal Chem 66(6):1485–1499
48. Mirocha CJ, Panthre SV, Pawlosky RJ, Hewetson DW (1986) Mass spectra of selected trichothecenes. In: Cole RJ (ed) Modern methods in the analysis and structure elucidation of mycotoxins. Academic Press, New York, pp 353–392
49. Mocarelli P, Brambilla P, Gerthoux PM, Patterson DG Jr, Needham LL Jr (1996) Change in sex ration with exposure to dioxin. Lancet 348:409
50. Morais S, Maquieeira A, Puchades R (1999) Immunofiltration: A new methodology for preconcentration and determination of organic pollutants. Anal Chem 71:1905–1909
51. Morris BA, Clifford MN (eds) (1985) Immunoassays for Food Analysis. Elsevier Applied Science, London
52. Morse PM (1999) Soaps and detergents. C&EN, February 1:35–48
53. Mullett W, Lai EPC, Yeung JM (1998) Immunoassay of fumonisins by a surface plasmon resonance biosensor. Anal Biochem 258:161–167
54. Mulvad G, Pederson HS, Hansen JC, Dewailly E et al (1996) Exposure of Greenlandic Inuit to organochlorines and heavy metals through the marine food-chain: an international study. Tot Environ 186:137–139
55. Nabok AV, Starodub NF, Ray AK, Hassan AK (2000a) Registration of immunoglobuline AB/AG reaction with planar polarisation interferometer. Biochemical and biomolecular sensing. In: Liberman RA (ed) Proceedings SPIE 4200, pp 1–9
56. Nabok AV, Starodub NF, Ray AK, Hassan AK (2000b) Enzyme/indicator optrodes for detection of heavy metal ions and pesticides. Biochemical and biomolecular sensing. In: Liberman RA (ed) Proceedings SPIE 4200, pp 32–41
57. Nabok AV, Tsargorodskaya A, Holloway A et al (2007a) Registration of T2 mycotoxin with total internal reflection ellipsometry and QCM impedance methods. Biosens Bioelectron 22:885–890
58. Nabok A, Tsargorodskaya A, Holloway A et al (2007b) Specific binding of large aggregates of amphiphilic molecules to the respective antibodies. Langmuir 23:8485–8490
59. Nabok A., Tsargorodskaya A., Mustafa M.K., et al (2009) Detection of low molecular weight toxins using optical phase detection techniques. Proc. of the Eurosensors XXIII conference. Procedia Chem www.elsevier.com/locate/procedia, pp 1–4
60. Ostrŷ V (1999) Micromycetes, mycotoxins and human health. Cas Lek Cesk 138(17): 515–521
61. Paepens C, De Saeger S, Sibanda L et al (2004) A flow-through enzyme immunoassay for the screening of fumonisins in maize. Anal Chem Acta 523:229–235
62. Peng S, Dong J, Yang J (1996) Studies on relationship between toxicity of trichothecene toxin T-2 and its structure. Chung Hua Yu Fang I Hsueh Tsa Chih 30(N 3):141–143
63. Powers KW, Brown SC (2006) Characterization of nanoscale particles for toxicological evaluation. Toxicol Sci 90(2):296–303

64. Pylypenko LN, Egorova AV, Pylypenko IV et al (2007) Investigation of toxic effect of patulin with the help of biosensorics systems. Food Sci Technol N1:35–38
65. Pylypenko IV, Pylypenko LN, Starodub NF (2010) Biosensors at the determination of mycoyoxins. In: Proccedingsof 4th International schentific-technical conference "Sensor electronics and microsystem technologies" (SMEST-4), Ukraine, Odessa, 28 June–02 July, p 227
66. Ramakrishna N, Lacey J, Candlish AA et al (1990) Monoclonal antibody-based enzyme linked immunosorbent assay of aflatoxin B1, T-2 toxin, and ochratoxin A in barley. J. Assoc Anal Chem 73(1):71–76
67. Romer TR (1984) Chromatographic techniques for mycotoxins. In: Lawrence JF (ed) Food constituents and food residues: their chromatographic determination. Marcel Dekker, New York, pp 393–415
68. Rosen RT, Rosen JD (1982) Presence of four Fusarium mycotoxins and synthetic material in yellow rain. Evidence for the use of chemical weapons in Laos. Biomed Mass Spectrom 9 (10):443–450
69. Rotter BA, Thompson BK, Prelusky DB, Trenholm HL (1991) Evaluation of potential interactions involving trichothecene mycotoxins using the chick embryotoxicity bioassay. Arch Environ Contam Toxicol 21(4):621–624
70. Sahoo SK, Parveen S, Panda JJ (2007) The present and future of nanotechnology in human health care. Nanomedicine: Nanotechnol Hum Healthc 3:20–31
71. Sarkisov A (1954) Ch. Mycotoxicoses. M., Publ. SIAL, 216 pp
72. Schoening MJ, Ronkel F, Crott M et al (1997) Miniaturization of potentiometric sensors using porous silicon microtechnology. Electrochim Acta 42:3185–3193
73. Solodkij MV, Melnichenko MM, Starodub MF, Shmyryeva OM (2008) Biosensor for the determination of the concentration of toxic substances with the help of nanostructured silicon. In: Proceedings of the of VI International scientific-technical conference, 21–23 Oct 2008, pp 160–163
74. Starodub NF, El'skaya AV, Piletsky SA, Lavryk NV (1993) Template sensors for low weight organic molecules based on SiO2 surfaces. Sens Actuat 708–710
75. Starodub NF, Fedorenko LL, Starodub VM, Dikij SP, Svechnicov SV (1996) Sens Actuat 35:44–47
76. Starodub VM, Fedorenko LL, Starodub NF (1998) Control of a myoglobin level in solution by the bioaffinic sensor based on the photoluminescence of porous silicon. Eurosensors XII. In: Proceedings of the European conference on solid-state transducers and 9-th UK Conference on sensors and their applications, Southampton, UK, 13–16 September, vol 2, pp 817–820
77. Starodub VM, Starodub NF (1999) Optical immune sensors for the monitoring protein substances in the air. Eurosensor XII. In: The 13th European conference on solid-state transducers, 12–15 Sept 1999, The Hague, The Netherlands, pp 181–184
78. Starodub VM, Fedorenko LL, Sisetskiy AP, Starodub NF (1999) Control of mioglobin level in an immune sensor based on the photoluminescence of porous silicon. Sens Actuat B 58:409–414
79. Starodub NF, Starodub VM, Demchenko AV et al (2004a) Creation of optical immune sensors for the determination of low weight toxic substances in environment. In: Proceedings of the International trade fair for optical and microtechnology products with conference, 25–27 May 2004. Nuremberg Exhibition Center, Germany, pp 335–340
80. Starodub NF, Demchenko A, Starodub VM et al (2004b) Optical immune sensors for the determination of low weight toxic substances in environment. Sens Electron Microsyst Technol Odessa 2004:150–151
81. Starodub NF, Pirogova LV, Starodub VM, Demchenko A (2005) Antibody immobilization on the metal and silicon surface. The use of self-assembled layer and specific receptors. Bioelectrochemistry 66:111–115
82. Starodub NF, Pilipenko LN, Egorova AV et al (2007) Analysis of mycotoxins: selection and preparation of samples. Biotechnology 1(1):106–115

expected approach is to build up optical fiber sensor with high surface-to-volume aspect ratio sensitive layer.

2 Materials/Methods

Blood is a bodily fluid which is supplied to all the cells of the body by means of the circulatory system. The role of the blood in the body is to transport oxygen and carbon dioxide, the supply of nutrients and disposal of end products of metabolism. In addition, the blood provides thermal regulation of the organism and self-healing mechanisms for preventing blood leakage from injured blood vessels. Blood is a heterogeneous solution, which consists of formed elements (cells) suspended in plasma, all of which have different characteristics and fulfill different functions in the organism. There are three types of the cellular components: red blood cells (RBC), white blood cells (WBC), and platelets (PLT), which in total account for about 35–50 % of the volume of whole blood, while the remainder is plasma [9–12].

Red blood cells (erythrocytes) are the most frequently occurring blood cells. There are approx. 5×10^6 μl^{1} red blood cells (RBCs) in normal human blood. They have a round shape of a biconcave disk, with an average diameter of 7.2–7.8 μm, thickness 1.4–2.1 μm, and an area of 130 μm^2 and a volume of 90 μm^3. Their index of refraction is 1.4 in the wavelength range 600–1100 nm [9]. They are very flexible and elastic, which allows them to squeeze through narrow blood vessels and protects against mechanical damage. Their task is to deliver oxygen from the lungs to all cells of the body and discharge carbon dioxide from them back to the lungs, where the gas exchange occurs. The ability to transport oxygen is enabled by the hemoglobin contained in them. Erythrocytes have no nucleus or mitochondria, they cannot divide, and cannot recover from damage. Thus they are formed by the body itself, which takes place in the bone marrow. Their average life is approx. 120 days [10–12].

White blood cells (leukocytes) area part of the body's immune system. Their key role is to protect the body against viruses, bacteria and other foreign bodies and remove dead erythrocytes. There are several types, which differ in structure and purpose: neutrophils, eosinophils, basophils, monocytes, macrophages and lymphocytes. They contain nuclei and have the ability to move. They neutralize the threat by phagocytosis (devouring) of pathogens. Their number is roughly 4×10^3–10×10^3 μl^{-1} [10]. They are formed like spheres with diameter of 8–22 μm [9].

Platelets (thrombocytes) are devoid of the cell nucleus are broken away portions of specific bone marrow cells of irregular shape. Platelets in the blood stream are biconvex, disk-like particles with diameter ranging between 2–4 μm. There are approx. $(2–4) \times 10^5$ μl^{-1} cells in human blood [9–12]. They are used to maintain hemostasis, or the prevention of blood leakage in the event of vascular injury. They contain clotting factors, which are released during platelet aggregation. Within seconds after the injury, platelets adhere to collagen fibrils to form a platelet plug which is sustained by the fibrin fibers, which prevents further the bleeding [10].

Blood plasma—is an extracellular scaffolding, which is mainly used to transport morphotic elements, nutrients and metabolic products. Plasma water constitutes approx. 90 % of plasma, where the rest is composed of proteins, lipids and other components in smaller quantities, including glucose, albumin, globulin, cholesterol, triglycerides, lipoproteins, hormones, minerals, electrolytes [10–12].

Hematocrit (Hct) is one of the basic parameters of the blood. It is defined as the ratio of the volume of RBCs to the total blood volume [9–12]:

$$\mathrm{Hct}\ [\%] = V_{RBC}/V_{Blood}, \tag{1}$$

where V_{RBC} is the volume of red blood cells, and V_{Blood} is the total volume of blood. This parameter is extremely important, because it provides the information on the amount of RBCs in the blood sample, which dominates all optical properties of blood samples. Elevated levels of Hct may be the result of overproduction of RBCs, which is a sign of the circulatory system diseases. This can cause blockages, which increases the risk of stroke and myocardial infarction [9–11].

The mean corpuscular volume of red blood cells (MCV) is defined as the ratio of the hematocrit to the total amount of RBCs, the Red Blood Cell Count (RBCC):

$$\mathrm{MCV}[fl] = Hct[\%]/RBCC\left[10^6/\mu l\right] \tag{2}$$

MCV determines the size of blood cells. The smaller sizes of blood cells can be caused by iron deficiency, while larger sizes are the evidence of deficiency of vitamin B12 or folic acid.

The mean corpuscular hemoglobin (MCH)—the average weight of hemoglobin in a RBC is defined as the ratio of total hemoglobin in a sample to the amount of RBC's:

$$\mathrm{MCH}[pg] = Hb[g/l]/RBCC[10^6/\mu l]. \tag{3}$$

The average weight of hemoglobin carried by the cell is determined by this parameter. Various disease states, and particularly of anemia caused by iron deficiency, reduce MCH.

The mean corpuscular hemoglobin concentration (MCHC) is an average concentration of hemoglobin in RBCs and is defined as the ratio of hemoglobin to hematocrit:

$$\mathrm{MCHC}[g/l] = Hb[g/l]/HCT[10^6/\mu l]. \tag{4}$$

The Hb saturation of RBCs increases in the states of overhydration, due to penetration of water into the cells. In contrast, dehydration is reflected in artificially low MCHC. The drop in MCHC is also typical for microcytic anemias.

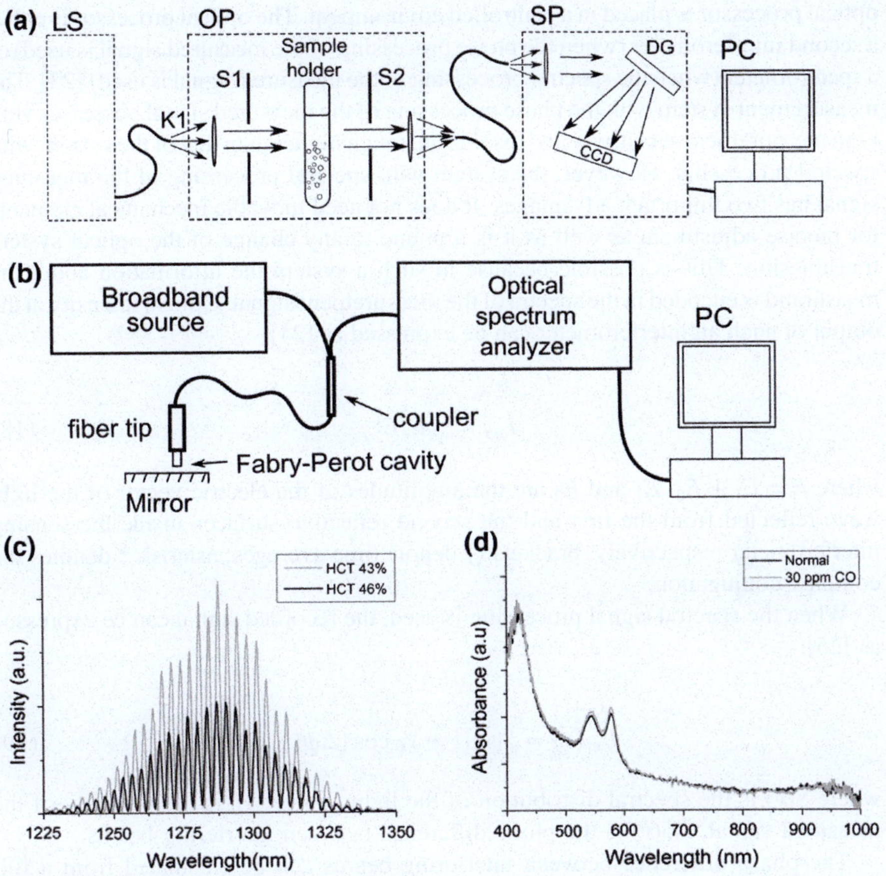

Fig. 1 Measurements systems and spectra: **a** absorption spectrometer set-up; **b** low-coherence interferometry set-up; **c** spectra of human blood samples with different Hct level measured by the low-coherence interferometry set-up; **d** absorption spectra of normal and CO-exposed whole blood samples

One of the methods allowing measurements of blood parameters, such as complex refractive index, is low-coherence interferometry. The light from the broadband source is transmitted to the sensing interferometer by the coupler and an optical fiber. At the sensing interferometer the amplitude of light is divided into two components and an optical path difference (OPD), which depends on the instantaneous value of the measurand, is introduced between them. The sensing interferometer is designed in such a way that a defined relationship exists between the optical path difference and the measurand. The signal from the sensing interferometer is transmitted back by the fiber to the optical processor. The optical processor consists of a second optical system, the output of which is a function of OPD generated at the sensing interferometer. The sensing interferometer is located inside the measurand field whilst the optical processor is placed in a controlled environment. The optical processor is either a second interferometer (when the phase processing of the measured signal is used) or a spectrometer (when the spectral processing of the measured signal is used) [23]. The measurement system with the phase processing of the measured signal possesses very high measurement sensitivity and resolution, much higher than that of the system with spectral processing. However, the system with spectral processing of the measured signal has two important advantages. It does not need movable mechanical elements for precise adjustment as well as it is immune to any change of the optical system transmission. This is possible because in such a system the information about the measurand is encoded in the spectra of the measurement signal. Optical intensity at the output of such an interferometer can be expressed as [24]:

$$I_{out} = \langle E_1 E_2^* \rangle, \tag{18}$$

where $E = E_1 + E_2$, E_1 and E_2 are the amplitudes of the electric vector of the light wave reflected from the first and the second reflective surfaces inside the sensing interferometer respectively, brackets $\langle\rangle$ denote time averages; asterisk * denotes the complex conjugation.

When the spectral signal processing is used, the recorded signal can be expressed as [25]:

$$I_{out}(v) = S(v)[1 + V_0 \cos(\Delta\phi(v))] \tag{19}$$

where S(v) is the spectral distribution of the light source; V_0 is the visibility of the measured signal, $\Delta\phi(v)$ is the phase difference between interfering beams.

The phase difference between interfering beams can be calculated from a following equation [26]:

$$\phi(v) = \frac{2\pi\, v\, \delta}{c} \tag{20}$$

where: δ is the optical path difference, c is the velocity of the light in vacuum.

If the light source exhibits a Gaussian spectrum, the normalized spectra pattern is predicted to be a cosine function modified by the Gaussian visibility profile. In the spectral domain signal processing the modulation frequency of the measurement signal gives information about the measurand. It can be noted that for $\Delta\phi = 0$ there is no spectral modulation. If the phase difference between the interfering beams varies from zero, the function takes the form of the cosine curve.

Low-coherence measurement system with spectral signal processing is not sensitive for any change of a transmission of the optical system. This is possible because in the system information about the measurand is encoded in the spectra of the measured signal. Therefore such a setup is the most convenient for the low-coherence blood samples measurements.

Low-coherence interferometer measurement set-up has been assembled from commercially available components Fig. 1b. Super luminescent diode Superlum Broadlighter S1300-G-I-20 with following optical parameters: λ_0 = 1290 nm, $\Delta\lambda$ = 50 nm, was used as a broadband light source. Signal was propagated through standard telecommunication fiber SMF-28. Optical Spectrum Analyzer Ando AQ6319 (wavelength resolution of 1 nm, wavelength accuracy of ±50 pm) was used as an optical processor. Measurements were carried out in a highly controlled conditions (especially temperature in laboratory room). In order to achieve satisfactory measurement precision for blood parameters monitoring, a large number of in vitro experiments with whole human blood were carried out. Blood samples with various parameters were collected from heathy volunteers via Gdansk Blood Donor Centre. Therefore, measurement range has been limited to blood parameters of healthy persons. Examination of numerous samples using low-coherence interferometer measurement set-up, allowed to determine relationship between registered signals and optical parameters. With the use of the described measurement system, the complex refractive index of numerous samples have been measured [27]. Changes in the real part of complex refractive index of measured sample have changed the modulation frequency of measured spectra, whilst the imaginary part influences the value of visibility of measured signal, as shown on Fig. 1c.

3 Blood CO Sensor Design

The main purpose of the new sensor is to shorten the process of diagnosis and continuous monitoring of intoxicated patients. Developed sensors will be dedicated to support diagnostics of critically ill patients in the emergency room at the hospital, small clinics and doctors' offices as well as fire engines and ambulances on the scene of the accidents. That will enable early and appropriate diagnosis, adequate treatment recommendations and specific antidotes selection.

The lab-on-a-chip will be realized in polymer, by employing state-of-the-art technologies such as soft lithography using the elastomer polydimethylsiloxane (PDMS), or by injection moulding of a thermoplastic material such as e.g. cyclic olefin copolymer (COC) or cyclic olefin polymer (COP). The latter solution will permit rapid production of a high number of chips. A solution where a thermoplastic and a soft polymer are combined will also be investigated, as well as possibilities for direct integration onto a Low Temperature Co-fired Ceramic package. Signal readout will be realized using commercial available data sampling hardware and software and carefully adapted for sensors needs.

A possible realization of the microfluidic package is shown in Fig. 2. The package will provide a housing for the silicon electrode chip (i.e. the Lab-on-a-chip component itself) with microfluidic channels ($\sim$50 μm × 400 μm: channel depth × width) for delivery and control of the sample liquid, thermal control, and electrical access to the electrode system of the silicon chip, as well as adequate optical access to the biosensing area.

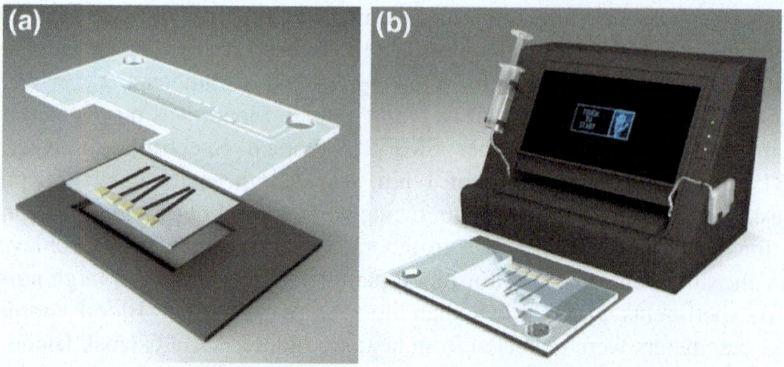

Fig. 2 Schematics of the proposed Lab-on-a-chip blood CO sensor: **a** disposable microfuidic chip; **b** a workstation housing optical and electronic parts of the system

The microfluidic sample delivery channel will ensure laminar flow in the sensing area of the silicon chip. Flow control will be achieved by an external syringe pump.

A particular challenge of the Lab-on-a-chip system is efficient integration of the electrical connection from the sensor chip to the external read-out electronics. Theoretical investigation of light interaction with blood in microfluidic channel of the sensor by Monte Carlo simulations will provide optimal parameters for the implementation of the optical fiber to the sensor. Nanotechnology will be used for the deposition of sensitive layers on the optical fibers surface. Reliability of the nano-sensor will be assessed by conducting a series of in vitro experiments on whole human blood with CO toxic levels measured in a reference analytical laboratory. Biostatistical analysis of measurement data will be used for validation of the sensor performance. Clinical trials will be conducted with the use of nano-sensor in development of approaches for diagnosis of "toxic effect of blood" on the brain microcirculation, which will contribute to the emergence of new knowledge about the resistance of cerebral vessels to hypoxia as a major factor, e.g. complicating stroke, post-birth brain injury, head trauma, brain edema.

4 Hollow Core Microstructured Waveguide Biosensors

Hollow core microstructured waveguides (MSWs) are novel unique materials for building highly sensitive sensors for a variety of biomedical applications including point-of-care medicine [28–30]. In addition to high sensitivity, basic advantages of MSW-based sensors are short response time and small size. The MSWs belong to photonic crystal waveguide family. Their hollow core diameter is a few times larger than the lattice period of the surrounding photonic crystal structures. The cross section of typical hollow MSW used for sensing is shown in Fig. 3a. The structure of the MSW, made from flint glass, consists of the hollow core of 150 μm in

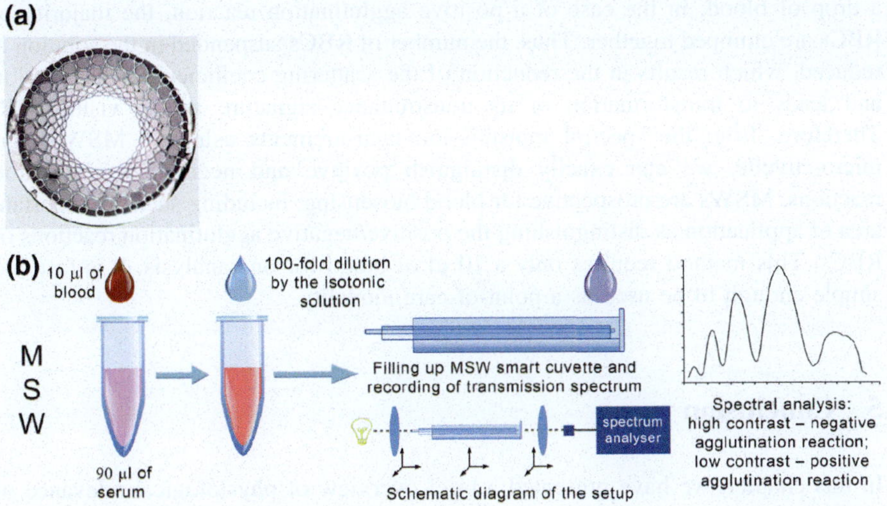

Fig. 3 MSW method for blood testing [33]: **a** electronic micrograph of the cross-section of hollow MSW (designed and manufactured by SPE "Nanostructured Glass Technology", Saratov, Russia); **b** protocol and equipment for blood testing using MSW smart cuvette and a portable spectrometer

diameter and five functional concentric layers of capillaries and an external buffer layer. Its outer diameter is around 1 mm.

The MSW possesses a few narrow and smooth transmission bands in the visible/NIR wavelength range (Fig. 3b, right). The location and number of transmission bands depend on the diameter of the MSW's core [28–33]. The unique features of MSW specify a high sensitivity to alterations of scattering, absorption, and refractivity of solutions filling up the MSW's core and channels in cladding. Usually, a short piece of MSW (a few centimeters in length) serves as a smart micro-cuvette filled up by a biological liquid sample tested in a portable fiber-optic spectrometer for providing spectral transmittance measurements.

For example, the 6 cm MSW with a hollow core diameter of 150 μm was utilized as a smart cuvette for blood typing [33]. Transmission spectra were measured using a setup containing a broadband light source (halogen lamp), fiber-optical elements for launching and collecting light, adjustable opto-mechanics providing a positional resolution of 1.25 μm, a spectrometer, Ocean Optics HR4000, operating in the visible/NIR range, and a PC (schematic diagram of the setup is present in Fig. 3b). An MSW-cuvette serves as a basic replaceable element of a biosensor for blood typing using a standard agglutinating serum technique. The protocol of the MSW technique for blood typing is present in Fig. 3b. A blood type (or group) is one of the most important characteristics of blood evaluated by the presence or absence of agglutinogens (antigens) on the surface of RBCs and antibodies in the blood plasma.

The new technique of blood typing based on hollow core MSW potentially allows one to decrease the percentage of mistakes [33]. After mixing the serum with a drop of blood, in the case of a positive agglutination reaction, the majority of RBCs are clumped together. Thus, the number of RBCs suspended in the solution is reduced, which results in the reduction of the scattering coefficient of the solution and leads to transformation of its transmittance signature in the visible/NIR. Therefore, from the spectral transmission measurements using an MSW smart micro-cuvette, we can exactly distinguish positive and negative agglutination reactions. MSWs are prospective for blood biosensing, including such an important area of application as distinguishing the positive/negative agglutination reactions of RBCs. This method requires only a 10 μl of blood for one analysis. It is fast and simple enough to be used as a point-of-care modality.

5 Conclusion

In this chapter, we have presented a brief overview of physiological relevance of blood carbon monoxide levels on poisoning, as well as the optical properties of blood. Low-coherence interferometer and absorption spectroscopy set-up which we designed and elaborated are described. Those systems enable to obtain the optical parameters of blood such as complex refractive index and absorption spectrum. This gives us opportunity to find relationship between optical and hematological parameters of blood, such as complex refractive index and hematocrit level. We show the concept of optoelectronic sensors for rapid detection of environmental toxicity in blood of poisoned people. The application of the sensor can shorten the time of diagnosis of poisoned patients. It will be dedicated to support diagnostics of all patients in severe clinical state, such as fire victims. We plan to use such method in the Emergency Departments, small clinics and doctors' offices. The final section focuses on the hollow core microstructured waveguide biosensors for applications in biomedical sensors. Such waveguide can be used for investigation of optical parameters (scattering, refractivity, etc.) of many biological liquids, such as blood.

Acknowledgments This study was partially supported by the National Science Center, Poland under the grant 2011/03/D/ST7/03540, DS Programs of the Faculty of Electronics, Telecommunications and Informatics, Gdańsk University of Technology as well as European Cooperation in Science and Technology (COST) Action BM1205. The work done by VVT was supported by grant No. 14-15-00186 of the Russian Science Foundation.

References

1. Raub JA, Mathieu-Nolf M, Hampson NB, Thom SR (2000) Carbon monoxide poisoning—a public health perspective. Toxicology 145:1–14
2. Nieścior M, Jackowska T (2013) Carbon monoxide intoxication. Postępy Nauk Medycznych 7:519–522

3. Ryter SW, Otterbein LE (2004) Carbon monoxide in biology and medicine. BioEssays 26(3):270–280
4. Federal Register (1994) National ambient air quality standards for carbon monoxide—final decision. Fed Regist 59, 38:906–938, 917
5. Torrance RW (1998) Transport of carbon monoxide in the body by hemoglobin, IUTOX VIIIth International Congress of Toxicology “Carbon monoxide: the unnoticed poison of the 21st Century”. Université de Bourgogne, France, pp 115–122
6. Prockop LD, Chichkova RI (2007) Carbon monoxide intoxication: An updated review. J Neurol Sci 262:122–130
7. Mutschler E, Geisslinger G, Kroemer HK, Ruth P, Schaefer-Korting M (2008) Arzneimittel-wirkungen—Lehrbuch der Pharmakologie und Toxikologie *(Effects of drugs—textbook of pharmacology and toxicology)*. Endo-Praxis 24(03):1077–1078
8. Sohn YH (2010) Carbon monoxide poisoning. In: Kompoliti K, Metman LV (eds) Encyclopedia of movement disorders. Academic Press, Oxford, pp 187–189
9. Handin R, Lux S, Stossel T (2003) Blood: principles & practice of hematology. Lippincott Williams & Wilkins, Philadelphia
10. Tuchin VV (2012) Dictionary of biomedical optics and biophotonics. SPIE Press, Bellingham
11. Greer J, Foerrster J, Rodgers G, Paraskevas F, Glader B, Arber D, Means R (2008) Wintrobe’s clinical hematology. Lippincott Williams & Wilkins, Philadelphia
12. Ciesla B (2007) Hematology in practice. F.A. Davis Company, Philadelphia
13. Born M, Wolf E (1993) Principles of optics. Pergamon Press, Oxford, pp 613–614
14. Tuchin VV (2006) Optical clearing of tissue and blood. SPIE Press, Bellingham
15. Tuchin VV (ed) (2002) Handbook of optical biomedical diagnostics. SPIE Press, Bellingham
16. Parson WW (2007) Modern optical spectroscopy with examples from biophysics and biochemistry. Springer, Berlin
17. Hollas JM (2004) Modern spectroscopy. Wiley, West Sussex
18. Zwart A, Buursma A, Van Kampen EJ, Zijlstra WG (1984) Multicomponent analysis of hemoglobin derivatives with reversed-optics spectrophotometer. Clin Chem 30:373–379
19. Zwart A, Van Kampen EJ, Zijlstra WG (1986) Results of routine determination of clinically significant hemoglobin derivatives by multicomponent analysis. Clin Chem 32:972–978
20. Roggan A, Friebel M, Dörschel K, Hahn A, Müller G (1999) Optical properties of circulating human blood in the wavelength range 400–2500 nm. J. Biomed. Opt. 4(1):36–46
21. Meinke M, Gersonde I, Friebel M, Helfmann J, Müller G (2005) Chemometric determination of blood parameters using visible–near-infrared spectra. Appl Spectrosc 59(6):826–835
22. Egan WJ, Brewer WE, Morgan SL (1999) Measurement of carboxyhemoglobin in forensic blood samples using UV-visible spectrometry and improved principal component regression. Appl Spectrosc 53(2):218–225
23. Grattan KTV, Meggit B (2000) Optical Fiber Sensor Technology. Kluwer Academic Publisher, Boston
24. Yu F (2002) Fiber optic sensors. Marcel Dekker, New York
25. Jędrzejewska-Szczerska M, Gnyba M, Kosmowski B (2011) Low-coherence fibre-optic interferometric sensors. Acta Phys Pol A 120(4):621–624
26. Jędrzejewska-Szczerska M, Kosmowski B, Hypszer R (2006) Shaping of coherence function of sources used in low-coherent measurement techniques. J Phys IV 137:103–106
27. Jędrzejewska-Szczerska M (2013) Measurement of complex refractive index of human blood by low-coherence interferometry. Eur Phys J Spec Top 222(9):2367–2372
28. Zheltikov AM (2011) Microstructure fibers in biophotonics. In: Popp J, Tuchin VV, Chiou A, Heinemann SH (eds) Handbook of Biophotonics: Vol. 2: Photonics for Health Care, Wiley-VCH Verlag GmbH & Co., KGaA, Weinheim, pp. 77–103
29. Skibina Y, Tuchin VV, Beloglazov VI, Steinmeyer G, Bethge J, Wedell R, Langhoff N (2011) Photonic crystal fibres in biomedical investigations. Quantum Electron 41(4):284–301

30. Skibina JS, Malinin AV, Zanishevskaya AA, Tuchin VV (2014) Photonic crystal waveguide sensing. In: Nikolelis DP, Varzakas T, Erdem A, Nikoleli G-P (eds) Portable biosensing of food toxicants and environmental pollutants, series in sensors. CRC Press, Boca Raton, pp 1–32
31. Malinin AV, Skibina Y, Tuchin VV, Chainikov MV, Beloglazov VI, Silokhin IYu, Zanishevskaya AA, Dubrovskii VA, Dolmashkin AA (2011) The use of hollow-core photonic crystal fibres as biological sensors. Quantum Electron 41(4):302–307
32. Zanishevskaya AA, Malinin AV, Tuchin VV, Skibina YS, Silokhin IY (2013) Photonic crystal waveguide biosensor. J Innovative Opt Health Sci 6(2):1350008-1-6
33. Zanishevskaya AA, Shuvalov AA, Skibina YS, Tuchin VV (2015) Blood typing using microstructured waveguide smart cuvette. J Biomed Opt 20(4):040503

Emerging Biosensor for Pesticide Detection

Ilaria Palchetti

Abstract Biosensors are considered interesting devices for pesticides monitoring. The peculiar characteristics of biosensors allow them to complement current screening and monitoring methods, especially when continuous, real-time, in situ monitoring is required. However, biosensors still faces to stability and detection sensitivity of the biomolecular recognition element. In the recent years, innovative catalytic biosensors have been proposed to solve this aspect. In this chapter, some of these innovative enzyme based biosensors will be reviewed.

Keywords Cholinesterase · Pesticides · Whole cells

1 Introduction: Role of Biosensors in Pesticide Detection

Monitoring of contaminants in the air, water and soil is an instrumental component in understanding and managing risks to human health and ecosystems [1]. Among other pollutants, pesticides are regarded as one of most dangerous environmental contaminant class because of their ability to accumulate and their long-term effects on living organisms. Pesticide is a term used in broad sense for organic toxic compounds used to control weeds, insects, fungi, and other pests. Currently, over 800 active ingredients, which belong to more than 100 substance classes, are present in a wide range of commercial pesticides [2]. Organochlorine, organophosphorus (OP), and organonitrogen pesticides are the most important groups. Among them, organochlorine pesticides are more toxic to many organisms and persist in the environment for more than 30 years; therefore, they are beginning to be withdrawn

I. Palchetti (✉)
Dipartimento di Chimica, Università degli Studi di Firenze, Via della Lastruccia 3, 50019 Sesto Fiorentino, FI, Italy
e-mail: ilaria.palchetti@unifi.it

D.P. Nikolelis and G.-P. Nikoleli (eds.), *Biosensors for Security and Bioterrorism Applications*, Advanced Sciences and Technologies for Security Applications,
DOI 10.1007/978-3-319-28926-7_20

and replaced by compounds that are biodegraded quickly. In place of organochlorine pesticides, OP pesticides have become the most popular pesticides because of their shorter environmental half-life, wide range of applications, and price [2, 3]. OPs are inhibitors of cholinesterase enzymes that are involved in the central nervous system of humans and insects leading to neurotransmitter acetylcholine buildup which interferes with muscular response, functioning of vital organs, and eventually death [3, 4]. Thus are extremely toxic compounds that have found use also as chemical warfare agents (nerve gases).

These issues call for fast and cost-effective analytical techniques to be used in extensive monitoring programs. Given this requirement as well as the time and cost involved in traditional chemical analysis of environmental samples (e.g. chromatographic methods and related hyphenated techniques), there is an expanding need of rapid and innovative methods. Biosensors appear well suited to complement standard analytical methods for a number of environmental monitoring applications [5, 6].

The main advantages offered by biosensor technology over conventional analytical techniques are fast and economical measurements, the possibility of miniaturization, portability, the possibility of continuous monitoring and, in some cases, the ability to measure pollutants in complex matrices with minimal sample preparation. Although many of the systems developed cannot compete with conventional analytical methods in terms of accuracy and reproducibility, they can be used by regulatory authorities and by industry to provide enough information for routine testing and screening of samples. The peculiar characteristics of biosensors allow these devices to complement current field screening and monitoring methods such as ELISA kit, especially when continuous, real-time, in situ monitoring is required [1]. Some excellent reviews summarized the recent progress in use of biosensors for environmental applications [7–13].

Nowadays, biosensor technology largely benefit from the results obtained in other fields of life and applied sciences, such as biotechnology, nanoengineering and nanotechnology. Lab on a chip technology increased a lot the possibility of obtaining easy-to-use and self-contained devices, minimizing the need for sample processing and analysis at a laboratory. Micro- and nanotechnology development continuously help in miniaturize the dimension of the devices.

Considering these facts, in this chapter we would like to give an overview of the enzyme biosensor used for pesticide monitoring, focusing on the innovative aspects. Biotechnologies help to increase the number of stable, sensitive and selective sensing elements that can be used for obtaining reliable biosensors. Thus, description of the attempts to increase stability and sensitivity of the bioreceptors are here reported. Generally, biosensors for pesticide detection can be classified into two major categories, enzyme-based biosensors and immunosensors, and in lesser extent as aptasensors and MIP-based biosensors. This chapter is focused on enzyme-based biosensors.

2 Enzyme-Biosensor: Definition, General Features and Classifications

In early 2000, two Divisions of the International Union of Pure and Applied Chemistry (IUPAC), namely Physical Chemistry and Analytical Chemistry prepared recommendations on the definition, classification and nomenclature related to electrochemical biosensors; these recommendations have been then extended to other types of biosensors [14]. Following these IUPAC recommendations, "a biosensor is defined as a self-contained integrated device, which is capable of providing specific quantitative or semi-quantitative analytical information using a biological recognition element (biochemical receptor) which is retained in direct spatial contact with an electrochemical transduction element. Because of their ability to be repeatedly calibrated, a biosensor should be clearly distinguished from a bioanalytical system, which requires additional processing steps, such as reagent addition. A device that is both disposable after one measurement, i.e. single use, and unable to monitor the analyte concentration continuously or after rapid and reproducible regeneration, should be designated a single use biosensor" [14].

Nowadays, the concept of biologically derived sensing element (i.e. synthetic chemical compounds that mimic the biological material in the development of biosensors) is fully accepted in the scientific community [15].

Biosensors are classified according to the biological specificity-conferring mechanism (Fig. 1) or, alternatively, to the mode of physico-chemical signal transduction. Biosensors may be further classified according to the analytes or reactions that they monitor: direct monitoring of analyte concentration or of reactions producing or consuming such analytes; alternatively, an indirect monitoring of inhibitor or activator of the biological recognition element (biochemical receptor) may be achieved [1, 16].

Regarding the transduction principles, biosensors can be classified as optical, electrochemical, mass, magnetic, calorimetric or micromechanical biosensors. Optical detection by fluorescence spectroscopy is a popular method, due largely to the ease with which biomolecule (especially nucleic acids) can be fluorescently labeled, the availability of many different fluorophores and quenchers, and the inherent capability for real-time multiplex detection [1]. Chemiluminescence is another optical technique widely used. A different optical transduction, based on evanescent wave device, can offer real time label-free optical detection. These biosensors rely on monitoring changes in surface optical properties (shift in resonance angle due to change in the interfacial refractive index) resulting from the surface binding reaction. Electrochemical devices have also proven very useful, due to inherent miniaturization and their compatibility with advanced microfabrication technology. Electrochemical detection usually involves monitoring a current response under controlled potential conditions. However other changes in electrochemical parameters such as capacitance, impedance and conductivity have been used. Another useful label-free detection scheme relies on the use of quartz crystal microbalance (QCM) transducers. QCM biosensors consist of an oscillating crystal

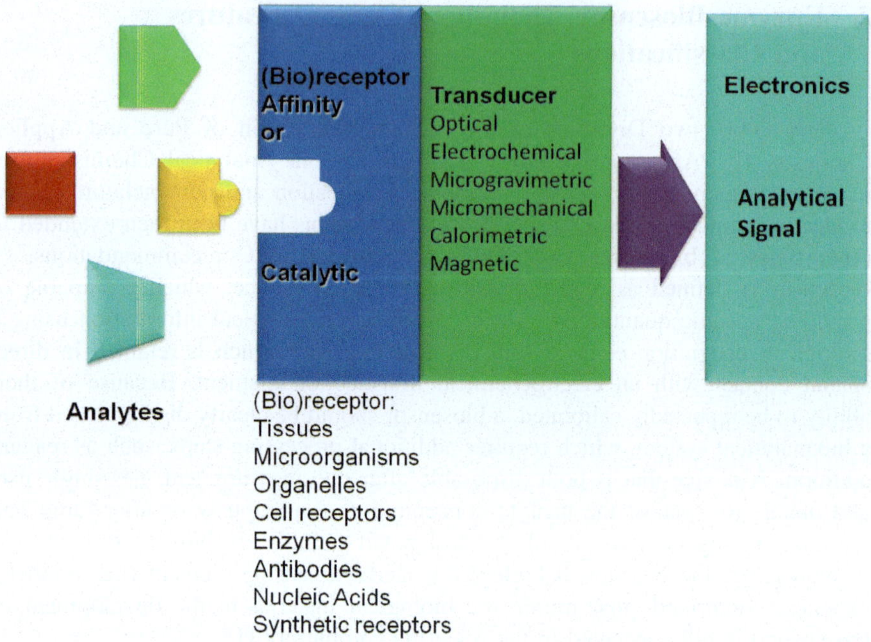

Fig. 1 Schematic representation of a biosensor. Following the different (bio)receptors, biosensors can be classified as catalytic (enzymes, tissues, etc.), affinity (antibodies, cell receptor, synthetic receptors, nucleic acids in genosensors and aptasensors, etc.) or based on complex cellular functions (whole cell, etc.). Biosensors can be also classified in accordance to the transducer (i.e. optical or electrochemical etc.). Reprinted with permission from Ref. [1]

with a bioreceptor immobilized on its surface. The increased mass, associated with the biorecognition reaction, results in a decrease of the oscillating frequency. Acoustic wave sensors used in thickness-shear mode with a liquid sample detect changes in a number of physical properties including mass, viscosity and charge density [1]. Micromechanical transduction are based on cantilevers and label-free biosensor capable of detecting biomolecular interactions via the bending of microfabricated cantilevers coated with bioreceptors were reported in literature [1]. Finally, in magneto-biosensor, magnetic label are used to detect magnetoresistance, giant magnetoresistive effect (GMR), spin-value GMR, and other parameters [1].

All these different types of transducers can be combined with enzymes or enzyme-based systems as the biochemical component, to form the largest and best-known category of biosensors. Enzymes (and all biological elements based on the enzymes contained in it, like tissues, cells, microrganisms) represent the class of what is now called "catalytic elements". L. C. Clark introduced the first example of enzyme-based biosensor. Professor Clark, in a Conference at a Symposium in the New York Academy of Sciences in 1962, described how "to make electrochemical sensors (pH, polarographic, potentiometric, or conductometric) more intelligent," that is by adding "enzyme" [17]. This first example was illustrated by entrapping the enzyme Glucose Oxidase (GOx) in a dialysis membrane over an oxygen sensor. Then subsequently in 1967 Updike and Hicks use the same term "enzyme electrode" to describe a similar device where again the enzyme GOx was immobilized in a polyacrylamide gel onto a surface of an oxygen electrode for the rapid and quantitative determination of glucose [18]. Another step forward in the technology was performed by Guilbault and Montalvo in 1969, with the use of a glass electrode coupled with urease to measure urea concentration by a potentiometric transducer [19]. During the 1970s, thermal [20] enzyme-based biosensors were proposed. In the same years, Lubbers and Opitz [21] demonstrated the concept to make an optical biosensor for alcohol by immobilizing alcohol oxidase on the end of a fiber optic oxygen sensor.

Nowadays, enzyme-based biosensors are among the most important commercially available category of biosensors. Yellow Springs Instrument Company, USA, launched the first glucose analyzer (known as Model 23A YSI analyzer) in 1975, which is employed till date as a standard in clinical diagnosis [22]. However, it was only during the 1980s that large-scale commercial success of enzyme biosensors for glucose monitoring was first achieved [23].

The mode of operation of an enzyme-biosensor can be summarized in a five-step process, including the transport of the substrate from the bulk of solution/matrix toward the enzyme layer, the diffusion of the substrate through the layer to the enzyme active site, the reaction at the enzyme active site, the diffusion of the product towards the transducer, and finally the conversion of the concentration of the product at this interface into an analytical signal by the transducer. The detailed description of the reaction occurring at the active site can be found in any standard biochemistry book, and here we will remind only the basic enzyme catalysis mechanism based on Michaelis-Menten equation [24, 25] and the K*m* Michaelis constant of the enzyme for the given substrate. Enzyme-based biosensors may be classified according to the analytes or reactions that they monitor [24]; for example, direct monitoring of analyte concentration or of reactions producing or consuming such analytes, is the direct mode; alternatively, an indirect monitoring of inhibitor or activator of the enzyme may be achieved. Biosensors based on the principle of enzyme inhibition have by now been applied for a wide range of significant analytes including pesticides [26–31]. These toxic analytes inhibit normal enzyme function. In general, the development of these biosensing systems relies on a quantitative measurement of the enzyme activity before and after exposure to a target analyte. Typically the percentage of inhibited enzyme (I%) that results after exposure to the inhibitor is quantitatively related to the inhibitor (i.e., analyte) concentration and the incubation time. Consequently, the residual enzyme activity is inversely related to the inhibitor concentration. The inhibition can be either reversible or result in an irreversible inactivation of the enzyme [25].

In the following paragraphs, some of the most common enzymes used in the development of biosensors for pesticide detection are described, highlighting the progress of biotechnologies in obtaining more stable and sensitive catalytic biosensors.

3 Emerging Biorecognition Elements for Pesticide Detection

3.1 *Esterase-Based Biosensors*

Since the pioneering work of Guilbault et al. [32] on cholinesterase (ChE) purification, inhibitor analysis, and assay development, biosensors based on the principle of enzyme inhibition have become an active research area. Typical examples of enzymes involved in environmental applications are acetylcholinesterase (AChE) and butyrylcholinesterase (BChE) for organophosphate (OP) and carbammate pesticide analysis. OP and carbamate pesticides selectively inhibit ChEs by blocking the serine in the active site through nucleophilic attack to produce a serine phosphoester (via phosphorylation). AChE enzymes are available from different sources, such as Electric eel (Ee), Bovine or Human erythrocytes, Horse serum and Human blood, insects (Drosophila). Generally, ChE enzymes isolated from insects are more sensitive than those extracted from other sources. ChE enzymes modified by the use of recombinant DNA technologies have also been reported [9, 33, 34]. These genetically modified AChE can decrease orders of magnitude the detection limit found for the use of AChE from wild type Drosophila melanogaster and Electric Eel, as reported in [9, 34, 35]. Valdés-Ramírez et al. reported on the use of a recombinant ChE for the highly sensitive detection of the organophosphorus insecticide dichlorvos. The enzyme was immobilized by entrapment in a photo-crosslinkable PVA-SbQ polymer on a screen-printed graphite electrode. The enzyme activity was estimated amperometrically at +100 mV versus Ag/AgCl by measuring the thiocholine produced by the enzymatic hydrolysis of the acetylthiocholine substrate using cobalt phthalocyanine as electron mediator. The pesticide was measured in the presence of 5 % acetonitrile without loss of enzyme activity. The best sensitivity was achieved with the genetically modified enzyme with a detection limit of 7×10^{-11} M as compared to 1×10^{-8} M with the wild type and 6×10^{-7} M with the Ee-isolated enzyme. The genetically modified enzyme biosensor was used to quantify dichlorvos in a sample of skin apple after extraction with acetonitrile [35].

The increased performance of biomolecular and genetic techniques leads to the development of new sources of enzyme production, also allowing better isolation and characterization steps [36]. Recently, the possibility of using the esterase 2 from Alicyclobacillus acidocaldarius (EST2), as the biological part of a colorimetric and fluorescence biosensor for paraoxon detection, demonstrating its characteristics of stability over time, reproducibility and sensitivity was reported [4, 37]. EST2 is a carboxylesterase belonging to the hormone sensitive lipase (HSL) family that includes several AChEs. This enzyme shows a very long time stability and an appropriate resistance and activity at different pH values and temperatures, as well as a good stability in the presence of low concentrations of organic solvents and detergents. Furthermore, the EST2 3D structure has been solved at 2.6 Å, giving the possibility of having a model structure for in silico studies such as molecular

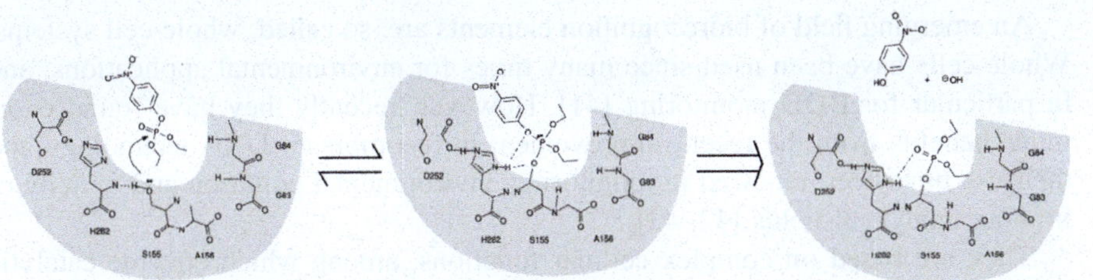

Fig. 2 Schematic representation of the EST2-paraoxon reaction, reprint with permission from Ref. [4]

docking predictions. EST2 interacts with paraoxon in a similar way to the substrate (Fig. 2), but the covalent intermediate of reaction is too stable to dissociate, so irreversibly inhibiting the enzyme. EST2's peculiarity in terms of stability, in addition to its sensitivity and selectivity toward phosphoryl OPs, like paraoxon and methyl paraoxon, makes this enzyme a candidate for use as a bio-receptor in biosensors for qualitative and quantitative OP detection.

3.2 OPH-Based Biosensors

The class of biosensors based on OPs as substrates of organophosphorus hydrolase (OPH), as opposed to inhibitors, is single step, simple, rapid, reversible and selective for OP [38].

The basis of the mechanism of an OPH-based biosensor is that OPH is able to hydrolyze a number of OP pesticides, such as paraoxon, parathion, coumaphos, diazinon, chlorpyrifos, and methyl parathion, [39], leading to production of organophosphorus acid and alcohol as a result of the cleavage of the P–O, P–F, P–S, or P–CN bonds. In many cases, the alcohol produced is chromophoric and/or electroactive. An amperometric biosensor for oganophosphorus (OP) pesticides based on a carbon nanotube (CNT)-modified transducer and an organophosphorus hydrolase (OPH) biocatalyst is described in [40]. A bilayer approach with the OPH layer atop of the CNT film was used for preparing the CNT/OPH biosensor. The CNT layer leads to a greatly improved anodic detection of the enzymatically generated p-nitrophenol product, including higher sensitivity and stability. The sensor performance was optimized with respect to the surface modification and operating conditions. Under the optimal conditions the biosensor was used to measure as low as 0.15 μM paraoxon and 0.8 μM methyl parathion with sensitivities of 25 and 6 nA/μM, respectively.

However, the lower sensitivity of potentiometric and optical biosensors and poor selectivity of amperometric biosensors over phenolic compounds limit their applications in environmental monitoring. Moreover, these enzyme electrodes required purified enzyme, which is a laborious, time-consuming, and costly effort.

An emerging field of biorecognition elements are, so called, whole-cell systems. Whole cells have been used since many times for environmental applications, and in particular for BOD monitoring [41]. However, recently they have found enormous benefits from the recent improvement in recombinant DNA technology and their use have been renewed in monitoring environmental pollution and toxicity or in other analytical fields [42, 43].

They are based on complex cellular functions, among which enzyme catalytic reactions play an important role. These biosensors are constructed through the fusion of promoters, responsive to the relevant environmental conditions, to easily monitored reporter genes. Depending on the choice of reporter gene, expression can be monitored by the production of color, light, fluorescence or electrochemical reactions. Although there are numerous examples of genetic modification to these whole cell types (bacteria, yeast, algae and tissue culture cells) genetically engineered bacteria are most often reported in cell-based biosensors [38]. Over the past decades, different types of bacterial cells have been incorporated into the transducers for making microbial biosensors for the detection of pesticides. Thus, using microorganisms as biorecognition element provides an ideal alternative to purified enzyme. As already mentioned, different kind of transducers have been coupled to these bioreceptors. An amperometric microbial biosensor for the direct measurement of OP is described in Fig. 3 [44]. The sensor is based on a carbon paste electrode containing genetically engineered cells expressing organophosphorus hydrolase (OPH) on the cell surface. OPH catalyzes the hydrolysis of organophosphorus pesticides with p-nitrophenyl substituent such as paraoxon, parathion and methyl parathion to p-nitrophenol. The later is detected anodically at the carbon transducer with the oxidation current being proportional to the nerve-agent concentration. The sensor sensitivity was optimized with respect to the buffer pH and loading of cells immobilized using paraoxon as substrate. The best sensitivity was obtained using a sensor constructed with 10 mg of wet cell weight per 100 mg of carbon paste and operating in pH 8.5 buffer. Using these conditions, the biosensor was used to measure as low as 0.2 μM paraoxon and 1 μM methyl

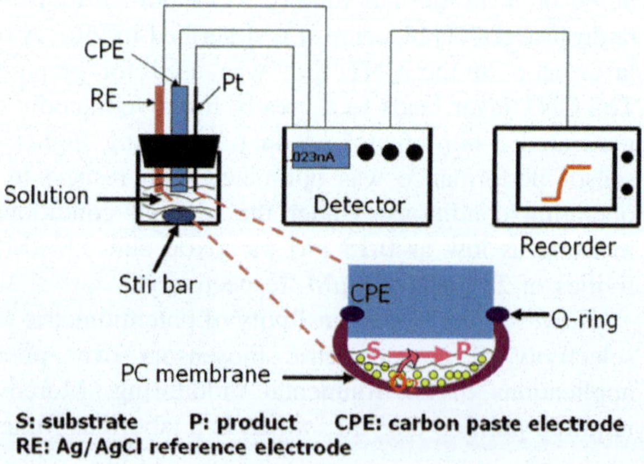

Fig. 3 Schematic of carbon paste electrode-based amperometric biosensor, reprinted with permission from Ref. [38]

parathion with very good sensitivity, excellent selectivity and reproducibility. The microbial biosensor had excellent storage stability, retaining 100 % of its original activity when stored at 4 °C for up to 45 days.

A microbial biosensor consisting of a dissolved oxygen electrode modified with the genetically engineered PNP-degrader Moraxella sp. displaying organophosphorus hydrolase (OPH) on the cell surface for sensitive, selective, rapid and direct determination of p-nitrophenyl (PNP)-substituted organophosphates (OPs) is reported in [39]. Surface-expressed OPH works in tandem with the PNP oxidation machinery of the Moraxella sp. to degrade PNP-substituted OPs and PNP simultaneously while consuming oxygen, that is proportional to the analyte concentration. The optimum performance was obtained by electrodes constructed using 0.35 mg dry weight of cell and operating at pH 7.5. Operating at optimum conditions the biosensor was able to measure as low as 0.1 μM (27.5 ppb) of paraoxon and had excellent selectivity against triazines, carbamates and OPs without PNP substitutent. The biosensor was stable for a week when stored at 4 °C. The applicability of the biosensor to measure OPs in lake water was demonstrated [39].

A yeast strain that co-displays OPH and enhanced green fluorescent protein (EGFP) on the cell surface using a Flo1p anchor system was used to detect OP in Ref. [45]. OP degradation releases protons and causes a change in pH. This pH change results in structural deformation of EGFP, which triggers quenching of its fluorescence, thereby making this cell useful for visual detection of OPs. Fluorescence microscopy confirmed the high-intensity fluorescence displayed by EGFP on the cell surface. The yeast strain possessed sufficient OPH hydrolytic activities for degrading OPs, as measured by incubation with 1 mM paraoxon for 24 h at 30 °C. In addition, with 20 mM paraoxon at 30 °C, fluorescence quenching of EGFP on the single yeast cell was observed within 40 s in a microchamber chip. Other examples are reported in Ref. [38].

3.3 *Other Enzymes*

Inhibition is the mechanism used also by tyrosinase based biosensor [46]. Tyrosinase oxidizes monophenols in two consecutive steps: first, the enzyme catalyzes the o-hydroxylation of monophenol to o-diphenol, which, in a second step, is oxidized to its corresponding o-quinone. Tyrosinase is inhibited by different compounds, such as carbamate pesticides and atrazine. Numerous electrochemical biosensors based on the inhibition of tyrosinase activity have been reported. Tyrosinase biosensors suffer from poor specificity since many substrates and inhibitors can interfere. The enzyme is inherently unstable, reducing the lifetime of the tyrosinase-based biosensors. However, tyrosinase can stand high temperatures and the organic solvents used to dissolve the pesticides [2, 3]. Even if genetic techniques have been already applied to produce more stable tyrosinases for industrial application, to the best of our knowledge these are not yet applied to biosensor development. Alkaline Phosphatase (ALP), Acid Phosphatase and

Peroxidase are other enzymes used in the development of biosensors for pesticides monitoring, due to the inhibition effect of many pesticides on these enzymes. However poor specificity is generally obtained. Other enzymes, instead, like paraoxanase (PON), a serum high-density lipoprotein-associated esterase which among other things provides some protection against OP poisoning and diisopropyl fluorophosphate (DFPase) found in squid nervous tissues, organophosphorus acid anhydrolase (OPAA), can use OP as substrate, and should be interesting enzyme to be studied in biosensor develpment.

4 Conclusions

Sensor technologies hold exceptional promise for providing critical information for continuous, real-time and in situ data collection. Simultaneous measurement of multiple agents (multiplexing) within a single device, has been also demonstrated. New sensing modalities have emerged from biotechnology, nanotechnology and nanoengineering that could be adapted and developed for environmental monitoring.

In addition to being self-contained, biosensors are capable of quantitative, continuous data capture in the field, without the need for sample processing and analysis at a laboratory. However, to be competitive with other existing technologies, these devices must be easy-to-use, portable, minimally inconvenient, rugged, and inexpensive to deploy. In our opinion, all these things are within the reach of existing manufacturing technologies and of immediate development. The real challenge, nowadays, is the possibility to obtain stable and selective (bio) sensing molecules. As reported in this chapter biotechnology can greatly help in reaching this goal.

References

1. Palchetti I, Mascini M (2011) Biosensor techniques for environmental monitoring. In: R.S. of Chemistry (ed) Nucleic Acid Biosens. Environ Pollut Monit
2. Liu S, Zheng Z, Li X (2013) Advances in pesticide biosensors: current status, challenges, and future perspectives. Anal Bioanal Chem 405:63–90
3. Sassolas B, Prieto-Simón J-L (2012) Marty, enw_18264. Am J Anal Chem 3:210–232
4. Carullo P, Cetrangolo GP, Mandrich L, Manco G, Febbraio F (2015) Fluorescence spectroscopy approaches for the development of a real-time organophosphate detection system using an enzymatic sensor. Sensors 15:3932
5. Verma N, Bhardwaj A (2015) Biosensor technology for pesticides—a review. Appl Biochem Biotechnol 175:3093–3119
6. Kumar P, Kim K-H, Deep A (2015) Recent advancements in sensing techniques based on functional materials for organophosphate pesticides. Biosens Bioelectron 70:469–481
7. Rogers KR (2006) Recent advances in biosensor techniques for environmental monitoring. Anal Chim Acta 568:222–231

8. Rodriguez-Mozaz S, Lopez de Alda M, Barceló D (2006) Biosensors as useful tools for environmental analysis and monitoring. Anal Bioanal Chem 386:1025–1041
9. Campàs M, Prieto-Simón B, Marty J-L (2009) A review of the use of genetically engineered enzymes in electrochemical biosensors. Semin Cell Dev Biol 20:3–9
10. Tudorache M, Bala C (2007) Biosensors based on screen-printing technology, and their applications in environmental and food analysis. Anal Bioanal Chem 388:565–578
11. Rodriguez-Mozaz S, de Alda MJL, Marco M-P, Barceló D (2005) Biosensors for environmental monitoring: A global perspective. Talanta 65:291–297
12. Ron EZ (2007) Biosensing environmental pollution. Curr Opin Biotechnol 18:252–256
13. Amine A, Arduini F, Moscone D, Palleschi G (2016) Recent advances in biosensors based on enzyme inhibition. Biosens Bioelectron 76:180–194
14. Thévenot DR, Toth K, Durst RA, Wilson GS (2001) Electrochemical biosensors: recommended definitions and classification1. Biosens Bioelectron 16:121–131
15. Turner APF (2013) Biosensors: sense and sensibility. Chem Soc Rev 42:3184–3196
16. Palchetti I, Mascini M, Escarpa A, González MC, López MÁ (2015) Electrochemical enzyme biosensors. In: Agricultural and food electroanalysis. Wiley, New York, 207–221
17. Clark LC Jr, Lyons C (1962) Electrode systems for continuous monitoring in cardiovascular surgery. Ann NY Acad Sci 102:29–45
18. Updike SJ, Hicks GP (1967) The enzyme electrode. Nature 214:986–988
19. Guilbault GG, Montalvo JG (1969) Urea-specific enzyme electrode. J Am Chem Soc 91:2164–2165
20. Danielsson B, Mosbach K (1974) An enzyme thermistor. Biocim Biophys Acta 364:140–145
21. Völkl KP, Opitz N, Lübbers DW (1980) Continuous measurementof concentration of alcohol using a fluorescence-photometric enzymatic method. Fresen Z Anal Chem 301:162–163
22. Vashist SK, Zheng D, Al-Rubeaan K, Luong JHT, Sheu F-S (2011) Technology behind commercial devices for blood glucose monitoring in diabetes management: a review. Anal Chim Acta 703:124–136
23. Newman JD, Turner APF (2005) Home blood glucose biosensors: a commercial perspective. Biosens Bioelectron 20:2435–2453
24. Eggins B (2002) Chemical sensors and biosensors. Wiley, Chichester
25. Tran MC (1993) Biosensors, Chapman & Hall, London
26. Cagnini A, Palchetti I, Mascini M, Turner APF (1995) Ruthenized screen-printed choline oxidase-based biosensors for measurement of anticholinesterase activity. Mikrochim Acta 121:155–166
27. Cagnini A, Palchetti I, Lionti I, Mascini M, Turner APF (1995) Disposable ruthenized screen-printed biosensors for pesticides monitoring. Sens Actuators B Chem 24:85–89
28. Hernandez S, Palchetti I, Mascini M (2000) Determination of anticholinesterase activity for pesticides monitoring using a thiocholine sensor. Int J Environ Anal Chem 78:263–278
29. Laschi S, Ogończyk D, Palchetti I, Mascini M (2007) Evaluation of pesticide-induced acetylcholinesterase inhibition by means of disposable carbon-modified electrochemical biosensors. Enzyme Microb Technol 40:485–489
30. Palchetti I, Cagnini A, Del Carlo M, Coppi C, Mascini M, Turner APF (1997) Determination of anticholinesterase pesticides in real samples using a disposable biosensor. Anal Chim Acta 337:315–321
31. Arduini F, Forchielli M, Amine A, Neagu D, Cacciotti I, Nanni F et al (2015) Screen-printed biosensor modified with carbon black nanoparticles for the determination of paraoxon based on the inhibition of butyrylcholinesterase. Microchim Acta 182:643–651
32. Guilbault GG, Sadar MH, Kuan SS, Casey D (1970) Enzymatic methods of analysis: Trace analysis of various pesticides with insect cholinesterases. Anal Chim Acta 52:75–82
33. de Oliveira PRB, Marques GS, dos Nunes TCR, Santos S, Andreescu J-L (2004) Marty, Comparative investigation between acetylcholinesterase obtained from commercial sources and genetically modified Drosophila melanogaster: Application in amperometric biosensors for methamidophos pesticide detection. Biosens Bioelectron 20:825–832

34. Campas M, Garibo D, Prieto-Simon B (2012) Novel nanobiotechnological concepts in electrochemical biosensors for the analysis of toxins. Analyst. 137:1055–1067
35. Valdés-Ramírez G, Fournier D, Ramírez-Silva MT, Marty J-L (2008) Sensitive amperometric biosensor for dichlorovos quantification: Application to detection of residues on apple skin. Talanta 74:741–746
36. Mandrich L, Manco G, Rossi M, Floris E, Jansen-van den Bosch T, Smit G et al (2006) Alicyclobacillus acidocaldarius thermophilic esterase EST2's activity in milk and cheese models. Appl Environ Microbiol 72:3191–3197
37. Febbraio F, Merone L, Cetrangolo GP, Rossi M, Nucci R, Manco G (2011) Thermostable esterase 2 from alicyclobacillus acidocaldarius as biosensor for the detection of organophosphate pesticides. Anal Chem 83:1530–1536
38. Mulchandani A (2011) Rajesh, microbial biosensors for organophosphate pesticides. Appl Biochem Biotechnol 165:687–699
39. Mulchandani P, Chen W, Mulchandani A (2006) Microbial biosensor for direct determination of nitrophenyl-substituted organophosphate nerve agents using genetically engineered Moraxella sp. Anal Chim Acta 568:217–221
40. Deo RP, Wang J, Block I, Mulchandani A, Joshi KA, Trojanowicz M et al (2005) Determination of organophosphate pesticides at a carbon nanotube/organophosphorus hydrolase electrochemical biosensor. Anal Chim Acta 530:185–189
41. Nakamura H (2010) Recent organic pollution and its biosensing methods. Anal Methods 2:430–444
42. Xia L, Liang B, Li L, Tang X, Palchetti I, Mascini M et al (2013) Direct energy conversion from xylose using xylose dehydrogenase surface displayed bacteria based enzymatic biofuel cell. Biosens Bioelectron 44:160–163
43. Liang B, Li L, Tang X, Lang Q, Wang H, Li F et al (2013) Microbial surface display of glucose dehydrogenase for amperometric glucose biosensor. Biosens Bioelectron 45:19–24
44. Mulchandani P, Chen W, Mulchandani A, Wang J, Chen L (2001) Amperometric microbial biosensor for direct determination of organophosphate pesticides using recombinant microorganism with surface expressed organophosphorus hydrolase. Biosens Bioelectron 16:433–437
45. Fukuda T, Tsuchiya K, Makishima H, Tsuchiyama K, Mulchandani A, Kuroda K et al (2010) Organophosphorus compound detection on a cell chip with yeast coexpressing hydrolase and eGFP. Biotechnol J 5:515–519
46. Vidal JC, Esteban S, Gil J, Castillo JR (2006) A comparative study of immobilization methods of a tyrosinase enzyme on electrodes and their application to the detection of dichlorvos organophosphorus insecticide. Talanta 68:791–799

visible wavelength region for typically applied metal nanoparticles. The rate of electric field enhancement is limited by the size of the nanoparticles.

2.2 *Optical Waveguide Based Biosensors*

Optical waveguides are structures with high refractive index capable to guide the propagation of electromagnetic waves by total internal reflection within the high index volume. Planar waveguides are typically created by forming layers of high refractive index materials between low index layers. At appropriate angles of incidence (or wavelength), light is reflected by the low refractive index surfaces, therefore, is entrapped in the waveguide. In the case of SPR the entire gold surface of the sensor serves as a guiding layer, allowing planar translational movement of plasmon excitation along the surface.

On the basis of recording changes in optical density, various sensor platforms can be designed. Optical effects created may include reflectance, interference or fluorescence spectroscopy, taking advantage of the evanescent field principle in case of a stripe or film waveguides, as well as mode coupling at or off resonance, using polarized or non-polarized light to observe phase shifts [16]. Optical parameters or effects to be measured include phase changes, coupling effects between waveguide structures, grating couplers, dual beam and mode beat interferometric signals. Besides the above techniques, ellipsometry can also be applied, making use of light of varying polarization states and exhibiting high sensitivity, or simpler techniques like multiple reflection interference (white light interference) at thin films can be mentioned. Most important measurement setups include various optical waveguides, interferometer based sensors, reflectometric interference spectroscopy (RIfS), total internal reflection fluorescence (TIRF), total internal reflection ellipsometry (TIRE), ring resonator, etc. Signals obtained through the optical waveguide may be directly measured, or can also be amplified by optical means, including interferometry using planar waveguides or optical ring resonator based sensors [7, 35].

Optical waveguide signals are typically affected by an adlayer built up on the waveguide layer, where it comes to contact with the analyte medium. Molecular processes going on in this adlayer may modify various characteristics of light propagating in the waveguide, including its intensity, phase, polarization state or its incoupling resonance angle, thus, all these optical characteristics can serve as signal transducing elements [16, 30, 35, 74]. A variety of optical methods have been developed for biosensing by evanescent wave optical devices in real time, without labels [94, 185], and gained importance in pharmaceutical and biomedical research [27, 28, 39, 96, 130, 137, 179], as well as in environmental and food safety [7, 11].

Waveguide sensors are based on the phenomenon of incident light incoupling into a dielectric layer by using a diffraction grating (first described in 1902 [178], developed further to support guided waves [70] and applied for sensors [156, 160]), and are mostly constructed in "normal symmetry"; consisting a high refractivity index layer between two low refractivity index layers (see above), where the bottom layer can be the surface of the glass (n ~ 1.5) substrate (chip). The waveguide film can be made of titanium dioxide (n ~ 2), and the upper layer can be formed by the aqueous analyte medium (n ~ 1.4) in direct contact with the waveguide. In such normal symmetry setups red or green laser entrapped in the waveguide may generate the evanescent field sensitive to molecular changes in the aqueous medium at a depth of 100–200 nm with an exponential decay towards the cover medium [35, 74]. Yet, waveguide generation is also possible by employing a "reverse symmetry" configuration [37, 74–77], where the setup is similarly a layered structure as in the case of the normal symmetry sensors, but the waveguide support is made of a lower refractivity index material (e.g. nanoporous silica, n = 1.193) than that of the covering medium. In such a setup the spatial distribution of the evanescent field is of reverse symmetry, allowing deeper (up to μm scale) penetration into the covering medium, and therefore, sensing in a thicker adlayer than in the case of normal symmetry waveguide sensors, which is particularly preferential when sensing interactions inside living cells. Employing multimode waveguides even several different penetrations depths can be simultaneously applied for monitoring cell refractive index variations at various depths [73].

The detection sensitivity of evanescent optical field based sensors can be enhanced by the use of interferometry. For this purpose, the incident light (laser) is usually split into two beams: the object beam and the reference beam. The object beam passes through the study object (e.g. cells) as in single-beam techniques, but upon interaction with it, it is not directly detected, but brought in interference with the reference beam, unaffected by any interactions with the object. Changes in light characteristics can be determined with improved sensitivity due to the interference, or alternatively a hologram may be created by the split beams. The most common utilization of the technique is seen by the use of Mach–Zehnder interferometry in optical biosensors [37, 89].

3 Optical Biosensors for Detection of Cells and Cell Refractive Index Changes

Biosensors can be applied for the detection of cells principally in two ways: (a) in real matrices serving as habitat or medium for naturally propagating cells or deliberately contaminated with cells and (b) in artificial matrices, where cells are reared, e.g. cell cultures. The first type is typical for prokaryotic cells, while the second can apply to both prokaryotic and eukaryotic cells. As related to the fight against bioterrorism, the first type is typically used in the sensitive detection of bioterrorism agents (e.g. *Bacillus anthracis*, *Variola* (smallpox), *Clostridium botulinum*, *Yersinia pestis*, *Francisella tularensis* or other microbial warfare agents developed from these and other pathogenic microorganisms and their toxins), posing even national security risks; while the second is more used for the identification and characterization of cytotoxic activites on various cell lines.

Applications utilize the principle of SPR [93, 164], LSPR [44, 72], as well as integrated optical waveguide methods [37, 94, 112].

3.1 Detection of Bacterial Cells by Biosensors

From the aspect of biosafety assurance against biowarfare agents, bacterial cell detection by biosensors has the highest significance [111]. Additional common types of such sensors are mostly applied in environmental or food safety assurance and in human or animal health related uses, but can also serve microbial research or toxicity monitoring purposes. Selective discrimination of bacterial strain types and detection of actual cell concentrations is of great importance in all areas of, clinical diagnostics, food safety and environmental sciences, as well as biothreat identification [142, 185, 191]. In turn, planar and fiber-optic optical waveguides are commonly suggested for the detection of various pathogens [109], and numerous commercialized technologies e.g., EPIC® by Corning Inc. (Corning, NY, USA), BIND™ by SRU Biosystems Inc. (Urbana-Champaign, IL, USA) and Zeptosens® (Bayer Technology Services GmbH, Leverkusen, Germany) are applicable for these purposes.

Optical biosensors of the highest importance in biosecurity applications include simple fiber optic laser intensity reduction sensor utilizing immobilized pathogen-specific short peptide ligands [4], an evanescent wave fiber-optic biosensor [161] and an SPR sensor [171, 53] for the detection of *B. anthracis* spores. It has to be noted, however, that numerous other biosensors formats (including electrochemical, piezoelectric, fluorescent, acoustic, PCR based, etc.) for these pathogens have been also developed. Equally important to pathogen detection is the quantitative determination of various protein type or small molecular weight toxins by suitable sensors [143]. The most commonly used optosensoric detection of prokaryotic organisms is perhaps that of *E. coli* strains, especially *E. coli* O157: H7, the causative agent of severe food-borne poisoning by its secreted secondary metabolite verotoxin that binds with high mortality consequences to receptors in human kidney, brain and gut cells. *E. coli* cells have been reported to be selectively determined by a two-way laser intensity reduction sensor measuring photosignal drop due to antibody capturing *E. coli* relative to an untreated reference channel, capable to detect as low as 45 cells per well surface [3]. Nonetheless, a range of different types of immunosensors, including SPR biosensors were also developed, utilizing target microorganism specific antibodies [85, 155, 173] and polyethylene glycol terminated alkanethiol mixed self-assembled monolayer [149]. The same sensors were also used for detection of *Salmonella* [173] and *Staphylococcus* [155] strains. A fiber optic waveguide biosensor was developed for detection of pathogenic *E. coli* O157:H7 from both aqueous media and beef meat [90], and waveguide capillary and its integrating waveguide immunosensor was developed for its detection as a water-borne pathogen [190]. A grating-coupled planar optical waveguide sensor of „reverse symmetry" was also developed for sensing bacteria, including *E. coli* K12 [75]. An electrochemical optical waveguide lightmode spectroscopy (EC-OWLS) sensor facilitating the adsorption of *E. coli* was also applied to direct microbiological measurements, and was indicated to differentiate between living and heat treated bacteria in the sample [153]. To assay detection sensitivity of *E. coli* O157:H7-specific fiber optic and planar waveguide immunosensor platforms, *E. coli* O157:H7 genetically modified to produce green fluorescent protein was used and indicated cell capture efficiency between 0.4 and 1.2 % [140].

Specificity for *E. coli* has been achieved through oligonucleotide recombination on the sensor surface as well, for which single stranded oligonucleotide probes were immobilized allowing binding of the complementary DNA stretch with high specificity. DNA biosensors were developed in the quartz crystal microbalance (QCM) format without [104] or with [101] nanoparticle amplification. In the former [104] the target specific oligonucleotide immobilized for subsequent recombination was the *lacZ* gene from *E. coli*, and in situ polymerase chain reaction (PCR) was used. Due to the amplification efficacy of PCR the method allowed outstanding limits of detection below 10 fg of genomic *E. coli* DNA and as low as a few viable *E. coli* cells per 100 ml of water. In the latter one [101] the single stranded DNA stretch was the *eaeA* gene from *E. coli* O157:H7, and streptavidin conjugated Fe_3O_4 nanoparticles (average diameter = 145 nm) were used as "mass enhancers" to amplify the frequency change. Such DNA based biosensors have gained increased attention over traditional diagnostic methods due to their fast and responsive operation and cost-effective design. Thus, a multi-analyte bacterial sensor using multi-walled carbon nanotubes direct dispersion in DNA solution [159] allowed detection of 32 pathogenic microorganisms, including *E. coli* O157:H7.

An integrated metal clad leaky waveguide biosensor was developed for the detection of bacteria, including spores of *Bacillus subtilis* var. *niger*, with a sensor platform fabricated to increase the overlap of the evanescent field extension from the sensor surface (reverse symmetry) with the bacteria bound at the sensor surface, allowing detection of 8×10^4 spores/ml [192]. OWLS biosensors were applied to study lactic acid bacteria, part of the normal human microflora [6, 110] with lactic acid bacteria cells immobilized on the sensor surface, and the in situ effects of various chemical stress factors (hydrogen peroxide, acetic acid, lactic acid, fructooligosaccharides from Jerusalem artichoke) were recorded. A waveguide immunosensor was developed [138] with an integrated microfluidic channel, enabling label-free detection of live cells of *Listeria monocytogenes*, a common food-borne microbial contaminant and causative agent of listerosis. SPR immunosensors were developed for *Campylobacter jejuni* [174] and for *S. typhimurium* [92], common food-borne pathogens, in poultry meat samples with an acceptable sensitivity (1×10^6 cfu/ml). A DNA based biosensor specific for *C. jejuni* [54] utilized the hippuricase gene (*hipO*) from the target bacterium immobilized on the sensor surface in both SPR and diffraction optics technology (DOT, dotLab) formats. A five layer structured metal clad planar optical waveguide immunosensor was described as a possible means for the detection of *Pseudomonas* and *Pseudomonas*-like bacteria [141]. A rapid and sensitive DNA strip sensor was constructed [188] based on gold nanoparticle-labeled oligonucleotide probes for the detection of *Acidovorax avenae* subsp. *citrulli* allowing both qualitative and semi-quantitative detection of the target microorganism.

Facilitation of binding of various bacterial strain types with different specificity to the sensing surfaces have been achieved by coating with proteins of selective binding affinity, other than antibodies, e.g. lectins and silicatein. Sensor surfaces treated with concanavalin A or agglutinin were found to create biomimetic surfaces to enhance binding of e.g., Gram-negative (*E. coli* DH5a and *Enterobacter cloacae*) or Gram-positive (*Bacillus subtilis*) bacteria and a yeast (*Saccharomyces cerevisiae*), or even for a mammalian cell line (HeLa) [180], as well as for the sulfate-reducing bacterium *Desulfovibrio caledoiensis* [170]. Silicatein modified *E. coli* BL21AI cells were immobilized on an OWLS platform, allowing in situ investigations of the cells to various chemical stress factors, including hydrogen peroxide, antibiotics (penicillin G, chloramphenicol) and pesticide active ingredient carbofuran [5].

A microfluidic fiber optic waveguide immunosensor (Raptor) was utilized for practical application, analysis of *Enterococci* in recreational water to avoid public exposure to pathogens of fecal origin and to monitor decontamination efficacy of on-site ultrafiltration. *Enterococci* were detected at concentrations above 10^6 cfu/l [97].

3.2 Detection of Eukaryotic Cells by Label-Free Biosensors

Effects on eukaryotic cells are typically tested in biomedical and life sciences on cultured cells isolated from biological tissues, and may serve as disease models. The same types of cellular sensors for the determination of cytotoxicity are highly applicable in detection of the effects of biowarfare agents or toxins of bioterrorist uses. A large number of cellular functions and processes are disease indicators, and such cellular interactions and molecular event within a single cell are readily recordable by microfabricated electrochemical cell based biosensors [172] and by optical biosensors [73, 168]. Cell phenotyping is a particularly important tool in clinical diagnostics [22, 45, 46, 164, 189]. In addition screening of receptor-ligand binding by receptor based biosensors [32] can facilitate determination of functional responses (e.g. cell proliferation). Cellular features detected are mostly major cell characteristics (e.g. shape, volume or optical properties) and functions (e.g. adhesion, flattening or globularization, proliferation), but may also include subcellular descriptors (e.g. protein/metabolite production, substance uptake or even effects of altered gene expression).

An early exploration of the utility of the OWLS method in cell biology studies [127] revealed the cellular processes of utmost importance on the sensor surface. When a cell suspension was seeded onto the surface, cells initially attached, subsequently flattened in shape, and underwent reorganization leading to spreading over the sensor surface. In other words, small and subsequently growing flat regions were formed at the initial points of contact, and only the shape of the cells modified, their number remained constant. Consequently, the change in the signal measured by the OWLS sensor over time resulted from the flattening and spreading of the cells on the sensor surface. This observation has been the basis of numerous later applications and data interpretation [59, 81, 106, 115, 126, 169]. In addition to cell adhesion to surfaces, sensor models have been applied to description of cell barrier functions and regulation, cell to cell communication *via* direct interactions or chemical communication, cell proliferation and migration, tissue regeneration and repair, inflammation and blood clotting, signaling *via* receptor activation by an agonist, as well as viral infection [80, 40, 44, 46, 47], and have been applied to receptor assays [41]. Evanescent field based optical biosensors, e.g. OWLS [9, 10, 44, 116, 126], photonic crystal biosensors [136], grating coupling interferometry (GCI) [120, 121] and resonant waveguide grating EPIC® biosensors [47] are particularly suitable to monitor surface adhesion processes as these techniques detect molecular interactions on the sensor surface in a 100–200 nm thick layer [156].

The utility of various label-free sensor technologies in pharmaceutical research and in cytotoxicity testing, including SPR, guided mode resonance, resonant waveguide grating sensors in evanescent wave sensing, as well as impedance based sensors, resonant acoustic profiling and bio-layer interferometry has been reviewed [25]. By combining technologies of polymer waveguides and patterned surfaces for cell immobilization, the evanescent-field technique was proposed for immobilization of living cells including neural cells [125], and label-free detection with possible involvement of biosensors have been targeted to cell signaling studies on ligands acting at 7-transmembrane receptors [133]. Resonant waveguide grating biosensors have been applied in biomedical analysis to probe cancer signaling [42, 43], and biosensor systems have been applied to receptor assays [41].

The use of SPR technique in cellular characterization is frequent. Using immobilized antigens on the sensor surface, T cells were efficiently immunophenotyped with SPR based sensor techniques including imaging function [129]. A multi-gold-array SPR biosensor using antibody-immobilized gold sensing spots of the sensor surface [38] was used to record SPR spectra of CD33 cells from the bone marrow of leukemia patients, and showed the utility of the cell-specific label-free analytical biosensor. A small size SPR sensor equipped with an optical fiber was applied to detect activation of immobilized RBL-2H3 mast cells [181]. Fourier transform infrared SPR was applied in monitoring intercellular and cell-substrate interactions in various epithelial cell lines including highly differentiated canine kidney cells, rat intestinal epithelial cells, as well as cancerous and poorly differentiated human epithelial cells (e.g. melanoma MEL 1106 and cervix carcinoma HeLa cells) on gold-coated prism surface coupled with optical microscopy and image processing [182–184]. As mentioned already, treatment of sensor sensing surfaces with concanavalin A or agglutinin based self-assembled lectin monolayers useful to facilitate binding of not only bacteria, but also a mammalian cell line (HeLa) [180]. In addition, cell binding enhancement by two lectins (agglutinin and asialofetuin) alone or in parallel in a two-channel microfluidic device was applied to SPR signal mapping [31, 60].

LSPR systems have proven their utility in clinical bioanalysis [50, 72, 165], and have been applied to biochemical protein interactions [12], plasma-enhanced enzyme-linked immunosorbent assay (ELISA) [19], and cell based assays [34]. Equipped with charge-coupled optical devices (CCDs), such as a CCD camera or CCD based spectrophotometer, the method allows LSPR imaging (LSPRi) that promises outstanding utility in cellular studies [128], and in visualization detection of cancer cell lines [78]. Thus, LSPRi detected secreted cellular substances from live cell lines, to map the fractional occupancy of surface-bound receptors at individual nanostructures with nanomolar sensitivity and a temporal resolution of 225 ms [128]. LSPR systems formed with ordered arrays of Au nanoparticles was applied to normal human epithelial and breast cancer cells. The method was able to distinguish cell concentrations (8.6×10^3–1.19×10^6 cells/ml) and allowed comparisons of cell shapes between the flat and higher confluency human-derived retinal pigment epithelial RPE-1 cells and breast cancer MCF-7 cells [99]. Another application allowed detection of cytokine (tumor necrosis factor (TNF)-R) secretion from human blood immune cells (THP-1 and CD45 cells) and determination of the functional response of the cells and the immune status of patient. Analysis required only minimal blood sample volume (3 μL) and a assay time [113]. Infrared applications of SPR are also expanding, e.g. for measuring D-glucose uptake by erythrocytes in suspension with SPR coupled with Fourier transform infrared spectrometry, demonstrating that the sensitivity of the SPR technique can be improved in the infrared range [55].

Yet, SPR is not the only technique applied in cell sensorics. A grating coupled interferometry biosensor was developed using on poly-L-lysine coated sensor surface to test whether microvesicles secreted by cultured human lymphoblast-like T cells bind to extracellular matrix molecules e.g. type I collagen or fibronectin [121]. OWLS was used to model cell-substrate interactions [9, 127, 193, 73] and to measure the attachment process of NE4C neuroectodermal stem cell-like cells [152]. A label-free photonic crystal based optical biosensor incorporated into the bottoms of the wells of 96-well microplates to detect cell attachment to these surfaces was utilized for assaying E6-1 human T cell leukemia cells and J45.1 Jurkat T cells. Immobilized antibodies specific to T cell antigens expressed on the cell surfaces provided sensor capability to distinguish between closely related cell lines [98]. Using CHO-K1 cells (cell lone isolated from Chinese hamster ovary) attached to well surfaces in a EPIC® system, loss in bound cell material upon cytoskeletal changes due to the effect of saponin, a known cytoskeleton modulator has been described [48, 49]. Cellular responses of native human epidermal carcinoma A431 cells to epinephrine, an agonist of β_2-adrenergic receptors expressed by these cells and adenylate cyclase activator forskolin were detected by a microfluidic resonant waveguide grating biosensor [186]. A microfluidic structure planar waveguide attenuated by metal particles fabricated for capturing cells from human blood allowed semi-quantitative determination of 100 cells/μl of blood [51].

An OWLS method, allowing allowed parallel phase-contrast microscopy visualization was used to study adhesion, morphology and metabolic state of anchorage-dependent baby hamster kidney (BHK) cells [81]. An impedance based (CellKey^{TM}) and two optical biosensors BIND® and EPIC® were shown to be able to detect G-protein-coupled receptor signaling through each of the three major classes of G-proteins (Gi-coupled, Gq-coupled and Gs-coupled) in various Chinese hamster ovary (CHO) cell lines expressing various G-protein types [124], demonstrating the utility of these sensoric methods in pharmaceutical research. A novel approach to biosensing of cellular processes is the use of light based detection techniques applied in the body using optical probes excited by and emitting light at ultraviolet and visible wavelengths allowing monitoring of physiological processes in the body in real-time with a high resolution [71, 132]. In a unique, in vivo biosensing application in real time, implantable waveguide sensors on hydrogel patches are used to guide light as sensor signal from the sensor properly positioned in the body of the test animal and to accommodate optogenetic cells (e.g. genetically altered HeLa cells producing green fluorescent protein in response of metal-induced cellular stress-related heat shock protein (HSP70) production in the cells, $<10^5$ cells). The implanted biosensor design may serve as the basis of a new generation of implantable systems for in-body optical sensing and therapy [21, 150].

An early and unique imaging application of evanescent field effect based sensorics is represented by scattering optical waveguide microscopy, allowing microscopic characterization with lateral resolution below 1 μm. The method utilizes the evanescent field of a guided optical wave in a thin film over the sensor surface, and image contrast is generated by the scattering intensity of the thin film sample within the evanescent field [158]. A fluorescence-enchanced waveguide microscopy method, waveguide excitation fluorescence microscopy has been further developed relying on the planar optical waveguide exciting fluorescence in the near interface region. The technique has been demonstrated to be of target sensitivity for fluorescence detection in the femtomolar range, good precision due to an order of magnitude improvement in the signal to noise ratio compared to conventional fluorescence microscopy, submicron optical resolution, high surface specificity and therefore, high utility in the dynamic and quantitative in situ characterization of heterologous phase biological surface processes in real time including focal adhesion formation in cell-surface interactions, e.g. fibroblast focal adhesion [59]. To enhance sensor performance and functionality, a Mach–Zehnder interferometer setup and other interferometry elements were applied in sensing devices, and were used for measuring the dry/wet mass of a single living cell [157].

Imaging applications include fluorescence in the range of optical signal detected as well. Waveguide evanescent field fluorescence microscopy [64, 67] was utilized to investigate plasma membranes of living osteoblast cells, imaging plasma membranes of the cells with high axial resolution, and allowing time-lapse imaging to investigate cell morphology, the effect of membrane disrupters (Triton X-100), resulting membrane solubilization was visualized, a three-stage solubilization model was proposed for liposomes and supported lipid bilayers [66, 69], the method was used on HEK293 cells [68] and to measure cell-substratum separation distance [63, 65].

4 Holographic Microscopy for Label-Free Visualization of Living Cells

In addition to evanescent field based optical waveguide microscopy [31, 38, 59, 60, 129, 158], cellular characteristics and processes can be visualized by laser interference, similar to the use in Mach-Zehnder interferometers, applied to optical microscopy. Light based techniques for tissue or cell imaging are precise, rapid and non-invasive methods of visualizing biological functions in living organisms [17], yet light transmission in in vitro, ex vivo or in vivo systems is severely limited by the absorption and scattering of light at ultraviolet, visible (including laser) and, to a lesser extent, near-infrared wavelengths by biological molecules and structures [20] and allow penetration of only a few centimeters or so into biological tissue [163]. As a consequence, traditional optical microscopy methods routinely used for following cells and cellular physiological processes remain to allow only limited magnification ($\sim$1 mm) and achievable depth of focus (3–5 mm). The latter limitation is even more problematic than the former, as conventional optical microscopes can thus, allow visualization of a very thin layer of the sample observed. A flow-through measuring chamber design [139] may facilitate sample throughput capacity, yet still allowing limited volumes measured. Digital holographic microscopy, a relatively novel, label-free, non-invasive, nondestructive and non-phototoxic method, surpassed this problem by recording all available diffraction information in a hologram, from which any particular image in the observed volume can be reconstructed subsequently by numeric simulation of wave propagation [13, 18, 24, 52, 102] with high precision. Digital holograms can be obtained using single or multiple light sources of different wavelengths, with or without an objective (lens) and possibly enhanced by interferometry (Fig. 2).

In one set of application, taking advantage of the 100-fold increase in the observed volume, digital holographic microscopy has been applied as a fluid monitoring system [56, 58] for water quality measurement, extended with advanced twin image elimination algorithms [117, 118] and triple wavelength illumination [88]. The system has been applied to automatized recognition of various aquatic microorganisms, including microalgae. Computerized evaluation of the captured images by machine learning algorithms analyzes and classifies the various algae types found in the water samples. Reconstructed objects in the observed volume segmented arbitrarily can be stored electronically and further processed with an adaptive model based classification algorithm allowing creation of a biological

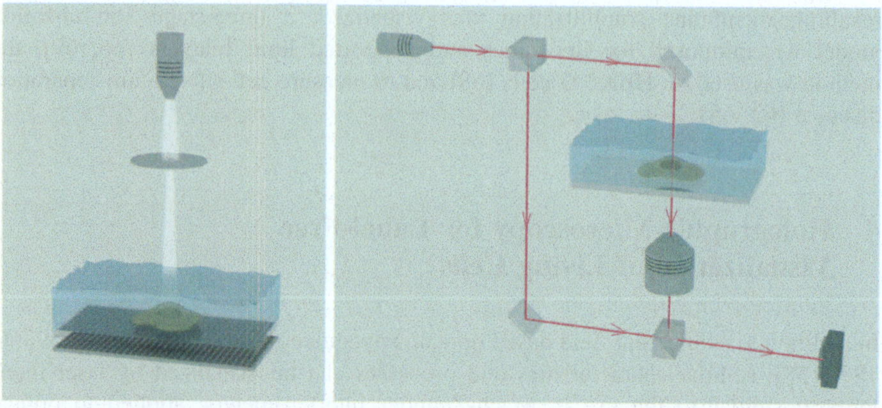

Fig. 2 Schematic view of digital holographic microscopy setups. A lens-free design (*left*) uses a light (not necessarily laser) source and an adjustable and a movable aperture. Light transmitting through the study object is detected by a sensor array plane below. With multiple light sources (at multiple wavelengths) 3D images are created by the sensor array. In a Mach-Zehnder interferometer based setup (*right*) the illuminating laser beam is split into two identical (object and reference) beams. The object beam transmitting through the study object (in this case living cells adhered to the surface) is reunited and interferred with the reference beam creating a hologram

microorganisms database. Color hologram reconstructions obtained by the system achieve at least 1 μm lateral resolution.

Digital holographic transmission microscopy was used in biological systems for the integral refractive index determination of living pancreas tumor cells (PaTu 8988S and 8988T cells) in cell culture medium. In addition to cell shape and thickness measurement, the method also demonstrated the strong cell cytoskeleton disruptive action of the marine toxin Latrunculin B, determined in dynamic assessment of cell morphological characteristics [87]. In another application, cell viability and cell morphology parameters in response to cytotoxic chemical agents $HgCl_2$, chloroquine and gambogic acid of commonly used tumor cell lines, CHO-K1 and HeLa cells were tested using digital holographic microscopy with a DHM T-1001 system with a motorized plate holder [91].

Digital holographic microscopy is also feasible without the use of a laser. Unique multi-view, multi-wavelength illumination based computational on-chip imaging platforms allowed entirely new three-dimensional (3D) imaging of cells [14, 114, 119, 134, 147], and on the basis of the technique, lens-free computational microscopy is rapidly emerging into fields including high-throughput biosensing [29], antibody-antigen, DNA-DNA and cellular interactions [103], medical diagnostics as well as microscopic cell analysis and cytometry [82, 135, 175] and tomography [83]. 3D imaging of various cells including microorganisms (such as *E. coli*) [107], plant pollen [15], sperm cells [145, 146, 148], white blood cells [144], skin tissue cell [95], human breast cancer cells with an optical system improved with synthetic aperture [100], even viruses [176], therefore, gained diagnostic and medical utility coupled to optoelectric tweezers [79] or even with remote (cell phone based) sensing [108, 167]. Visual efficacy of imaging micron-scale fluorescent objects was improved by the use of micro-reflectors for fluorescent light (self-assembled polyethylene glycol droplets) that enhance excitation efficiency. The method also utilizes the waveguide principle, as the polyethylene glycol droplets partially redirect of the emitted fluorescent light, obtained without the use of any fluorescent labeling, towards the detector due to internal reflections at the liquid-air interface of the meniscus [57].

The resolution and contrast of digital holographic microscopy can be enhanced by interferometric amplification, e.g. Mach-Zehnder interferometry, when the illuminating light (e.g. HeNe laser at 635 nm wavelength) is split into an object beam and a reference beam. The object beam upon illumination of the object is re-joined and interfered with the reference beam creating a hologram. This interference pattern is recorded on a digital sensor. Focusing within the hologram is possible to any point without any mechanical movement by iteratively created images any time after the actual recording. This holographic visualization technique can be applied to refractometry of various microscopic objects, including living cells [61, 84, 86, 8], allowing both qualitative and quantitative measurements of living cells over time, visualization of cell integrity, and observation of ongoing cell-morphological changes including the processes of cell differentiation, cell growth and cell death. Along with experimental setups, a commercial digital holographic microscope, HoloMonitor M4 is also available, allowing cytometric time-lapse microscopy created from image sequences of cultured cells recorded over long time periods.

In another application example by a recent cytotoxicity study [151], holographic transmission microscopy was successfully used to measure cell morphology parameters as descriptors of cell viability. Thus, holographic microscopy using HoloMonitor M4 (PHI AB, Lund, Sweden) was shown to be applicable to quantitatively determine the effect of various agrochemicals (herbicide active ingredient glyphosate, its formulated herbicide preparation and its adjuvant surfactant polyethoxylated tallowamine used in the formulated preparation studied) on a neuroectodermal cell line. Quantitative visualization of average cell thickness and area, calculated using software HoloStudio M4 (PHI AB, Lund, Sweden), indicated effects on the cells in close correlation with those determined in colorimetric cytotoxicity tests, indicating, besides the toxicological significance, the utility of the digital holographic monitoring technique in toxicology (Fig. 3).

A similar study [162] on the effects of methylglyoxal and edaravone on brain endothelial cells essential in the blood-brain barrier. Methylglyoxal, a strongly oxidative and glycolative glycotoxin, exerted a cytotoxic action on the endothelial cells causing these typically elongated and flat cells to form globular forms with increased cell thickness. Treatment with a neuroprotective free radical scavenger edaravone, active ingredient of the recently developed Japanese pharmaceutical preparation Radicut, prevented such cytotoxicity of methylglyoxal, as seen in unchanged cell morphology during the treatment period. An additional cell

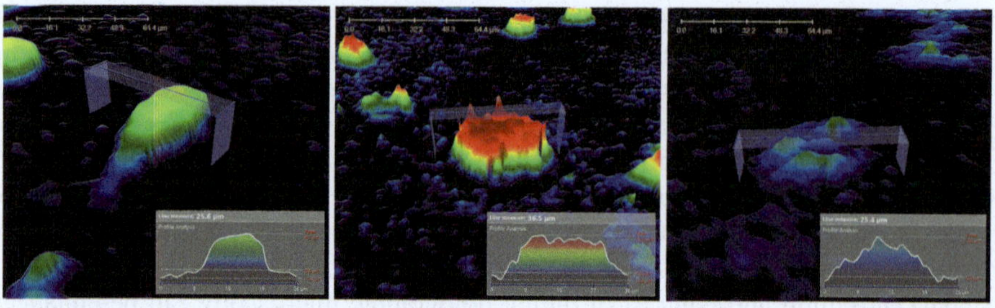

Fig. 3 Cells under toxic effect visualized with phase-contrast holographic microscopy. Time dependence of cytotoxicity on neuroectodermal (NE4C) cells exposed to 0.1 % of formulated herbicide preparation Roundup (containing active ingredient glyphosate and adjuvant polyethoylated tallowamines). Cells of typically elongated form origically (*left*) form uneven globular shape in 5 min (*middle*) and undergo rapid disintegration in 10 min (*right*)

morphological parameter, surface area similarly indicated the protective combined effect, while cell volumes remained unchanged.

A recent study evaluated the effects of polyphenolic bioactive components in green tea on HeLa cancer cells using a heavily miniaturized and incubator proof holographic microscopy [123]. HoloMonitor M4 was applied for in situ monitoring of altered adhering of preosteoblast cells on nanostructured titanate surfaces.

5 Conclusions

Cell based analytical biosensors and cell imaging techniques gained dual significance in respect of biothreat identification: through specific detection of certain microorganisms or cell types with the use of sensors utilizing specificity or recognition elements to detect given microorganisms on the one hand, and through the use of given cell types or lines in these techniques to detect cytotoxicity on the other hand. As microbial and chemical contamination can take place at numerous vulnerable points, rapid diagnostic tests are required to control the risks by such accidental or deliberate contamination, and label-free, evanescent optical field based sensors are excellent tools for detection of pathogen transmission, biotoxin occurrence and effect determination.

Optical biosensors for pathogen detection include laser intensity reduction sensors, SPR and LSPR, planar and fiber-optic (capillary) optical waveguides, OWLS, metal clad leaky waveguide biosensor, and their commercialized optical waveguide sensor technologies e.g. EPIC®, BIND™, Zeptosens®, OWLS210 (Microvacuum Ltd., Budapest, Hungary), Creoptix™ WAVE (Creoptix AG, Wädenswil, Switzerland); and application forms in immunosensor and bioaffinity (e.g. lectins and silicatein) or PCR based applications. Effects on various cell types and lines are typically tested with cell based optical sensors to determine cytotoxicity of bioactive agents (e.g. biothreat agents) including SPR, resonant waveguide grating biosensors, OWLS, grating coupling interferometry (GCI), resonant waveguide grating (EPIC®) biosensors, photonic crystal biosensors and bio-layer interferometry. Cell imaging techniques, applied alone or coupled with optosensors, include scattering optical waveguide microscopy, fluorescence-enchanced waveguide microscopy (waveguide excitation fluorescence microscopy), as well as digital holographic microscopy applied in single- or multi-wavelength illumination, even in lens-free formats, and enhanced by Mach-Zender interferometry e.g. HoloMonitor M4.

Acknowledgments Research was funded by projects "Development of optical bio-sensors for detection of bio-toxins" (NUKR.SFPP 984637) in the NATO SPS Programme, "Mechanism-related teratogenic, hormone modulant and other toxicological effects of veterinary and agricultural surfactants" (OTKA K109865) by the Hungarian Research Fund, and by the Momentum Program (Lendület) of the Hungarian Academy of Sciences.

References

1. Abbas A, Linman MJ, Cheng Q (2011) New trends in instrumental design for surface plasmon resonance-based biosensors. Biosens Bioelectron 26:1815–1824
2. Abdulhalim I, Zourob M, Lakhtakia A (2007) Overview of optical biosensing techniques. In: Marks RS, Cullen DC, Karube I, Lowe CR, Weetall HH (eds) Handbook of biosensors and biochips. Wiley, London, pp 1–34
3. Acharya G, Chang C-L, Savran C (2006) An optical biosensor for rapid and label-free detection of cells. J Am Chem Soc 128:3862–3863
4. Acharya G, Doorneweerd DD, Chang C-L, Henne WA, Low PS, Savran CA (2007) Label-free optical detection of anthrax-causing spores. J Am Chem Soc 129:732–733
5. Adányi N, Bori Z, Szendrő I, Erdélyi K, Wang X, Schröder HC, Müller WEG (2013) Bacterial sensors based on biosilica immobilization for label-free OWLS detection. New Biotechnol 30(5):493–499
6. Adányi N, Németh E, Halász A, Szendrő I, Váradi M (2006) Application of electrochemical optical waveguide lightmode spectroscopy (EC-OWLS) for studying the effect of different stress factors on lactic acid bacteria. Anal Chim Acta 573–574:41–47
7. Adányi N, Majer-Baranyi K, Székács A (2015) Evanescent field effect based nanobiosensors for agro-environmental and food safety. In: Grumezescu AM (ed) Nanotechnology in food industry. Springer, Berlin (in press)
8. Alm K, El-Schich Z, Falck Miniotis M, Gjörloff Wingren A, Janicke B, Oredsson S (2013) Cells and holograms—Holograms and digital holographic microscopy as a tool to study the morphology of living cells. In: Mihaylova E (ed) Holography—Basic Principles and Contemporary Applications. InTech, Rijeka, pp 335–351
9. Aref A, Horvath R, McColl J, Ramsden JJ (2010) Optical monitoring of stem cellsubstratum interactions. J Biomed Opt 14:010501
10. Aref A, Horvath R, Ramsden JJ (2010) Spreading kinetics for quantifying cell state during stem cell differentiation. J Biol Phys Chem 10:1–7
11. Bata-Vidács I, Adányi N, Beczner J, Farkas J, Székács A (2013) Nanotechnology and microbial food safety. In: Mendez-Vilas A (ed) Microbial pathogens and strategies for combating them: science, technology and education. Formatex, Zubaran, pp 155–159

12. Bellapadrona G, Tesler AB, Grunstein D, Hossain LH, Kikkeri R, Seeberger PH, Vaskevich A, Rubinstein I (2012) Optimization of localized surface plasmon resonance transducers for studying carbohydrate-protein interactions. Anal Chem 84:232–240
13. Bergoend I, Colomb T, Pavillon N, Emery Y, Depeursinge C (2009) Depth-of-field extension and 3D reconstruction in digital holographic microscopy. Proc SPIE 7390:73901C-1
14. Bishara W, Hongying Zhu H, Ozcan A (2010) Holographic opto-fluidic microscopy. Opt Express 18(26):27499
15. Bishara W, Sikora U, Mudanyali O, Su T-W, Yaglidere O, Luckhart S, Ozcan A (2011) Holographic pixel super-resolution in portable lensless on-chip microscopyusing a fiber-optic array. Lab Chip 11:1276–1279
16. Brecht A, Gauglitz G (1995) Optical probes and transducers. Biosens Bioelectron 10: 923–936
17. Bremer C, Ntziachristos V, Weissleder R (2003) Optical-based molecular imaging: contrast agents and potential medical applications. Eur Radiol 13:231–243
18. Carl D, Kemper B, Wernicke G, Bally GV (2004) Parameter optimized digital holographic microscope for high resolution living cell analysis. Appl Opt 43:6536–6544
19. Chen S, Svedendahl M, Van Duyne RP, Käll M (2011) Plasmon-enhanced colorimetric ELISA with single molecule sensitivity. Nano Lett 11:1826–1830
20. Cheong W-F, Prahl SA, Welch AJ (1990) A review of the optical properties of biological tissues. IEEE J Quantum Elect 26:2166–2185
21. Choi M, Choi JW, Kim S, Nizamoglu S, Hahn SK, Yun SH (2013) Light-guiding hydrogels for cell-based sensing and optogenetic synthesis *in vivo*. Nat Photon 7:987–994
22. Citartan M, Gopinath SC, Tominaga J, Tang TH (2013) Label-free methods of reporting biomolecular interactions by optical biosensors. Analyst 138:3576–3592
23. Codex Alimentarius (2001) Principles and guidelines for the conduct of microbiological risk assessment. In: Food hygiene baic texts. Food and Agricultural Organizaton/World Health Organization, Rome
24. Colomb T, Pavillon N, Kühn J, Cuche E, Depeursinge C, Emery Y (2010) Extended depth-of-focus by digital holographic microscopy. Opt Lett 35(11):1840–1842
25. Comley J (2004) Label-free detection: New biosensors facilitate broader range of drug discovery applications. Drug Discov World Winter 2004(5):63–74
26. Conroy PJ, Hearty S, Leonard P, O'Kennedy RJ (2009) Antibody production, design and use for biosensor-based applications. Semin Cell Dev Biol 20:10–26
27. Cooper MA (2006) Optical biosensors: where next and how soon? Drug Discovery Today 11 (23–24):1061–1067
28. Cooper MA (2009) Signal transduction profiling using label-free biosensors. J Recept Signal Transduction 29(3–4):224–233
29. Coskun AF, Cetin AE, Galarreta BC, Alvarez DA, Altug H, Ozcan A (2014) Lensfree optofluidic plasmonic sensor for real-time and label-free monitoring of molecular binding events over a wide field-of-view. Sci Rep 4:6789
30. Cunningham B, Li P, Lin B, Pepper J (2002) Colorimetric resonant reflection as a direct biochemical assay technique. Sens Actuators B 81(2–3):316–328
31. D'Agata R, Grasso G, Iacono G, Spoto G, Vecchio G (2006) Lectin recognition of a new SOD mimic bioconjugate studied with surface plasmon resonance imaging. Org Biomol Chem 4:610–612
32. de Jong LAA, Uges DRA, Franke JP, Bischoff R (2005) Receptor-ligand binding assays: technologies and applications. J Chromatogr B 829:1–25
33. Duffus JH (1993) Glossary for chemists of terms used in toxicology. Pure Appl Chem 65:2003–2122
34. Endo T, Yamamura S, Tamiya E (2008) Label-free cell-based assay using localized surface plasmon resonance biosensor. Anal Chim Acta 614:182–189
35. Erdélyi K, Frutos AG, Ramsden JJ, Szendrő I, Voirin G (2008) Grating-based optical biosensors. In: Marks RS, Cullen DC, Karube I, Lowe CR Weetall HH (eds) Handbook of biosensors and biochips. Wiley, New York, pp 569–586

36. Commission European (2000) First Report on the Harmonisation of Risk Assessment Procedures (SSC 2000). European Commission, Brussels
37. Fan X, White IM, Shopova SI, Zhu H, Suter JD, Sun Y (2008) Sensitive optical biosensors for unlabeled targets: a review. Anal Chim Acta 620:8–26
38. Fang XY, Liu CL, Cheng XL, Wang YL, Yang YC (2011) A spectral imaging array biosensor and its application in detection of leukemia cell. Sens Actuat B-Chem 156:760–764
39. Fang Y (2006) Label-free cell-based assays with optical biosensors in drug discovery. Assay Drug Devel Technol 4(5):583–595
40. Fang Y (2007) Non-invasive optical biosensor for probing cell signaling. Sensors 7(10):2316–2329
41. Fang Y (2010) Label-free receptor assays. Drug Discov Today Technol 7(1):e5–e11
42. Fang Y (2010) Probing cancer signaling with resonant waveguide grating biosensors. Exp Opin Drug Discov 5(12):1237–1248
43. Fang Y (2010) Resonant waveguide grating biosensor for microarrays. In: Zourob M, Lakhtakia A (eds) Optical guided-wave chemical and biosensors II. Springer, Berlin, pp 27–42
44. Fang Y (2011) Label-free biosensors for cell biology. Int J Electrochem (2011: Article ID 460850, 16 p)
45. Fang Y (2013) Biosensors: On the origin of label-free cell phenotypic profiles of drug-target interactions. J Biochip Tissue Chip 3:e126
46. Fang Y (2013) Troubleshooting and deconvoluting label-free cell phenotypic assays in drug discovery. J Pharmacol Toxicol Meth 67:69–81
47. Fang Y, Ferrie AM, Fontaine NH, Mauro J, Balakrishnan J (2006) Resonant waveguide grating biosensor for living cell sensing. Biophys J 91(5):1925–1940
48. Fang Y, Ferrie AM, Li G (2005) Probing cytoskeleton modulation by optical biosensors. FEBS Lett 579:4175–4180
49. Fang Y, Ferrie AM, Tran E (2009) Resonant waveguide grating biosensor for whole-cell GPCR assays. In: Leifert WR (ed) G Protein-coupled receptors in drug discovery. Methods in molecular biology 552 (pp 239–252). Humana Press, New York
50. Feng J, Siu VS, Pacifici D (2012) Nanoscale plasmonic interferometers for multispectral, high-throughput biochemical sensing. Nano Lett 12:602–609
51. Garcia D, Ghansah I, LeBlanc J, Butte MJ (2012) Counting cells with a low-cost integrated microfluidics-waveguide sensor. Biomicrofluidics 6:014115
52. Garcia-Sucerquia J, Xu W, Jericho S, Klages P, Jericho M, Kreuzer H (2006) Digital in-line holographic microscopy. Appl Optics 45(5):836–850
53. Ghosh N, Gupta G, Boopathi M, Pal V, Singh AK, Gopalan N, Goe AK (2013) Surface plasmon resonance biosensor for detection of *Bacillus anthracis*, the causative agent of anthrax from soil samples targeting protective antigen. Indian J Microbiol 53(1):48–55
54. Gnanaprakasa TJ, Oyarzabal OA, Olsen EV, Pedrosa VA, Simonian AL (2011) Tethered DNA scaffolds on optical sensor platforms for detection of hipo gene from *Campylobacter jejuni*. Sens Actuat B-Chem 156:304–311
55. Golosovsky M, Lirtsman V, Yashunsky V, Davidov D, Aroeti B (2009) Midinfrared surface-plasmon resonance: a novel biophysical tool for studying living cells. J Appl Phys 105:102036
56. Göröcs Z, Kiss M, Tóth V, Orzó L, Tőkés S (2010) Multicolor digital holographic microscope (DHM) for biological purposes. Proc SPIE 7568:75681P
57. Göröcs Z, McLeod E, Ozcan A (2015) Enhanced light collection in fluorescence microscopy using self-assembled micro-reflectors. Sci Rep 5:10999
58. Göröcs Z, Orzó L, Kiss M, Tóth V, Tőkés S (2010) In-line color digital holographic microscope for water quality measurements. Proc SPIE 7376:737614
59. Grandin HM, Städtler B, Textor M, Vörös J (2006) Waveguide excitation fluorescence microscopy: a new tool for sensing and imaging the biointerface. Biosens Bioelectron 21:1476–1482

60. Grasso G, D'Agata R, Zanoli L, Spoto G (2009) Microfluidic networks for surface plasmon resonance. Microchem J 93:82–86
61. Gustafsson M, Sebesta M (2004) Refractometry of microscopic objects with digital holography. Appl Opt 43:4796–4801
62. Haes AJ, Van Duyne RP (2004) A unified view of propagating and localized surface plasmon resonance biosensors. Anal Bioanal Chem 379:920–930
63. Hassanzadeh A, Armstrong S, Dixon SJ, Mittler S (2009) Multimode waveguide evanescent field fluorescence microscopy: Measurement of cell-substratum separation distance. Appl Phys Lett 94:033503
64. Hassanzadeh A, Azami D (2014) Waveguide evanescent field fluorescence microscopy: theoretical investigation of optical pressure on a cell. J Nanophoton 8:083076
65. Hassanzadeh A, Ma HK, Armstrong S, Dixon SJ, Sims SM, Mittler S (2009) Waveguide evanescent field fluorescence microscopy: from cell-substratum distances to kinetic cell behavior. Proc SPIE 2009:73220A
66. Hassanzadeh A, Ma HK, Dixon SJ, Mittler S (2012) Visualization of the solubilization process of the plasma membrane of a living cell by waveguide evanescent field fluorescence microscopy. J Biomed Opt 17(7):076025
67. Hassanzadeh A, Mittler S (2011) Waveguide evanescent field fluorescence microscopy: high contrast imaging of a domain forming mixed lipid Langmuir-Blodgett monolayer mimicking lung surfactant. J Biomed Opt 16(4):046022
68. Hassanzadeh A, Nitsche M, Armstrong S, Nabavi N, Harrison R, Dixon SJ, Langbein U, Mittler S (2010) Optical waveguides formed by silver ion exchange in Schott SG11 glass for waveguide evanescent field fluorescence microscopy: evanescent images of HEK293 cells. J Biomed Opt 15(3):036018
69. Hassanzadeh A, Nitsche M, Mittler S, Armstrong S, Dixon J, Langbein U (2008) Waveguide evanescent field fluorescence microscopy: Thin film fluorescence intensities and its application in cell biology. Appl Phys Lett 92:233503
70. Hessel A, Oliner AA (1965) A new theory of Wood's anomalies on optical gratings. Appl Opt 4(10):1275–1297
71. Hilderbrand SA, Weissleder R (2010) Near-infrared fluorescence: application to *in vivo* molecular imaging. Curr Opin Chem Biol 14:71–79
72. Hong Y, Huh Y-M, Yoon DS, Yang J (2012) Nanobiosensors based on localized surface plasmon resonance for biomarker detection. J Nanomater (Article ID 759830, 13 p)
73. Horvath R, Cottier K, Pedersen HC, Ramsden JJ (2008) Multidepth screening of living cells using optical waveguides. Biosens Bioelectron 24(4):799–804
74. Horvath R, Lindvold LR, Larsen NB (2002) Reverse-symmetry waveguides: theory and fabrication. Appl Phys B 74:383–393
75. Horvath R, Pedersen HC, Skivesen N, Selmeczi D, Larsen NB (2003) Optical waveguide sensor for on-line monitoring of bacteria. Opt Lett 28:1233–1235
76. Horvath R, Pedersen HC, Skivesen N, Selmeczi D, Larsen NB (2005) Monitoring of living cell attachment and spreading using reverse symmetry waveguide sensing. Appl Phys Lett 86:071101
77. Horvath R, Pedersen HC, Skivesen N, Svanberg C, Larsen NB (2005) Fabrication of reverse symmetry polymer waveguide sensor chips on nanoporous substrates using dip-floating. J Micromech Microeng 15:1260–1264
78. Hu R, Yong K-T, Roy I, Ding H, He S, Prasad PN (2009) Metallic nanostructures as localized plasmon resonance enhanced scattering probes for multiplex dark-field targeted imaging of cancer cells. J Phys Chem C 113:2676–2684
79. Huang K-W, Su T-W, Ozcan A, Chiou P-Y (2013) Optoelectronic tweezers integrated with lensfree holographic microscopyfor wide-field interactive cell and particle manipulation on a chip. Lab Chip 13:2278–2284
80. Hug TS, Prenosil JE, Maier P, Morbidelli M (2002) Optical waveguide lightmode spectroscopy (OWLS) to monitor cell proliferation quantitatively. Biotechnol Bioeng 80:213–221

81. Hug TS, Prenosil JE, Morbidelli M (2001) Optical waveguide lightmode spectroscopy as a new method to study adhesion of anchorage-dependent cells as an indicator of metabolic state. Biosens Bioelectron 16:865–874
82. Isikman SO, Greenbaum A, Lee M, Bishara W, Mudanyali O, Su T-W, Ozcan A (2012) Lensfree computational microscopy tools for cell and tissue imaging at the point-of-care and in low-resource settings. Anal Cell Path 35:229–247
83. Isikman SO, Greenbaum A, Luo W, Coskun AF, Ozcan A (2012) Giga-pixel lensfree holographic microscopy and tomography using color image sensors. PLoS ONE 7(9):e45044
84. Joseph S, Gineste J-M, Whelan M, Newport D (2010) A heterodyne Mach-Zehnder Interferometer employing static and dynamic phase demodulation techniques for live-cell imaging. Proc SPIE 7554:75540P
85. Joung CK, Kim HN, Lim MC, Jeon TJ, Kim HY, Kim YR (2013) A nanoporous membrane based impedimetric immunosensor for label free detection of pathogenic bacteria in whole milk. Biosens Bioelectron 44:210–215
86. Kaul RA, Mahlmann DM, Loosen P (2010) Mach-Zehnder interference microscopy optically records electrically stimulated cellular activity in unstained nerve cells. J Microsc 240(1):60–74
87. Kemper B, Carl D, Schnekenburger J, Bredebusch I, Schäfer M, Domschke W, von Bally G (2006) Investigation of living pancreas tumor cells by digital holographic microscopy. J Biomed Optics 11(3):034005
88. Kiss M, Nagy B, Lakatos P, Göröcs Z, Tőkés S, Wittner B, Orzó L (2014) Special multicolor illumination and numerical tilt correction in volumetric digital holographic microscopy. Opt Express 22:7559–7573
89. Kozma P, Kehl F, Ehrentreich-Förster E, Stamm C, Bier FF (2014) Integrated planar optical waveguide interferometer biosensors: A comparative review. Biosens Bioelectron 58:287–307
90. Kramer MF, Tims TB, DeMarco DR, Lim DV (2002) Recovery of *Escherichia coli* 0157:H7 from fiber optic waveguides used for rapid biosensor detection. J Rapid Meth Autom Microbiol 10:93–106
91. Kühn J, Shaffer E, Mena J, Breton B, Parent J, Rappaz B, Chambon M, Emery Y, Magistretti P, Depeursinge C, Marquet P, Turcatti G (2013) Label-free cytotoxicity screening assay by digital holographic microscopy. Assay Drug Devel Technol 11(2):101–107
92. Lan Y, Wang S, Yin Y, Hoffmann WC (2008) Using a surface plasmon resonance biosensor for rapid detection of *Salmonella typhimurium* in chicken carcass. J Bionic Eng 5:239–246
93. Lechuga LM (2000) Optical sensors based on evanescent field sensing Part I. Surface plasmon resonance sensors. Quim Anal 19:54–60
94. Lechuga LM (2000) Optical sensors based on evanescent field sensing Part II. Integrated optical sensors. Quim Anal 19:61–67
95. Lee M, Yaglidere O, Ozcan A (2011) Field-portable reflection and transmission microscopy based on lensless holography. Biomed Opt Express 2(9):2721
96. Lee PH (2009) Label-free optical biosensor: a tool for G protein coupled receptors pharmacology profiling and inverse agonists identification. J Recept Sign Transduct 29(3–4):146–153
97. Leskinen SD, Lim DV (2008) Rapid ultrafiltration concentration and biosensor detection of enterococci from large volumes of Florida recreational water. Appl Environ Microbiol 74 (15):4792–4798
98. Lin B, Li P, Cunningham BT (2006) A label-free biosensor-based cell attachment assay for characterization of cell surface molecules. Sens Actuat B-Chem 114:559–564
99. Liu F, Wong MM, Chiu SK, Lin H, Ho JC, Pang SW (2014) Effects of nanoparticle size and cell type on high sensitivity cell detection using a localized surface plasmon resonance biosensor. Biosens Bioelectron 55:141–148
100. Luo W, Greenbaum A, Zhang Y, Ozcan A (2015) Synthetic aperture-based on-chip microscopy. Sci Appl 4:e261
101. Mao X, Yang L, Su X, Li Y (2006) Nanoparticles amplification based quartz crystal microbalance DNA sensor for detection of *E. coli O157:H7*. Biosens Bioelectron 7:1178–1185

102. Marquet P, Rappaz B, Magistretti PJ, Cuche E, Emery Y, Colomb T, Depeursinge C (2005) Digital holographic microscopy: a noninvasive contrast imaging technique allowing quantitative visualization of living cells with sub wavelength accuracy. Opt Lett 30:468–470
103. McLeod E, Ozcan A (2014) Nano-imaging enabled via self-assembly. Nano Today 9: 560–573
104. Mo XT, Zhou YP, Lei H, Deng L (2002) Microbalance-DNA probe method for the detection of specific bacteria in water. Enzyme Microb Technol 5:583–589
105. Moskovits M (1985) Surface-enhanced spectroscopy. Rev Mod Phys 57:783–826
106. Mrksich M (2000) Model organic surfaces for mechanistic studies of cell adhesion. Chem Soc Rev 29(4):267–273
107. Mudanyali O, Bishara W, Ozcan A (2011) Lensfree super-resolution holographic microscopy using wetting films on a chip. Opt Express 19(18):17378
108. Mudanyali O, Tseng D, Oh C, Isikman SO, Sencan I, Bishara W, Oztoprak C, Seo S, Khademhosseini B, Ozcan A (2010) Compact, light-weight and cost-effective microscope based on lensless incoherent holography for telemedicine applications. Lab Chip 10: 1417–1428
109. Mukundan H, Anderson AS, Grace WK, Grace KM, Hartman N, Martinez JS, Swanson BI (2009) Waveguide-based biosensors for pathogen detection. Sensors 9:5783–5809
110. Németh E, Adányi N, Halász A, Váradi M, Szendrő I (2007) Real-time study of the effect of different stress factors on lactic acid bacteria by electrochemical optical waveguide lightmode spectroscopy. Biomol Eng 24:631–637
111. Nikolelis DP (ed) (2012) Portable biosensors: weapons against bioterrorism. NATO science for peace and security series a: chemistry and biology. Springer Verlag, Dordrecht
112. Nirschl M, Reuter F, Vörös J (2011) Review of transducer principles for label-free biomolecular interaction analysis. Biosensors 1:70–91
113. Oh B-R, Huang N-T, Chen W, Seo JH, Chen P, Cornell TT, Shanley TP, Fu J, Kurabayashi K (2014) Integrated nanoplasmonic sensing for cellular functional immunoanalysis using human blood. ACS Nano 8(3):2667–2676
114. Oh C, Serhan SO, Khademhosseinieh B, Ozcan A (2010) On-chip differential interference contrast microscopy using lensless digital holography. Opt Express 18(5):4717
115. Orgovan N, Peter B, Sz Bősze, Ramsden JJ, Szabó B, Horvath R (2014) Dependence of cancer cell adhesion kinetics on integrin ligand surface density measured by a high-throughput label-free resonant waveguide grating biosensor. Sci Reports 4:4034
116. Orgovan N, Salánki R, Sándor N, Bajtay Z, Erdei A, Szabó B, Horvath R (2013) In-situ and label-free optical monitoring of the adhesion and spreading of primary monocytes isolated from human blood: dependence on serum concentration levels. Biosens Bioelectron 54:339–344
117. Orzó L, Göröcs Z, Fehér A, Tőkés S (2013) In-line hologram segmentation for volumetric samples. Appl Optics 52(1):A45–A55
118. Orzó L, Wittner B, Tőkés S (2013b) High speed water monitoring systems based on digital holographic microscopy. In: Computer Science and Information Technologies (CSIT), pp. 1–9
119. Ozcan A, Isikman S, Mudanyali O, Tseng D, Sencan I (2010) Lensfree on-chip holography facilitates novel microscopy applications. SPIE Newsroom, 2010 May 19:00294
120. Patko D, Cottier K, Hamori A, HorvatH R (2012) Single beam grating coupled interferometry: high resolution miniaturized label-free sensor for plate based parallel screening. Opt Express 20:23162
121. Patko D, Gyorgy B, Nemeth A, Szabó-Taylor KE, Kittel A, Buzas EI, Horvath R (2013) Label-free optical monitoring of surface adhesion of extracellular vesicles by grating coupled interferometry. Sens Actuat B-Chem 188:697–701
122. Patko D, Hamori A, Cottier K, Kurunczi S, Horvath R (2011) Label free biosensing using Grating Coupled Interferometry. Eur Biophys J with Biophys Lett 40:230–231
123. Peter B, Nador J, Juhasz K, Dobos A, Körösi L, Székács I, Patko D, Horvath R (2015) Incubator proof miniaturized Holomonitor to *in situ* monitor cancer cells exposed to green tea polyphenol and preosteoblast cells adhering on nanostructured titanate surfaces: validity of the measured parameters and their corrections. J Biomed Opt 20(6):067002, 10 p

124. Peters MF, Vaillancourt F, Heroux M, Valiquette M, Scott CW (2010) Comparing label-free biosensors for pharmacological screening with cell-based functional assays. ASSAY Drug Devel Technol 8(2):219–227
125. Rabus DG, Welle A, R. Seger A, Ichihashi Y, Bruendel M, Hieb J, Isaacson M (2006) Determination of living cell characteristics and behavior using biophotonic methods. Proc of SPIE 6329:63290H
126. Ramsden JJ, Horvath R (2009) Optical biosensors for cell adhesion. J Recept Sign Transduct Res 29:211–223
127. Ramsden JJ, Li SY, Heinzle E, Prenosil JE (1995) Optical method for measurement of number and shape of attached cells in real time. Cytom 19:97–102
128. Raphael MP, Christodoulides JA, Delehanty JB, Long JP, Pehrsson PE, Byers JM (2013) Quantitative LSPR imaging for biosensing with single nanostructure resolution. Biophys J 104:30–36
129. Rice JM, Stern LJ, Guignon EF, Lawrence DA, Lynes MA (2012) Antigen-specific T cell phenotyping microarrays using grating coupled surface plasmon resonance imaging and surface plasmon coupled emission. Biosens Bioelectron 31:264–269
130. Rich RL, Myszka DG (2006) Survey of the year 2005 commercial optical biosensor literature. J Mol Recogn 19(6):478–534
131. Rivet C, Lee H, Hirsch A, Hamilton S, Lu H (2011) Microfluidics for medical diagnostics and biosensors. Chem Eng Sci 66:1490–1507
132. Rizzo MA, Davidson MW, Piston DW (2010) Fluorescent protein tracking and detection. In: Goldman RD, Swedlow JR, Spector DL (eds) Live cell imaging: a laboratory manual, 2nd edn. Cold Spring Harbor Laboratory Press, New York, pp 3–34
133. Rocheville M, Jerman JC (2009) 7TM pharmacology measured by label-free: a holistic approach to cell signaling. Curr Opinion Pharmacol 9(5):643–649
134. Sencan I, Coskun AF, Sikora U, Ozcan A (2014) Spectral demultiplexing in holographic and fluorescent on-chip microscopy. Sci Reports 4:3760
135. Seo S, Su T-W, Tseng DK, Erlinger A, Ozcan A (2009) Lensfree holographic imaging for on-chip cytometry and diagnostics. Lab Chip 9:777–787
136. Shamah SM, Cunningham BT (2011) Label-free cell-based assays using photonic crystal optical biosensors. Analyst 136:1090–102
137. Shiau AK, Massari ME, Ozbal CC (2008) Back to basics: label-free technologies for small molecule screening. Comb Chem High T Scr 11(3):231–237
138. Sim JH, Kwak YH, Choi CH, Paek S-H, Park SS, Seo S (2012) A birefringent waveguide biosensor platform for label-free live cell detection of Listeria monocytogenes. Sens Actuat B-Chem 173:752–759
139. Simonnet C, Groisman A (2005) Two-dimensional hydrodynamic focusing in a simple microfluidic device. Appl Phys Lett 87:114104
140. Simpson-Stroot JM, Kearns EA, Stroot PG, Magaña S, Lim DV (2008) Monitoring biosensor capture efficiencies: Development of a model using GFP-expressing *Escherichia coli* O157: H7. J Microbiol Meth 72:29–37
141. Singh V, Kumar D (2009) Theoretical modeling of a metal-clad planar waveguide based biosensor for the detection of pseudomonas-like bacteria. Progr Electromagn Res M 6: 167–184
142. Spichiger-Keller UE (1998) Chemical sensors and biosensors for medical and biological applications. Wiley-VCH, Weinheim
143. Starodub NF, Ogorodniichuk YA, Sitnik YA, Slishik NF (2012) Biosensors for the control of some toxins, viral and microbial infections to prevent actions of bioterrorists. In: Nikolelis DP (ed) Portable chemical sensors. NATO science for peace and security series a: chemistry and biology. Springer, Dordrecht, pp 95–117
144. Stybayeva G, Mudanyali O, Seo S, Silangcruz J, Macal M, Ramanculov E, Dandekar S, Erlinger A, Ozcan A, Revzin A (2010) Lensfree holographic imaging of antibody microarrays for high-throughput detection of leukocyte numbers and function. Anal Chem 82(9):3736–3744

145. Su T, Ozcan A (2013) On-Chip holographic microscopy and its application for automated semen analysis. In: Shaked NT, Zalevsky Z, Satterwhite LL (eds) Biomedical optical phase microscopy and nanoscopy. Elsevier, Amsterdam, pp 153–171
146. Su T-W, Choi I, Feng J, Huang K, McLeod E, Ozcan A (2013) Sperm trajectories form chiral ribbons. Sci Rep 3:1664
147. Su T-W, Isikman SO, Bishara W, Tseng D, Erlinger A, Ozcan A (2010) Multi-angle lensless digital holography for depth resolved imaging on a chip. Opt Express 18(9):9690
148. Su TW, Xue L, Ozcan A (2012) High-throughput lensfree 3D tracking of human sperms reveals rare statistics of helical trajectories. Proc Natl Aacad Sci USA 109:16018–16022
149. Subramanian A (2006) A mixed self-assembled monolayer-based surface plasmon immunosensor for detection of *E. coli O157H7*. Biosens Bioelectron 7:998–1006
150. Sykes EA, Albanese A, Chan WCW (2013) Implantable waveguides. Nat Photon 7:940–941
151. Székács I, Fejes Á, Sz Klátyik, Takács E, Patkó D, Pomóthy J, Mörtl M, Horvath R, Madarász E, Darvas B, Székács A (2014) Environmental and toxicological impacts of glyphosate with its formulating adjuvant. Intl J Biol Food Vet Agric Food Engineer 8(3):213–218
152. Székács I, Pál I, Zs Környei, Szendrő I, Madarász E (2010) Initial cell attachment: events and kinetics of surface adhesion of NE-4C neural stem cells. Eur Cells Mater 20(3):252
153. Szendrő I, Erdélyi K, Puskás Z, Fábián M, Adányi N, Somogyi K (2012) Development and experiments with conductive oxide nanofilm coated planar waveguide sensors. Nanopages 7:17–24
154. Taitt CR, Anderson GP, Ligler FS (2005) Evanescent wave fluorescence biosensors. Biosens Bioelectron 20:2470–2487
155. Tan F, Leung PHM, Liud Z, Zhang Y, Xiao L, Ye W, Zhang X, Yi L, Yang M (2011) A PDMS microfluidic impedance immunosensor for *E. coli O157:H7* and *Staphylococcus aureus* detection via antibody-immobilized nanoporous membrane. Sensor Actuat B-Chem 159:328–335
156. Teifenthaler K, Lukosz W (1989) Sensitivity of grating couplers as integrated-optical chemical sensors. J Opt Soc Am 6(2):209–220
157. Testa G, Persichettia G, Zenib L, Sarroc PM, Berninia R (2013) Optofluidics: a new tool for sensing. Proc SPIE 8794:879402
158. Thoma F, Langbein U, Mittler-Neher S (1997) Waveguide scattering microscopy. Opt Commun 134:16–20
159. Thuy NT, Tam PD, Tuan MA, Le AT, Tam LT, Thu VV, Hieu NV, Chien ND (2012) Detection of 32 pathogenic microorganisms using biosensor based on multi-walled carbon nanotubes dispersed in DNA solution. Curr Appl Phys 12:1553–1560
160. Tiefenthaler K, Lukosz W (1985) Grating couplers as integrated optical humidity and gas sensors. Thin Solid Films 126(3–4):205–211
161. Tims TB, Lim DV (2004) Rapid detection of *Bacillus anthracis* spores directly from powders with an evanescent wave fiber-optic biosensor. J Microbiol Methods 59(1):127–130
162. Tóth AE, Walter FR, Bocsik A, Sántha P, Veszelka S, Nagy L, Puskás LG, Couraud PO, Takata F, Dohgu S, Kataoka Y, Deli MA (2014) Edaravone protects against methylglyoxal-induced barrier damage in human brain endothelial cells. PLoS ONE 9(7): e100152
163. Tromberg BJ, Shah N, Lanning R, Cerussi A, Espinoza J, Pham T, Svaasand L, Butler J (2000) Non-invasive *in vivo* characterization of breast tumors using photon migration spectroscopy. Neoplasia 2:26–40
164. Tung Y-C, Huang N-T, Oh B-R, Patra B, Pan C-C, Qiu T, Chu PK, Zhang W, Kurabayashi K (2012) Optofluidic detection for cellular phenotyping. Lab Chip 12: 3552–3565
165. Unser S, Bruzas I, He J, Sagle L (2015) Localized surface plasmon resonance biosensing: current challenges and approaches. Sensors 15:15684–15716
166. van Leeuwen CJ, Vermeire TG (eds) (2007) Risk assessment of chemicals. In: An introduction 2nd edn. Springer, Dordrecht

domain of mycotoxins, their occurrence and detection in food and feed. We have centralized our discussion to the biosensor designs described for OTA and AF detection, as these biosensors architects provide general platform, and are also applied for the detection of other mycotoxins.

2 Mycotoxins

Mycotoxins are low molecular weight natural molecules produced as secondary metabolite by filamentous fungi. They have been associated with different diseases, and can affect agricultural crops during crops growth, harvest, storage and processing. Their presence can cause auto-immune illness, allergenic properties, and some of them are teratogenic, carcinogenic, mutagenic, etc. [3]. The mycotoxigenic potential depends on species and strains of fungus, composition of matrix and environmental factors such as temperature and moisture [4]. *Fusarium*, *Aspergillus*, and *Penicillium* are the most abundant moulds that produce mycotoxins and contaminate human foods and animal feeds [5]. Among more than 300 reported different mycotoxins, only a small number is of particular interest [6]. Ochratoxins, aflatoxins, fumonisins, zearalenone, trichothecenes, and tremorgenic toxins occur in food and feed at significant levels, and therefore require special concern for monitoring (Table 1).

Table 1 Mycotoxins, fungal species, effect on animals and classification

Mycotoxins	Examples of producing-fungi	Effects on animals	Classification (IARC)
Aflatoxins	*A. flavus A. parasiticus*	Teratogenic and carcinogenic effects, hepatotoxic and Hepathocarcinegenic	1,2B
Ochratoxin A	*A. ochraceus, P. viridicatum P. verrucosum*	Nervous and gangrenous syndromes, Nephrotoxicity, porcine nephropathy, immune suppression	2B
Patulin	*Byssochlamys, Aspergillus Penicillium*	Neurotoxic, immunotoxic, mutagenic, carcinogenic, and genotoxic effects	3
Fumonisins	*F. verticilliodes, F. prolifiratum*	Pulmonary edema, Hepatotoxicity, nephrotoxicity, leukoencephalomalacia in horses	3
Zearalenone	*F.graminearum*	Estrogenic effects, Atrophy of ovaries and testicles, abortion	2B
Trichothecenes	*F.graminearum*	Immunologic effects, hematological changes, dermatitis, oral lesions, edema, hemorrhages of intestinal tissues	3

2.1 *Aflatoxins*

The aflatoxins (A-flavus-toxins) were isolated and characterized for the first time after the death of more than 100,000 turkeys in England, due to the consumption of a mold-contaminated Brazilian peanut meal [7]. They are difuranocoumarin derivatives produced by a polypeptide pathway by many strains of *Aspergillus flavus, Aspargillus parasiticus* [8], *and Aspergillus nomius* [9]. Moreover *Aspergillums pseudotamarii, Aspergillus ochraceoroseus, Aspergillus rambellii, Aspergillus toxicarius, Emericella astellata, Emericella olivicola and Emericella venezuelensis* [10] are also responsible for such impacts under some special environmental conditions that include high temperatures (13–40 °C) and high relative humidity (85 %). Among the 20 different types of aflatoxins, the most extremely toxic, mutagenic and carcinogenic type are aflatoxin B1 (AFB1), aflatoxin B2 (AFB2), aflatoxin G1 (AFG2) and aflatoxin G2 (AFG2) [11], in addition to aflatoxins M1 (AFM1) and M2 (AFM2). The later are hydroxylated metabolites of AFB1 and AFB2 [12] (Fig. 1). These aflatoxins are heterocyclic compounds (Figs. 2, 3 and 4) and they are classified according to their different fluorescing properties. AFB1 and AFB2 fluoresce in blue; AFG1 and AFG2 fluoresce yellow-green under ultraviolet light [13, 14]. The toxicity varies from a toxin to another, i.e. AFB1 is the most potent natural carcinogen aflatoxin produced by toxigenic strains among all aflatoxins [15], whereas AFM1 is hepatotoxic but not carcinogenic [16].

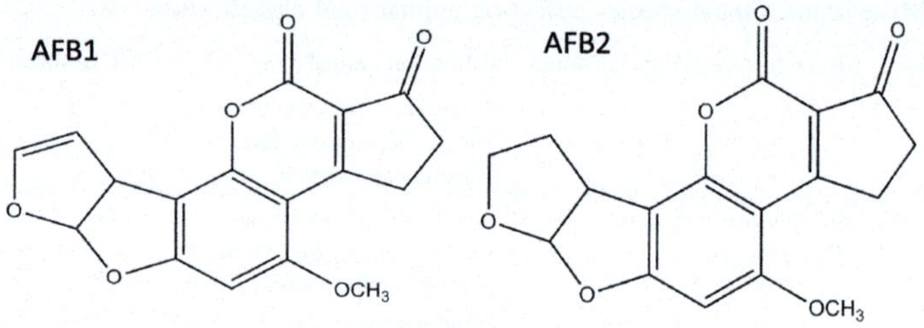

Fig. 1 Chemical structure of aflatoxin B (AFB1 and AFB2)

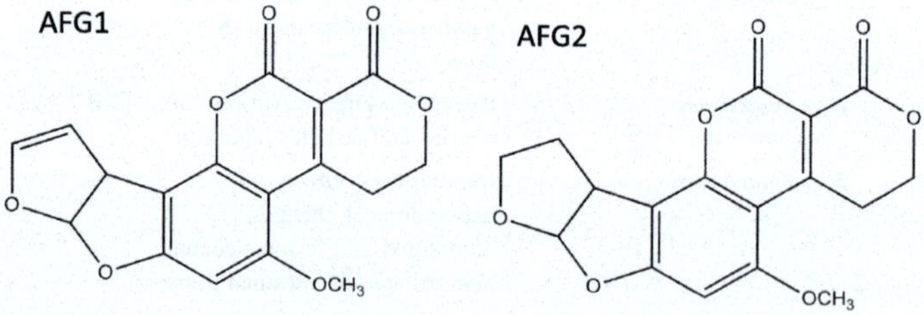

Fig. 2 Chemical structure of aflatoxin G (AFG1 and AFG2)

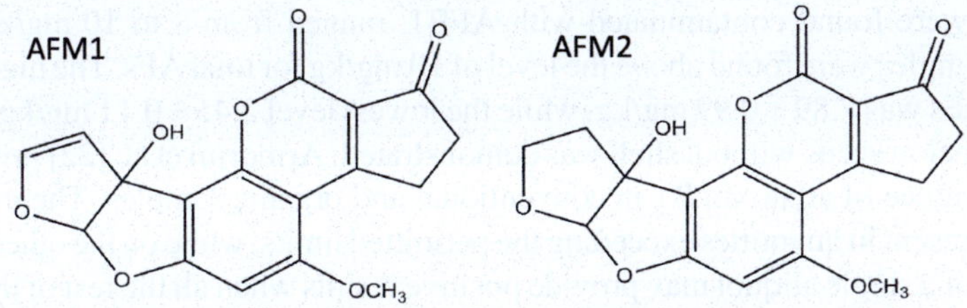

Fig. 3 Chemical structure of aflatoxin M (AFM1 and AFM2)

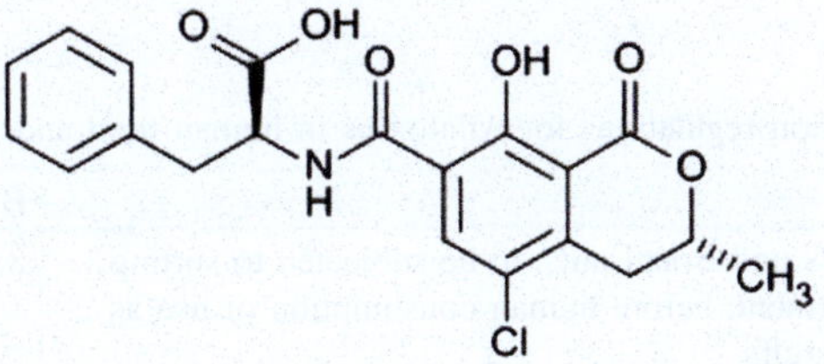

Fig. 4 OTA chemical structure

2.1.1 Aflatoxins in Foods and Feeds

Aflatoxins contaminate several food and animal feeds, to ensure food safety, maximum aflatoxin levels (reported in Table 2) were set within the European Commission Regulation No. 1881/2006. Maize is among the most commonly affected grains with the AF contamination. Kos et al. [17] studied the influence of weather conditions on the AFs production. They analyzed maize samples under two different weather conditions. The extremely hot and dry weather conditions followed by drought in 2012 caused AFs presence in 137 (68.5 %) of samples in the concentration range from 1.01 to 86.1 mg/kg with the mean level of 36.3 mg/kg. It was concludes that extremely hot and dry weather contributes to mold "hot spots" growth on grain and mycotoxins production. Dimitrieska-Stojkovic et al. [18] reported that AFM_1 is a main source of contamination in raw milk with calculated carry-over rates of AFB_1 to AFM_1 in the range from 0.22 to 3.74 %. This study was based on the analysis of thousands of milk samples and corresponsive feed samples from maize and complementary feed for dairy cattle. In the same context, Battacone et al. [19] administrated AFB_1 and detected AFM_1 after 3–6 h, with a total disappearance after 3.5 days [26]. Red pepper is among the most consumed spices in the world, Aydin, Emin Erkan, Baokaya and Ciftcioglu [20] demonstrated that the AFB1 levels in 18 % of red pepper samples were higher than the legal limits of European Commission in 2002 (>5 g/kg) with the highest level close to 40.9 μg/kg. In Pakistan, Masood et al. [21] showed the AFs contamination of over 307 samples of dry fruits and edible nuts (including dried plums, dates, dried apricot, raisins, almonds, walnut with shell and without shell, peanuts with shell and without shell, dried figs, watermelon seeds, melon seeds, pistachio with shell and without shell, pine nuts and cashew nut) from Northern areas and Khyber Pakhtunkhwa, Pakistan. 75 (24 %)

samples were found contaminated with AFB1, ranged from 8 to 10 mg/kg and 41 (13 %) samples were found above the level of 10 mg/kg for total AFs. The highest level of total AFs was 7.89 ± 0.99 mg/kg, while the lowest level 2.45 ± 0.11 mg/kg of AFB1 in watermelon seeds without shell was demonstrated. Armorini et al. [22] investigated the occurrence of aflatoxin B1 in conventional and organic samples. The mycotoxin may be present in quantities exceeding the permitted limits, while on the other hand the analysis of a single aliquot may provide positive results when all the rest of the sample are not contaminated with the mycotoxin. In conclusion, the selection of the test sample for laboratory analysis is a very crucial step to get a reliable result in ascertaining the toxin contaminations.

Table 2 European union regulations for Aflatoxins in human food and feed MLs (μg/kg)

Food	B1	B1,B2,G1,G2	M1
Groundnuts, Hazelnuts and Brazil nuts, to be subjected to sorting, or other physical treatment, before human consumption or use as an ingredient in foodstuffs	8.0	15.0	–
Almonds, pistachios and apricot kernels to be subjected to sorting, or other physical treatment, before human consumption or use as an ingredient in foodstuffs	12.0	10.0	–
Almonds, pistachios and apricot kernels, intended for direct human consumption or use as an ingredient in foodstuff	8.0	10.0	–
Dried figs	6,0	10,0	–
Tree nuts, Maize and rice to be subjected to sorting, or other physical treatment, before human consumption or use as an ingredient in foodstuff	5.0	10.0	–
Following species of spices: *Capsicum spp.* (dried fruits, chilli powder, cayenne and paprika)			
Piper spp. (fruits, white and black pepper)			
Myristica fragrans (nutmeg)			
Zingiber officinale (ginger)			
Curcuma longa (turmeric)			
Mixtures of spices containing one or more of the abovementioned spices			
All cereals and all products derived from cereals, including processed cereal products dried fruit, hazelnuts and Brazil nuts, intended for direct human consumption or use as an ingredient in foodstuffs	2.0	4.0	–
Milk and milk-based products			
Feed (exception below)	–	–	0.05
Complementary feedstuffs for cattle, sheep, goats	50		
Complementary feedstuffs for pig and poultry	50		
Complete feedstuff for pig and poultry, groundnuts, palm kernel, cottonseed maize and product derived from	30 20		
Complete feed for lambs and calves	10		
Complete dairy feed	5		

2.1.2 Effects on Human and Animals Health

The harmful effects of aflatoxins have been demonstrated in several studies; Oliveira et al. [23] classified AFB1 as a natural carcinogen that acts by forming adducts with guanine, and reacting with lungs RNA and proteins causing liver and lungs cancer. Additionally, it can also induce RNA damage, lesions, and inactivation of the ATR/Chk1 pathway of RNA damages repairing in the HCT-116 colonic cell line human colon carcinoma [24]. According to a study by Hirst et al. [25] AFB1 is the inciting agent in the development of malignant hepatomas in 5–28 % of cases, by deleting mutations in the P53 tumor-suppressing gene and activation of dominant oncogenes [26]. Besides, AFB1 is involved in promoting liver tumor growth when co-infecting with fumonisine [27], increasing the hepatic lipid levels and organ weight when co-infecting with OTA [28]. It may also lead to 60 times increase the relative risk for developing liver cancer for HBV (hepatitis B) patients when co-infected with HBV [26]. Periaca et al. [29] reported many worldwide aflatoxicosis outbreaks including R eye syndrome, liver, brain, kidney and blood issues. Immunologically, AFB_1 is involved in phagocytic activity suppression, T-lymphocyte function and number reduction, thymic aplasia induction [26] as well as cell-mediated immune responses suppression in many farm animals such as pigs and poultry [30]. Smith et al. [31] showed that feeding boilers with a high level (3.5 mg/Kg) of a mixture of aflatoxins, including 79 % AFB1, 16 % AFB2, 4 % AFG1 and 1 % AFG2, induced a body weight loss and an increase in the liver and kidney weights. Jend et al. [32] presented that a high level of AF in poultry feed induces salmonellosis, coccidiosis, hepatitis and infectious bursal diseases. Moreover, injection of AF alone or with OTA increases mortality or impaired development in chicken embryos [33].

2.2 *Ochratoxins*

Ochratoxins are secondary metabolites produced by several *Aspergillus* and *Penicillium* genera, chemically described as 3,4-dihydromethylisocoumarin deriva [34]. Ochratoxin A is the most prevalent and relevant fungal toxin of this group, while ochratoxins B and C are of lesser importance. Ochratoxin A (Fig. 1) (C20H18ClNO6, 1 N-{[(3R)-5-chloro-8-hydroxy-3-methyl-1-oxo-3,4-dihydro-1H-isochromen-7-yl]carbonyl}-L-phenylalanine) is a fluorescent compound with a molecular weight of 403.82 g/mol [35]. Van der Meurve et al. isolated it from *Aspergillus ochraceus* [36]. It's *a* secondary metabolite produced by fungi from the genera *Aspergillus* and a single *Penicillium* species: *P. verrucosum* [37]. The major species producing OTA are *Aspergillus ochraceus*, *Aspergillus* species and *Aspergillus carbonarius* [38]. Other fungi such as *Aspergillus niger* are producing OTA in some commodities or geographic areas [39]. Its presence is related to health effects such as nephrotoxic, genotoxic, neurotoxic, mutagenic, imunotoxic, embiotoxic and teratogenic effects and possible carcinogenicity [40, 41]. This mycotoxin has also been involved in chronic karyomegalic interstitial nephropathology, chronic interstitial nephropathy [42], end-stage renal failure, nephrosis and urothelial cancer [43].

2.2.1 Ochratoxin A in Foods and Feeds

OTA contaminates foodstuffs such as cereals, dried fruits, cocoa beans, coffee beans, peanuts, poultry eggs, milk and wine [44] and the maximum residue levels (MRLs) fixed by European regulations for OTA in several foods and beverage are reported in Table 3. Because of the diversity of products contaminated by fungus and ochratoxins, it is difficult to describe the visible aspect of the fungus producing ochratoxins, except for some species. For example, *A. ochraceus* is visible in yellowish-tan, *Penicillium* in blue-green, and black for *A. niger*. For the grain with no visible aspect of contamination, a musty odor is an indication of ochratoxins presence [34]. The humidity, moisture and warm temperatures are the most prosperous condition for fungus and mycotoxins occurrence, therefore a good storage conditions must be applied and rigorous testing must be carried out.

To assess OTA occurrence in European special wine, Valero et al. [45] studied 121 representative special wines from Europe, mostly from zone with a hot and dry climate during summer, and warm and wet during autumn. They concluded that 60 % of the total wine samples contained OTA with 27 % higher than 2 μg/L, with a maximum level up to 15.62 μg/L. They also concluded that wines made from grapes dried in cool and dry chambers contained higher OTA levels than those made from grapes dried in hot and dry chambers. The uncontaminated samples represent wines that generally made in colder region with climates suitable for noble rot rather than summer rot, which shows that the prevalence of OTA in some wines is dependent on the environmental and fermentation conditions. Scodamore [46] concluded that drying the grain up to 15 % before storage, and maintaining water activity around 0.8 can preserve the grains from mold infection. In the same context, Jorsengen et al. [47] studied the precipitation effect on grains contaminations level, and reported that the harvest from the year characterized by normal to wet climate, generates higher average contamination level than the year considered dry to very dry. They also suggested that to prevent the OTA production during storage, the grain must be dry

Table 3 European union regulations for Ochratoxin A in food maximum levels (μg/Kg)

Food	OTA
Soluble coffee, dried vine fruit (currants, raisins and sultanas)	10
Unprocessed cereals, roasted coffee beans and ground roasted coffee, excluding soluble coffee	5
All products derived from unprocessed cereals intended for direct human consumption	3
Wine, fruit wine, all beverage from grape	2
Baby foods for infants and young children	0.5

to reduce water activity. The type of agriculture practices used in crop production is decisive, since samples from organic rice and cereal grain presented OTA level much higher than samples provided from conventional farms. This difference is due to the limited use of chemical products like fertilizers and fungicides in organic crop [48]. Duarte et al. [49] reported OTA contamination in several cereals samples from all over the world. In Tunisia, the occurrence of OTA in wheat, barley and sorghum were respectively 38, 40 and 43. Cereals grains in UK represented less contamination levels, with 16.25 % for wheat, 15.9 % for barley and no contamination for oat. In addition to its presence in food and feed, OTA occur in house speck collected from the heating ducts in a household, and dirt from a crawl space [50].

2.2.2 Effects on Human and Animals Health

Ochratoxin A is the most redundant and prevalent among ochratoxins. Up to now, it has been presented with teratogenic, nephrotoxic, and carcinogenic effect, and to be immunosuppressive in several experimental studies [51]. The National Toxicology Program (NTP) in the Thirteenth Report on Carcinogens has narrated it as "Reasonably anticipated to be human carcinogens" [52]. It has been shown as a potential causative agent of the Balkan Endemic Nephropathology disease, [53] especially due to the similarity of symptoms with the ochratoxin A-induced porcine nephropathy [54]. OTA has been also correlated with testicular cancer in several European countries, by inducing adducts in testicular DNA in animals. The consumption of OTA contaminated foods during both pregnancy and childhood may result in testicular DNA lesions transformed in puberty into testicular cancer [55]. In addition, Appelgren and Arora [56] conducted an auto-radiographic study on mice and concluded that [14] C-labelled OTA crosses the placental barrier preferentially on day 9 of pregnancy.

OTA metabolic disposition in human is not well known, except that OTA has a long serum half-life in blood of 840 h (35 days), due to serum macromolecules binding. It has been detected in human breast milk, urine, blood, and even in breast cancer, gastric mucosa, renal cell carcinomas and skin biopsy samples [57]. Liver elimination of OTA is retained by protein carriers that pick the toxin from the bonded macromolecules into the bile, then uptaken by the tubule cells to be secreted into urine [58]. In case of swine, OTA seems to have an additive effect on weight gain [59]. Glavitis and Vanyi [60] in (1995) showed that OTA could cause uncoordinated movement, faintness, anorexia, an increase in the total body weight and urination due to water retention. Stove and colleagues investigated kidneys of effected pigs and observed that the kidneys were enlarged and pale, presenting vascular lesions, damaged proximal tubules, renal fibrosis, denser, with hyalinisation and sclerosis of glomeruli and slightly undulating surface [61].

2.3 Other Mycotoxins

In addition to OTs and AFs, there are several other potentially toxic mycotoxins such as patulin, zearalenone, fumonisins … etc. These toxins have been found in foods and/or feeds and present a high potential for causing severe diseases in human and animals.

Patulin is an unsaturated heterocyclic lactone (Fig. 5) [62] produced by fungi belonging to *Penicillium, Aspergillus* and *Byssochlamys* genera [63], contaminating vegetables, fruits and particularly apple and apple products, which are more subject to the accumulation of this toxin [64]. Numerous studies illustrated neurotoxic, immunotoxic, mutagenic [65], carcinogenic [66], and genotoxic [67] effects of patulin residues on human health. Due to its high toxicity, the European Union (EU) fixed its maximum levels up to 50 μg/L in apple juice and cider, 25 μg/kg in solid apple, and 10 μg/L in infants and young children products [68, 69].

Zearalenone is a non-steroidal, estrogenic mycotoxin produced by several *Fusarium* species (*F. semitectum F. culmorum, F. equiseti, F. graminearum, F. cerealis, F. crookwellense and F. sporotrichioides,*) [70] in temperate and warm countries. Chemically defined as 6-(10- hydroxy-6-oxo-trans-1-undecenyl)-b-resorcyclic acid lactone (C18H22O5, MW: 318.36, CAS 17924-92-4); [71] (EFSA, 2004), this mycotoxin contaminates corn, wheat, barley, sorghum and rye [3]. The infected grains demonstrated a pink color associated to a pink pigment produced by the same strains [18]. Zearalenone shows oestrogen-like activity in certain farms animals, essentially sheep, pigs and cattle [1]. Its alcoholic metabolites (α-zearalenol and β- zearalenol) are also known to be oestrogenic [72].

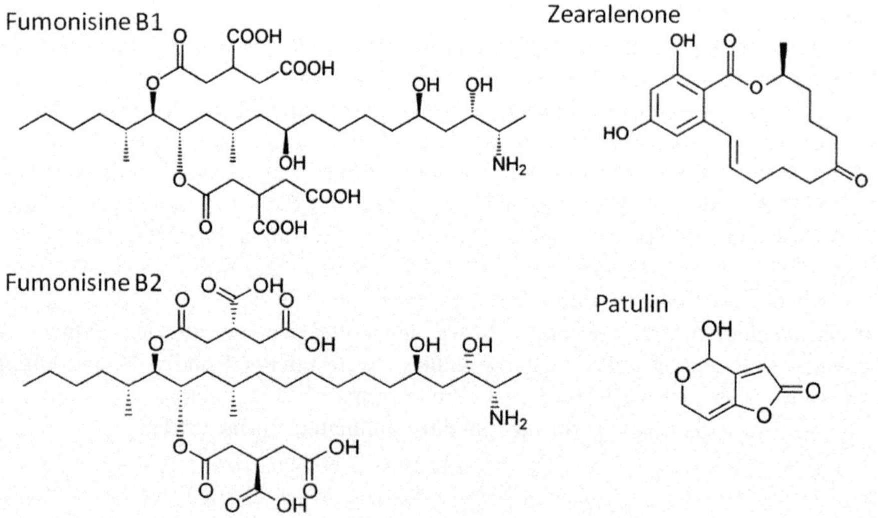

Fig. 5 Chemical structure of patulin, zearalenone, fumonisines B1 and B2

Fumonisins are a group of non-fluorescent structurally related mycotoxins including fumonisin B1 (FB1) and fumonisin B2 (FB2) [73] (Fig. 5). They are mainly produced by some *Fusarium* species, such as *Fusarium verticillioides* and *fusarium proliferatum* [74] that typically contaminates maize and maize-based products worldwide in addition to sorghum and rice. FB1 and FB2 are the most abundant and the most toxic among all. They are cancer promoting, and have been associated to a wide range of diseases in animals, such as leukoencephalomalacia in horses, pulmonary edema and hydrothorax in wines, and liver cancer in rats [75]. They are suspected to increase the risk of oesophagus cancer [76] and classified as possibly carcinogens to humans as class 2B carcinogens, [77] according to the International Agency for Research on Cancer (IARC) (Table 1).

3 Electrochemical Biosensors

Because of the wide spread occurrence of mycotoxins and the serious economic and health threat associated with them, it has been very crucial to develop ultrasensitive detection devices in order to ensure food safety and prevent risks in agro-food and environmental sector. For these purposes, many biosensors based on different bio-recognition elements and using different analyzing technique have been reported.

Electrochemical biosensors present many advantages comparing to classic analytical methods. They are simple, highly sensitive, allow trace level detection, enable a very rapid analysis time, easy to use and could be integrated into miniaturized systems [78, 79]. They are based on potentiometric, amperometric, conductometric and impedimetric detection methodologies. In this chapter, we will give an overview of recent advancements in the domain of electrochemical biosensors and their applications in mycotoxins detection for food safety. We will summarize them according to their recognition elements as enzymatic sensors, immunosensors, aptasensors and peptide based sensors.

3.1 Enzymatic Biosensors

Enzymatic biosensors developed for mycotoxins detection are based on either catalytic activity or enzyme inhibition. The first one is based on the detection of the products of the reaction when the toxin acts as substrate. It relies on the detection of the ochratoxin A based on enzymatic reaction with horseradish peroxidase (HRP) enzyme. It was shown that enzymatic reaction of HRP with OTA produced oxidized OTA that can be monitored through their electrochemical signal by amperometric analysis. The enzymatic biosensor could be constructed following the immobilization of enzyme directly onto the surface of screen-printed electrode or through polypyrrole matrix [80]. This approach allows selective and sensitive determination of OTA within concentration range of 23 to 200 nM, and detection of OTA in beer has been demonstrated [81] (Fig. 6).

The enzymatic inhibition produced by some mycotoxins has been exploited in the fabrication of enzymatic biosensors. This is the case of aflatoxin B1 that inhibits the catalytic activity of various acetylcolinesterases with a non-competitive inhibition pattern [82]. The toxin has been demonstrated to inhibit reversibly the enzyme by binding to the peripheral site located at the entrance of the active site where the binding constant was estimated to be 0.3 μM [83]. Various biosensors of aflatoxins have been developed, based on the inhibition of acetylcholinesterase catalytic activity [84]. The choline esterase hydrolyses the substrate of enzyme acetylcoline to choline and acetic acid (Fig. 7).

Potentiometric biosensor are based on the measuring the charge variation following catalytic reaction that results in a pH change. This has been achieved through field effect transistor where aflatoxin has been detected in the concentration range of 0.2 to 40 μg/ml, and the method has been demonstrated with real samples [85].

Amperometric biosensors are based on measuring the activity of the electroactive species produced after the catalytic reaction. As the reactant and products involved in the enzymatic reaction of acetylcolineesterase are not electroactive, the biosensor could be developed through bienzymatic reaction in which the first enzymatic reaction is coupled to choline oxidase which oxidases the choline to betaine with the formation of hydrogen peroxide that could be monitored by amperometric measurement (Fig. 7). In order to reduce the applied potential and to enhance the sensitivity of detection, redox mediators such as Prussian blue [86] or *cobalt-phthalocyanine* are integrated in the sensor design [82] (Fig. 8).

The main problem of the enzymatic biosensors is the lack of selectivity, as other toxins or molecules from the sample might also react or inhibit the enzyme activity. This area of research is still under progress and promising proof of concept has been demonstrated.

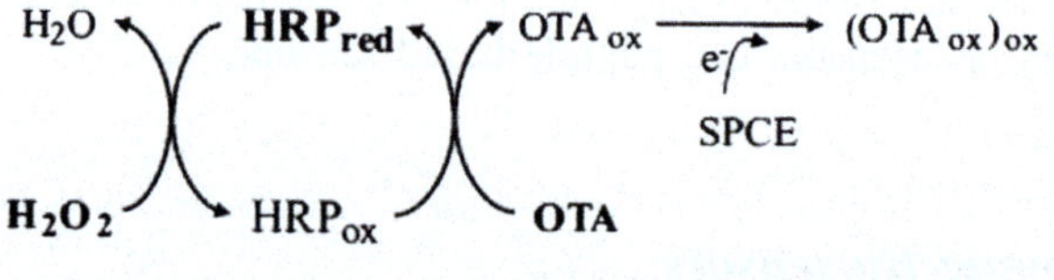

Fig. 6 Mechanism of OTA detection through enzymatic reaction with HRP [81]

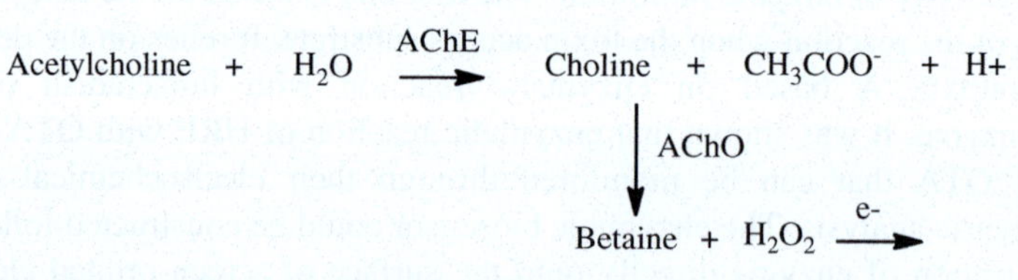

Fig. 7 The mechanism of bienzymatic biosensor for amperometric detection

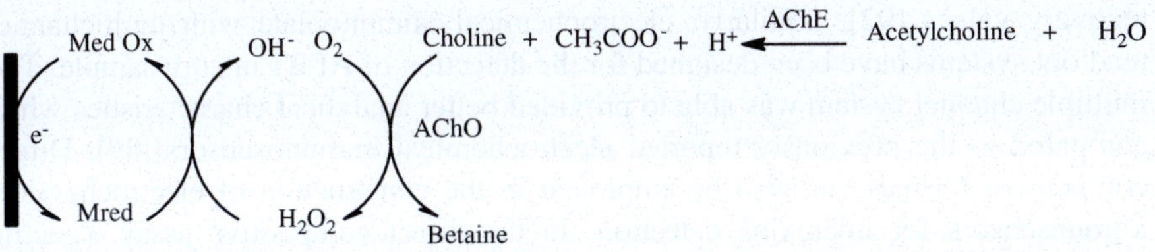

Fig. 8 Schematic representation of the bienzymatic reaction for AFB1 determination measured with redox mediator

3.2 *Electrochemical Immunosensors*

Various electrochemical detection methodologies such as potentiometry, amperometry and conductometry have been integrated in the design of electrochemical immunosensors for the detection of mycotoxins. Among all, amperometric immunosensors are the most commonly used methods due to their good sensitivity, cost effectiveness and the possibility of miniaturization [87]. Different electrodes materials including platinum, gold and carbon in its various forms have been used as transducer platform in the fabrication of electrochemical immunosensors. To overcome the problems associated with conventional types of transducer platforms, screen printed technology has attracted an increasing interest with the characteristics of mass production, low cost fabrication and single drop analysis [88]. The combination of electrochemical detection methodologies with the advances in sensor technologies and the suitability for in situ applications make this approach one of the most commonly used analysis system in the field of electrochemical immunosensors.

3.2.1 Electrochemical Immunosensors for Aflatoxins

In order to achieve high sensitivity and with a requirement to design single use disposable methodologies, various formats of electrochemical immunosensors have been designed and successfully applied for the detection of AFB1 and AFM1 in food samples such as corn, barley and milk [89, 90]. Aflatoxins are low molecular weight analytes, hence sandwich type formats are not ideal candidate to design electrochemical immunosensing platforms. The assays based on the competition formats are usually preferred in the fabrication of electrochemical immunosensors. For example, indirect competitive electrochemical immunosensors based on the disposable screen printed electrodes have been extensively reported in the literature. The specificities of the designed assays are demonstrated by assessing the cross reactivity of the specific integrated antibody against the other closely related non specific aflatoxins. It has been reported that specific antibodies provide a vital way to distinguish different types of aflatoxins [91]. In this context, an electrochemical immunosensor was developed for the detection of AFB1 in barley sample which was able to distinguish AFB1 from other toxins with low matrix effect and good recovery values [92]. Similarly, electrochemical immunoplate with multichannel read out systems have been designed for the detection of AFB1 in corn sample. The multiple channel system was able to provided better analytical characteristics when compared to the previously reported electrochemical immunosensors [89]. Direct competitive formats can also be employed in the construction of electrochemical immunsensors for aflatoxins detection. In the direct competitive assay, specific antibody is immobilized on the electrochemical transducer platform, and the competition is performed between the labeled and non labeled analytes to bind with the immobilized bioreceptor. Such a design has been successfully applied for the detection of AFM1 in milk sample [93]. Flow injection electrochemical immunosensors have also been reported to monitor aflatoxins in food samples. These automated systems are characterized with low cost instrumentation, fast analysis time and improved reproducibility. The better reproducibility is associated to the automation of the system, as this helps to minimize the possible human errors linked to the batch and manual systems. Such an automated electrochemical sensing system has been applied to the detection of AFM1 in milk samples. The obtained results were comparable to those obtained with HPLC methods with the additional advantages of automation, less analysis time and low cost system [94].

3.2.2 Electrochemical Immunosensors for Ochratoxin A

Due to the small size of OTA, competition bases format offers the potential for rapid and quantitative analysis of OTA. Sandwich assays are considered to undergo the problems of less sensitivity and irreproducibility in case of small size molecules [95]. The first electrochemical immunsensor reported for OTA analysis was based on direct competitive format, and enzyme labels were used on the secondary antibody with screen printed carbon electrode as transducer platform. Differential pulse voltammetry based electrochemical immunosensors was applied to detect OTA in wheat samples [96]. Since then, our group has developed many disposable screen printed electrochemical immunosensors for the monitoring of OTA in wine [97] and wheat samples [98]. Because of the complexities associated with the use of label reagents, efforts are focused to design label free electrochemical sensing methodologies. In this context, electrochemical impedance spectroscopy has been emerged as a powerful technique to develop label free immunosensors. OTA specific antibody is immobilized on the electrode surface and changes in the electron transfer resistance in response to target analyte concentrations are monitored to obtain a calibration curve. Various formats of impedimetric immunosensors have been reported in the literature for the detection of OTA [99], but most of them have not been demonstrated with real sample analysis. Array based electrochemical immunosensors also offer an attractive alternative for simultaneous detection of many analytes, and have been applied for the detection of OTA along with other food toxins [100].

3.3 Aptasensors

Since the first report of aptamers by the end of the 19th century and with the emergence of the Systematic Evolution of Ligands by Exponential enrichment (SELEX) process early 20th century, numerous studies have reported the use of aptamers in biosensor field for food and environmental monitoring.

To detect AFM1 in milk, recently, Istambouli et al. [101], have developed an impedimetric aptasensor based on the immobilization of a hexaethylenehlycol-modified 21-mer oligonucleotide on a carbon screen-printed electrode, after activation of the sensing surface with diazonium salt. This approach permitted to detect AFM_1 in buffer in the range of 2–150 ng/L with a LOD = 1.15 ng/L, and in filtered milk from 20 to 1000 ng/Kg, in agreement with the maximum levels of AFM1 permitted by the European Union for adults (50 ng/Kg) and infants (25 ng/Kg) milk and dairy products.

In the same context, Nguyen and co-workers [102] fabricated an electrochemical Fe_3O_4/polyaniline-based aptasensor capable of detecting AFM1 in the range of 6–60 ng/L, with a detection limit of 1.98 ng/L. This aptasensor was based on the polymerization of the Fe_3O_4 incorporated polyaniline (Fe_3O_4/PANi) on inter-digitized electrode (IDE) as sensitive film on which aptamer was immobilized. Magnetic nanoparticles were used as signal amplification element. The AFB1 detection was performed via electrochemical signal change acquired by cyclic and square wave voltammetry. Dinçkaya et al. [103] developed a three step DNA based biosensor; a first layer of self-assembled monolayer (SAM) of cysteamine was deposited on gold electrodes, followed by a gold nanoparticles layer to immobilize a thiol-modified single stranded DNA (ss-HSDNA) probe. The assembly processes were monitored with the help of electrochemical impedance spectroscopy (EIS) and cyclic voltammetry (CV) techniques, and the method permitted a linear response to AFM1 in the range of 1–14 ng/mL with a standard deviation of ±0.36 ng/mL.

Evtugyn and colleagues [104] reported the fabrication of an electrochemical aptasensor based on neutral red modified polycarboxylated macrocyclic ligands "calixarens" on which AFB_1 specific DNA aptamer was covalently immobilized. The interaction with the target was monitored with cyclic voltametry (CV) and electrochemical impedance spectroscopy (EIS), and a negligible variation was observed with a LOD of 0.1 nM for CV and 0.05 nM for EIS. This method allowed the detection of AFB1 in white wine, soy sauce, peanuts and cashew nuts with a recovery of 80–100 %.

Castillo et al. [105] fabricated a sensitive electrochemical aptamer-based sensor for the direct detection of mycotoxin AFB1. They covered gold electrode with cystamine, then immobilized Poly (amidoamine) dendrimers of fourth generation (PAMAM G4) and subsequently employed for attachment of single stranded amino-modified DNA aptamers specific to AFB1. The different steps of the aptasensor construction were monitored with cyclic voltammetry (CV) and electrochemical impedance spectroscopy (EIS). This aptasensors allowed to detect AFB1 in buffer between 0.1 and 10 nM, with a LOD = 0.40 ± 0.03 nM and was characterized with comparable results in treated certified contaminated peanuts extract as well as in spiked peanuts snacks with a recovery of 96–120 %.

The specificity of this aptasensor to AFB1 was investigated in the presence of other interfering toxins.

Electrochemical aptasensors have been widely used in OTA sensing including label and label-free detection methods. For labeled modes, several biomarkers such as quantum dots (CdS, PbS), methylene blue (MB), ferrocenes (Fc), enzyme (horseradish peroxidase) and gold nanoparticles have been extensively employed in aptasensing design. The label free mode is mainly based on EIS detection. Some recent advances for OTA detection are summarized in Table 4.

Huang and colleague [96] optimized a signal-amplified electrochemical aptasensor for ultrasensitive OTA detection based on signal enhancement with rolling circle amplification (RCA). This aptasensor demonstrated a very low LOD of 0.065 pg/mL and a 10000 fold extended dynamic range. Many other researchers have demonstrated OTA aptasensor based on amplified signal. Bulbul et al. [79] reported an ultrasensitive electrochemical aptasensor based on the synergistic contribution of a nanoceria (nCe) tag and graphene oxide (GO) used to immobilize the aptamer which capture OTA via a specific competitive mechanism between the free and the nCe labeled one. This aptasensor exhibited a linear response to OTA in the range 0.15−180 nM with a detection limit of 0.1 nM. Xie et al. [108] used an outstanding DNA amplification procedure: loop-mediated isothermal amplification (LAMP) whose products were combined with methylene blue (MB) and analyzed using differential pulse voltammograms (DPV). This amplification allowed to detect OTA at a range of 0.005–50 nM with a LOD of 0.3 pM.

Tong et al. [111] developed a different strategy to enhance OTA sensing. They used a single-stranded DNA specific exonuclease which released the OTA associated DNA from the complimentary Fc labeled DNA, digested it, and liberated the OTA to re-engage it in the detection process and resulted in signal-on format. On the other hand, Kuang and colleague [110] demonstrated a "signal-off" electrochemical aptasensor based on immobilizing of MB labeled aptamer to the electrode surface by complementary pairing to DNA. Wu et al. [105], also demonstrated a "signal-off" one step aptasensor, based on the conformational change of aptamer after binding to OTA inducing remoteness of the MB modified side of the aptamer from the electrode surface and loss of the signal was monitored.

Mishra et al. [107] demonstrated one of the label free impedimetric aptasensors for OTA detection and quantification. It is based on immobilizing anti-OTA-aptamer using diazonium coupling reaction and demonstrates a limit of detection of 0.15 ng/mL with a good recovery percentage in cocoa beans.

3.4 *Peptide Based Sensors*

Since last few years, researchers are interested in replacing natural biorecognition elements with artificial receptors that are robust, more stable and cheap to produce, easy to modify, and are able to bind to specific chemical functions and/or molecular structures which makes them very suitable candidate for the development of

Table 4 Electrochemical aptasensors for OTA detection

Technique	Assay format	Detection range	LOD	Matrix	Reference	Recovery
DPV/EIS	Au-DNA1/DNA2/GNPs-DNA3/DNA4	2.5 Pm–2.5 nM	5 pM	Wine	Yang et al. [78]	90–95 %
EIS	Au-Apta/MB	0.1 pg/mL– 1 μg/mL	0.095 pg/mL	Red wine	Wu et al. [106]	94–102.3 %
DPV	SPCE-Diaz/Apta/BSA	0.15–5 ng/mL	0.07 ng/mL	Cocoa beans	Mishra et al. [107]	82.1–85 %
EIS	SPEC-Apta	0.15–2.5 ng/mL	0.15 ng/mL	Cocoa beans	Mishra et al. [108]	91–95 %
DPV/EIS	GCE-AuNPs/CaptureDNA/HT/Apta	0.001–50 nM	0.3 pM	Wine	Wie et al. [109]	97.1–108 %
Amperometric	SPEC-MBs(OTA) SPEC-MBs(Apta)	–	0.05 μg/L 0.06 μg/L	Beer	Rhouati et al. [110]	97.5–100 %
CV	GCEDBA1/DNA2(Apta)/DNA3-Au	0.1–20 ng/mL	30 pg/mL	Wine	Kuang et al. [111]	95–110 %
DPV	GCE-Au-S/Fc-DNA/Apta	0.005–10.0 ng/mL	1.0 pg/mL	Wheat	Tong et al. [112]	90–108 %
EIS	ITO-SA/PANI-DNA/Apta	0.1 ng/mL 0.01 μg/mL 1–25 μg/mL	0.1 ng/mL	–	Prabhakar et al. [113]	–
CA	AuHAP(DNA)/MCH/Avindin-HRP	1.0–20 ng/mL	0.4 pg/mL	Wheat	Zhang et al. [114]	95.8–105.4 %
CV/EIS	SPCE-NH2-PEG-Apta	0.12–5.5 ng/L 1.0–8.5 ng/L	0.12 ng/L 1 ng/L	Beer	Akhtar el al. [115]	–
CV/EIS	SPCE-Ar-NO_2-p/Ar-Eth/H	1.25 ng/L– 500 ng/L	0.25 ng/L	Beer	Akhtar el al. [116]	–
DPV	SPCE-MBs-Apta	0.78–8.74 ng/mL	0.07 ng/mL	Wheat	Bonel et al. [117]	102–104 %
CV/EIS/DPV	Au captureDNA/MCH/RCA	0.1 pg/mL– 5 μg/mL	0.065 pg/mL	Wine	Huang et al. [118]	96–108 %

CV cyclic voltammetry
EIS electrochemical impedance spectroscopy
DPV differential pulse voltammetry
CA chronoamperometry

biosensors [119]. Tozzi and co-workers have synthesized and identified a hexapeptide which showed a good affinity (Keq = 3.4 × 104 M^{-1}) towards the mycotoxin ochratoxin A (OTA) detection. This hexapeptide has been used in an immunoaffinity-like stationary phase to develop a solid-phase extraction method for the quantification of OTA in wine samples. Several different wine samples fortified with OTA showed recovery of 94.7 and 98.4 % [120].

Bazin et al. [121] elaborated a sensing method that was based on the use of peptides to detect OTA. The peptide NFO_4 was selected because of its high affinity toward OTA compared to other peptides. It was used and validated in performing a peptide based competitive ELISA assay, and discriminated a concentration of 2 µg/L of OTA in red wine without pre-treatment of the sample. The same peptide was immobilized via a histidine tag onto 3D porous chitosan foam pre-adsorbed with Zn^{2+} [122]. This study revealed that the histidine tag modification and immobilization has improved the OTA detection with the finding that the N-terminus is 2.5 times more efficient rather than C-terminus.

4 Conclusion

By summarizing recent progress in developing analytical methodologies, we have demonstrated that electrochemical biosensors can be used as highly sensitive and selective alternatives to conventional analytical methods for the detection of mycotoxins. The small size and simplicity of biomolecules, added to their easy functionality, enable their oriented immobilization, increasing the efficiency in the binding of targets and minimizing the non-specific adsorptions, making electrochemical biosensors as ideal candidate for mycotoxin detection.

In comparison to other immunosensors, electrochemical aptasensors offer several advantages such as extended stability, low production cost and easy and diverse modification sites. The use of screen printed electrodes as transducer in combination with aptamer as biorecognition element allows to miniaturized the system, and improve the reliability, speed and cost-effectiveness of the analysis. Broad ranges of sensing formats are being studied in order to apply the developed strategies to other. Recently, researchers have explored nanomaterials of vital interests with advancement in the field of nanotechnology. With this progress, nanomaterials are extensively integrated in the field of aptasensors for diverse applications. For example, our group has recently employed nanoceria particles and carboxy fluorescent particle as signal generating probes to replace the commonly used labels in the design of aptasensors [79, 123]. Future research may focus on the portability and cost effectiveness of the affinity sensor, making them highly suitable for field applications.

References

1. Zain ME (2011) J Saudi Chem Soc 15:129–144
2. Hussein HS, Brasel JM (2001) Toxicol 167:101–134
3. CAST (2013) Mycotoxins: risks in plant, animal and human systems. Task Force Report N° 139. Council for Agricultural Science and Technology, Ames
4. Fernandez-Cruz ML, Mansilla ML, Tadeo JL (2010) JARE 1:113–122
5. Binder EM (2007) Anim Feed Sci Technol 133:149–166
6. Miller JD (1995) JD. J Stored Prod Res 1:1–16
7. Turner NW, Subramanyam S, Piletsky SA (2009) Anal Chim Acta 632:168–180
8. Shapira R, Paster N, Menasherov M, Eyal O, Mett A, Meiron T, Kuttin E, Salomon R (1997) Appl Environ Microbiol 63:990–995
9. Agriopoulou S, Koliadima A, Karaiskakis G, Kapolos J (2016) Food Control 61:221–226
10. Reiter E, Zentek J, Razzazi E (2009) Mol Nutr Food Res 53:508–524
11. Inan F, Pala M, Doymaz I J (2007) J Stored Prod Res 43:425–429
12. Giray B, Girgin B, Engin AB, Aydin S, Sahin G (2007) Food Control 18:23–29
13. Saegeant (1963) KCRBAAR chemistry and origin of aflatoxins. Chemical Ind, London, pp 53–55
14. D'Mello JPF, MacDonald AMC (1997) Anim Feed Sci Tech 69:553–557
15. Squire RA (1981) Science 214:877–880
16. Carnaghan RBA, Hartley RD, O'Kelly J (1963) Nature 200:1101–1102
17. Kos J, Mastilovic J, Janic Hajnal E, Saric B (2013) Food Control 34:31–34
18. Dimitrieska-Stojkovic E, Stojanovska-Dimzoska B, Ilievska G, Uzunov R, Stojković G, Hajrulai-Musliu Z, Jankuloski D (2016) Food Control 59:201–206
19. Battacone G, Nudda A, Rassu SPG, Decandia M, Pulina G (2012) J Diary Sci 92(10): 4997–5004
20. Aydin, Emin Erkan M, Baskaya R, Ciftcioglu G (2007) Food Control 18:1015–1018
21. Masood M, Iqbal SZ, Asi MR, Malik N (2015) Food Control 55:62–65
22. Armorini S, Altafini A, Zaghini A, Roncada P (2015) Food Control 50:858–863
23. Oliveira PA, Colaço A, Chaves R, Guedes-Pinto H, De-La-Cruz PLF, Lopez C (2007) Ann Bras Acad Sci 79(4):593–616
24. De Ruyck K, De Boevre M, Huybrechts I, De Saeger S (2015) Rev Mutat Res 766:32–41
25. Hurst JM, Sterdman OJ (2011) J Gen Microbiol 48(3):329–355
26. Walliams JH, Phillips TD, Jolly PE, Stiles JK, Jolly CM, Aggarwal D (2004) Am J Clin Nutr 80:1006–1022
27. Carlson DB, Williams DE, Spitsbergen JE, Frank Ross P, Bacon CW, Meredith FI, Riley RT (2001) Toxicol Appl Pharmacol 172(1):29–36
28. Grenier B, Oswald IP (2011) World Mycotoxin J 4(3):285–313
29. Peraica M, Radic B, Lucic A, Pavlović M (1999) Bull World Health Org 77(9):754–766
30. Ali MV, Mohiuddin SM, Reddy AC (1978) Indian Vet J 71:801–817
31. Smith EE, Kubena LF, Braithwaite CE, Harvey RB, Phillips TD, Reine AH (1992) Poult Sci 71:1136–1144
32. Jand SK, Singh PP, Singh A (1995) Ind J Anim Sci 65:1063–1067
33. Erdington TS, Harvey RB, Kubena LF (1995) Bull Environ Contam Toxicol 36:331–326
34. Cole RJ, Cox RH (1981) Handbook of toxic and fungal metabolites. Academic press, New York
35. Richard JL (2007) Int J Food Microbiol 119:3–10
36. Van Der Merwe KJ, Steyn PS, Fourie L, Scott De B, Theron JJ (1965) Nature (London) 205:112–113
37. Meri K, Marika J, Aldo R (2005) Int J Food Microbiol 99:207–214
38. Benford D, Boyle C, Dekant W, Fuchs E, Gaylor DW, Hard G, McGregory DB, Pitt JI, Plestina R, Shephard G, Solfrizzo M, Verger PJP, Walker R (2006) Ochratoxin A safety

evaluation of certain mycotoxins in food. WHO Food Additives Series 47. FAO Food and Nutrition Paper, vol. 74. WHO Geneva, Switzerland, 281–415
39. Tjamos SE, Antoniou PP, Azantizidou A (2004) J Phytopathol 152:250–255
40. Darte SC, Pena A, Lino CM (2009) Microchem J 93:115–120
41. Ghali R, Hmaissia-Khalifa K, Ghorbel H (2008) Food Control 19:921–924
42. Maaroufi K, Zakhama A, Baudrimont I (1999) Human Exp Toxicol 18:410–415
43. Wafa EW, Yahya RS, Sobh MA, Eraky I, el-Baz M, el-Gayar HA, Betbeder AM, Creppoy EE (1998) Human Exp Toxicol 17(2):124–9
44. Weidenborner M (2001) Encyclopedia of food mycotoxins. Springler, Berlin
45. Valero, SM, Ramos AJ (2008) Food Chem 108:593–599
46. Scudamore KA (2005) Food Addit Contam 22(1):17–25
47. Jorgensen K, Rasmussen G, Thorup I (1996) Food Addit Contam 13:95–104
48. Gonzales L, Juan C, Soriano JM, Molto JC (2006) J Manes Int J Food Microbiol 107:223–227
49. Duarte SC, Pena A, Lino CM (2010) Food Microbiol 27:187–198
50. Richard JL, Plattner RD, May J, Liska SL (1999) Mycopathologia 146(2):99–103
51. Selected mycotoxins: chratoxins, trichothecenes, ergot. Report of an Expert Committee (1990) World Health Organization
52. http://ntp.niehs.nih.gov/ntp/roc/content/listed_substances_508.pdf
53. Pfohl-Leszkowicz T, Petkova-Bocharova IN, Chernozemsky MC (2002) Food Addit Contam 19(3):282–302
54. Krogh P (1976) Nordisk Veterinaermedicin 28(9):452–458
55. Schwartz GG (2002) Cancer Causes Control 13:91–100
56. Appelgren LE, Arora RG (1983) Food Chem Toxicol 21(5):563–568
57. Hope JH, Hope BE (2012) J Environ Sci Health (Article ID 835059, 10 p)
58. Reddy L, Bhoola K (2010) Toxins 2:771–779
59. Lindemann MD, Blodgett DJ, Kornegay ET (1993) J Animal Science 71(1):171–178
60. Glavitis R, Vanyi A (1995) Magyar Allatorvosak Lapja 50:407–420
61. Stoev SD, Hald B, Mantle PG (1998) The Veterinary Record
62. Bueno D, Istamboulie G, Munõz O, Marty J-L (2015) Appl Spectrosc Rev 50:1–47
63. Pennacchio A, Varriale MG, Esposito M, Staiano S (2015) D'Auria. Anal Biochem 481:55–59
64. Reddy KRN, Spadaro D, Lore A, Gullino ML, Garibaldi A (2010) Mycotoxin Research 26 (4):257–265
65. Zouaoui N, Sbaii N, Bacha H, Abid-Essefi S (2015) Food Control 51:356–360
66. Osswald H, Frank HK, Komitowski D, Winter H (1978) Food Cosmetics Toxicology 16 (3):243–247
67. Donmez-Altuntas H, Gokalp-Yildiz P, Bitgen N, Hamurcu Z (2013) Mycotoxin Research 29 (2):63–70
68. Beretta B, Gaiaschi A, Galli CL, Restani P (2000) Food Addit Contam 17:399–406
69. Yuanyuan Z, Boqiang L, Shiping T (2015) Int J Food Microbiol 206:102–108
70. Zinedine JM, Soriano JC, Molto J (2007) Manes. Food Chem Toxicol 45:1–18
71. Opinion of the scientific committee on food on fusarium toxins (2000) PART 2: Zearalenone (ZEA). http://europa.eu.int/comm/food/fs/sc/scf/index_en.html
72. Cheeke PR (1998) Mycotoxins in cereal grains and supplement. In: Cheeke PR (ed) Natural toxicants in feeds, forages, ansPoisonous plants. Interstate Publishers Inc, Danville, pp 87–136
73. Lino CM, Silva LJG, Pena ALS, Silveira MI (2006) Anal Bioanal Chem 384:1214–1220
74. Renjie L, Bu T, Minhao P, Yingchao L, Dong J (2015) Food Control 50:838–842
75. WFO Marasas V, Miller JD, Riley R (1999) Mycotoxins of growing interest, Fumonisins. In: Third Joint FAO/WHO/UNEP International, Conference on Mycotoxins Tunis, Tunisia, pp 3–6
76. Pagliuca G, Zironi E, Ceccolini A, Matera R, Serrazanetti GP, Piva A (2005) J Chromatogr B 819:97–103
77. IARC (International Agency for Research on Cancer) (2002) Monographs on the evaluation of carcinogenic risks to humans. In: IARC (ed) Some naturally occurring substances; food items and constituents, heterocyclic aromatic amines and mycotoxins, vol 56. Geneve, pp 489–521

78. Lê HQA, Sauriat-Dorizon H, Korri-Youssoufi H (2010) Anal Chim Acta 674:1–8
79. Bulbul G, Hayat A, Andreescu S (2015) Nanoscale 7:13230–13238
80. Alonso-Lomillo MA, Domínguez-Renedo O, Ferreira-Gonçalvez L, Arcos-Martinez MJ (2014) Biosens Bioelectron 25:1333–1337
81. Alonso-Lomillo MA, Domínguez-Renedo O, Roman LDT, Arcos-Martinez MJ (2011) Anal Chim Acta 688:49–53
82. Arduini F, Errico I, Amine AA, Micheli L, Palleschi G, Moscone D (2007) Anal Chem 79:3409–3415
83. Hansmann T, Sanson B, Stojan J, Weik M, Marty J-L, Fournier D (2009) Biosens Bioelectron 24:2119–2124
84. Arduini F, Amine A, Moscone D, Palleschi G (2010) Microchim Acta 170:193–214
85. Stepurska KV, Soldatkin OO, Arkhypova VM, Soldatkin AP, Lagarde F, Jaffrezic-Renault N, Dzyadevych SV (2015) Talanta 44:1079–1084
86. Ben Rejeb FA, Arvinte A, Amine A, Gargouri M, Micheli L, Bala C, Moscone D, Palleschi G (2009) Biosens Bioelectron 24:1962–1968
87. Ricci G, Volpe L, Micheli G, Palleschi G (2007) Anal Chim Acta 605:111–129
88. Ricci F, Adornetto G, Palleschi G (2012) Electrochim Acta 84:74–83
89. Paniel N, Radoi A, Marty J-L (2010) Sensors 10:9439–9448
90. Piermarini S, Micheli L, Ammida NHS, Palleschi G, Moscone D (2007) Biosens Bioelectron 22:1434–1440
91. Radoi MT, Prieto-Simon B, Marty J-L (2008) Talanta 77:138–143
92. Ammida NHS, Micheli L, Piermarini S, Moscone D, Palleschi G (2006) Anal Lett 39: 1559–1572
93. Micheli L, Grecco R, Badea M, Moscone D, Palleschi G (2005) Biosens Bioelectron 21: 588–596
94. Badea M, Micheli L, Messia MC, Candigliota T, Marconi E, Mottram T, Velasco-Garcia M, Moscone D, Palleschi G (2004) Anal Chim Acta 520:141–148
95. Hayat NP, Rhouati A, Marty J-L, Barthelmebs L (2012) Food Control 26:401–415
96. Alarcon SH, Micheli L, Palleschi G, Compagnone D (2004) Anal Lett 37:1545–1558
97. Prieto-Simon B, Campas M, Marty JL, Noguer T (2008) Biosens Bioelectron 23:995–1002 b
98. Bonel L, Vidal JC, Duato P, Castillo JR (2010) Anal Methods 2:335–341
99. Radi AE, Munoz-Berbel X, Lates V, Marty J-L (2009) Biosens Bioelectron 24:1888–1892
100. Sauceda-Friebe JC, Karsunke XYZ, Vazac S, Biselli S, Niessner R, Knopp D (2011) Anal Chim Acta 689:234–242
101. Istamboulié G, Paniel N, Zara L, Reguillo Granados L, Barthelmebs L, Noguer T (2016) Talanta 146:464–469
102. Nguyen BH, Tran LD, Do QP, Nguyen HL, Tran NH, Nguyen PX (2013) Mater Sci Eng C Mater Biol Appl 33:2229–2234
103. Dinçkaya E, Kınık O, Sezgintürk MK, Altug C, Akkoca A (2011) Biosens Bioelectron 26(9):3806–3811
104. Evtugyn G, Porfireva A, Stepanova V, Sitdikov R, Stoikov I, Nikolelis D, Hianik T (2014) Electroanalysis 26:2100–2109
105. Castillo G, Spinella K, Poturnayova A, Šnejdárková M, Mosiello L, Hianik T (2015) Food Control 52:9–18
106. Chu H, Wu J, Mei Z, Deng Y, Xue F, Zheng L (2012) W Chen Anal Chim Acta 753:27–31
107. Mishra RK, Hayat A, Catanante G, Istamboulie G, Marty J-L (2016) Food Chem 192:799–804
108. Mishra RK, Hayat A, Catanante G, Ocana C, Marty J-L (2015) Anal Chim Acta 889:106–112
109. Xie S, Chai Y, Yuan Y, Bai L, Yuan R (2014) Biosens Bioelectron 55:324–329
110. Rhouati A, Hayat A, Bueno Hernandez D, Meraihi Z, Munoz-Guerrero R, Marty J-L (2013) Sens Actuators B 176:1160–1166
111. Kuang H, Chen W, Xu D, Xu L, Zhu Y, Liu L, Chu H, Peng C, Xu C, Zhu S (2010) Biosens Bioelectron 26:710–716

112. Tong P, Zhang L, Xu JJ, Chen H-Y (2011) Biosens Bioelectron 29:97–101
113. Prabhakar N, Matharu Z, Malhotra BD (2011) Biosens Bioelectron 26:4006–4011
114. Zhang J, Chen J, Zhang X, Zeng Z, Chen M, Wang S (2012) Electrochem Com 25:5–7
115. Hayat A, Andreescu S, Marty J-L (2013) Biosens Bioelectron 45:168–173
116. Hayat A, Sassolas A, Marty J-L, Radi AE (2013) Talanta 103:14–19
117. Bonel L, Vidal JC, Duato P, Castillo JR (2011) Biosens Bioelectron 26:3254–3259
118. Huang L, Wu J, Zheng L, Qian H, Xue F, Wu Y, Pan D, Adeloju SB, Chen W (2013) Anal Chem 85:10842–10849
119. Tohill IE (2011) Emerging bio-sensing methods for mycotoxins analysis. In: Determining Mycotoxins and Mycotoxigenic Fungi in Food and Feed. Elsevier, 456 p
120. Giraudi G, Anfossi L, Baggiani C (2007) J Chromatogr A 1175:174–180
121. Bazin I, Andreotti N, Ibn Hadj Hassine A (2013) Biosens Bioelectron 40:240–246
122. Soleri R, Demey H, Tria SA, Guiseppi-Elie A, IBN Had Hassine A, Gonzalez C, Bazin I (2015) Biosens Bioelectron 67:634–641
123. Hayat A, Mishra R, Catanante G, Marty J-L (2015) Anal Bioanal Chem 407(2015): 7815–7822

Comparative Studies on Optical Biosensors for Detection of Bio-Toxins

Alexei Nabok

Abstract A number of optical bio-sensing methods were reviewed with their principles and main characteristics outlined. The advantages and disadvantages of optical methods were discussed in a view of their application in detection of bio-toxins. A case study presented the comparative analysis of results in detection of mycotoxins obtained with the method of total internal reflection ellipsometry. The future prospects of optical biosensing technologies were discussed with the main focus on development of portable and highly sensitive biosensors suitable for *in-field* analysis.

Keywords Optical biosensors · SPR · Ellipsometry · Interferometry · Planar waveguides · Mycotoxins

1 Introduction: The Role of Optical Methods in Detection of Bio-Toxins

Detection of bio-toxins of different origins (either produced naturally by different bio-organisms or synthetic) is of great interest nowadays for healthcare, environmental safety, and security. Traditional analytical methods, such as chromatography and mass spectroscopy, are quite capable of both identification and quantification of various bio-toxins in very small concentrations down to ppt range. However those laboratory-based, high-tech methods are expensive and require efforts of specially trained personnel, thus leading to long waiting time and high cost of analysis. Traditional bio-sensing approach based on ELISA immunoassay is more affordable though still laboratory-based and require specialised equipment, highly trained technicians, and expensive chemicals; yet the sensitivity of detection is in ppb range at best.

A. Nabok (✉)
Materials and Engineering Research Institute, Sheffield Hallam University, Sheffield S1 1WB, UK
e-mail: A.Nabok@shu.ac.uk

D.P. Nikolelis and G.-P. Nikoleli (eds.), *Biosensors for Security and Bioterrorism Applications*, Advanced Sciences and Technologies for Security Applications, DOI 10.1007/978-3-319-28926-7_23

These days, the focus of the development of chemical and bio-sensors has shifted towards portable (ideally hand-held) devices suitable for in-field (or point of care) analysis and at the same time highly sensitive and capable of detection of single molecules of analytes of interest. From the point of view of portability and simplicity of design, nothing can be better than electrochemical sensors; that is why electrochemical sensors dominate the bio-sensing market. Their sensitivity is reasonably high though the selectivity is rather poor, so that the identification of analytes is the main problem for electrochemical sensors. The most common sensitive elements in electrochemical sensors are enzymes which are not directly react with bio-toxins but rather inhibited. The identification of toxins is therefore required the use of sensor array approach with quite complex statistical data analysis and thus unreliable outcome. Since most of bio-toxins are rather small molecules, the use of gravimetric sensors, i.e. QCM and SAW sensors, is limited. Optical immuno-sensors are perhaps the most suitable for detection of bio-toxins since they combine high sensitivity of optical transducers with high selectivity of immune reactions.

The main purpose of this work is to compare several optical bio-sensing technologies on their suitability for detection of bio-toxins and to suggest the ways for future development of sensor devices.

2 Review of Existing Optical Bio-Sensing Technologies

Before going into description of different bio-sensing technologies, we have to outline the parameters on which all these methods will be compared. The sensitivity of detection can be defined as $S = \Delta R/\Delta C$, where ΔR is the changes of the sensor response and ΔC is the variation of the analyte concentration. Since the majority of optical bio-sensing methods are based on detection of changes in refractive index in the sensitive molecular layer, without mentioning specific analytes, the sensitivity can be expressed as $S = \Delta R/\mathrm{RIU}$ (RIU standing for refractive index unity). Very often instead of sensitivity, we tend to quote the low detection limit (LDL) which is defined as minimal concentration of analytes detected. Obviously LDL is inversely proportional to sensitivity, but also depends on signal to noise ratio for the response. The units for LDL could be in molar concentrations (mM, μM, nM, pM, etc.), or weight of analyte per ml of solution (μg/ml, ng/ml, pg/ml, etc.), which are equivalent to ppm, ppb, and ppt, respectively. Other important characteristics are the level of sensor recovery and the characteristic times of response and recovery. These parameters are related to the kinetics of the response and recovery and depend on the adsorption and desorption rates (k_a, and k_d) as well as on their ratios known as association constant ($K_A = k_a/k_d$) and affinity constant $K_D = 1/K_A = k_d/k_a$. The latter two parameters describe well the specificity of binding the molecule of analyte to receptor. For example, in the case of highly specific immune reactions K_A reaches the level of 10^6–10^8 Mol^{-1}.

Optical immuno-sensing technologies can be split into two categories, namely luminescence (fluorescence) sensors and label-free sensors. In the first one, the sensitive elements, such as proteins, antibodies, enzymes, nano-particles are conjugated with the fluorescent labels; binding analyte molecules to such receptors causes luminescence (fluorescence) or it's quenching. As result the response can be easily visualised either by naked eye or with a suitable photodetector. Fluorescence-based sensing technologies are very popular and particularly suitable for high throughput analysis. The example could be the method of ELISA [1] where the immune reaction can be quantified simultaneously in a large number of channels (typically 96). ELISA was established as a standard bio-sensing method in analytical laboratories, and other bio-sensing methods are commonly compared with it. The sensitivity of ELISA is reasonably high, typically below the ng/ml level. The main drawback of this method is the use of expensive chemicals and long time of preparation for analysis; yet it is a laboratory based method not suitable for *in-field* analysis.

The main focus of this review is on label-free optical methods. Majority of these methods are based on the phenomenon of evanescent field or wave which appear as electromagnetic wave propagating along the interface between two materials with different refractive indices when the light enter the material with lower refractive index at total internal reflection condition; the amplitude of evanescent is exponentially decaying away from the interface. Optical waveguides are based on this principle with the light propagating along the material of higher refractive index (core) when sandwiched between materials of lower refractive index (cladding).

The waveguides could be of different geometries from standard cylindrical in optical fibres to rectangular in planar waveguides and optical slabs. There are a number of sensing technologies based on optical fibres with the simplest one having a tip of the optical fibre coated with sensitive material. The intensity of the reflected light carries the information on interaction between the receptor and analyte molecules. Although such sensors are simple and could be handy for remote in-field analysis, the sensitivity of such optical probes is rather poor, typically in ppm range and below.

More advanced optic fibres sensors utilize diffraction gratings written either in the core or cladding resulting in Bragg-Grating (BG) sensors [2] and long period grating (LPG) sensors [3], respectively. These sensors were found application in remote sensing though the sensitivity of detection has never been high. Regenerating of the sensing surface could be another problem which limits their use for detection of bio-toxins. Sensing applications of the waveguides of planar geometry are very promising and will be discussed later.

The phenomenon of surface plasmon resonance (SPR) constitutes the basis of the most popular optical biosensing technology [4]. Surface plasmon resonance takes place in thin metal (Ag, Au, Cu) films (around 40 nm) deposited on glass. In a classical Kretschmann SPR geometry the light is coupled into thin metal film through the prism (Fig. 1a). When the angle of incidence exceeds the critical angle

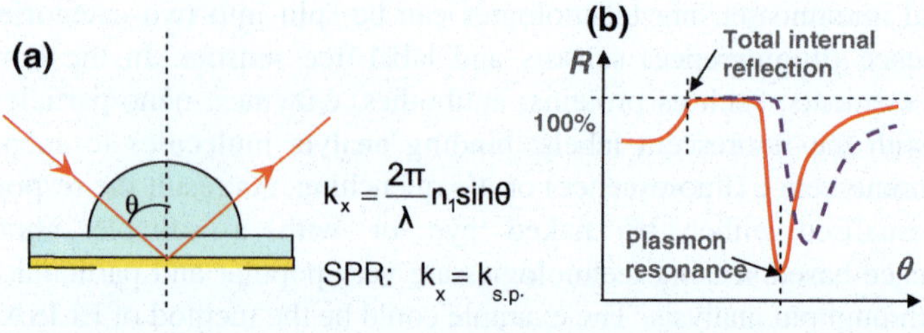

Fig. 1 Kretchamann configuration of SPR (**a**); Typical SPR curve (**b**)

of total internal reflection, almost 100 % of light is reflected and the evanescent wave is propagating along the surface. However, if the k_x vector of the evanescent wave matches the $k_{s.p.}$ vector of surface plasmons in thin metal films, the energy is transferred to plasmons and the reflection intensity drops down.

The dependence of refection intensity against the angle of incidence showed a dip which is called surface plasmon resonance or shortly SRP (Fig. 1b). Since the evanescent wave penetrates into the medium beyond the metal films any molecular adsorption on the surface of metal will cause a shift of SPR peak to high angles, which could be calibrated in terms of molecular concentrations. The effect of SPR is behind the most popular optical bio-sensing technology which was implemented in a large number of commercial instrumentations with BIACORE the most known. The designs are varied: instead of varying the angle of incidence, the wavelength of incident light can be varied which results in a spectroscopic SPR, portable SPR instruments based on this principle were developed. Different types of prisms (triangular, semi-cylindrical) can used for light coupling; instead of a prism, the diffraction grating can be used. The light spot could be widened resulting in SPR imaging instrument [5]. The sensitivity of SPR detection depends very much on the configuration used. The highest sensitivity of 10^{-9} RIU quoted by BIACORE was achieved with the use of special gold plates coated with a 10 nm thick porous layer of dextrane providing the large sensing surface and also containing the reference channel. Otherwise, typical sensitivity of majority of commercial SPR instruments is 3–4 orders of magnitude smaller.

Ellipsometry is another popular optical analytical method based on detection of changes in polarization of light upon its reflection from the investigated surface [4, 6]. The state of polarization of light is fully described by two ellipsometric parameters, namely Ψ and Δ which represent respectively the ratio of amplitudes and the phase shift between p- and s- components of polarized light, $\Psi = A_p/A_s$, $\Delta = \varphi_p - \varphi_s$. Such changes of Ψ and Δ are related to the optical parameters of the reflected surface through Fresnel equations. The presence of any thin film on the reflected surface, which could be a layer of adsorbed molecules, affects the values of Ψ and Δ. By solving Fresnel equations using appropriate software, it is possible to calculate the values of the thickness and refractive index of the adsorbed molecular layer from experimentally recorded values of Ψ and Δ, therefore the ellipsometry can be utilised for detecting different chemical reactions on the surface, e.g. for sensing. The ellipsometry is a very sensitive analytical tool which is typically used for optical characterisation of various thin films and coatings with the 0.01 nm accuracy of thickness detection and 10^{-5} for refractive index.

Typically ellipsometric instruments comprise the light source, polarizer, compensator, analyser, and photodetector. There are a number of commercial ellipsometric instruments available from basic monochromatic and fixed angle instruments to automated variable-angle, spectroscopic instruments, among them the most known are J.A. Woollam and Jobin Yvon instruments. Despite high sensitivity the ellipsometry in its traditional configuration is hardly used for sensing applications. For sensing purposes, a cell has to be attached to the instrument allowing injection different substances (both gaseous and liquid). The fact that the light travels trough the cell appeared to be the main obstacle for sensing application. The cells can be relatively easily used for gas sensing, but not for sensing in liquids. Different values of refractive index of liquids affect the performance; the liquid medium may also absorb or scatter light. The commercial cells are usually of high volume (tens of millilitres) which is definitely not suitable for bio-sensing which typically operate with microlitres quantities. This problem was overcome about 10 years ago, by combining the ellipsometry and SPR [7, 8] and the new method of total internal reflection ellipsometry (TIRE) was established. The method of TIRE was further developed in the research group of A. Nabok and appeared to be particularly suitable for detection low molecular weight analytes such as mycotoxins [9–13]. The experimental TIRE set-up shown in Fig. 2a is based on a spectroscopic J.A. Woolam M2000 ellipsometric instrument with an addition of a 68° prism in optical contact (via index matching fluid) with gold coated glass slide. A 0.2 ml PTFE cell is sealed against the gold surface via silicon rubber O-ring; the cell has inlet and outlet tubes allowing injection of different solutions. The polarized light is coupled into the gold coated glass slide through the 68° prism which provides total internal reflection conditions on the interface between glass and water. Typical TIRE spectra of Ψ and Δ recorded on bare gold film of 25 nm thick deposited on glass with the cell filled with water are shown in Fig. 2b. As one can see, the spectrum of

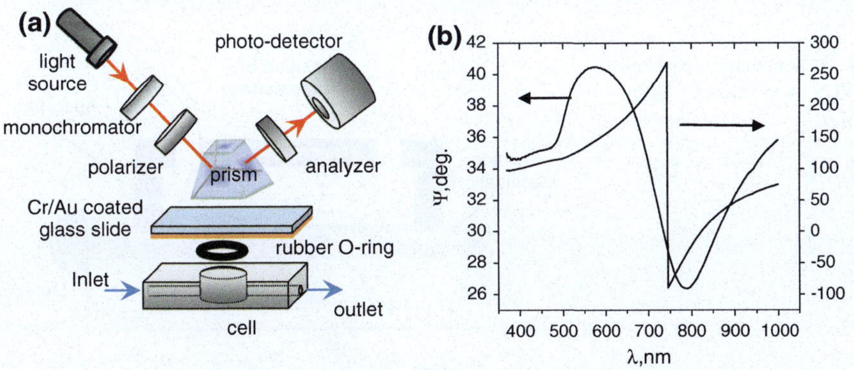

Fig. 2 TIRE experimental set-up (**a**); Typical spectra of Ψ and Δ recorded on bare gold (**b**)

Ψ resembles the classical SRP curve with the maximal signal at about 550 nm wavelength corresponding to total internal reflection conditions and the minimum corresponding to plasmon resonance. It is SPR indeed with the only difference of using a mixture of p- and s- polarised components in contrast to traditional SPR which used p-polarized light only. The spectrum of Δ however represents a new quantity of phase shift between p- and s- components of polarized light, which doesn't exist in SPR. The spectrum of Δ in Fig. 2b shows a phase drop near the resonance, which actually a characteristic of particular instrument used (J.A. Woollam, M200), but it is very sensitive to adsorption of molecules on gold surface and can be calibrated and used as a sensing response. The modelling showed 10 times better sensitivity of the parameter Δ that that of Ψ to small changes of thickness and refractive index in the adsorbed molecular layer [11]. Therefore, for sensing purposes, it better to use the spectra of Δ; the method of TIRE can be called as phase SPR. In both SPR and ellipsometry, the fitting of experimental data to Fresnels equations allows finding the optical parameters of thin molecular films, such as thickness (d) and complex refractive index $N = n - jk$ (where n is refractive index and k is extinction coefficient). However, the limitation of SPR and ellipsometry is that for transparent dielectric films ($k = 0$, $n > 1$) with the thicknesses less than 10 nm it is impossible to find simultaneously the values of d and n. Traditionally in SPR, all changes in the sensing layer are associated with the refractive index, which is fully justified for BIACORE instruments or similar using a porous layer of dextrane with a constant thickness, but it is not always the case. In our experiments using TIRE, for example, it is more logical to "fix" the refractive index and associate all changes with the layer thickness [11]. This is obviously an approximation but it is very close to reality, since majority of biochemicals used have more or less the same refractive index of 1.42 in the visible spectral range [14].

Another very promising method based on evanescent field principle is called optical waveguide lightmode spectroscopy (OWLS) [15–18]. The schematic diagram of OWLS is shown in Fig. 3. It is similar to SPR with the light coupled to a slab waveguide using diffraction grating. The angles of coupling depend on the type of light polarization used p- and s- (often referred in literature as TM and TE,

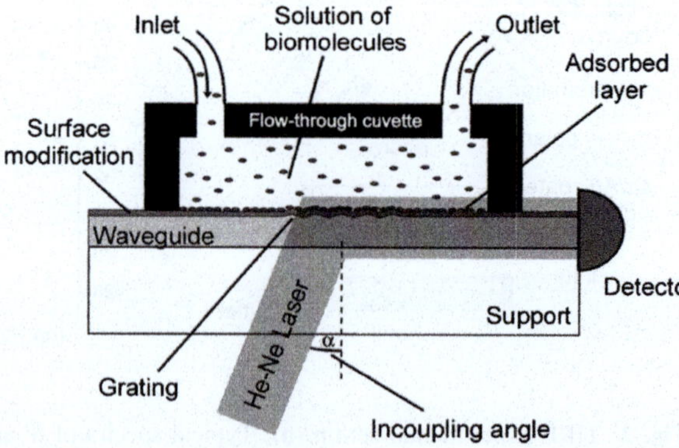

Fig. 3 Schematic diagram of OWLS experimental set-up [15]

respectively). In the method of OWLS, these angles are recorded continuously by rocking the chip by a small angle ±7°, and the two coupling intensity peaks are recoded at the angles of α_s and $\alpha_{p.}$.

If the adsorption of molecules occurs on the top of the waveguide (a flow injection cell is attached for this purpose) positions of both peaks α_s and α_p shift to higher angles. Because s-polarized light is less affected (as compared to p-polarization) by changes in the adsorbed molecular layer difference between coupling angles $\Delta = \alpha_p - \alpha_p$ can be used as a sensor response and calibrated in concentrations of adsorbed molecules. The sensitivity of OWLS is similar to SPR and is in the range of 10^{-6} RIU. An additional advantage of OWLS, is that the exact solution of mode equations [15, 16] allows the simultaneous determination of *n* and *d* of adsorbed molecular layer. A relative disadvantage of OWLS is rather bulky and expensive experimental set-up, so the OWLS is laboratory-based equipment.

It is well known that interferometry is a very sensitive analytical tool in optics; these advantages of interferometry have been exploited in chemical and bio-sensing development. A very successful commercial biosensing instrument was based on dual polarization interferometer (DPI) [19, 20]. The idea of this instrument is quite simple and based on interference of two waves propagating in adjacent slabs and the formation of the interference pattern as shown schematically in Fig. 4. Since the upper waveguiding slab is exposed to the environment, the molecular adsorption affecting the propagating wave causes the shift of the interference pattern which can be quantified in the concentrations of adsorbed molecules. The sensitivity of this method is claimed to be of 10^{-7} RIU which is comparable with the best SPR achievements. The interpretation of the results is not model-dependent (as compared to SPR or ellipsometry), so the outcomes can be easily interpreted as changes in refractive index or thickness, or optical density of molecular layer. The instrument is however quite expensive, bulky, and obviously laboratory based.

Perhaps, the most promising interferomeric scheme for bio-sensing applications is Mach Zehnder interferometer [21], a well-known optoelectronic device used for

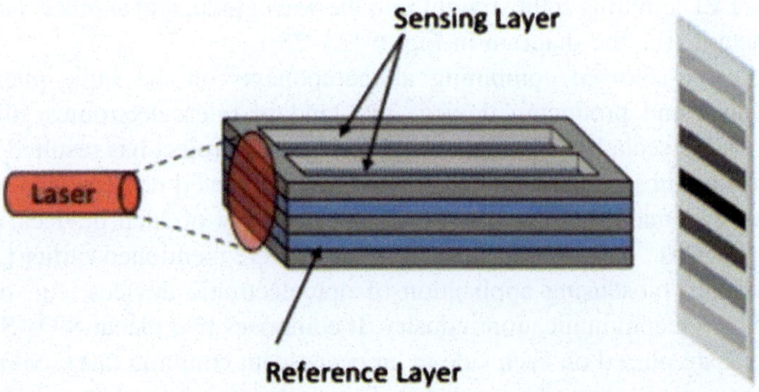

Fig. 4 Schematic diagram of DPI sensor

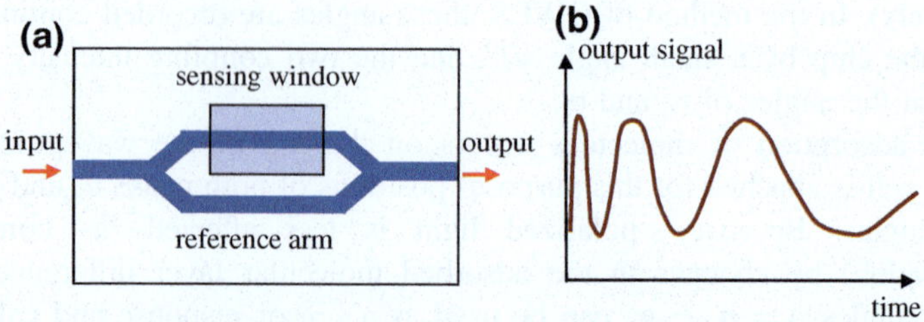

Fig. 5 The scheme of Mach-Zehnder interferometer (**a**); Multiperiodic output signal (**b**)

mixing and modulating optical signals. It based on a planar waveguide splitting into two arms which then merge together again (see Fig. 5a).

An opening in the cladding in one of the arms (called as a sensing window) acts as the sensing element, while the second arm serves as a reference. The output of MZ interferometer shown in Fig. 5b is a multi-periodic signal proportional to the phase shift between the two arms: $R \approx Sin(\varphi_s - \varphi_r)$, where indices s and r stand for "sensing" and "reference" arms. This multi-periodic signal has to be converted to a voltage proportional to the phase shift. For a while, MZ interferometers were bulky optical equipment, typically built on optical benches and used lasers. Relatively recently, miniaturised MZ devices were fabricated commercially using silicon microelectronics technology for telecommunication. Scientist working in the area of chemical and bio-sensing found these devices extremely useful. The technical problems of connecting optical fibres to MZ planar devices were solved successfully using Telecom fabrication facilities.

It was reported recently on the development of commercial multi-channel bio-sensor comprising dozens of MZ interferometers and ring-resonators (another optoelectronic device to be explained later [22]), integrated LED light sources, and micro-fluidic system of liquid sample delivery.

In the last few years, MZ-based biosensors were developed further using a concept of fully integrated MZ-devices based on SiO_2-Si_3N_4-SiO_2 planar waveguide built on Si wafer with two p-n junctions on each end, one acting as an avalanche LED emitting light straight into the waveguide, and another one acting as photo-detector (see the diagram in Fig. 6 [23–25]).

Such design allowed combining all components on the same platform, e.g. silicon wafer, and producing devices by standard microelectronics silicon technology. A large scale EU FP7 collaborative research project has resulted in a range of biosensing chips comprising an array of MZ integrated devices data acquisition electronics and microfluidics. The reported sensitivity of such devices was in the range of 10^{-7}–10^{-9} RIU. Optical ring resonators were mentioned earlier [22]; this is another fruitful biosensing application of optoelectronic devices, e.g. band filter, produced by telecommunication industry. It comprises two planar SiO_2-Si_3N_4-SiO_2 waveguides are placed on each side in an evanescent coupling distance from a ring made of the same waveguiding material. The electromagnetic wave coupled into the ring forms standing waves with characteristic resonances at $\lambda = 2\pi r/n$, where r is the

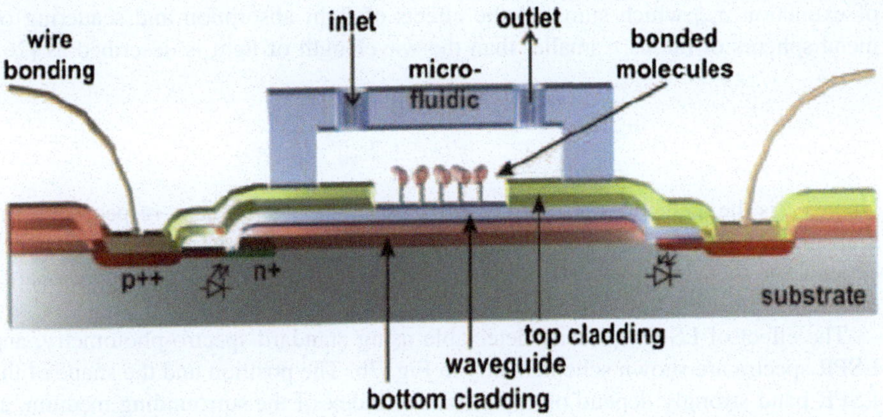

Fig. 6 Fully integrated MZ interferometer sensor [24]

ring radius and *n* is integer number. Such band-filter devices can be utilized for bio-sensing applications if the sensing window is etched on one of the ring, with another one serving as reference. The sensitivity in the range of 10^{-6} RIU is achievable.

Another very promising biosensing development appeared in the last decade is based on the phenomenon of localized surface plasmon resonance (LSPR). It is quite remarkable that the phenomenon of LSPR has been predicted theoretically by Mie in 1908 [26], then been deserted for nearly a century, and now appeared with a bang due to the rise of material nanotechnologies. The effect of LSPR is caused by oscillations of surface electrons typically in metal nano-structures, such as nano-particles of dimensions comparable with the wavelength of light. Illustration of this phenomenon is given in Fig. 7a. Such oscillations have a characteristic resonance frequency (wavelength) in visible optical range which depends on the type and geometry of metal nano-structures. According to Mie theory, the cross-section

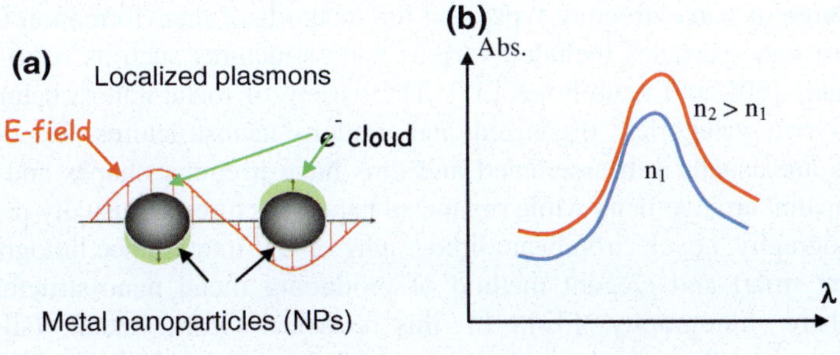

Fig. 7 Interaction of EM wave with surface plasmons in metal nanoparticles (**a**); Spectral shift of LSPR band due to changes in refractive index of the medium (**b**)

of extinction σ_{ext} which sums up the effects of light absorption and scattering on metal spheres of radius r smaller than the wavelength of light is described as [26]:

$$\sigma_{ext} = \frac{12\omega\pi r^3}{c} \frac{\varepsilon_m^{3/2}\varepsilon''(\omega)}{[\varepsilon'(\omega) + 2\varepsilon_m]^2 + \varepsilon''(\omega)^2}$$

where ε_m is the dielectric permittivity of the medium; ε' and ε'' are, respectively, the real and imaginary parts of the dielectric permittivity of metal particles which depend on the circular frequency (ω) of electromagnetic wave. The resonance conditions are achieved when $\varepsilon'(\omega) = -2\varepsilon_m$.

The effect of LSPR is easily detectable using standard spectro-photometry, and LSPR spectra are shown schematically in Fig. 7b. The position and the shape of the LSPR band strongly depend on the refractive index of the surrounding medium, as it obvious from Mie formula. As has been shown more precisely in [27] the refractive index of a medium (n_m) causes the shift of the LSPR peak (λ_{LSPR}) to longer wavelengths in respect to the plasmon peak in bulk material (λ_0): $\lambda_{LSPR} = \lambda_0\sqrt{2n_m^2 + 1}$. This constitutes the main principle of LSPR-based sensors; the molecules adsorbed on the surface of metal nanostructures alter the local refractive index in the vicinity of metal nano-particles thus causing the changes of the absorption band which then can be quantified and calibrated against the concentration of the adsorbed molecules.

The practicality of the LSPR is largely related to the type of metal nano-structures and technology of their fabrications. First of all, metals must be highly conductive which gives an obvious choice of Au, Ag, Cu, etc. Among those metals, gold is the most popular since it combines high conductivity with chemical inertness and well-developed thiol chemistry of its surface modification for immobilization of proteins. Historically, gold nano-particles were perhaps the first type of nano-structures used for LSPR sensing applications [28]. There are a number of well-established recipes of fabrication of gold nano-particles which researches can easily reproduce in the lab to make gold-nanoparticles of particular sizes. Also, nano-particles of gold (as well as some other metals) of various sizes are commercially available these days.

The range of nano-structure types and the methods of their formation is vast and apart from nano-particles included various nano-structures such as nano-rods [29], nano-islands [30], and nano-holes [31]. The variety of metal-nanostructures can be split into two categories: organized and random nanostructures. Random nano-structures are usually self-assembled and thus have irregular shapes and sizes and random spatial arrangement, while organized nano-structures are usually produced by nano-lithography, i.e. electron beam lithography or UV interference lithography [32].

A very smart and elegant method of producing metal nano-structures is the micro-sphere lithography [33]. In this classical work metal (silver) was electro-deposited through the gaps in a monolayer of closely packed polysterene spheres, and the resulted triangular nano-islands arranged in hexagonal 2D lattice were revealed after removing microspheres (Fig. 8).

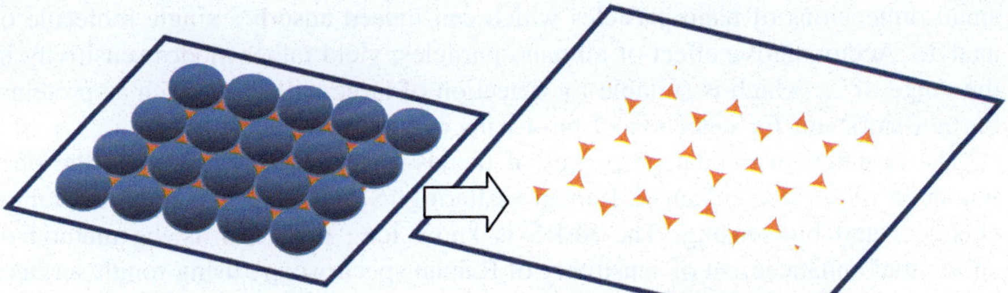

Fig. 8 Formation of metal nanostructures by microsphere lithography [33]

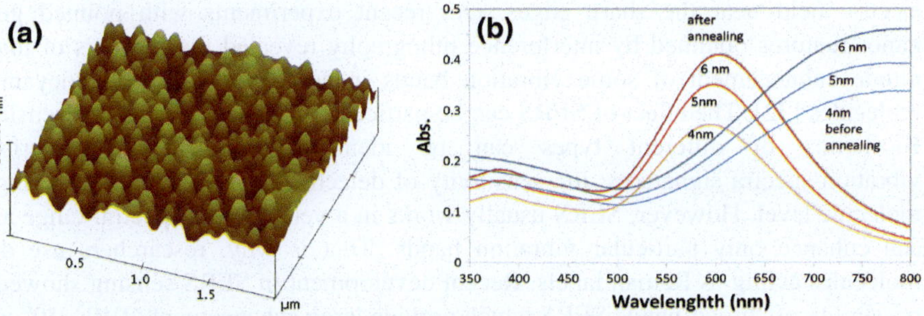

Fig. 9 AFM image of 4 nm Au film thermally evaporated on glass (**a**); UV-vis absorption spectra of thin Cr/Au films before and after annealing (**b**)

However, the simplest technology of making metal nano-structures is annealing of thin metal films. Figure 9a shows the formation of nano-islands structure after annealing at 480 °C for 1–2 h a 4 nm thick gold film thermally evaporated on glass (2 nm layer of Cr was evaporated first to provide good adhesion). The respective absorption spectra in Fig. 9b show the transition from a classical Drude dispersion to a pronounced LSPR band at about 560 nm as a result of annealing.

The formation of nano-islands by de-wetting of continuous thin metal films at high temperatures has been studied systematically by Rubinstein and his colleagues. The refractive index sensitivity depends on the thickness and the size of metal nano-structures, and typically found to be in the range of 50–150 nm/RIU [34].

The spectra of alongated nanoparticles or nano-structures of more complex geometries revealed multiple LSPR peaks which still can be explained using classical electrodynamic theory [35]. For example, the rectangular particles of different aspect ratio yield three LSP peaks which correspond to depolarization factors along all three dimensions, length, width, and height [36]. Increasing of the aspect ratio results in further red shift of the third peak (at the longest wavelength) while the first two peaks remained the same. A very high sensitivity of LSPR down to single molecule level was claimed. This is however a trivial fact considering small dimensions of nano-particles which can indeed adsorb a single molecule of analyte. Accumulative effect of all nano-particles, yield rather modes sensitivity in the range of … which is suitable for detection of large molecules such as proteins, but not sufficient for detection of bio-toxins of low molecular weight.

The research in optical properties of metals nano-structures revealed the phenomenon of surface enhanced Raman scattering (SERS) which may revolutionise chemical and bio-sensing. The SERS is know long time ago as the method of substantial enhancement of sensitivity of Raman spectroscopy using rough surfaces [37] due to scattering of the electromagnetic waves and the associated local enhancement of electric field. Certain types of metal nanostructures, such as pointed towards each other triangles formed by micro-sphere lithography, appeared to show a huge enhancement of Raman signal up to 10^9 times [37]. The reason behind such enhancement is the presence of so-called "hot spots" having much large values of electric field near the sharp edges. Our recent experiments with pointed gold nanostructures obtained by interference lithography revealed the 6 orders of magnitude enhancement of some vibration bands in adsorbed metal-phthalocyanine molecules [32]. The effect of SERS can be particularly useful for chemical sensing. Bio-toxins of different types can be identified by their characteristic vibration-spectra signatures; the sensitivity of detection could truly go to a single molecule level. However, SERS usually works in a very narrow spectral range and can enhance only particular vibration bands. That is why, researchers use dye molecules acting as Raman labels. Recent development in SERS sensing showed a possibility of "broad-band" SERS which provide large enhancement of 10^5–10^6 in a wide spectra range covering practically all IR spectra [38]. This was achieved using a specially designed and fabricated with electron beam lithography SERS nano-antenna having multiple hot-spots of different resonant frequencies.

3 Case Study: Detection of Bio-Toxins Using TIRE

The main problem of optical detection of bio-toxins is their relatively small molecular weight, so that only the most sensitive analytical tools can be utilised for this task. The detection of mycotoxins was (and still is) a very important problem related to food and feed safety, as well as to security. There are huge varieties of fungi releasing different mycotoxins some of which are toxic, carcinogenic, endocrine disruptive agents, so that the contamination of agriculture products with mycotoxins as well as derivative food and feed constitute a real danger. The above hazardous properties together with the simplicity of mycotoxins production make them listed as substances of potential terrorism activities and biological warfare. In this research we focused on detection of three mycotoxins, namely T2, zearalenone, and aflatoxin B1 in the immunoassays with respective antibodies. Initially tried method of SPR appeared to be not sensitive enough for detection of mycotoxins, therefore the recently developed method of TIRE was selected as the main biosensing technique.

For immobilization of antibodies on the surface of gold, the method of electrostatic deposition, which has been developed in our research group, was used [4]. Standard microscopic glass slides were coated with 25 nm thick gold film with a 3 nm underlayer of chromium providing good adhesion of gold to glass. In order to increase the negative surface charge, gold coated slides were treated overnight in 100 mM solution of mercapto-ethylene-sulfonate sodium salt. The following depositions steps were carried out in the reaction cell of TIRE experimental set-up shown in Fig. 2. The spectra of Ψ and Δ were recorded in the same standard Tris-HCl buffer solution (pH 7.5) after each adsorption or binding step. In addition, the kinetics of molecular adsorption or binding has been studied by recording Ψ and Δ spectra many times after a certain time interval, and then plotting time dependences of Ψ and Δ at particular wavelengths. The first step was the deposition of polycationic layer of poly-allylamine hydrochloride (PAH) from its 1 mg/ml solution in deionised water. Then a layer of protein A was electrostatically adsorbed from its solution in Tris-HCl buffer, pH 7.5–8. Then antibodies to respective mycotoxins were immobilized on the surface. IgG based antibodies having a binding site to protein A in the second domain were oriented on the surface with the fab-fragments pointing towards the solution. Immune binding of the mentioned above mycotoxins to their specific antibodies immobilised on the surface was carried out by consecutive injection of small amounts of toxin into the cell starting form the smallest concentrations.

In the example of T2 immunoassay, the kinetics of the immune binding of T2 molecules to specific antibodies was studied by TIRE dynamic scans during injection of T2 into the cell. The kinetics data analysis protocol was based on differential equation of adsorption on single binding cites [11–13]. The experimental time dependencies of Ψ or Δ were fitted to exponential function and the time constant (τ) was evaluated. Such procedure was repeated for different concentrations (C) of mycotoxins, and a linear dependence of $1/\tau = k_a C + k_d$ versus C was plotted. The rates of adsorption (k_a) and desorption (k_d) were found, respectively, as the gradient and intercept of this dependence. Then the value of the association constant can be found as $K_A = k_a/k_d$. The obtained values of K_A for all three mycotoxins studied were in the range of 10^7 Mol^{-1} which is typical for highly specific immune binding between mycotoxins and respective antibodies.

Typical spectra of Δ recorded after each adsorption (binding) step shown in Fig. 10 demonstrate the "red" spectral shift which is proportional to the size (or mass) of the adsorbed molecules. For sensing purposes, it would be sufficient to just calibrate the shift of Δ spectra in concentrations of mycotoxins. The detection limit for mycotoxins in the range of 0.1 ng/ml is very good for direct immunoassay. However, we were also interested in the physical meaning of the response, so that the fitting of TIRE data was carried out using an upside-down four-layer ellipsometric model consisting of BK7 glass (ambient), Cr/Au film, molecular layer, and water (substrate) [11].

Dispersions functions for n and k for all materials used were taken from J.A. Woollam database. The effective thickness (d) and dispersions of $n(\lambda)$ and $k(\lambda)$ for Cr/Au layer were found first from TIRE spectra recorded on bare gold layer, then

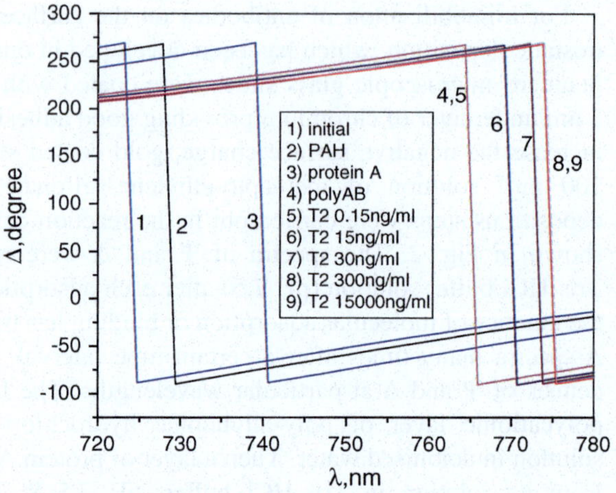

Fig. 10 TIRE spectra of D recorded on after each adsorption (binding) step [10]

these parameters were kept fixed at further fittings. The molecular layer was modelled by Cauchy layer $n = A + B/\lambda^2 + C/\lambda^4$ where the parameters of $A = 1.39$, $B = 0.01$, and $C = 0$ were fixed giving the value of $n = 1.42$ at 633 nm, also $k = 0$. Therefore all changes taken place in the molecular layer were associated with the thickness d.

The resulted data were presented in Fig. 11 as calibration curves, i.e. dependences of the thickness increment vs concentration of mycotoxins, for all three mycotoxins studied. As one can see, all mycotoxins show similar sigmoid type calibrations typical for immune reactions with LDL of 0.1 ng/ml. 10 times lower LDL (0.01 ng/ml) was achieved for detection of Zearalenone using competitive immunoassay [12]. It was interesting to note that the thickness increments of 4.5,

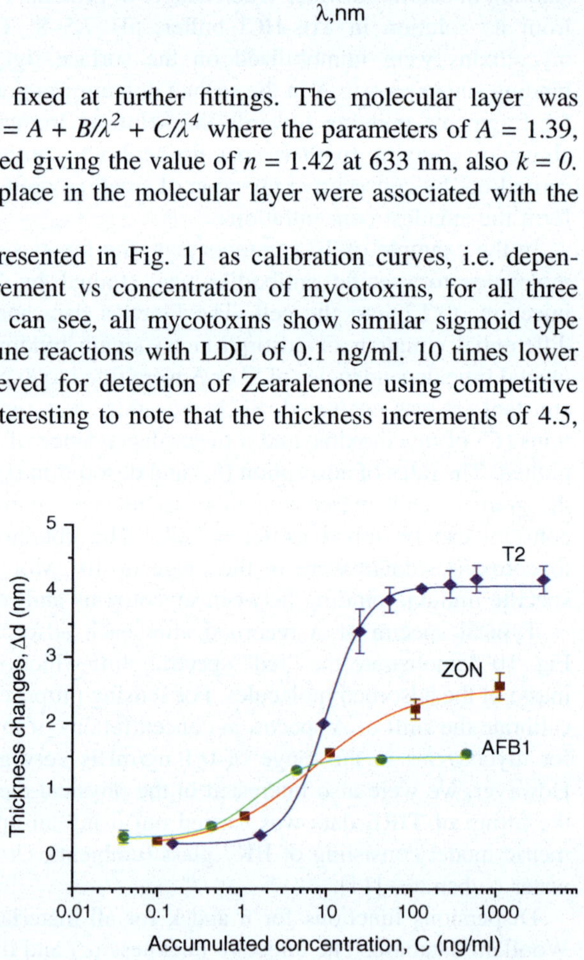

Fig. 11 The TIRE calibration curves for three mycotoxins studied

2.5, and 1.5 nm for T2, ZON, and AFT, respectively, are larger than actual dimensions of corresponding mycotoxins, especially considering that the obtained thickness values are effective because of signal averaging over a large area (15 mm^2) of a light beam spot. This could be a result of aggregation of hydrophobic mycotoxin molecules in water solutions.

The concept of molecular aggregation has been proved when studying immune binding of nonylphenol to respective monoclonal antibodies using a similar TIRE experimental protocol [39]. It appeared that the maximal thickness increment in calibration curve reached 25 nm, a huge thickness nearly 20 times larger than the length of nonylphenol molecule. Also a substantial mass gain was recorded with QCM, and the formation of large aggregates was observed with AFM. The experimental results obtained supported the idea of the formation of micelles of amphiphilic nonylphenol molecules when the initial stock solution of nonylphenol in acetonitrile was diluted with water. Such large aggregates can still bind specifically to antibodies with the association constant in the range of 10^6 Mol^{-1}; the modelling showed that the nonylphenol micelles having rather a "flat pancake" shape can bind to several anaibodies simultaneously [40]. Similarly, specific binding of molecular aggregates (though of much smaller sizes) to antibodies may take place in the case of hydrophobic mycotoxin molecules. This may give an additional boost in sensitivity as compared to adsorption of single molecules.

The method of TIRE has been successfully utilized for detection of some other bio-toxins, namely herbicides pesticides simazine and atrazine [9] and microcystin LR released by bacterial algae [41]. Therefore, the method of TIRE is particularly attractive for detection of low molecular weight analytes as most of bio-toxins are. The sensitivity in sub-ppb level is achievable with TIRE.

4 Conclusions and Future Prospects for Bio-Toxins' Sensor Development

The main outcome of the analysis of literature as well as from our own experience was that the interferometric sensors offer much lower LDL for bio-toxins in pMol or even in fMol range, three orders of magnitude better as compared to more traditional optical methods of SPR, OWLS, TIRE. This is because of multiple reflections of light in planar waveguide structures having large differences between the core and cladding. Scaling down sensing devices to a hand-held type is another major issue in sensing development. From that point of view, MZ interferometers and ring resonators together with microfluidic system produced with the use of advance telecommunication technological facilities is very promising. Fully integrated MZ devices recently developed within EU FP7 project are perhaps the most attractive from the point of view of miniaturization. Such devices can be integrated together with microfluidic and signal acquisition system into a single silicon

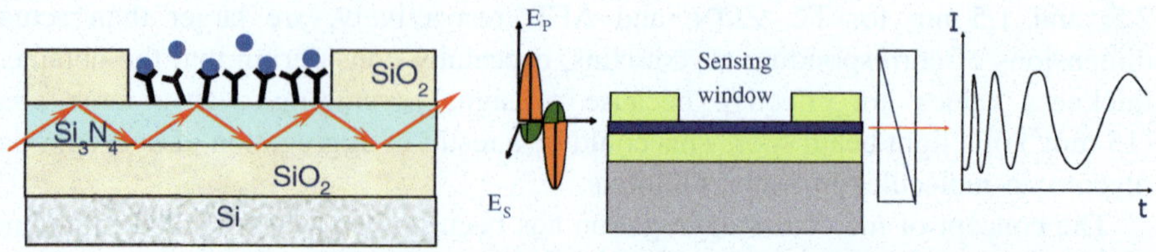

Fig. 12 The design and experimental set-up of planar waveguide-based PPI sensor

chip. This may result in development of highly sensitive and portable (mobile phone size) sensors for *in-field* analysis with the grate benefits for agriculture, food industry, medical testing, environmental safety, and security.

In addition to existing planar waveguide designs, we would like to develop (within ongoing NATO SPS project) a novel type of interferometric sensor devices based the principle of planar polarization interferometry (PPI) shown in Fig. 12. A polarized light coupled into the waveguide experiences a large number of reflections (up to 5000 per mm) causing a substantial phase shift between *p*- and *s*-polarization components (the latter one less affected by the medium serves as a reference). The resulted output signal is multi-periodic similarly to MZ interferometer.

Preliminary experiments [42] and modelling [12] showed a possibility of sensitivity enhancement of minimum 1000 times as compared to the TIRE method, so that the LDL of detection bio-toxins may go down to pg/ml or even lower.

Acknowledgment This work was financially supported by NATO project, SPS(NUKR.SFPP 984637).

References

1. Lequin R M (2005) Enzyme Immunoassay (EIA)/Enzyme-Linked Immunosorbent Assay (ELISA). Clin Chem 51(12):2415–2418
2. Rao Y J (1999) Recent progress in applications in fibre Bragg grating sensors. Opt Lasers Eng 31:297–324
3. James S W, Tatam R P (2003) Optical fibre long-period grating sensors: characteristics and application. Meas Sci Technol 14:R49–R61
4. Nabok A (2005) Organic and inorganic nanostructures. Artech House, Boston
5. Brockman J M, Nelson B P, Corn R M (2000) Surface plasmon imaging of ultrathin organic films. Ann Rev Phys Chem 51:41–63
6. Azzam R M A, Bashara N M (1992) Ellipsometry and polarized light. North Holland, Amsterdam
7. Westphal P, Bornmann A (2002) Biomolecular detection by surface plasmon enhanced ellipsometry. Sens Actuators B 84:278–282
8. Arwin H, Poksinski M, Johansen K (2004) Total internal reflection ellipsometry: principles and applications. Appl Opt 43:3028–3036

9. Nabok A, Tsargorodskaya A, Hassan A K, Starodub N F (2005) Total internal reflection ellipsometry and SPR detection of low molecular weight environmental toxins. Appl Surf Sci 246(4):381–386
10. Nabok A, Tsargorodskaya A, Holloway A, Starodub N F, Gojster O (2007) Registration of T-2 mycotoxin with total internal reflection ellipsometry and QCM impedance methods. Biosens Bioelectron 22(6):885–890
11. Nabok A, Tsargorodskaya A (2008) The method of total internal reflection ellipsometry for thin film characterisation and sensing. Thin Solid Films 516(24):8993–9001
12. Nabok A, Tsargorodskaya A, Mustafa M K, Szekacs I, Starodub N F, Szekacs A (2011) Detection of low molecular weight toxins using optical phase detection techniques. Sens Actuators B Chem 154(2):232–237
13. Nabok A, Mustafa M K, Tsargorodskaya A, Starodub N F (2011) Detection of aflatoxin B1 with a label free ellipsometry immunosensor. BioNanoScience 1(1):38–45
14. Székács A, Adányi N, Székács I, Majer-Baranyi K, Szendrő I (2009) Optical waveguide lightmode spectroscopy immunosensors for environmental monitoring. Appl Opt 48:B151–B158
15. Voros J, Ramsden J J, Gsucs G, Szendro I, De Paul S M, Textor M, Spenser N D (2002) Optical grating couple biosensor. Biomaterials 23:3699–3710
16. Voros J (2004) The density and refractive index od adsorbing protein layers. Biophys J 87:553–561
17. Label-free immunosensor for herbicide trifluralin detection, OWLS Application notes No-002, http://www.OWLS-sensors.com
18. Label-free immunosensor for Aflatoxin B1, OWLS Application notes No-006. http://www.OWLS-sensors.com
19. Cross G, Reeves A A, Brand S, Popplewell J F, Peel L L, Swann M J, Freeman N J (2003) A new quantitative optical biosensor for protein characterisation. Biosens Bioelectron 19 (4):383–390
20. Cross G, Reeves A A, Brand S, Swann M J, Peel L L, Freeman N J, Lu J R (2004) The metrics of surface adsorbed small molecules on the Yung's fringe dual-slab waveguide interferometer. J Phys D Appl Phys 37:74–80
21. Luff B J, Wikinson J S, Piehler J, Hollenbach U, Ingenhoff J, Fabricius N (1998) Integrated optical Mach-Zehnder biosensor. J Lightwave Technol 16(4):583–592
22. Sun Y, Fan X (2011) Optical ring resonators for biochemical and chemical sensing. Anal Bioanal Chem 399:205–211
23. Misiakos K, Kakabakos S E, Petrou P S, Ruf H H (2004) A monolithic silicon optoelectronic transducer as a real-time affinity biosensor. Anal Chem 76:1366–1373
24. Kitsara M, Misiakos K, Raptis I, Makarona E (2010) Integrated optical frequency-resolved Mach-Zehnder interferometers for label-free affinity sensing. Opt Express 18:8193–8206
25. Misiakos K, Raptis I, Makarona E, Botsialas A, Salapatas A, Oikonomou P, Psarouli A, Petrou PS, Kakabakos SE, Tukkiniemi K, Sopanen M, Jobst G (2014) All silicon monolithic Mach-Zehnder interferometer as a refractive index and bio-chemical sensor. Opt Express 22 (22):26803–26813
26. Mie G (1908) Beitge zur optic trüber medien, speziell kolloidaler metallösungen. Ann Phys 330(3):377–445
27. Hong Y, Huh Y-M, Yoon D S, Yang J (2012) Nanobiosensors based on localized surface plasmon resonance for biomarker detection, Hindawi Publishing Co. J Nanomater 759830. doi:10.1155/2012/759830
28. Zhao J, Zhang X, Yonzon C R, Haes A J, Van Duyne R P (2006) Localized surface plasmon resonance biosensors. Nanomedicine 1(2):1029–1034
29. Lee K-S, El-Sayed M (2005) Dependence of the enhanced optical scattering efficiency relative to that of absorption for gold metal nanorods on aspect ratio, size, and- cup shape, and medium refractive index. J Phys Chem 109(43):20331–20338
30. Vaskevich A, Rubistein I (2013) Nanoplasmonic sensors. Springer, Integrated Analytical System, pp 333–368

31. Jonsson M P, Dahlin A B, Jonsson P, Hook F (2008) Nanoplasmonic biosensing with focus on short-range ordered nanoholes in thin metal films. Biointerphases. J Biomater Biol Interphases 3:FD30. doi:10.1116/1.3027483
32. Tsargorodska A, El Zubir O, Darroch B, Cartron M L, Basova T, Hunter C N, Nabok A V, Leggett G J (2014) Fast, simple, combinatorial routes to the fabrication of reusable, plasmonically active gold nanostructures by interferometric lithography of self- assembled monolayers. ACS Nano 8(8):7858–7869
33. Jensen T R, Duval M L, Kelly K L, Lazarides A A, Schatz G C, Van Duyne R P (1999) Single nanosphere lithography: effect of external dielectric medium on the surface plasmon resonance spectrum of a periodic array of silver nanoparticles. J Phys Chem B 103(45):9846–9853
34. Karakouz T, Holder D, Goomanovsky M, Vaskevich A, Rubinstein I (2009) Morphology and refractive index sensitivity of gold island films. Chem Mater 21:5875–5885
35. Gans R (1912) Ann Phys (Weinheim, Ger.) 103:9846
36. Petryaeva E, Krull U J (2011) Localized surface plasmon resonance: nanostructures, bioassays and siosensing—a review. Anal Chim Acta 706:8–24
37. Larkin P J (2005) IR and Raman Spectroscopy. Jones and Bartlett Publishers Inc, Burlington
38. Aouani H, Rahmani M, Šípová H, Torres V, Hegnerová K, Beruete M, Homola J, Hong M, Navarro-Cía M, Maier S A (2013) Plasmonic nanoantennas for multispectral surface-enhanced spectroscopies. J Phys Chem C 117:18620–18626
39. Nabok A, Tsargorodskaya A, Holloway A, Starodub NF, Demchenko A (2007) Specific binding of large aggregates of amphiphilic molecules to respective antibodies. Langmuir 23 (16):8485–8490
40. Lishchuk S, Tsargorodskaya A, Nabok A (2008) The model of alkylphenol micelles bound to respective antibodies on the solid surface. Colloids Surf A 324:117–121
41. Al-Ammar R, Nabok A, Hashim A, Smith T (2015) Microcystin-LR produced by bacterialalgae: Optical detection and purification of contaminated substances. Sens Actuators B Chem 209:1070–1076
42. Starodub N F, Nabok A V, Starodub V M, Ray A K, Hassan A K (2001) Immobilisation of biocomponents for immune optical sensors. Ukrainian Bio Chem J 73:55–64

The manufacturer's authorised representative in the EU is Springer
Nature Customer Service Centre GmbH, Europaplatz 3, 69115 Heidelberg,
Germany. If you have any concerns regarding our products, please
contact ProductSafety@springernature.com

Printed and bound by CPI Group (UK) Ltd, Croydon, CR0 4YY
03/10/2026
02227493-0003